# Computer Simulation of Liquids

M. P. ALLEN

*H. H. Wills Physics Laboratory*
*University of Bristol*

and

D. J. TILDESLEY

*Department of Chemistry*
*Imperial College of Science, Technology and Medicine, London*

CLARENDON PRESS · OXFORD

**OXFORD**
UNIVERSITY PRESS

Great Clarendon Street, Oxford OX2 6DP

Oxford University Press is a department of the University of Oxford.
It furthers the University's objective of excellence in research, scholarship,
and education by publishing worldwide in

Oxford New York

Auckland Cape Town Dar es Salaam Hong Kong Karachi Kuala Lumpur
Madrid Melbourne Mexico City Nairobi New Delhi Shanghai Taipei Toronto

With offices in

Argentina Austria Brazil Chile Czech Republic France Greece Guatemala
Hungary Italy Japan South Korea Poland Portugal Singapore Switzerland
Thailand Turkey Ukraine Vietnam

Published in the United States
by Oxford University Press Inc., New York



First published 1987
First published in paperback (with corrections) 1989
Reprinted 1990 (twice), 1991, 1992, 1993, 1994, 1996,
1997, 1999, 2000, 2001, 2002, 2003, 2004, 2005



British Library Cataloguing in Publication Data

Allen, M. P.
Computer simulation of liquids.
1. Liquids—Simulation methods
2. Digital computer simulation
I. Title II. Tildesley, D. J.
530.4'2'0724 QC145.2

Library of Congress Cataloging in Publication Data

Allen, M. P.
Computer simulation of liquids.
1. Liquids—Mathematical models.
2. Liquids—Data processing.
3. Molecular dynamics.
4. Monto Carlo method.
I. Tildesley, D. J.
II. Title.
QC145.2.A43 1987 532'.00724 87-1555

ISBN 0 19 855645 4 (Pbk)

Printed in Great Britain by
Arrowsmith, Bristol

# COMPUTER SIMULATION OF LIQUIDS

*To*

Diane and Pauline

# PREFACE

This is a 'how-to-do-it' book for people who want to use computers to simulate the behaviour of atomic and molecular liquids. We hope that it will be useful to first-year graduate students, research workers in industry and academia, and to teachers and lecturers who want to use the computer to illustrate the way liquids behave.

Getting started is the main barrier to writing a simulation program. Few people begin their research into liquids by sitting down and composing a program from scratch. Yet these programs are not inherently complicated: there are just a few pitfalls to be avoided. In the past, many simulation programs have been handed down from one research group to another and from one generation of students to the next. Indeed, with a trained eye, it is possible to trace many programs back to one of the handful of groups working in the field 20 years ago. Technical details such as methods for improving the speed of the program or for avoiding common mistakes are often buried in the appendices of publications or passed on by word of mouth. In the first six chapters of this book, we have tried to gather together these details and to present a clear account of the techniques, namely Monte Carlo and molecular dynamics. The hope is that a graduate student could use these chapters to write his own program.

The field of computer simulation has enjoyed rapid advances in the last five years. Smart Monte Carlo sampling techniques have been introduced and tested, and the molecular dynamics method has been extended to simulate various ensembles. The techniques have been merged into a new field of stochastic simulations and extended to cover quantum-mechanical as well as classical systems. A book on simulation would be incomplete without some mention of these advances and we have tackled them in Chapters 7 to 10. Chapter 11 contains a brief account of some interesting problems to which the methods have been applied. Our choices in this chapter are subjective and our coverage far from exhaustive. The aim is to give the reader a taste rather than a feast. Finally we have included examples of computer code to illustrate points made in the text, and have provided a wide selection of useful routines which are available on-line from two sources. We have not attempted to tackle the important areas of solid state simulation and protein molecular mechanics. The techniques discussed in this book are useful in these fields, but additionally much weight is given to energy minimization rather than the simulation of systems at non-zero temperatures. The vast field of lattice dynamics is discussed in many other texts.

Both of us were fortunate in that we had expert guidance when starting work in the field, and we would like to take this opportunity to thank P. Schofield (Harwell) and W. B. Streett (Cornell), who set us on the right road some years ago. This book was largely written and created at the Physical Chemistry Laboratory, Oxford, where both of us have spent a large part of our research careers. We owe a great debt of gratitude to the head of department, J. S. Rowlinson, who has provided us with continuous encouragement and support in this venture, as well as a meticulous criticism of early versions of the manuscript. We would also like to thank our friends and colleagues in the Physics department at Bristol and the Chemistry department at Southampton for their help and encouragement, and we are indebted to many colleagues, who in discussions at conferences and workshops, particularly those organized by CCP5 and CECAM, have helped to form our ideas. We cannot mention all by name, but should say that conversations with D. Frenkel and P. A. Madden have been especially helpful. We would also like to thank M. Gillan and J. P. Ryckaert, who made useful comments on certain chapters, and I. R. McDonald who read and commented on the completed manuscript. We are grateful for the assistance of Mrs L. Hayes, at Oxford University Computing Service, where the original Microfiche was produced. Lastly, we thank Taylor and Francis for allowing us to reproduce diagrams from *Molecular Physics* and *Advances in Physics*, and ICL and Cray Research (UK) for the photographs in Fig. 1.1. Detailed acknowledgements appear in the text.

Books are not written without a lot of family support. One of us (DJT) wants to thank the Oaks and the Sibleys of Bicester for their hospitality during many weekends in the last three years. Our wives, Diane and Pauline, have suffered in silence during our frequent disappearances, and given us their unflagging support during the whole project. We owe them a great deal.

*Bristol* M. P. A.
*Southampton* D. J. T.
*May 1986*

# CONTENTS

# LIST OF SYMBOLS

**Latin Alphabet**

| | | |
|---|---|---|
| $a$ | atom index | (1.3.3) |
| $\mathbf{a}$ | molecular acceleration | (3.2) |
| $A$ | Helmholtz free energy | (2.2) |
| $\mathscr{A}$ | general dynamic variable | (2.1) |
| $\mathbf{A}$ | rotation matrix | (3.3.1) |
| $\mathscr{A}$ | set of dynamic variables | (8.1) |
| $b$ | atom index | (1.3.3) |
| $\mathbf{b}$ | time derivative of acceleration | (3.2) |
| $B$ | second virial coefficient | (1.4.3) |
| $\mathscr{B}$ | general dynamic variable | (2.3) |
| $c_{\mathscr{A}\mathscr{B}}(t)$ | normalized time correlation function | (2.7) |
| $c(r)$ | direct pair correlation function | (6.5.2) |
| $C_{\mathscr{A}\mathscr{B}}(t)$ | un-normalized time correlation function | (2.7) |
| $C_P$ | constant-pressure heat capacity | (2.5) |
| $C_V$ | constant-volume heat capacity | (2.5) |
| $d$ | spatial dimensionality | (5.5) |
| $d_{ab}$ | intramolecular bond length | (5.5.2) |
| $\mathbf{d}_{ia}$ | atom position relative to molecular centre of mass | (3.3) |
| $D$ | diffusion coefficient | (2.7) |
| $\mathbf{D}_{ij}$ | pair diffusion matrix | (9.4) |
| $\mathscr{D}^l_{mm'}$ | Wigner rotation matrix | (2.6) |
| $\mathbf{e}_i$ | molecular axis unit vector | (1.3.3) |
| $E$ | total internal energy | (2.2) |
| $\mathscr{E}$ | electric field | (5.5.3) |
| $\mathbf{f}$ | force on molecule | (2.4) |
| $\mathbf{f}_{ij}$ | force on molecule $i$ due to $j$ | (2.4) |
| $\mathscr{F}$ | Fermi function | (7.2.3) |
| $\mathscr{F}$ | applied field | (8.1) |
| $g(r)$ | pair distribution function. | (2.6) |
| $g(r_{ij}, \mathbf{\Omega}_i, \mathbf{\Omega}_j)$ | molecular pair distribution function | (2.6) |
| $g_{ab}(r_{ab})$ | site–site distribution function | (2.6) |
| $g_{ll'm}(r)$ | spherical harmonic coefficients of pair distribution function | (2.6) |
| $g_l$ | angular correlation parameter | (2.6) |
| $\mathbf{g}$ | constraint force | (3.4) |
| $G$ | Gibbs free energy | (2.2) |
| $G(r, t)$ | van Hove function | (2.7) |
| $\hbar = h/2\pi$ | Planck's constant | (2.2) |
| $h(r)$ | total pair correlation function | (6.5.2) |
| $H$ | enthalpy | (2.2) |
| $\mathscr{H}$ | Hamiltonian | (1.3.1) |
| $i$ | molecule index | (1.3.1) |
| $I$ | molecular moment of inertia | (2.9) |
| $I_{xx}, I_{yy}, I_{zz}$ | principal components of inertia tensor | (2.9) |

| | | |
|---|---|---|
| $I(k, t)$ | intermediate scattering function | (2.7) |
| $j$ | molecule index | (1.3.2) |
| $\mathbf{j}$ | particle current | (8.2) |
| $\mathbf{j}^{\epsilon}$ | energy current | (2.7) |
| $\mathbf{J}$ | total angular momentum | (3.6.1) |
| $k$ | generalized coordinate or molecule index | (1.3.2) |
| $k_B$ | Boltzmann's constant | (1.2) |
| $\mathbf{k}$ | wave vector | (1.5.2) |
| $\mathscr{K}$ | kinetic energy | (1.3.1) |
| $l$ | cell length | (5.3.2) |
| $L$ | box length | (1.5.2) |
| $L$ | Liouville operator (always found as $iL$) | (2.1) |
| $\mathbf{L}$ | angular momentum | (3.1) |
| $\mathscr{L}$ | Lagrangian | (3.1) |
| $m$ | molecular mass | (1.3.1) |
| $m$ | possible outcome or state label | (4.3) |
| $\mathbf{M}$ | memory function matrix | (9.2) |
| $n$ | possible outcome or state label | (4.3) |
| $N$ | number of molecules | (1.3.1) |
| $O$ | octopole moment | (1.3.3) |
| $\mathbf{p}$ | molecular momentum | (1.3.1) |
| $P$ | pressure | (2.4) |
| $P_l$ | Legendre polynomial | (2.6) |
| $\mathbf{P}$ | total linear momentum | (2.2) |
| $\mathscr{P}$ | instantaneous pressure | (2.4) |
| $\boldsymbol{\mathscr{P}}$ | instantaneous pressure tensor | (2.7) |
| $\mathbb{P}$ | projection operator | (9.2) |
| $\mathbf{q}$ | generalized coordinates | (1.3.1) |
| $Q$ | quadrupole moment | (1.3.3) |
| $Q$ | partition function | (2.1) |
| $\mathbf{Q}$ | quaternion | (3.3.1) |
| $\mathbb{Q}$ | projection operator | (9.2) |
| $r$ | molecular separation | (1.3.2) |
| $\mathbf{r}$ | molecular position | (1.3.1) |
| $\mathbf{r}_{ij}$ | position of molecule $i$ relative to $j$ ($\mathbf{r}_i - \mathbf{r}_j$) | (1.3.2) |
| $\mathbf{r}_{ab}$ | site–site vector (short for $\mathbf{r}_{ia} - \mathbf{r}_{jb}$) | (2.4) |
| $\mathscr{R}$ | region of space | (4.4) |
| $s$ | statistical inefficiency | (6.4.1) |
| $s$ | scaled time variable | (7.4.2) |
| $\mathbf{s}$ | scaled molecular position | (4.5) |
| $S$ | entropy | (2.2) |
| $S(k)$ | structure factor | (2.6) |
| $S(k, \omega)$ | dynamic structure factor | (2.7) |
| $\mathbf{S}$ | total intrinsic angular momentum (spin) | (3.6.1) |
| $\mathscr{S}$ | action integral | (10.2) |
| $t$ | time | (2.1) |
| $t_{\mathscr{A}}$ | correlation or relaxation time | (2.7) |
| $T$ | temperature | (1.2) |

| | | |
|---|---|---|
| $\mathcal{T}$ | instantaneous kinetic temperature | (2.4) |
| $u(r)$ | pseudo-potential | (10.4) |
| $v(r)$ | pair potential | (1.3.2) |
| $\mathbf{v}$ | molecular velocity | (2.7) |
| $\mathbf{v}_{ij}$ | velocity of molecule $i$ relative to $j$ ($\mathbf{v}_i - \mathbf{v}_j$) | (3.6.1) |
| $V$ | volume | (1.2) |
| $\mathcal{V}$ | total potential energy | (1.3.1) |
| $w(r)$ | pair virial function | (2.4) |
| $W$ | weighting function | (7.2.2) |
| $\mathcal{W}$ | total virial function | (2.4) |
| $x$ | Cartesian coordinate | (1.3.1) |
| $x(r)$ | pair hypervirial function | (2.5) |
| $\mathcal{X}$ | total hypervirial function | (2.5) |
| $y$ | Cartesian coordinate | (1.3.1) |
| $Y_{lm}$ | spherical harmonic function | (2.6) |
| $z$ | Cartesian coordinate | (1.3.1) |
| $z$ | charge | (1.3.2) |
| $z$ | activity | (4.6) |
| $Z$ | configuration integral | (2.2) |

**Greek alphabet**

| | | |
|---|---|---|
| $\alpha$ | Cartesian component ($x$, $y$, $z$) | (1.3.1) |
| $\alpha_P$ | thermal expansion coefficient | (2.5) |
| $\boldsymbol{\alpha}$ | underlying stochastic transition matrix | (4.3) |
| $\beta$ | Cartesian component ($x$, $y$, $z$) | (1.3.1) |
| $\beta$ | inverse temperature ($1/k_BT$) | (2.3) |
| $\beta_S$ | adiabatic compressibility | (2.5) |
| $\beta_T$ | isothermal compressibility | (2.5) |
| $\gamma$ | a general transport coefficient | (2.7) |
| $\gamma_s$ | surface tension | (11.2) |
| $\gamma_V$ | thermal pressure coefficient | (2.5) |
| $\gamma_{ij}$ | angle between molecular axis vectors | (2.6) |
| $\boldsymbol{\Gamma}$ | point in phase space | (2.1) |
| $\delta(\ldots)$ | Dirac delta function | (2.2) |
| $\delta t$ | time step | (3.2) |
| $\delta\mathcal{A}$ | deviation of $\mathcal{A}$ from ensemble average | (2.3) |
| $\varepsilon$ | energy parameter in pair potentials | (1.3.2) |
| $\varepsilon_i$ | energy per molecule | (2.7) |
| $\varepsilon$ | relative permittivity (dielectric constant) | (5.5.2) |
| $\varepsilon_0$ | permittivity of free space | (1.3.2) |
| $\zeta$ | random number | (G.3) |
| $\eta$ | shear viscosity | (2.7) |
| $\eta_V$ | bulk viscosity | (2.7) |
| $\theta$ | Euler angle | (3.3.1) |
| $\theta$ | bond bending angle | (1.3.3) |
| $\theta(x)$ | unit step function | (6.5.3) |
| $\kappa$ | inertia parameter | (3.6.1) |

| | | |
|---|---|---|
| $\kappa$ | inverse of charge screening length | (5.5.2) |
| $\lambda$ | Lagrange undetermined multiplier | (3.4) |
| $\lambda_T$ | thermal conductivity | (2.7) |
| $\Lambda$ | thermal de Broglie wavelength | (2.2) |
| $\mu$ | chemical potential | (2.2) |
| $\boldsymbol{\mu}$ | molecular dipole moment | (5.5.2) |
| $\nu$ | exponent in soft-sphere potential | (1.3.2) |
| $\nu$ | discrete frequency index | (6.3.2) |
| $\xi$ | random number in range (0, 1) | (G.2) |
| $\xi$ | friction coefficient | (9.3) |
| $\xi(\mathbf{r}, \mathbf{p})$ | dynamical friction coefficient | (7.4.3) |
| $\boldsymbol{\pi}$ | stochastic transition matrix | (4.3) |
| $\rho$ | number density | (2.1) |
| $\rho(k)$ | spatial Fourier transform of number density | (2.6) |
| $\rho(\boldsymbol{\Gamma})$ | phase space distribution function | (2.1) |
| $\rho(\ldots)$ | general probability distribution function | (4.2.2) |
| $\boldsymbol{\rho}$ | set of all possible probabilities | (4.3) |
| $\sigma$ | length parameter in pair potentials | (1.3.2) |
| $\sigma(\mathscr{A})$ | RMS fluctuation for dynamical variable $\mathscr{A}$ | (2.3) |
| $\tau$ | discrete time or trial index | (2.1) |
| $\tau_{\mathscr{A}}$ | discrete correlation 'time' | (6.4.1) |
| $\boldsymbol{\tau}$ | torque acting on a molecule | (3.3) |
| $\phi$ | Euler angle | (3.3.1) |
| $\phi$ | bond torsional angle | (1.3.3) |
| $\chi$ | constraint condition | (3.4) |
| $\chi(\mathbf{r}, \mathbf{p})$ | dynamic scaling factor in constant-pressure simulations | (7.5.2) |
| $\psi$ | Euler angle | (3.3.1) |
| $\psi(\mathbf{r}, t)$ | wavepacket | (10.3) |
| $\Psi$ | thermodynamic potential | (2.1) |
| $\Psi(\mathbf{r}, t)$ | many-body wavefunction | (10.1) |
| $\omega$ | frequency | (D.1) |
| $\boldsymbol{\omega}$ | molecular angular velocity | (2.7) |
| $\boldsymbol{\Omega}$ | molecular orientation | (1.3.1) |
| $\boldsymbol{\Omega}$ | frequency matrix (always found as $i\boldsymbol{\Omega}$) | (9.2) |

**Subscripts and superscripts**

| | | |
|---|---|---|
| $\mathbf{r}_{ia}$ | denotes position of atom $a$ in molecule $i$ | (1.3.3) |
| $r_{i\alpha}$ | denotes $\alpha$ component ( $= x, y, z$) of position vector | (1.3.1) |
| $\parallel$ | denotes parallel or longitudinal component | (2.7) |
| $\perp$ | denotes perpendicular or transverse component | (2.7) |
| * | denotes reduced variables or complex conjugate | (B.1) |
| id | denotes ideal gas part | (2.2) |
| ex | denotes excess part | (2.2) |
| cl | denotes classical variable | (2.9) |
| qu | denotes quantum variable | (2.9) |
| p | denotes predicted values | (3.2) |
| c | denotes corrected values | (3.2) |

| | | |
|---|---|---|
| T | denotes matrix transpose | (7.5.4) |
| b | denotes body-fixed variable | (3.3.1) |
| s | denotes space-fixed variable | (3.3.1) |

**Special conventions**

| | | |
|---|---|---|
| $\nabla_{\mathbf{r}}$ | gradient with respect to molecular positions | (2.1) |
| $\nabla_{\mathbf{p}}$ | gradient with respect to molecular momenta | (2.1) |
| $\mathrm{d}/\mathrm{d}t$ | total derivative with respect to time | (2.1) |
| $\partial/\partial t$ | partial derivative with respect to time | (2.1) |
| $\dot{\mathbf{r}}$, $\ddot{\mathbf{r}}$ etc. | single and double time derivatives | (2.1) |
| $\langle\ldots\rangle_{\mathrm{trials}}$ | MC trial or step-by-step average | (4.2.2) |
| $\langle\ldots\rangle_{\mathrm{time}}$ | time average | (2.1) |
| $\langle\ldots\rangle_{\mathrm{ens}}$ | general ensemble average | (2.1) |
| $\langle\ldots\rangle_{\mathrm{ne}}$ | non-equilibrium ensemble average | (8.1) |
| $\langle\ldots\rangle_{\mathrm{W}}$ | weighted average | (7.2.2) |

# 1
# INTRODUCTION

## 1.1 A short history of computer simulation

What is a liquid? As you read this book, you may be mixing up, drinking down, sailing on, or sinking in, a liquid. Liquids flow, although they may be very viscous. They may be transparent, or they may scatter light strongly. Liquids may be found in bulk, or in the form of tiny droplets. They may be vaporized or frozen. Life as we know it probably evolved in the liquid phase, and our bodies are kept alive by chemical reactions occurring in liquids. There are many fascinating details of liquid-like behaviour, covering thermodynamics, structure, and motion. Why do liquids behave like this?

The study of the liquid state of matter has a long and rich history, from both the theoretical and experimental standpoints. From early observations of Brownian motion to recent neutron scattering experiments, experimentalists have worked to improve the understanding of the structure and particle dynamics that characterize liquids. At the same time, theoreticians have tried to construct simple models which explain how liquids behave. In this book, we concentrate exclusively on molecular models of liquids, and their analysis by computer simulation. For excellent accounts of the current status of liquid science, the reader should consult the standard references [Barker and Henderson 1976; Rowlinson and Swinton 1982; Hansen and McDonald 1986].

Early models of liquids [Morrell and Hildebrand 1936] involved the physical manipulation and analysis of the packing of a large number of gelatine balls, representing the molecules; this resulted in a surprisingly good three-dimensional picture of the structure of a liquid, or perhaps a random glass, and later applications of the technique have been described [Bernal and King 1968]. Even today, there is some interest in the study of assemblies of metal ball bearings, kept in motion by mechanical vibration [Pierański, Malecki, Kuczynski, and Wojciechowski 1978]. However, the use of large numbers of physical objects to represent molecules can be very time-consuming, there are obvious limitations on the types of interactions between them, and the effects of gravity can never be eliminated. The natural extension of this approach is to use a mathematical, rather than a physical, model, and to perform the analysis by computer.

It is now over 30 years since the first computer simulation of a liquid was carried out at the Los Alamos National Laboratories in the United States [Metropolis, Rosenbluth, Rosenbluth, Teller, and Teller 1953]. The Los Alamos computer, called MANIAC, was at that time one of the most powerful available; it is a measure of the recent rapid advance in computer technology that microcomputers of comparable power are now available to the general public at modest cost. Modern computers range from the relatively cheap, but powerful, single-user workstation to the extremely fast and expensive mainframe, as exemplified in Fig. 1.1. Rapid development of computer hardware is currently under way, with the introduction of specialized features, such as pipeline and array processors, and totally new architectures, such as the dataflow approach. Computer simulation is possible on most machines, and we provide an overview of some widely available computers, and computing languages, as they relate to simulation, in Appendix A.

The very earliest work [Metropolis *et al.* 1953] laid the foundations of modern 'Monte Carlo' simulation (so-called because of the role that random numbers play in the method). The precise technique employed in this study is still widely used, and is referred to simply as 'Metropolis Monte Carlo'; we will use the abbreviation 'MC'. The original models were highly idealized representations of molecules, such as hard spheres and disks, but, within a few years MC simulations were carried out on the Lennard-Jones interaction potential [Wood and Parker 1957] (see Section 1.3). This made it possible to compare data obtained from experiments on, for example, liquid argon, with the computer-generated thermodynamic data derived from a model.

A different technique is required to obtain the dynamic properties of many-particle systems. Molecular dynamics (MD) is the term used to describe the solution of the classical equations of motion (Newton's equations) for a set of molecules. This was first accomplished, for a system of hard spheres, by Alder and Wainwright [1957, 1959]. In this case, the particles move at constant velocity between perfectly elastic collisions, and it is possible to solve the dynamic problem without making any approximations, within the limits imposed by machine accuracy. It was several years before a successful attempt was made to solve the equations of motion for a set of Lennard-Jones particles [Rahman 1964]. Here, an approximate, step-by-step procedure is needed, since the forces change continuously as the particles move. Since that time, the properties of the Lennard-Jones model have been thoroughly investigated [Verlet 1967, 1968; Nicolas, Gubbins, Streett, and Tildesley 1979].

After this initial groundwork on atomic systems, computer simulation developed rapidly. An early attempt to model a diatomic molecular liquid [Harp and Berne 1968; Berne and Harp 1970] using molecular dynamics was quickly followed by two ambitious attempts to model liquid water, first by MC [Barker and Watts 1969], and then by MD [Rahman and Stillinger 1971]. Water remains one of the most interesting and difficult liquids to study [Stillinger 1975, 1980; Wood 1979; Morse and Rice 1982]. Small rigid molecules [Barojas, Levesque, and Quentrec 1973], flexible hydrocarbons [Ryckaert and Bellemans 1975] and even large molecules such as proteins [McCammon, Gelin, and Karplus 1977] have all been objects of study in recent years. Computer simulation has been used to improve our understanding of phase transitions and behaviour at interfaces [Lee, Barker, and Pound 1974; Chapela, Saville, Thompson, and Rowlinson 1977; Frenkel and McTague 1980]. We shall be looking in detail at these developments in the last

**Fig. 1.1** Two modern computers. (a) The PERQ computer, marketed in the UK by ICL: a single-user graphics workstation capable of fast numerical calculations. (b) The CRAY 1-S computer: a supercomputer which uses pipeline processing to perform outstandingly fast numerical calculations.

chapter of this book. The techniques of computer simulation have also advanced, with the introduction of 'non-equilibrium' methods of measuring transport coefficients [Lees and Edwards 1972; Hoover and Ashurst 1975; Ciccotti, Jacucci, and McDonald 1979], the development of 'stochastic dynamics' methods [Turq, Lantelme, and Friedman 1977], and the incorporation of quantum mechanical effects [Corbin and Singer 1982; Ceperley and Kalos 1986]. Again, these will be dealt with in the later chapters. First, we turn to the questions: What is computer simulation? How does it work? What can it tell us?

## 1.2 Computer simulation: motivation and applications

Some problems in statistical mechanics are exactly soluble. By this, we mean that a complete specification of the microscopic properties of a system (such as the Hamiltonian of an idealized model like the perfect gas or the Einstein crystal) leads directly, and perhaps easily, to a set of interesting results or macroscopic properties (such as an equation of state like $PV = Nk_BT$). There are only a handful of non-trivial, exactly soluble problems in statistical mechanics [Baxter 1982]; the two-dimensional Ising model is a famous example.

Some problems in statistical mechanics, while not being exactly soluble, succumb readily to analysis based on a straightforward approximation scheme. Computers may have an incidental, calculational, part to play in such work, for example in the evaluation of cluster integrals in the virial expansion for dilute, imperfect gases. The problem is that, like the virial expansion, many 'straightforward' approximation schemes simply do not work when applied to liquids. For some liquid properties, it may not even be clear how to begin constructing an approximate theory in a reasonable way. The more difficult and interesting the problem, the more desirable it becomes to have exact results available, both to test existing approximation methods and to point the way towards new approaches. It is also important to be able to do this without necessarily introducing the additional question of how closely a particular model (which may be very idealized) mimics a real liquid, although this may also be a matter of interest.

Computer simulations have a valuable role to play in providing essentially exact results for problems in statistical mechanics which would otherwise only be soluble by approximate methods, or might be quite intractable. In this sense, computer simulation is a test of theories and, historically, simulations have indeed discriminated between well-founded approaches (such as integral equation theories [Hansen and McDonald 1986]) and ideas that are plausible but, in the event, less successful (such as the old cell theories of liquids [Lennard-Jones and Devonshire 1939a, 1939b]). The results of computer simulations may also be compared with those of real experiments. In the first place, this is a test of the underlying model used in a computer simulation.

Eventually, if the model is a good one, the simulator hopes to offer insights to the experimentalist, and assist in the interpretation of new results. This dual role of simulation, as a bridge between models and theoretical predictions on the one hand, and between models and experimental results on the other, is illustrated in Fig. 1.2. Because of this connecting role, and the way in which simulations are conducted and analysed, these techniques are often termed 'computer experiments'.

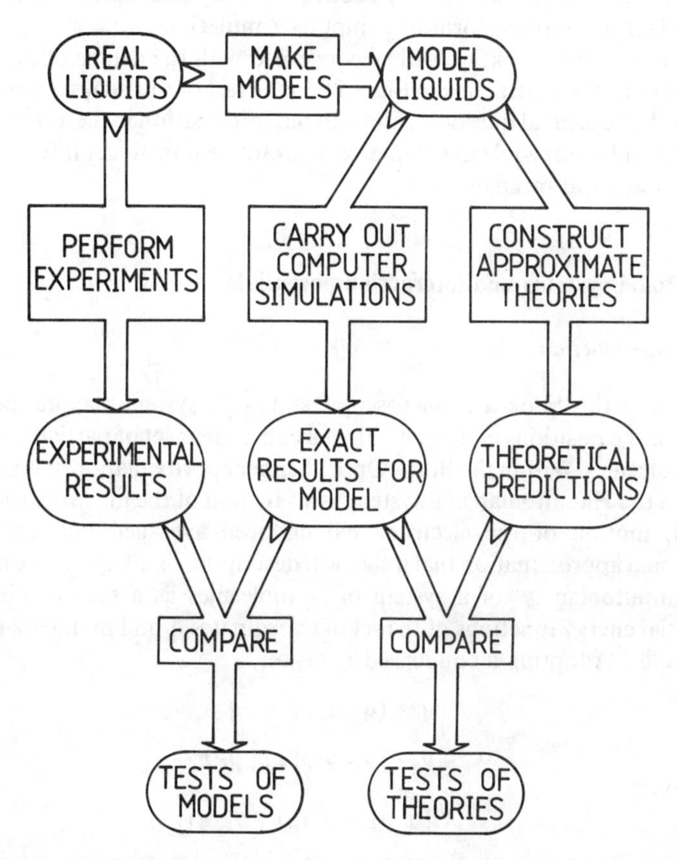


**Fig. 1.2** The connection between experiment, theory, and computer simulation.

Computer simulation provides a direct route from the microscopic details of a system (the masses of the atoms, the interactions between them, molecular geometry etc.) to macroscopic properties of experimental interest (the equation of state, transport coefficients, structural order parameters, and so on). As well as being of academic interest, this type of information is technologically useful. It may be difficult or impossible to carry out experiments under extremes of temperature and pressure, while a computer simulation of the material in, say, a shock wave, a high-temperature plasma, a nuclear reactor, or a planetary core, would be perfectly feasible. Quite subtle details of molecular motion and structure, for example in heterogeneous catalysis, fast ion conduction, or enzyme action, are difficult to probe experimentally, but can be extracted readily from a computer simulation. Finally, while the speed of molecular events is itself an experimental difficulty, it presents no hindrance to the simulator. A wide range of physical phenomena, from the molecular scale to the galactic [Hockney and Eastwood 1981], may be studied using some form of computer simulation.

In most of this book, we will be concerned with the details of carrying out simulations (the central box in Fig. 1.2). In the rest of this chapter, however, we deal with the general question of how to put information in (i.e. how to define a model of a liquid) while in Chapter 2 we examine how to get information out (using statistical mechanics).

## 1.3 Model systems and interaction potentials

### *1.3.1 Introduction*

In most of this book, the microscopic state of a system may be specified in terms of the positions and momenta of a constituent set of particles: the atoms and molecules. Within the Born–Oppenheimer approximation, it is possible to express the Hamiltonian of a system as a function of the nuclear variables, the (rapid) motion of the electrons having been averaged out. Making the additional approximation that a classical description is adequate, we may write the Hamiltonian $\mathscr{H}$ of a system of $N$ molecules as a sum of kinetic and potential energy functions of the set of coordinates $\mathbf{q}_i$ and momenta $\mathbf{p}_i$ of each molecule $i$. Adopting a condensed notation

$$\mathbf{q} = (\mathbf{q}_1, \mathbf{q}_2, \ldots, \mathbf{q}_N) \tag{1.1a}$$

$$\mathbf{p} = (\mathbf{p}_1, \mathbf{p}_2, \ldots, \mathbf{p}_N) \tag{1.1b}$$

we have

$$\mathscr{H}(\mathbf{q}, \mathbf{p}) = \mathscr{K}(\mathbf{p}) + \mathscr{V}(\mathbf{q}). \tag{1.2}$$

The generalized coordinates $\mathbf{q}$ may simply be the set of Cartesian coordinates $\mathbf{r}_i$ of each atom (or nucleus) in the system, but, as we shall see, it is sometimes useful to treat molecules as rigid bodies, in which case $\mathbf{q}$ will consist of the Cartesian coordinates of each molecular centre of mass together with a set of variables $\mathbf{\Omega}_i$ that specify molecular orientation. In any case, $\mathbf{p}$ stands for the appropriate set of conjugate momenta. Usually, the kinetic energy $\mathscr{K}$ takes the form

$$\mathscr{K} = \sum_{i=1}^{N} \sum_{\alpha} p_{i\alpha}^2 / 2m_i \tag{1.3}$$

where $m_i$ is the molecular mass, and the index $\alpha$ runs over the different $(x, y, z)$ components of the momentum of molecule $i$. The potential energy $\mathscr{V}$ contains the interesting information regarding intermolecular interactions: assuming that $\mathscr{V}$ is fairly sensibly behaved, it will be possible to construct, from $\mathscr{H}$, an equation of motion (in Hamiltonian, Lagrangian, or Newtonian form) which governs the entire time-evolution of the system and all its mechanical properties [Goldstein 1980]. Solution of this equation will generally involve calculating, from $\mathscr{V}$, the forces $\mathbf{f}_i$, and torques $\boldsymbol{\tau}_i$, acting on the molecules (see Chapter 3). The Hamiltonian also dictates the equilibrium distribution function for molecular positions and momenta (see Chapter 2). Thus, generally, it is $\mathscr{H}$ (or $\mathscr{V}$) which is the basic input to a computer simulation program. The approach used almost universally in computer simulation is to break up the potential energy into terms involving pairs, triplets, etc. of molecules. In the following sections we shall consider this in detail.

Before leaving this section, we should mention briefly somewhat different approaches to the calculation of $\mathscr{V}$. In these developments, the distribution of electrons in the system is not modelled by an effective potential $\mathscr{V}(\mathbf{q})$, but is treated by a form of density functional theory. In one approach, the electron density is represented by an extension of the electron gas theory [LeSar and Gordon 1982, 1983; LeSar 1984]. In another, electronic degrees of freedom are explicitly included in the description, and the electrons are allowed to relax during the course of the simulation by a process known as 'simulated annealing' [Car and Parrinello 1985]. Both these methods avoid the division of $\mathscr{V}$ into pairwise and higher terms. They seem promising for future simulations of solids and liquids.

### *1.3.2 Atomic systems*

Consider first the simple case of a system containing $N$ atoms. The potential energy may be divided into terms depending on the coordinates of individual atoms, pairs, triplets etc.:

$$\mathscr{V} = \sum_i v_1(\mathbf{r}_i) + \sum_i \sum_{j>i} v_2(\mathbf{r}_i, \mathbf{r}_j) + \sum_i \sum_{j>i} \sum_{k>j>i} v_3(\mathbf{r}_i, \mathbf{r}_j, \mathbf{r}_k) + \ldots \tag{1.4}$$

The $\sum_i \sum_{j>i}$ notation indicates a summation over all distinct pairs $i$ and $j$ without counting any pair twice (i.e. as $ij$ and $ji$); the same care must be taken for triplets etc. The first term in eqn (1.4), $v_1(\mathbf{r}_i)$, represents the effect of an external field (including, for example, the container walls) on the system. The remaining terms represent particle interactions. The second term, $v_2$, the pair potential, is the most important. The pair potential depends only on the magnitude of the pair separation $r_{ij} = |\mathbf{r}_i - \mathbf{r}_j|$, so it may be written $v_2(r_{ij})$. Figure 1.3 shows one of the more recent estimates for the pair potential between two argon atoms, as

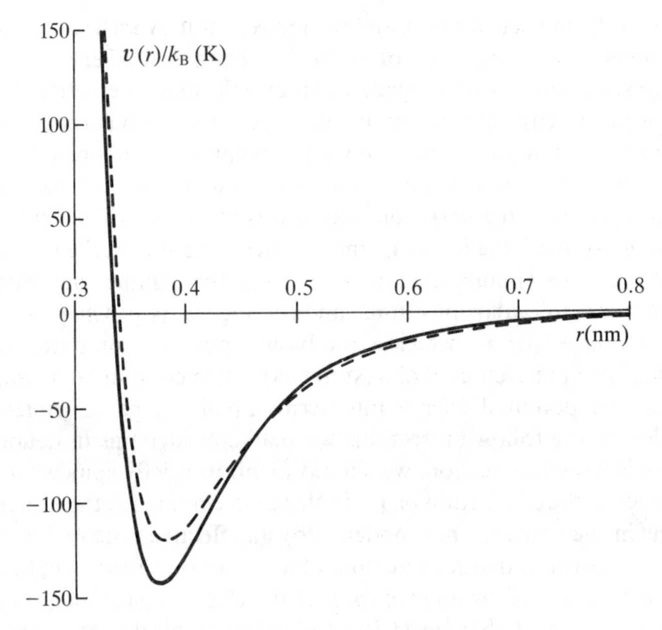


**Fig. 1.3** Argon pair potentials. We illustrate the BBMS pair potential for argon (solid line) [Maitland and Smith 1971]. The BFW potential [Barker *et al.* 1971] is numerically very similar. Also shown is the Lennard-Jones 12–6 effective pair potential (dashed line) used in computer simulations of liquid argon.

a function of separation [Bobetic and Barker 1970; Barker, Fisher, and Watts 1971; Maitland and Smith 1971]. This 'BBMS' potential was derived by considering a large quantity of experimental data, including molecular beam scattering, spectroscopy of the argon dimer, inversion of the temperature-dependence of the second virial coefficient, and solid-state properties, together with theoretical calculations of the long-range contributions [Maitland, Rigby, Smith, and Wakeham 1981]. The potential is also consistent with current estimates of transport coefficients in the gas phase.

The BBMS potential shows the typical features of intermolecular interactions. There is an attractive tail at large separations, essentially due to correlation between the electron clouds surrounding the atoms ('van der Waals' or 'London' dispersion). In addition, for charged species, Coulombic terms would be present. There is a negative well, responsible for cohesion in condensed phases. Finally, there is a steeply rising repulsive wall at short distances, due to non-bonded overlap between the electron clouds.

The $v_3$ term in eqn (1.4), involving triplets of molecules, is undoubtedly significant at liquid densities. Estimates of the magnitudes of the leading, triple–dipole, three-body contribution [Axilrod and Teller 1943] have been made for inert gases in their solid-state f.c.c. lattices [Doran and Zucker 1971; Barker 1976]. It is found that up to 10 per cent of the lattice energy of argon (and more in the case of more polarizable species) may be due to these non-additive terms in the potential; we may expect the same order of magnitude to hold in the liquid phase. Four-body (and higher) terms in eqn (1.4) are expected to be small in comparison with $v_2$ and $v_3$.

Despite the size of three-body terms in the potential, they are only rarely included in computer simulations [Barker *et al.* 1971; Monson, Rigby, and Steele 1983]. This is because, as we shall see shortly, the calculation of any quantity involving a sum over triplets of molecules will be very time-consuming on a computer. Fortunately, the pairwise approximation gives a remarkably good description of liquid properties because the average three-body effects can be partially included by defining an 'effective' pair potential. To do this, we rewrite eqn (1.4) in the form

$$\mathscr{V} \approx \sum_i v_1(\mathbf{r}_i) + \sum_i \sum_{j>i} v_2^{\text{eff}}(r_{ij}). \tag{1.5}$$

The pair potentials appearing in computer simulations are generally to be regarded as effective pair potentials of this kind, representing all the many-body effects; for simplicity, we will just use the notation $v(r_{ij})$ or $v(r)$. A consequence of this approximation is that the effective pair potential needed to reproduce experimental data may turn out to depend on the density, temperature etc., while the true two-body potential $v_2(r_{ij})$ of course does not.

Now we turn to the simpler, more idealized, pair potentials commonly used in computer simulations. These reflect the salient features of real interactions in a general, often empirical, way. Illustrated with the BBMS argon potential in Fig. 1.3 is a simple Lennard-Jones 12–6 potential

$$v^{\text{LJ}}(r) = 4\varepsilon((\sigma/r)^{12} - (\sigma/r)^6) \tag{1.6}$$

which provides a reasonable description of the properties of argon, via computer simulation, if the parameters $\varepsilon$ and $\sigma$ are chosen appropriately. The potential has a long-range attractive tail of the form $-1/r^6$, a negative well of depth $\varepsilon$, and a steeply rising repulsive wall at distances less than $r \sim \sigma$. The well-depth is often quoted in units of temperature as $\varepsilon/k_B$, where $k_B$ is Boltzmann's constant; values of $\varepsilon/k_B \approx 120\,\text{K}$ and $\sigma \approx 0.34\,\text{nm}$ provide reasonable agreement with the experimental properties of liquid argon. Once again, we must emphasize that these are not the values which would apply to an isolated pair of argon atoms, as is clear from Fig. 1.3

For the purposes of investigating general properties of liquids, and for comparison with theory, highly idealized pair potentials may be of value. In Fig. 1.4, we illustrate three forms which, although unrealistic, are very simple and convenient to use in computer simulation and in liquid-state theory. These are: the hard-sphere potential

$$v^{\text{HS}}(r) = \begin{cases} \infty & (r < \sigma) \\ 0 & (\sigma \leqslant r) \end{cases} \tag{1.7}$$

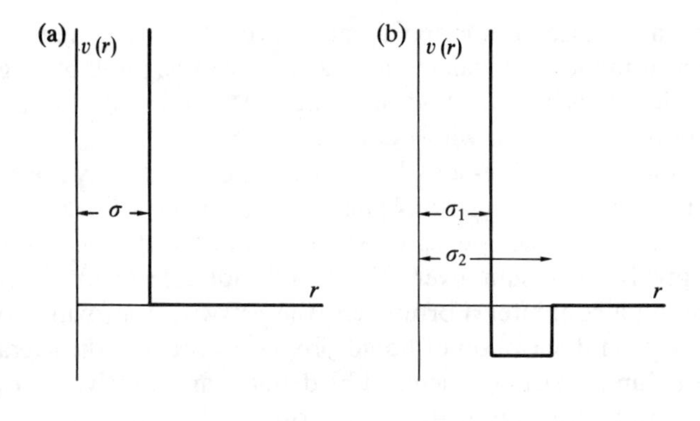


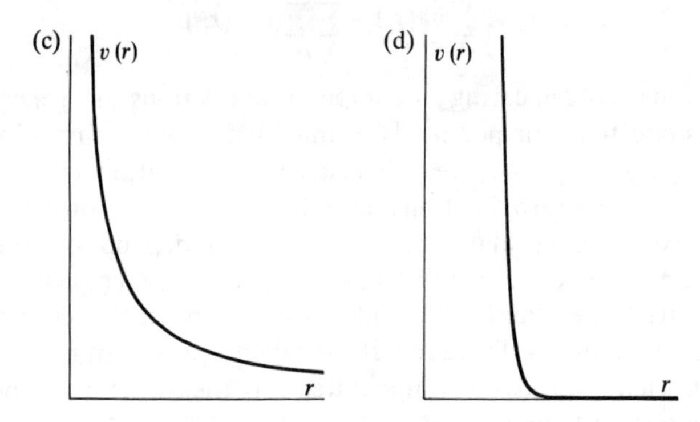


**Fig. 1.4** Idealized pair potentials. (a) The hard-sphere potential; (b) The square-well potential; (c) The soft-sphere potential with repulsion parameter $\nu = 1$; (d) The soft-sphere potential with repulsion parameter $\nu = 12$.

the square-well potential

$$v^{\mathrm{SW}}(r) = \begin{cases} \infty & (r < \sigma_1) \\ -\varepsilon & (\sigma_1 \leqslant r < \sigma_2) \\ 0 & (\sigma_2 \leqslant r) \end{cases} \tag{1.8}$$

and the soft-sphere potential

$$v^{\mathrm{SS}}(r) = \varepsilon(\sigma/r)^{\nu} = ar^{-\nu} \tag{1.9}$$

where $\nu$ is a parameter, often chosen to be an integer. The soft-sphere potential becomes progressively 'harder' as $\nu$ is increased. Soft-sphere potentials contain no attractive part.

It is often useful to divide more realistic potentials into separate attractive and repulsive components, and the separation proposed by Weeks, Chandler, and Andersen [1971] involves splitting the potential at the minimum. For the Lennard-Jones potential, the repulsive and attractive parts are thus

$$v^{\mathrm{RLJ}}(r) = \begin{cases} v^{\mathrm{LJ}}(r) + \varepsilon & r < r_{\min} \\ 0 & r_{\min} \leqslant r \end{cases} \tag{1.10a}$$

$$v^{\mathrm{ALJ}}(r) = \begin{cases} -\varepsilon & r < r_{\min} \\ v^{\mathrm{LJ}}(r) & r_{\min} \leqslant r \end{cases} \tag{1.10b}$$

where $r_{\min} = 2^{1/6}\sigma$. This separation is illustrated in Fig. 1.5. In perturbation theory [Weeks *et al.* 1971], a hypothetical fluid of molecules interacting via the repulsive potential $v^{\mathrm{RLJ}}$ is treated as a reference system and the attractive part $v^{\mathrm{ALJ}}$ is the perturbation. It should be noted that the potential $v^{\mathrm{RLJ}}(r)$ is significantly harder than the inverse 12th power soft-sphere potential, which is sometimes thought of as the 'repulsive' part of $v^{\mathrm{LJ}}(r)$.

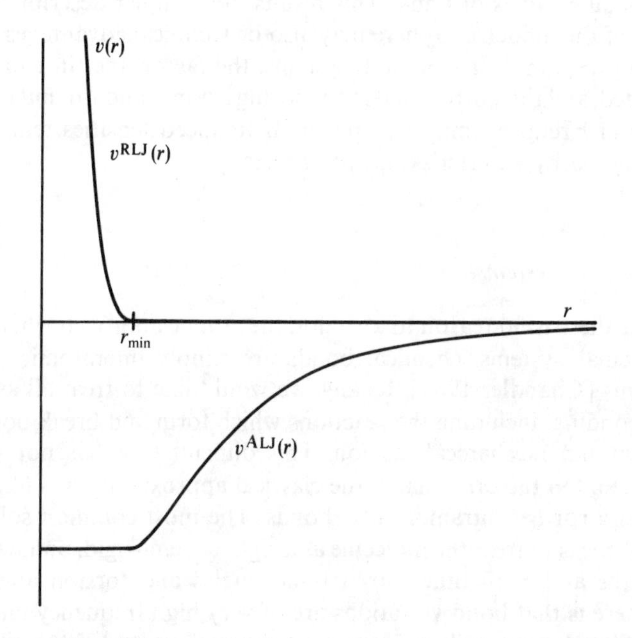


**Fig. 1.5** The separation of the Lennard-Jones potential into attractive and repulsive components.

For ions, of course, these potentials are not sufficient to represent the long-range interactions. A simple approach is to supplement one of the above pair potentials with the Coulomb charge–charge interaction

$$v^{zz}(r_{ij}) = \frac{z_i z_j}{4\pi\varepsilon_0 r_{ij}} \tag{1.11}$$

where $z_i$, $z_j$ are the charges on ions $i$ and $j$ and $\varepsilon_0$ is the permittivity of free space (not to be confused with $\varepsilon$ in eqns (1.6)–(1.10)).

For ionic systems, induction interactions are important: the ionic charge induces a dipole on a neighbouring ion. This term is not pairwise additive and hence is difficult to include in a simulation. The shell model is a crude attempt to take this ion polarizability into account [Dixon and Sangster 1976]. Each ion is represented as a core surrounded by a shell. Part of the ionic charge is located on the shell and the rest in the core. This division is always arranged so that the shell charge is negative (it represents the electronic cloud). The interactions between ions are just sums of the Coulombic shell–shell, core–core, and shell–core contributions. The shell and core of a given ion are coupled by a harmonic spring potential. The shells are taken to have zero mass. During a simulation, their positions are adjusted iteratively to zero the net force acting on each shell: this process makes the simulations very expensive.

When a potential depends upon just a few parameters, such as $\varepsilon$ and $\sigma$ above, it may be possible to choose an appropriate set of units in which these parameters take values of unity. This results in a simpler description of the properties of the model, and there may also be technical advantages within a simulation program. For Coulomb systems, the factor $4\pi\varepsilon_0$ in eqn (1.11) is often omitted, and this corresponds to choosing a non-standard unit of charge. We discuss such reduced units in Appendix B. Reduced densities, temperatures etc. are denoted by an asterisk, i.e. $\rho^*$, $T^*$ etc.

### *1.3.3 Molecular systems*

In principle there is no reason to abandon the atomic approach when dealing with molecular systems: chemical bonds are simply interatomic potential energy terms [Chandler 1982]. Ideally, we would like to treat all aspects of chemical bonding, including the reactions which form and break bonds, in a proper quantum mechanical fashion. This difficult task has not yet been accomplished. On the other hand, the classical approximation is likely to be seriously in error for intramolecular bonds. The most common solution to these problems is to treat the molecule as a rigid or semi-rigid, unit, with fixed bond lengths and, sometimes, fixed bond angles and torsion angles. The rationale here is that bond vibrations are of very high frequency (and hence difficult to handle, certainly in a classical simulation) but of low amplitude (therefore being unimportant for many liquid properties). Thus, a diatomic molecule with a strongly binding interatomic potential energy surface might be replaced by a dumb-bell with a rigid interatomic bond.

The interaction between the nuclei and electronic charge clouds of a pair of molecules $i$ and $j$ is clearly a complicated function of relative positions $\mathbf{r}_i$, $\mathbf{r}_j$ and orientations $\mathbf{\Omega}_i$, $\mathbf{\Omega}_j$ [Gray and Gubbins 1984]. One way of modelling a molecule is to concentrate on the positions and sizes of the constituent atoms [Eyring 1932]. The much simplified 'atom–atom' or 'site–site' approximation for diatomic molecules is illustrated in Fig. 1.6. The total interaction is a sum of

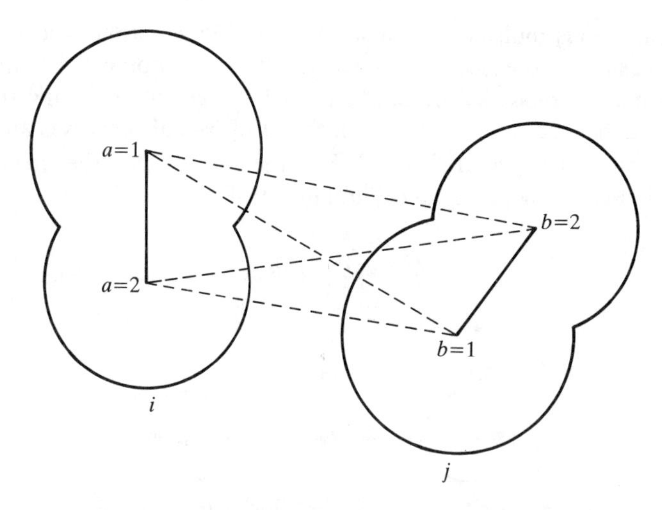


**Fig. 1.6** An atom–atom model of a diatomic molecule.

pairwise contributions from distinct sites $a$ in molecule $i$, at position $\mathbf{r}_{ia}$, and $b$ in molecule $j$, at position $\mathbf{r}_{jb}$

$$v(\mathbf{r}_{ij}, \mathbf{\Omega}_i, \mathbf{\Omega}_j) = \sum_a \sum_b v_{ab}(r_{ab}). \tag{1.12}$$

Here $a, b$ take the values $1, 2$, $v_{ab}$ is the pair potential acting between sites $a$ and $b$, and $r_{ab}$ is shorthand for the inter-site separation $r_{ab} = |\mathbf{r}_{ia} - \mathbf{r}_{jb}|$.

The interaction sites are usually centred, more or less, on the positions of the nuclei in the real molecule, so as to represent the basic effects of molecular 'shape'. A very simple extension of the hard-sphere model is to consider a diatomic composed of two hard spheres fused together [Streett and Tildesley 1976], but more realistic models involve continuous potentials. Thus, nitrogen, fluorine, chlorine etc. have been depicted as two 'Lennard-Jones atoms' separated by a fixed bond length [Barojas *et al.* 1973; Cheung and Powles 1975; Singer, Taylor, and Singer 1977]. Similar approaches apply to polyatomic molecules.

The description of the molecular charge distribution may be improved somewhat by incorporating point multipole moments at the centre of charge [Streett and Tildesley 1977]. These multipoles may be equal to the known (isolated molecule) values, or may be 'effective' values chosen simply to yield a better description of the liquid structure and thermodynamic properties. It is now generally accepted that such a multipole expansion is not rapidly convergent. A promising alternative approach for ionic and polar systems, is to use a set of fictitious 'partial charges' distributed 'in a physically reasonable way' around the molecule so as to reproduce the known multipole moments [Murthy, O'Shea, and McDonald 1983], and a further refinement is to distribute fictitious multipoles in a similar way [Price, Stone, and Alderton 1984]. For example, the electrostatic part of the interaction between nitrogen molecules may be modelled using five partial charges placed along the axis, while, for methane, a tetrahedral arrangement of partial charges is appropriate. These are illustrated in Fig. 1.7. For the case of $N_2$, the quadrupole moment $Q$ is given by [Gray and Gubbins 1984]

$$Q = \sum_{a=1}^{5} z_a r_{az}^2 \tag{1.13}$$

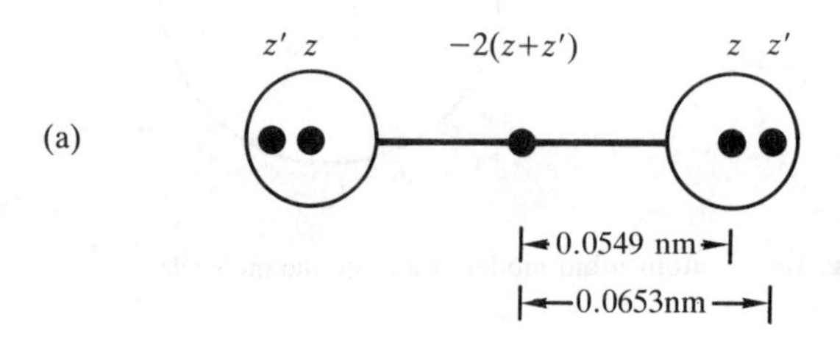


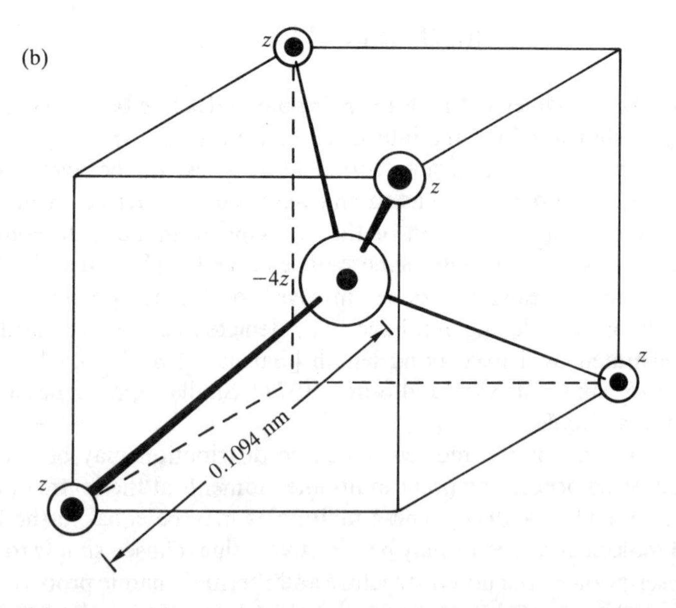


**Fig. 1.7** Partial charge models. (a) A five-charge model for $N_2$. There is one charge at the bond centre, two at the positions of the nuclei, and two more displaced beyond the nuclei. Typical values are (in units of the magnitude of the electronic charge) $z = +5.2366$, $z' = -4.0469$, giving $Q = -4.67 \times 10^{-40}\ \mathrm{C\,m^2}$ [Murthy *et al.* 1983]. (b) A five-charge model for $CH_4$. There is one charge at the centre, and four others at the positions of the hydrogen nuclei. A typical value is $z = 0.143$ giving $O = 5.77 \times 10^{-50}\ \mathrm{C\,m^3}$ [Righini, Maki, and Klein 1981].

with similar expressions for the higher multipoles (all the odd ones vanish for $N_2$). The first non-vanishing moment for methane is the octopole

$$O = \frac{5}{2} \sum_{a=1}^{5} z_a r_{ax} r_{ay} r_{az} \tag{1.14}$$

in the coordinate system of Fig. 1.7. The aim of all these approaches is to approximate the complete charge distribution in the molecule. In a calculation of the potential energy, the interaction between partial charges on different molecules would be summed in the same way as the other site–site interactions.

For large rigid models, a substantial number of sites would be required to model the repulsive core. For example, a crude model of the nematogen quinquaphenyl, which represented each of the five benzene rings as a single Lennard-Jones site, would necessitate 25 site–site interactions between each pair of molecules; sites based on each carbon atom would be more realistic but extremely expensive. An alternative type of intermolecular potential, introduced by Corner [1948], involves a single site–site interaction between a pair of molecules, characterized by energy and length parameters that depend on the relative orientation of the molecules. A version of this family of molecular potentials that has been used in computer simulation studies is the Gaussian overlap model generalized to a Lennard-Jones form [Berne and Pechukas 1972]. The basic potential acting between two linear molecules is the Lennard-Jones interaction, eqn (1.6), with the angular dependence of $\varepsilon$ and $\sigma$ determined by considering the overlap of two ellipsoidal Gaussian functions (representing the electron clouds of the molecules). The energy parameter is written

$$\varepsilon(\mathbf{\Omega}_i, \mathbf{\Omega}_j) = \varepsilon_{\mathrm{sph}}(1 - \chi^2(\mathbf{e}_i \cdot \mathbf{e}_j)^2)^{-1/2} \tag{1.15}$$

where $\varepsilon_{\mathrm{sph}}$ is a constant and $\mathbf{e}_i$, $\mathbf{e}_j$ are unit vectors describing the orientation of the molecules $i$ and $j$. $\chi$ is an anisotropy parameter determined by the length of the major and minor axes of the electron cloud ellipsoid

$$\chi = (\sigma^{\parallel 2} - \sigma^{\perp 2})/(\sigma^{\parallel 2} + \sigma^{\perp 2}). \tag{1.16}$$

The length parameter is given by

$$\sigma(r_{ij}, \mathbf{\Omega}_i, \mathbf{\Omega}_j) = \sigma_{\mathrm{sph}}\left[1 - \frac{1}{2}\chi\left(\frac{(\mathbf{r}_{ij} \cdot \mathbf{e}_i + \mathbf{r}_{ij} \cdot \mathbf{e}_j)^2}{r_{ij}^2(1 + \chi \mathbf{e}_i \cdot \mathbf{e}_j)} + \frac{(\mathbf{r}_{ij} \cdot \mathbf{e}_i - \mathbf{r}_{ij} \cdot \mathbf{e}_j)^2}{r_{ij}^2(1 - \chi \mathbf{e}_i \cdot \mathbf{e}_j)}\right)\right]^{-1/2} \tag{1.17}$$

where $\sigma_{\mathrm{sph}}$ is a constant. In certain respects, this form of the overlap potential is unrealistic, and it has been extended to make $\varepsilon$ also dependent upon $\mathbf{r}_{ij}$ [Gay and Berne 1981]. The extended potential can be parameterized to mimic a linear site–site potential, and should be particularly useful in the simulation of nematic liquid crystals.

For larger molecules it may not be reasonable to 'fix' all the internal degrees of freedom. In particular, torsional motion about bonds, which gives rise to conformational interconversion in, for example, alkanes, cannot in general be neglected (since these motions involve energy changes comparable with normal thermal energies). An early simulation of n-butane, $CH_3CH_2CH_2CH_3$ [Ryckaert and Bellemans 1975; Maréchal and Ryckaert 1983], provides a good example of the way in which these features are incorporated in a simple model. Butane can be represented as a four-centre molecule, with fixed bond lengths and bond bending angles, derived from known experimental (structural) data (see Fig. 1.8). A very common simplifying feature is built into this model: whole groups of atoms, such as $CH_3$ and $CH_2$, are condensed into spherically symmetric effective 'united atoms'. In fact, for butane, the interactions between such groups may be represented quite well by the ubiquitous Lennard-Jones potential, with empirically chosen parameters. In a simulation, the $C_1$–$C_2$, $C_2$–$C_3$ and $C_3$–$C_4$ bond lengths are held fixed by a method of constraints which will be described in detail in Chapter 3. The angles $\theta$ and $\theta'$ may be fixed by additionally constraining the $C_1$–$C_3$ and $C_2$–$C_4$ distances, i.e. by introducing 'phantom bonds'. If this is done, just one internal degree of freedom, namely the rotation about the $C_2$–$C_3$ bond, measured by the angle $\phi$, is left unconstrained; for each molecule, an extra term in the potential energy, $v^{\text{torsion}}(\phi)$, periodic in $\phi$, appears in the hamiltonian. This potential would have a minimum at a value of $\phi$ corresponding to the *trans* conformer of butane, and secondary minima at the *gauche* conformations. It is easy to see how this approach may be extended to much larger flexible molecules. The consequences of constraining bond lengths and angles will be treated in more detail in Chapters 2–4.

As the molecular model becomes more complicated, so too do the expressions for the potential energy, forces, and torques, due to molecular interactions. In Appendix C, we give some examples of these formulae, for rigid and flexible molecules, interacting via site–site pairwise potentials, including multipolar terms. We also show how to derive the forces from a simple three-body potential.

### *1.3.4 Lattice systems*

We may also consider the consequences of removing rather than adding degrees of freedom to the molecular model. In a crystal, molecular translation is severely restricted, while rotational motion (in plastic crystals for instance) may still occur. A simplified model of this situation may be devised, in which the molecular centres of mass are fixed at their equilibrium crystal lattice sites, and the potential energy is written solely as a function of molecular orientations. Such models are frequently of theoretical, rather than practical, interest, and accordingly the interactions are often of a very idealized form: the molecules may be represented as point multipoles for example, and interac-

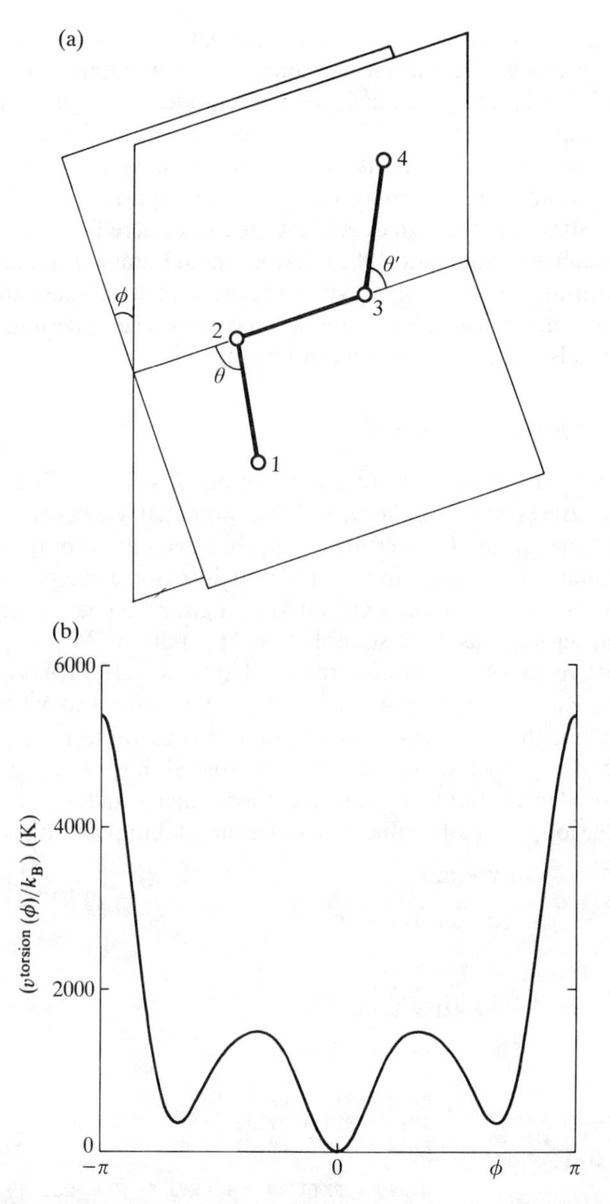


**Fig. 1.8** (a) A model of butane [Ryckaert and Bellemans 1975]. (b) The torsional potential [Marechal and Ryckaert 1983].

tions may even be restricted to nearest neighbours only [O'Shea 1978; Nosé, Kataoka, Okada, and Yamamoto 1981]. Ultimately, this leads us to the spin models of theoretical physics, as typified by the Heisenberg, Ising, and Potts models. These models are really attempts to deal with a simple quantum mechanical Hamiltonian for a solid-state lattice system, rather than the classical equations of motion for a liquid. However, because of its correspondence with the lattice gas model, the Ising model is still of some interest in classical liquid state theory. There has been a substantial amount of work involving Monte Carlo simulation of such spin systems, which we must regrettably omit from a book of this size. The importance of these idealized models in statistical mechanics is illustrated elsewhere [see e.g. Binder 1984, 1986; Toda, Kubo, and Saito 1983]. Lattice model simulations, however, have been useful in the study of polymer chains, and we discuss this briefly in Chapter 4. Paradoxically, lattice models have also been useful in the study of liquid crystals, which we mention in Chapter 11.

### *1.3.5 Calculating the potential*

This is an appropriate point to introduce our first piece of computer code, which illustrates the calculation of the potential energy in a system of Lennard-Jones atoms. Converting the algebraic equations of this chapter into a form suitable for the computer is a straightforward exercise in FORmula TRANslation, for which the FORTRAN programming language has historically been regarded as most suitable (see Appendix A). We suppose that the coordinate vectors of our atoms are stored in three FORTRAN arrays RX (I), RY (I) and RZ (I), with the particle index I varying from 1 to $N$ (the number of particles). For the Lennard-Jones potential it is useful to have precomputed the value of $\sigma^2$, which is stored in the variable SIGSQ. The potential energy will be stored in a variable V, which is zeroed initially, and is then accumulated in a double loop over all distinct pairs of atoms, taking care to count each pair only once.

```
        V = 0.0

        DO 100 I = 1, N - 1

           RXI = RX(I)
           RYI = RY(I)
           RZI = RZ(I)

           DO 99 J = I + 1, N

              RXIJ = RXI - RX(J)
              RYIJ = RYI - RY(J)
              RZIJ = RZI - RZ(J)

              RIJSQ = RXIJ ** 2 + RYIJ ** 2 + RZIJ ** 2
              SR2   = SIGSQ / RIJSQ
              SR6   = SR2 * SR2 * SR2
              SR12  = SR6 ** 2
              V     = V + SR12 - SR6

99         CONTINUE

100     CONTINUE

        V = 4.0 * EPSLON * V
```

Some measures have been taken here to avoid unnecessary use of computer time. The factor $4\varepsilon$ (4.0 *EPSLON in FORTRAN), which appears in every pair potential term, is multiplied in once, at the very end, rather than many times within the crucial 'inner loop' over index J. We have used temporary variables RXI, RYI, and RZI so that we do not have to make a large number of array references in this inner loop. Other, more subtle points (such as whether it may be faster to compute the square of a number by using the exponentiation operation ** or by multiplying the number by itself) are discussed in Appendix A. The more general questions of time-saving tricks in this part of the program are addressed in Chapter 5. The extension of this type of double loop to deal with other forms of the pair potential, and to compute forces in addition to potential terms, is straightforward, and examples will be given in later chapters. For molecular systems, the same general principles apply, but additional loops over the different sites or atoms in a molecule may be needed. For example, consider the site–site diatomic model of eqn (1.12) and Fig. 1.6. If the coordinates of site $a$ in molecule $i$ are stored in array elements RX (I, A), RY (I, A), RZ (I, A), then the intermolecular interactions might be computed as follows:

```
        DO 100 A = 1, 2

           DO 99 B = 1, 2

              DO 98 I = 1, N - 1

                 DO 97 J = I + 1, N

                    RXAB = RX(I,A) - RX(J,B)
                    RYAB = RY(I,A) - RY(J,B)
                    RZAB = RZ(I,A) - RZ(J,B)

                    ... calculate ia-jb interaction ...

97               CONTINUE

98            CONTINUE

99         CONTINUE

100     CONTINUE
```

This use of doubly dimensioned arrays may not be efficient on some machines, but is quite convenient. Note that, apart from the dependence of the loop over J on the index I, the order of nesting is a matter of choice. Here, we have placed a loop over molecular indices innermost; assuming that $N$ is relatively large, the vectorization of this loop on a pipeline machine will result in a great increase in speed of execution. Simulations of molecular systems may also involve the calculation of intramolecular energies, which, for site–site potentials, will necessitate a triple summation (over I, A, and B).

The above examples are essentially summations over pairs of interaction sites in the system. Any calculation of three-body interactions will, of course, entail triple summations of the kind

```
      DO 100 I = 1, N - 2
         DO 99 J = I + 1, N - 1
            DO 98 K = J + 1, N
               ... calculate i-j-k interaction ...
98          CONTINUE
99       CONTINUE
100   CONTINUE
```

Because all distinct triplets are examined, this will be much more time consuming than the summations described above. Even for pairwise-additive potentials, the energy or force calculation is the most expensive part of a computer simulation. We will return to this crucial section of the program in Chapter 5.

## 1.4 Constructing an intermolecular potential

### *1.4.1 Introduction*

There are essentially two stages in setting up a realistic simulation of a given system. The first is 'getting started' by constructing a first guess at a potential model. This should be a reasonable model of the system, and allow some preliminary simulations to be carried out. The second is to use the simulation results to refine the potential model in a systematic way, repeating the process several times if necessary. We consider the two phases in turn.

### *1.4.2 Building the model potential*

To illustrate the process of building up an intermolecular potential, we begin by considering a small molecule, such as $N_2$, OCS, or $CH_4$, which can be modelled using the interaction site potentials discussed in Section 1.3. The essential features of this model will be an anisotropic repulsive core, to represent the shape, an anisotropic dispersion interaction, and some partial charges to model the permanent electrostatic effects. This crude effective pair potential can then be refined by using it to calculate properties of the gas, liquid, and solid, and comparing with experiment.

Each short-range site–site interaction can be modelled using a Lennard-Jones potential. Suitable energy and length parameters for interactions between pairs of identical atoms in different molecules are available from a number of simulation studies. Some of these are given in Table 1.1.

**Table 1.1.** *Atom–atom interaction parameters*

| Atom | Source | $\varepsilon/k_B$(K) | $\sigma$(nm) |
|---|---|---|---|
| H | [Murad and Gubbins 1978] | 8.6 | 0.281 |
| He | [Maitland *et al.* 1981] | 10.2 | 0.228 |
| C | [Tildesley and Madden 1981] | 51.2 | 0.335 |
| N | [Cheung and Powles 1975] | 37.3 | 0.331 |
| O | [English and Venables 1974] | 61.6 | 0.295 |
| F | [Singer *et al.* 1977] | 52.8 | 0.283 |
| Ne | [Maitland *et al.* 1981] | 47.0 | 0.272 |
| S | [Tildesley and Madden, 1981] | 183.0 | 0.352 |
| Cl | [Singer *et al.* 1977] | 173.5 | 0.335 |
| Ar | [Maitland *et al.* 1981] | 119.8 | 0.341 |
| Br | [Singer *et al.* 1977] | 257.2 | 0.354 |
| Kr | [Maitland *et al.* 1981] | 164.0 | 0.383 |

The energy parameter $\varepsilon$ increases with atomic number as the polarizability goes up; $\sigma$ also increases down a group of the periodic table, but decreases from left to right across a period with the increasing nuclear charge. For elements which do not appear in Table 1.1, a guide to $\varepsilon$ and $\sigma$ might be provided by the polarizability and van der Waals radius respectively. These values are only intended as a reasonable first guess: they take no regard of chemical environment and are not designed to be transferable. For example, the carbon atom parameters in $CS_2$ given in the table are quite different from the values appropriate to a carbon atom in graphite [Crowell 1958]. Interactions between unlike atoms in different molecules can be approximated using the venerable Lorentz–Berthelot mixing rules. For example, in $CS_2$ the cross terms are given by

$$\sigma_{CS} = \tfrac{1}{2}[\sigma_{CC} + \sigma_{SS}] \tag{1.18}$$

and

$$\varepsilon_{CS} = [\varepsilon_{CC}\,\varepsilon_{SS}]^{1/2}\,. \tag{1.19}$$

In tackling larger molecules, it may be necessary to model several atoms as a unified site. We have seen this for butane in Section 1.3, and a similar approach has been used in a model of benzene [Evans and Watts 1976]. There are also complete sets of transferable potential parameters available for aromatic and aliphatic hydrocarbons [Williams 1965, 1967], and for hydrogen-bonded liquids [Jorgensen 1981], which use the site–site approach. In the case of the Williams potentials, an exponential repulsion rather than Lennard-Jones power law is used. The specification of an interaction site model is made complete by defining the positions of the sites within the molecule. Normally, these are located at the positions of the nuclei, with the bond lengths obtained from a standard source [CRC 1984].

The site–site Lennard-Jones potentials include an anisotropic dispersion which has the correct $r^{-6}$ radial dependence at long range. However, this is not the exact result for the anisotropic dispersion from second order perturbation theory. The correct formula, in an appropriate functional form for use in a simulation, is given by Burgos, Murthy, and Righini [1982]. Its implementation requires an estimate of the polarizability and polarizability anisotropy of the molecule.

The most convenient way of representing electrostatic interactions is through partial charges as discussed in Section 1.3. To minimize the calculation of site–site distances, they can be made to coincide with the Lennard-Jones sites, but this is not always desirable or possible; the only physical constraint on partial charge positions is that they should not lie outside the repulsive core region, since the potential might then diverge if molecules came too close. The magnitudes of the charges can be chosen to duplicate the known gas phase electrostatic moments [Gray and Gubbins 1984, Appendix D]. Alternatively, the moments may be taken as adjustable parameters. For example, in a simple three-site model of $N_2$ representing only the quadrupole–quadrupole interaction, the best agreement with condensed phase properties is obtained with charges giving a quadrupole 10–15 per cent lower than the gas phase value [Murthy, Singer, Klein, and McDonald 1980]. However, a sensible strategy is to begin with the gas phase values, and alter the repulsive core parameters $\varepsilon$ and $\sigma$ before changing the partial charges.

### *1.4.3 Adjusting the model potential*

The first-guess potential can be used to calculate a number of properties in the gas, liquid, and solid phases; comparison of these results with experiment may be used to refine the potential, and the cycle can be repeated if necessary. The second virial coefficient is given by

$$B(T) = -\frac{2\pi}{\Omega^2}\int_0^\infty r_{ij}^2\,\mathrm{d}r_{ij}\int \mathrm{d}\boldsymbol{\Omega}_i\int \mathrm{d}\boldsymbol{\Omega}_j \exp\left(-v(r_{ij},\boldsymbol{\Omega}_i,\boldsymbol{\Omega}_j)/k_\mathrm{B}T\right)-1 \tag{1.20}$$

where $\Omega = 4\pi$ for a linear molecule and $\Omega = 8\pi^2$ for a non-linear one. This multidimensional integral (four-dimensional for a linear molecule and six-dimensional for a non-linear one) is easily calculated using a non-product algorithm [Murad 1978]. Experimental values of $B(T)$ have been compiled by Dymond and Smith [1980]. Trial and error adjustment of the Lennard-Jones $\varepsilon$ and $\sigma$ parameters should be carried out, with any bond lengths and partial charges held fixed, so as to produce the closest match with the experimental

$B(T)$. This will produce an improved potential, but still one that is based on pair properties.

The next step is to carry out a series of computer simulations of the liquid state, as described in Chapters 3 and 4. The densities and temperatures of the simulations should be chosen to be close to the orthobaric curve of the real system, i.e. the liquid–vapour coexistence line. The output from these simulations, particularly the total internal energy and the pressure, may be compared with the experimental values. The coexisting pressures are readily available [Rowlinson and Swinton 1982], and the internal energy can be obtained approximately from the known latent heat of evaporation. The energy parameters $\varepsilon$ are adjusted to give a good fit to the internal energies along the orthobaric curve, and the length parameters $\sigma$ altered to fit the pressures. If no satisfactory fit is obtained at this stage, the partial charges may be adjusted.

Although the solid state is not the province of this book, it offers a sensitive test of any potential model. Using the experimentally observed crystal structure, and the refined potential model, the lattice energy at zero temperature can be compared with the experimental value (remembering to add a correction for quantum zero-point motion). In addition, the lattice parameters corresponding to the minimum energy for the model solid can be compared with the values obtained by diffraction, and also lattice dynamics calculations [Neto, Righini, Califano, and Walmsley 1978] used to obtain phonons, librational modes, and dispersion curves of the model solid. Finally, we can ask if the experimental crystal structure is indeed the minimum energy structure for our potential. These constitute severe tests of our model-building skills.

## 1.5 Studying small systems

### *1.5.1 Introduction*

Computer simulations are usually performed on a small number of molecules, $10 \leqslant N \leqslant 10\,000$. The size of the system is limited by the available storage on the host computer, and, more crucially, by the speed of execution of the program. The time taken for a double loop used to evaluate the forces or potential energy is proportional to $N^2$. Special techniques (see Chapter 5) may reduce this dependence to $\mathcal{O}(N)$, for very large systems, but the force/energy loop almost inevitably dictates the overall speed, and, clearly, smaller systems will always be less expensive. If we are interested in the properties of a very small liquid drop, or a microcrystal, then the simulation will be straightforward. The cohesive forces between molecules may be sufficient to hold the system together unaided during the course of a simulation; otherwise our set of $N$ molecules may be confined by a potential representing a container, which prevents them from drifting apart (see Chapter 11). These arrangements, however, are not satisfactory for the simulation of bulk liquids. A major obstacle to such a simulation is the large fraction of molecules which lie on the surface of any small sample; for 1000 molecules arranged in a $10 \times 10 \times 10$ cube, no less than 488 molecules appear on the cube faces. Whether or not the cube is surrounded by a containing wall, molecules on the surface will experience quite different forces from molecules in the bulk.

### *1.5.2 Periodic boundary conditions*

The problem of surface effects can be overcome by implementing periodic boundary conditions [Born and von Karman 1912]. The cubic box is replicated throughout space to form an infinite lattice. In the course of the simulation, as a molecule moves in the original box, its periodic image in each of the neighbouring boxes moves in exactly the same way. Thus, as a molecule leaves the central box, one of its images will enter through the opposite face. There are no walls at the boundary of the central box, and no surface molecules. This box simply forms a convenient axis system for measuring the coordinates of the $N$ molecules. A two-dimensional version of such a periodic system is shown in Fig. 1.9. The duplicate boxes are labelled A, B, C, etc., in an

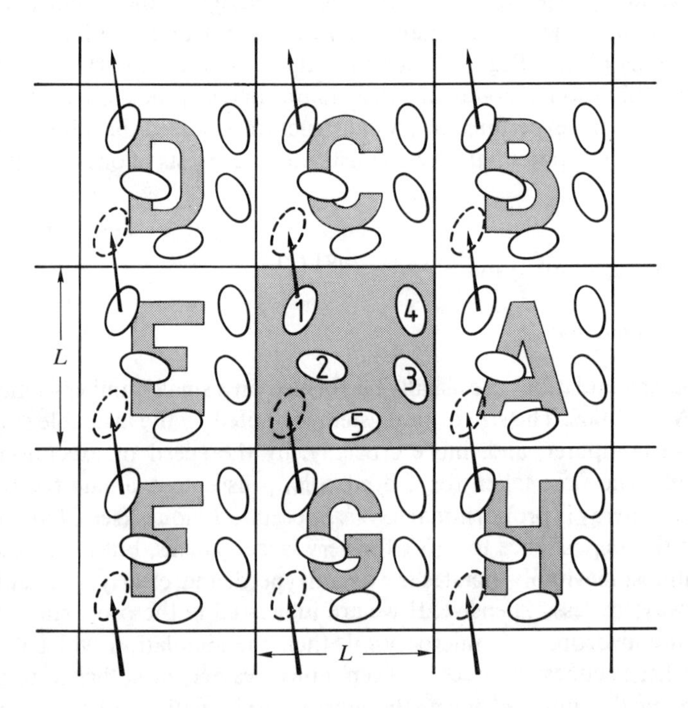


**Fig. 1.9** A two-dimensional periodic system. Molecules can enter and leave each box across each of the four edges. In a three-dimensional example, molecules would be free to cross any of the six cube faces.

arbitrary fashion. As particle 1 moves through a boundary, its images, $1_A$, $1_B$, etc. (where the subscript specifies in which box the image lies) move across their corresponding boundaries. The number density in the central box (and hence in the entire system) is conserved. It is not necessary to store the coordinates of all the images in a simulation (an infinite number!), just the molecules in the central box. When a molecule leaves the box by crossing a boundary, attention may be switched to the image just entering. It is sometimes useful to picture the basic simulation box (in our two-dimensional example) as being rolled up to form the surface of a three-dimensional torus or doughnut, when there is no need to consider an infinite number of replicas of the system, nor any image particles. This correctly represents the topology of the system, if not the geometry. A similar analogy exists for a three-dimensional periodic system, but this is more difficult to visualize!

It is important to ask if the properties of a small, infinitely periodic, system, and the macroscopic system which it represents, are the same. This will depend both on the range of the intermolecular potential and the phenomenon under investigation. For a fluid of Lennard-Jones atoms, it should be possible to perform a simulation in a cubic box of side $L \approx 6\sigma$, without a particle being able to 'sense' the symmetry of the periodic lattice. If the potential is long ranged (i.e. $v(r) \sim r^{-\nu}$ where $\nu$ is less than the dimensionality of the system) there will be a substantial interaction between a particle and its own images in neighbouring boxes, and consequently the symmetry of the cell structure is imposed on a fluid which is in reality isotropic. The methods used to cope with long-range potentials, for example in the simulation of charged ions ($v(r) \sim r^{-1}$) and dipolar molecules ($v(r) \sim r^{-3}$), are discussed in Chapter 5. Recent work has shown that, even in the case of short-range potentials, the periodic boundary conditions can induce anisotropies in the fluid structure [Mandell 1976; Impey, Madden, and Tildesley 1981]. These effects are pronounced for small system sizes ($N \approx 100$) and for properties such as the $g_2$ light scattering factor (see Chapter 2), which has a substantial long-range contribution. Pratt and Haan [1981] have developed theoretical methods for investigating the effects of boundary conditions on equilibrium properties.

The use of periodic boundary conditions inhibits the occurrence of long-wavelength fluctuations. For a cube of side $L$, the periodicity will suppress any density waves with a wavelength greater than $L$. Thus, it would not be possible to simulate a liquid close to the gas–liquid critical point, where the range of critical fluctuations is macroscopic. Furthermore, transitions which are known to be first order often exhibit the characteristics of higher order transitions when modelled in a small box because of the suppression of fluctuations. Examples are the nematic to isotropic transition in liquid crystals [Luckhurst and Simpson 1982] and the solid to plastic crystal transition for $N_2$ adsorbed on graphite [Mouritsen and Berlinsky 1982]. The same limitations apply to the simulation of long-wavelength phonons in model solids, where, in addition, the cell periodicity picks out a discrete set of available wave-vectors (i.e.

$\mathbf{k} = (k_x, k_y, k_z)2\pi/L$, where $k_x, k_y, k_z$ are integers) in the first Brillouin zone [Klein and Weis 1977]. Periodic boundary conditions have also been shown to affect the rate at which a simulated liquid nucleates and forms a solid or glass when it is rapidly cooled [Honeycutt and Andersen 1984].

Despite the above remarks, the common experience in simulation work is that periodic boundary conditions have little effect on the equilibrium thermodynamic properties and structures of fluids away from phase transitions and where the interactions are short-ranged. It is always sensible to check that this is true for each model studied. If the resources are available, it should be standard practice to increase the number of molecules (and the box size, so as to maintain constant density) and rerun the simulations.

The cubic box has been used almost exclusively in computer simulation studies because of its geometrical simplicity. Of the four remaining semi-regular space-filling polyhedra, the rhombic dodecahedron [Wang and Krumhansl 1972] and the truncated octahedron [Adams 1979, 1980] have also been studied. These boxes are illustrated in Fig. 1.10. They are more nearly spherical than the cube, which may be useful for simulating liquids, whose structure is spatially isotropic. In addition, for a given number density, the distance between periodic images is larger than in the cube. This property is useful in calculating distribution functions and structure factors (see Chapters 2 and 6).

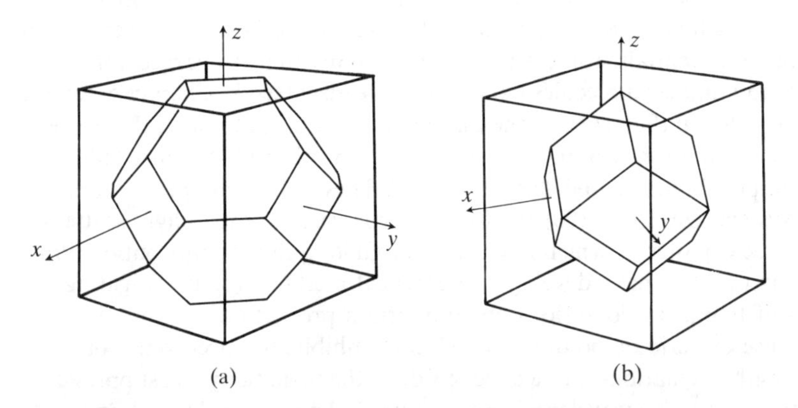


**Fig. 1.10** Non-cubic simulation boxes. (a) The truncated octahedron and its containing cube; (b) the rhombic dodecahedron and its containing cube. The axes are those used in microfiche F.1.

So far, we have tacitly assumed that there is no external potential, i.e. no $v_1$ term in eqns (1.4), (1.5). If such a potential is present, then either it must have the same periodicity as the simulation box, or the periodic boundaries must be abandoned. In some cases, it is not appropriate to employ periodic boundary conditions in each of the three coordinate directions. In the simulation of $CH_4$ on graphite [Severin and Tildesley 1980] the simulation box, shown in Fig. 1.11, is periodic in the plane of the surface. In the $z$-direction, the graphite surface

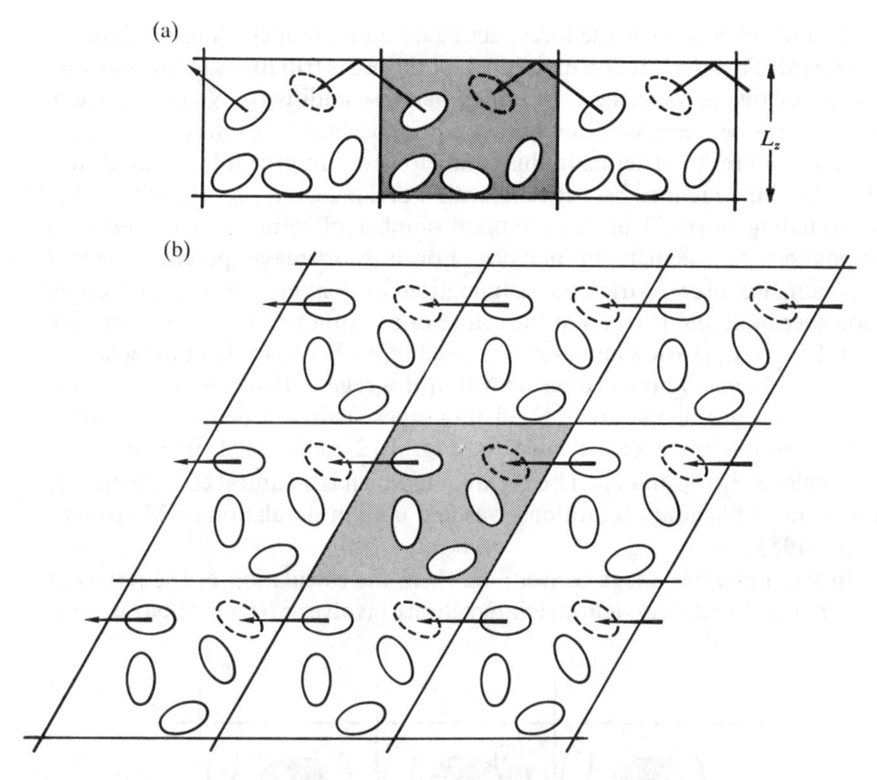


**Fig. 1.11** Periodic boundaries used in the simulation of adsorption [Severin and Tildesley 1980]. (a) A side view of the box. There is a reflecting boundary at a height $L_z$. (b) A top view, showing the rhombic shape (i.e. the same geometry as the underlying graphite lattice).

forms the lower boundary of the box, and the bulk of the adsorbate is in the region just above the graphite. Any molecule in the gas above the surface is confined by reversing its velocity should it cross a plane at a height $L_z$ above the surface. If $L_z$ is sufficiently large, this reflecting boundary will not influence the behaviour of the adsorbed monolayer. In the plane of the surface, the shape of the periodic box is a rhombus of side $L$. This conforms to the symmetry of the underlying graphite. Similar boxes have been used in the simulation of the electrical double layer [Torrie and Valleau 1979], of the liquid–vapour surface [Chapela *et al.* 1977], and of fluids in small pores [Subramanian and Davis 1979].

### *1.5.3 Potential truncation*

Now we must turn to the question of calculating properties of systems subject to periodic boundary conditions. The heart of the MC and MD programs involves the calculation of the potential energy of a particular configuration, and, in the case of MD, the forces acting on all molecules. Consider how we would calculate the force on molecule 1, or those contributions to the potential energy involving molecule 1, assuming pairwise additivity. We must include interactions between molecule 1 and every other molecule $i$ in the simulation box. There are $N-1$ terms in this sum. However, in principle, we must also include all interactions between molecule 1 and images $i_A$, $i_B$, etc. lying in the surrounding boxes. This is an infinite number of terms, and of course is impossible to calculate in practice. For a short-range potential energy function, we may restrict this summation by making an approximation. Consider molecule 1 to rest at the centre of a region which has the same size and shape as the basic simulation box (see Fig. 1.12). Molecule 1 interacts with all the molecules whose centres lie within this region, that is with the closest periodic images of the other $N-1$ molecules. This is called the 'minimum image convention': for example, in Fig. 1.12 molecule 1 interacts with molecules 2, $3_E$, $4_E$ and $5_C$. This technique, which is a natural consequence of the periodic boundary conditions, was first used in simulation by Metropolis *et al.* [1953].

In the minimum image convention, then, the calculation of the potential energy due to pairwise-additive interactions involves $\frac{1}{2}N(N-1)$ terms. This

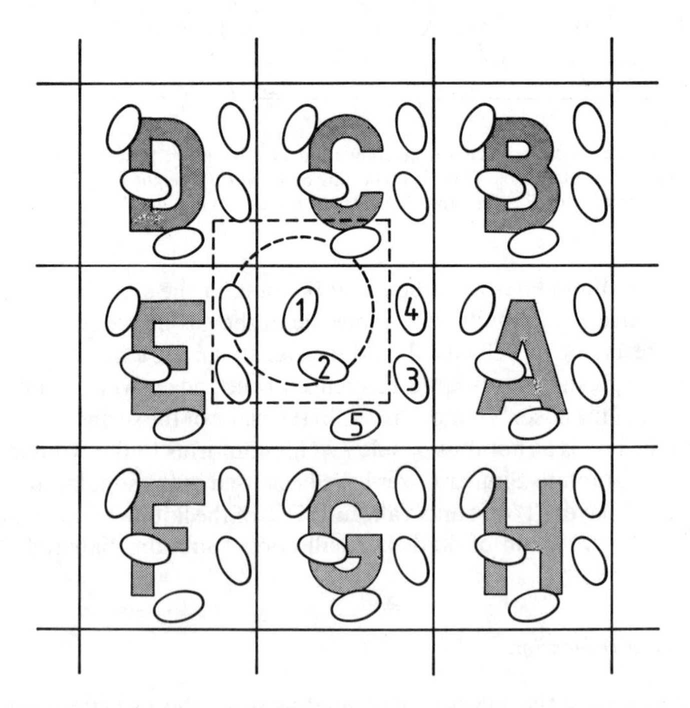


**Fig. 1.12** The minimum image convention in a two-dimensional system. The central box contains five molecules. The 'box' constructed with molecule 1 at its centre also contains five molecules. The dashed circle represents a potential cutoff.

may still be a very substantial calculation for a system of (say) 1000 particles. A further approximation significantly improves this situation. The largest contribution to the potential and forces comes from neighbours close to the molecule of interest, and for short-range forces we normally apply a spherical cutoff. This means setting the pair potential $v(r)$ to zero for $r \geqslant r_c$, where $r_c$ is the cutoff distance. The dashed circle in Fig. 1.12 represents a cutoff, and in this case molecules 2 and $4_E$ contribute to the force on 1, since their centres lie inside the cutoff, whereas molecules $3_E$ and $5_C$ do not contribute. In a cubic simulation box of side $L$, the number of neighbours explicitly considered is reduced by a factor of approximately $4\pi r_c^3/3L^3$, and this may be a substantial saving. The introduction of a spherical cutoff should be a small perturbation, and the cutoff distance should be sufficiently large to ensure this. As an example, in the simulation of Lennard-Jones atoms, the value of the pair potential at the boundary of a cutoff sphere of typical radius $r_c = 2.5\sigma$ is just 1.6 per cent of the well depth. Of course, the penalty of applying a spherical cutoff is that the thermodynamic (and other) properties of the model fluid will no longer be exactly the same as for (say) the non-truncated, Lennard-Jones fluid. As we shall see in Chapter 2, it is possible to apply long-range corrections to such results, so as to recover, approximately, the desired information.

The cutoff distance must be no greater than $\frac{1}{2}$L for consistency with the minimum image convention. In the non-cubic simulation boxes of Fig. 1.10, for a given density and number of particles, $r_c$ may take somewhat larger values than in the cubic case. Looked at another way, an advantage of non-cubic boundary conditions is that they permit simulations with a given cutoff distance and density to be conducted using fewer particles. As an example, a simulation in a cubic box, with $r_c$ set equal to $\frac{1}{2}L$, might involve $N = 256$ molecules; taking the same density, the same cutoff could be used in a simulation of $N = 197$ molecules in a truncated octahedron, or just $N = 181$ molecules in a rhombic dodecahedron.

### *1.5.4 Computer code for periodic boundaries*

How do we handle periodic boundaries and the minimum image convention, in a simulation program? Let us assume that, initially, the $N$ molecules in the simulation lie within a cubic box of side $L$, with the origin at its centre, i.e. all coordinates lie in the range $(-\frac{1}{2}L, \frac{1}{2}L)$. As the simulation proceeds, these molecules move about the infinite periodic system. When a molecule leaves the box by crossing one of the boundaries, it is usual to switch attention to the image molecule entering the box, by simply adding $L$ to, or subtracting $L$ from, the appropriate coordinate. One simple way to do this uses a FORTRAN IF statement to test the positions immediately after the molecules have been moved (whether by MC or MD):

```
IF ( RX(I) .GT.  BOXL2 ) RX(I) = RX(I) - BOXL
IF ( RX(I) .LT. -BOXL2 ) RX(I) = RX(I) + BOXL
```

Here, BOXL is a variable containing the box length $L$, and BOXL2 is just half the box length. Similar statements are applied to the $y$ and $z$ coordinates. An alternative to the IF statement is to use FORTRAN arithmetic functions to calculate the correct number of box lengths to be added or subtracted:

```
RX(I) = RX(I) - BOXL * ANINT ( RX(I) / BOXL )
```

The function ANINT(X) returns the nearest integer to X, converting the result back to type REAL; thus ANINT $(-0.49)$ has the value 0.0, whereas ANINT $(-0.51)$ is $-1.0$. By using these methods, we always have available the coordinates of the $N$ molecules that currently lie in the 'central' box. It is not strictly necessary to do this; we could, instead, use uncorrected coordinates, and follow the motion of the $N$ molecules that were in the central box at the start of the simulation. Indeed, as we shall see in Chapters 2 and 6, for calculation of transport coefficients it may be most desirable to have a set of uncorrected positions on hand. If it is decided to do this, however, care must be taken that the minimum image convention is correctly applied, so as to work out the vector between the two closest images of a pair of molecules, no matter how many 'boxes' apart they may be.

The minimum image convention may be coded in the same way as the periodic boundary adjustments. Of the two methods mentioned above, the arithmetic formula is usually preferable, being simpler; the use of IF statements inside the inner loop, particularly on pipeline machines, is to be avoided (see Appendix A). Immediately after calculating a pair separation vector, the following statements should be applied:

```
RXIJ = RXIJ - BOXL * ANINT ( RXIJ / BOXL )
RYIJ = RYIJ - BOXL * ANINT ( RYIJ / BOXL )
RZIJ = RZIJ - BOXL * ANINT ( RZIJ / BOXL )
```

The above code is guaranteed to yield the minimum image vector, no matter how many 'box lengths' apart the original images may be.

The calculation of minimum image distances is simplified by the use of reduced units: the length of the box is taken to define the fundamental unit of length in the simulation. Some workers define $L = 1$, others prefer to take $L = 2$. By setting $L = 1$, with particle coordinates nominally in the range $(-\frac{1}{2}, +\frac{1}{2})$, the minimum image correction above becomes

```
RXIJ = RXIJ - ANINT ( RXIJ )
RYIJ = RYIJ - ANINT ( RYIJ )
RZIJ = RZIJ - ANINT ( RZIJ )
```

which is simpler, and faster, than the code for a general box length. This approach is an alternative to the use of the pair potential to define reduced units as discussed in Appendix B, and is more generally applicable. For this reason a simulation box of unit length is adopted in some of the examples given in this book and on the attached microfiche.

There are several alternative ways of coding the minimum image corrections, some of which rely on the images being in the same, central box (i.e. on the periodic boundary correction being applied whenever the molecules move.) Some of these methods, for cubic boxes, are discussed in Appendix A. We have also mentioned the possibility of conducting simulations in non-cubic periodic boundary conditions. The FORTRAN code for implementing the minimum image correction in the cases of the truncated octahedron and the rhombic dodecahedron are given in program F.1 [see also Adams, 1983a; Smith 1983]. The code for the rhombic dodecahedron is a little more complicated than the code for the truncated octahedron, and the gain small, so that the truncated octahedron is preferable. We also give on the microfiche the code for computing minimum image corrections in the two-dimensional rhombic box often used in surface simulations.

Now we turn to the implementation of a spherical cutoff, i.e. we wish to set the pair potential (and all forces) to zero if the pair separation lies outside some distance $r_c$. It is easy to compute the square of the particle separation $r_{ij}$ and, rather than waste time taking the square root of this quantity, it is fastest to compare this with the square of $r_c$, which might be computed earlier and stored in a FORTRAN variable RCUTSQ. After computing the minimum image intermolecular vector, the following statements would be employed:

```
RIJSQ = RXIJ ** 2 + RYIJ ** 2 + RZIJ ** 2

IF ( RIJSQ .LT. RCUTSQ ) THEN

   ... compute i-j interaction ...
   ... accumulate energy and forces ...

ENDIF
```

In a large system, it may be worthwhile to apply separate tests for the $x$, $y$, and $z$ directions or some similar scheme.

```
IF ( ABS ( RXIJ ) .LT. RCUT ) THEN

   IF ( ABS ( RYIJ ) .LT. RCUT ) THEN

      IF ( ABS ( RZIJ ) .LT. RCUT ) THEN

         RIJSQ = RXIJ ** 2 + RYIJ ** 2 + RZIJ ** 2

         IF ( RIJSQ .LT. RCUTSQ ) THEN

            ... compute i-j interaction ...
            ... accumulate energy and forces ...

         ENDIF

      ENDIF

   ENDIF

ENDIF
```

The time saved in dropping out of this part of the program at any early stage must be weighed against the overheads of extra calculation and testing. A different approach is needed on a pipeline machine, since the IF statements may prevent vectorization of the inner loop. In this situation, it is generally simplest to compute all the minimum image interactions, and then set to zero the potential energy (and forces) arising from pairs separated by distances greater than $r_c$. The extra work involved here is more than offset by the speed increase on vectorization. Within the inner loop, this is simply achieved by setting the inverse squared separation $1/r_{ij}^2$ to zero, where appropriate, before calculating energies, forces etc. as functions of this quantity. The following code performs this task on the CRAY 1-S [Fincham and Ralston 1981]

```
RIJSQ  = RXIJ ** 2 + RYIJ ** 2 + RZIJ ** 2
RIJSQI = 1.0 / RIJSQ
RIJSQI = CVMGP ( RIJSQI, 0.0, RCUTSQ - RIJSQ )

... compute i-j interaction ...
... as functions of RIJSQI  ...
```

The function CVMGP(A, B, C) is a vector merge statement which returns the value A if C is non-negative and the value B otherwise. Note that no time is saved by using a spherical cutoff in this way on a pipeline machine. The only reason for implementing the spherical cutoff in this case is so that the usual long-range corrections may be applied to the simulation results (see Chapter 2). In Chapter 5 we discuss the more complicated time-saving tricks used in the simulations of large systems.

### *1.5.5 Spherical boundary conditions*

Before leaving this section, we should mention an alternative to the standard periodic boundary conditions for simulating bulk liquids. A two-dimensional system may be embedded in the surface of a sphere without introducing any physical boundaries [Hansen, Levesque, and Weis 1979] and the idea may be extended to consider a three-dimensional system as being the surface of a hypersphere [Kratky 1980; Kratky and Schreiner 1982]. The spherical or hyperspherical system is finite: it cannot be considered as part of an infinitely repeating periodic system. In this case, non-Euclidean geometry is an unavoidable complication, and distances between particles are typically measured along the great circle geodesics joining them. However, the effects of the curved geometry will decrease as the system size increases, and such 'spherical boundary conditions' are expected to be a valid method of simulating bulk liquids. Interesting differences from the standard periodic boundary conditions, particularly close to any solid–liquid phase transition, will result from the different topology. Periodic boundaries will be biased in favour of the formation of a solid with a lattice structure which matches the simulation box; in general, a periodic lattice is not consistent with spherical boundaries, and so the liquid state will be favoured in most simulations using this technique.

# 2

# STATISTICAL MECHANICS

## 2.1 Sampling from ensembles

Computer simulation generates information at the microscopic level (atomic and molecular positions, velocities etc.) and the conversion of this very detailed information into macroscopic terms (pressure, internal energy etc.) is the province of statistical mechanics. It is not our aim to provide a text in this field, since many excellent sources are available [Hill 1956; McQuarrie 1976; Landau and Lifshitz 1980; Friedman 1985; Hansen and McDonald 1986; Chandler, 1987]. In this chapter, our aim is to summarize those aspects of the subject which are of most interest to the computer simulator.

Let us consider, for simplicity, a one-component macroscopic system; extension to a multicomponent system is straightforward. The thermodynamic state of such a system is usually defined by a small set of parameters (such as the number of particles $N$, the temperature $T$, and the pressure $P$). Other thermodynamic properties (density $\rho$, chemical potential $\mu$, heat capacity $C_V$ etc.) may be derived through knowledge of the equations of state and the fundamental equations of thermodynamics. Even quantities such as the diffusion coefficient $D$, the shear viscosity $\eta$, and the structure factor $S(k)$ are state functions: although they clearly say something about the microscopic structure and dynamics of the system, their values are completely dictated by the few variables (e.g. $NPT$) characterizing the thermodynamic state, not by the very many atomic positions and momenta that define the instantaneous mechanical state. These positions and momenta can be thought of as coordinates in a multidimensional space: phase space. For a system of $N$ atoms, this space has $6N$ dimensions. Let us use the abbreviation $\mathbf{\Gamma}$ for a particular point in phase space, and suppose that we can write the instantaneous value of some property $\mathscr{A}$ (it might be the potential energy) as a function $\mathscr{A}(\mathbf{\Gamma})$. The system evolves in time, so that $\mathbf{\Gamma}$, and hence $\mathscr{A}(\mathbf{\Gamma})$ will change. It is reasonable to assume that the experimentally observable 'macroscopic' property $\mathscr{A}_{\text{obs}}$ is really the time average of $\mathscr{A}(\mathbf{\Gamma})$ taken over a long time interval:

$$\mathscr{A}_{\text{obs}} = \langle \mathscr{A} \rangle_{\text{time}} = \langle \mathscr{A}(\mathbf{\Gamma}(t)) \rangle_{\text{time}} = \lim_{t_{\text{obs}} \to \infty} \frac{1}{t_{\text{obs}}} \int_0^{t_{\text{obs}}} \mathscr{A}(\mathbf{\Gamma}(t))\, \mathrm{d}t\,. \tag{2.1}$$

The equations governing this time evolution, Newton's equations of motion in a simple classical system, are of course well known. They are just a system of ordinary differential equations: solving them on a computer, to a desired accuracy, is a practical proposition for, say, 1000 particles, although not for a truly macroscopic number (e.g. $10^{23}$). So far as the calculation of time averages is concerned, we clearly cannot hope to extend the integration of eqn (2.1) to infinite time, but might be satisfied to average over a long finite time $t_{obs}$. This is exactly what we do in a molecular dynamics simulation. In fact, the equations of motion are usually solved on a step-by-step basis, i.e. a large finite number $\tau_{obs}$ of time steps, of length $\delta t = t_{obs}/\tau_{obs}$ are taken. In this case, we may rewrite eqn (2.1) in the form

$$\mathscr{A}_{obs} = \langle \mathscr{A} \rangle_{time} = \frac{1}{\tau_{obs}} \sum_{\tau=1}^{\tau_{obs}} \mathscr{A}(\mathbf{\Gamma}(\tau)) . \tag{2.2}$$

In the summation, $\tau$ simply stands for an index running over the succession of time steps. This analogy between the discrete $\tau$ and the continuous $t$ is useful, even when, as we shall see in other examples, $\tau$ does not correspond to the passage of time in any physical sense.

The practical questions regarding the method are whether or not a sufficient region of phase space is explored by the system trajectory to yield satisfactory time averages within a feasible amount of computer time, and whether thermodynamic consistency can be attained between simulations with identical macroscopic parameters (density, energy etc.) but different initial conditions (atomic positions and velocities). The answers to these questions are that such simulation runs are indeed within the power of modern computers, and that thermodynamically consistent results for liquid state properties can indeed be obtained, provided that attention is paid to the selection of initial conditions. We will turn to the technical details of the method in Chapter 3.

The calculation of time averages by MD is not the approach to thermodynamic properties implicit in conventional statistical mechanics. Because of the complexity of the time evolution of $\mathscr{A}(\mathbf{\Gamma}(t))$ for large numbers of molecules, Gibbs suggested replacing the time average by the ensemble average. Here, we regard an ensemble as a collection of points $\mathbf{\Gamma}$ in phase space. The points are distributed according to a probability density $\rho(\mathbf{\Gamma})$. This function is determined by the chosen fixed macroscopic parameters ($NPT$, $NVT$ etc.), so we use the notation $\rho_{NPT}$, $\rho_{NVT}$, or, in general, $\rho_{ens}$. Each point represents a typical system at any particular instant of time. Each system evolves in time, according to the usual mechanical equations of motion, quite independently of the other systems. Consequently, in general, the phase space density $\rho_{ens}(\mathbf{\Gamma})$ will change with time. However, no systems are destroyed or created during this evolution, and Liouville's theorem, which is essentially a conservation law for probability density, states that $\mathrm{d}\rho/\mathrm{d}t = 0$ where $\mathrm{d}/\mathrm{d}t$ denotes the total derivative with respect to time (following a state $\mathbf{\Gamma}$ as it moves). As an example, consider a set of $N$ atoms with Cartesian coordinates $\mathbf{r}_i$ and momenta $\mathbf{p}_i$, in the classical approximation. The total time derivative is

$$\frac{\mathrm{d}}{\mathrm{d}t} = \frac{\partial}{\partial t} + \sum_i \dot{\mathbf{r}}_i \cdot \nabla_{\mathbf{r}_i} + \sum_i \dot{\mathbf{p}}_i \cdot \nabla_{\mathbf{p}_i} \tag{2.3a}$$

$$= \frac{\partial}{\partial t} + \dot{\mathbf{r}} \cdot \nabla_{\mathbf{r}} + \dot{\mathbf{p}} \cdot \nabla_{\mathbf{p}} . \tag{2.3b}$$

In eqn (2.3a), $\partial/\partial t$ represents differentiation, with respect to time, of a function, $\nabla_{\mathbf{r}_i}$ and $\nabla_{\mathbf{p}_i}$ are derivatives with respect to atomic position and momentum respectively, and $\dot{\mathbf{r}}_i$, $\dot{\mathbf{p}}_i$ signify the time derivatives of the position and momentum. Equation (2.3b) is the same equation written in a more compact way, and the equation may be further condensed by defining the Liouville operator $L$

$$iL = \left( \sum_i \dot{\mathbf{r}}_i \cdot \nabla_{\mathbf{r}_i} + \sum_i \dot{\mathbf{p}}_i \cdot \nabla_{\mathbf{p}_i} \right) = (\dot{\mathbf{r}} \cdot \nabla_{\mathbf{r}} + \dot{\mathbf{p}} \cdot \nabla_{\mathbf{p}}) \tag{2.4}$$

so that $\mathrm{d}/\mathrm{d}t = \partial/\partial t + iL$ and, using Liouville's theorem, we may write

$$\frac{\partial \rho_{\mathrm{ens}}(\mathbf{\Gamma}, t)}{\partial t} = -iL\, \rho_{\mathrm{ens}}(\mathbf{\Gamma}, t) . \tag{2.5}$$

This equation tells us that the rate of change of $\rho_{\mathrm{ens}}$ at a particular fixed point in phase space is related to the flows into and out of that point. This equation has a formal solution

$$\rho_{\mathrm{ens}}(\mathbf{\Gamma}, t) = \exp(-iLt)\, \rho_{\mathrm{ens}}(\mathbf{\Gamma}, 0) \tag{2.6}$$

where the exponential of an operator really means a series expansion

$$\exp(-iLt) = 1 - iLt - \tfrac{1}{2}L^2 t^2 + \ldots . \tag{2.7}$$

The equation of motion of a function like $\mathscr{A}(\mathbf{\Gamma})$, which does not depend explicitly on time, takes a conjugate form [McQuarrie 1976]:

$$\dot{\mathscr{A}}(\mathbf{\Gamma}(t)) = iL\, \mathscr{A}(\mathbf{\Gamma}(t)) \tag{2.8}$$

or

$$\mathscr{A}(\mathbf{\Gamma}(t)) = \exp(iLt)\, \mathscr{A}(\mathbf{\Gamma}(0)). \tag{2.9}$$

To be quite clear: in eqns (2.5) and (2.6) we consider the time-dependence of $\rho_{\mathrm{ens}}$ at a fixed point $\mathbf{\Gamma}$ in phase space; in eqns (2.8) and (2.9), $\mathscr{A}(\mathbf{\Gamma})$ is time-dependent because we are following the time evolution $\mathbf{\Gamma}(t)$ along a trajectory. This relationship is analogous to that between the Schrödinger and Heisenberg pictures in quantum mechanics.

If $\rho_{\mathrm{ens}}(\mathbf{\Gamma})$ represents an equilibrium ensemble, then its time-dependence completely vanishes, $\partial \rho_{\mathrm{ens}}/\partial t = 0$. The system evolution then becomes quite special. As each system leaves a particular state $\mathbf{\Gamma}(\tau)$ and moves on to the next, $\mathbf{\Gamma}(\tau + 1)$, another system arrives from state $\mathbf{\Gamma}(\tau - 1)$ to replace it. The motion resembles a long and convoluted conga line at a crowded party (see Fig. 2.1). There might be several such processions, each passing through different regions of phase space. However, if there is just one trajectory that passes through all the points in phase space for which $\rho_{\mathrm{ens}}$ is non-zero (i.e. the

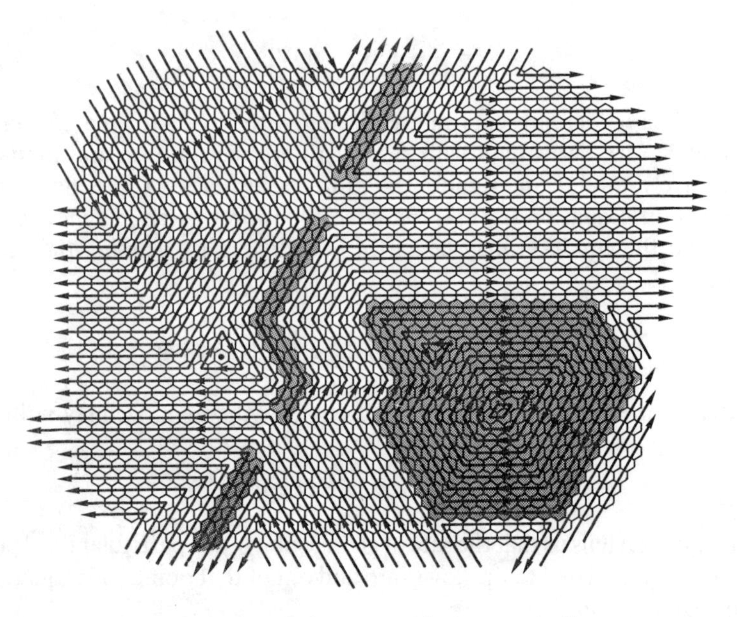

**Fig. 2.1** A schematic representation of phase space. The hexagonal cells represent state points (**q**, **p**). In an ergodic system, all the trajectories here would be different sections of a single long trajectory. A substantial region of cyclical trajectories, and a barrier region leading to bottlenecks, are shaded.

procession forms a single, very long, closed circuit) then each system will eventually visit all the state points. Such a system is termed 'ergodic' and the time taken to complete a cycle (the Poincaré recurrence time) is immeasurably long for a many-particle system (and for many parties as well it seems). Nevertheless, in principle, the fact that individual systems are moving around the circuit becomes unimportant: the average properties of our partygoers, for example, could be deduced from a snapshot of the event, rather than by following the complete experiences of one individual through the entire proceedings. This corresponds to replacing the time average in eqn (2.1) by an average taken over all the members of the ensemble, 'frozen' at a particular time:

$$\mathscr{A}_{\mathrm{obs}} = \langle \mathscr{A} \rangle_{\mathrm{ens}} = \langle \mathscr{A} | \rho_{\mathrm{ens}} \rangle = \sum_{\mathbf{\Gamma}} \mathscr{A}(\mathbf{\Gamma}) \rho_{\mathrm{ens}}(\mathbf{\Gamma}) . \tag{2.10}$$

The $\langle \mathscr{A} | \rho \rangle$ notation reminds us of the dependence of the average on both $\mathscr{A}$ and $\rho$: this is important when taking a thermodynamic derivative of $\mathscr{A}_{\mathrm{obs}}$ (we must differentiate both parts) or when considering time-dependent properties (when the Schrödinger/Heisenberg analogy may be exploited). Actually, we will be concerned with the practical question of efficient and thorough sampling of phase space, which is not quite the same as the rigorous definition of ergodicity [for a fuller discussion, see Tolman 1938]. In terms of our analogy of conga lines, there should not be a preponderance of independent closed circuits ('cliques') in which individuals can become trapped and fail fully to sample the available space (this is important in parties as well as in simulations). An MD simulation which started in the shaded cyclic region of Fig. 2.1, for example, would be disastrous. On the other hand, small non-ergodic regions are less likely to be dangerous and more likely to be recognized if they are unfortunately selected as starting points for a simulation. In a similar way, regions of phase space which act as barriers and cause bottlenecks through which only a few trajectories pass (see Fig. 2.1) can result in poor sampling by the relatively short simulation runs carried out in practice, even if the system is technically ergodic.

It is sometimes convenient to use, in place of $\rho_{ens}(\mathbf{\Gamma})$, a 'weight' function $w_{ens}(\mathbf{\Gamma})$, which satisfies the following equations:

$$\rho_{ens}(\mathbf{\Gamma}) = Q_{ens}^{-1} w_{ens}(\mathbf{\Gamma}) \tag{2.11}$$

$$Q_{ens} = \sum_{\mathbf{\Gamma}} w_{ens}(\mathbf{\Gamma}) \tag{2.12}$$

$$\langle \mathscr{A} \rangle_{ens} = \sum_{\mathbf{\Gamma}} w_{ens}(\mathbf{\Gamma}) \mathscr{A}(\mathbf{\Gamma}) / \sum_{\mathbf{\Gamma}} w_{ens}(\mathbf{\Gamma}) \,. \tag{2.13}$$

The weight function is essentially a non-normalized form of $\rho_{ens}(\mathbf{\Gamma})$, with the partition function $Q_{ens}$ (also called the sum over states) acting as the normalizing factor. Both $w_{ens}(\mathbf{\Gamma})$ and $Q_{ens}$ can contain an arbitrary multiplicative constant, whose choice corresponds to the definition of a zero of entropy. $Q_{ens}$ is simply a function of the macroscopic properties defining the ensemble, and connection with classical thermodynamics is made by defining a thermodynamic potential $\Psi_{ens}$ [see e.g. McQuarrie 1976]

$$\Psi_{ens} = -\ln Q_{ens} \,. \tag{2.14}$$

This is the function that has a minimum value at thermodynamic equilibrium (e.g. the negative of the entropy $S$ for a system at constant $NVE$, the Gibbs function $G$ for a constant-$NPT$ system).

Throughout the above discussion, although we have occasionally used the language of classical mechanics, we have assumed that the states $\mathbf{\Gamma}$ are discrete (e.g. a set of quantum numbers) and that we may sum over them. If the system were enclosed in a container, there would be a countably infinite set of quantum states. In the classical approximation, $\mathbf{\Gamma}$ represents the set of (continuously variable) particle positions and momenta, and we should replace the summation by a classical phase space integral. $w_{ens}$ and $Q_{ens}$ are then usually defined with appropriate factors included to make them dimensionless, and to match up with the usual semiclassical 'coarse-grained' phase space volume elements. On a computer, of course, all numbers are held to a finite precision and so, technically, positions and momenta are represented by discrete, not continuous, variables; we now have a countable and finite set of states. We assume that the distinction between this case and the classical limit is of no practical importance, and will use whichever representation is most convenient.

One conceivable approach to the computation of thermodynamic quantities, therefore, would be a direct evaluation of $Q_{\mathrm{ens}}$, for a particular ensemble, using eqn (2.12). This summation, over all possible states, is not feasible for many-particle systems: there are too many states, most of which have a very low weight due to non-physical overlaps between the repulsive cores of the molecules, rendering them unimportant. We would like to conduct the summation so as to exclude this large number of irrelevant states, and include only those with a high probability. Unfortunately, it is generally not possible to estimate $Q_{\mathrm{ens}}$ directly in this way. However, the underlying idea, that of generating (somehow) a set of states in phase space that are sampled from the complete set in accordance with the probability density $\rho_{\mathrm{ens}}(\mathbf{\Gamma})$, is central to the Monte Carlo technique.

We proceed by analogy with molecular dynamics in the sense that the ensemble average of eqn (2.13) is replaced by a trajectory average like eqn (2.2). Newton's equations generate a succession of states in accordance with the distribution function $\rho_{NVE}$ for the constant-$NVE$ or microcanonical ensemble. Suppose we wish to investigate other ensembles; experiments in the laboratory, for example, are frequently performed under conditions of constant temperature and pressure, while it is often very convenient to consider inhomogeneous systems at constant chemical potential. For each such case, let us invent a kind of equation of motion, i.e. a means of generating, from one state point $\mathbf{\Gamma}(\tau)$, a succeeding state point $\mathbf{\Gamma}(\tau+1)$. This recipe need have no physical interpretation, and it could be entirely deterministic or could involve a stochastic, random, element. It might be derived by modifying the true equations of motion in some way, or it may have no relation whatever with normal dynamics.

To be useful, this prescription should satisfy some sensible conditions:

(a) the probability density $\rho_{\mathrm{ens}}(\mathbf{\Gamma})$ for the ensemble of interest should not change as the system evolves;
(b) any 'reasonable' starting distribution $\rho(\mathbf{\Gamma})$ should tend to this stationary solution as the simulation proceeds;
(c) we should be able to argue that ergodicity holds, even though we cannot hope to prove this for realistic systems.

If these conditions are satisfied, then we should be able to generate, from an initial state, a succession of state points which, in the long term, are sampled in accordance with the desired probability density $\rho_{\mathrm{ens}}(\mathbf{\Gamma})$. In these circumstances, the ensemble average will be equal to a kind of 'time average':

$$\mathscr{A}_{\mathrm{obs}} = \langle \mathscr{A} \rangle_{\mathrm{ens}} = \frac{1}{\tau_{\mathrm{obs}}} \sum_{\tau=1}^{\tau_{\mathrm{obs}}} \mathscr{A}(\mathbf{\Gamma}(\tau)). \tag{2.15}$$

Here $\tau$ is an index running over the succession of $\tau_{obs}$ states or trials generated by our prescription; in a practical simulation, $\tau_{obs}$ would be a large finite number. This is exactly what we do in Monte Carlo simulations. The trick, of course, lies in the generation of the trajectory through phase space, and the different recipes for different ensembles will be discussed in Chapter 4. In general, because only a finite number of states can be generated in any one simulation, Monte Carlo results are subject to the same questions of initial condition effects and satisfactory phase space exploration as are molecular dynamics results.

## 2.2 Common statistical ensembles

Let us consider four ensembles in common use: the microcanonical, or constant-$NVE$, ensemble just mentioned, the canonical, or constant-$NVT$, ensemble, the isothermal–isobaric constant-$NPT$ ensemble, and the grand canonical constant-$\mu VT$ ensemble. For each ensemble, the aforementioned thermodynamic variables are specified, i.e. fixed. Other thermodynamic quantities must be determined by ensemble averaging and, for any particular state point, the instantaneous values of the appropriate phase function will deviate from this average value, i.e. fluctuations occur.

The probability density for the microcanonical ensemble is proportional to

$$\delta(\mathscr{H}(\mathbf{\Gamma}) - E)$$

where $\mathbf{\Gamma}$ represents the set of particle positions and momenta (or quantum numbers), and $\mathscr{H}(\mathbf{\Gamma})$ is the Hamiltonian. The delta function selects out those states of an $N$-particle system in a container of volume $V$ that have the desired energy $E$. When the set of states is discrete, $\delta$ is just the Kronecker delta, taking values of 0 or 1; when the states are continuous, $\delta$ is the Dirac delta function. The microcanonical partition function may be written:

$$Q_{NVE} = \sum_{\mathbf{\Gamma}} \delta(\mathscr{H}(\mathbf{\Gamma}) - E) \tag{2.16}$$

where the summation takes due note of indistinguishability of particles. In the quasi-classical expression for $Q_{NVE}$, for an atomic system, the indistinguishability is handled using a factor of $1/N!$

$$Q_{NVE} = \frac{1}{N!}\frac{1}{h^{3N}}\int \mathrm{d}\mathbf{r}\mathrm{d}\mathbf{p}\, \delta(\mathscr{H}(\mathbf{r},\mathbf{p}) - E). \tag{2.17}$$

Here, $\int \mathrm{d}\mathbf{r}\mathrm{d}\mathbf{p}$ stands for integration over all $6N$ phase space coordinates. The appropriate thermodynamic potential is the negative of the entropy

$$-S/k_{\mathrm{B}} = -\ln Q_{NVE}. \tag{2.18}$$

The factor involving Planck's constant $h$ in eqn (2.17) corresponds to the usual zero of entropy for the ideal gas (the Sackur–Tetrode equation).

For a classical system, Newton's equations of motion conserve energy and so provide a suitable method (but not the only method [Severin, Freasier, Hamer, Jolly, and Nordholm 1978; Creutz 1983]) for generating a succession of state points sampled from this ensemble, as discussed in the previous section. In fact, for a system not subjected to external forces, these equations also conserve total linear momentum **P**, and so molecular dynamics probes a subset of the microcanonical ensemble, namely the constant-$NVE\mathbf{P}$ ensemble (for technical reasons, as we shall see in Chapter 3, total angular momentum is not conserved in most MD simulations). Since it is easy to transform into the centre-of-mass frame, the choice of **P** is not crucial, and zero momentum is usually chosen for convenience. Differences between the constant-$NVE$ and constant-$NVE\mathbf{P}$ ensembles are minor: for the latter, an additional three constraints exist in that only $(N-1)$ particle momenta are actually independent of each other.

The density for the canonical ensemble is proportional to

$$\exp(-\mathscr{H}(\mathbf{\Gamma})/k_{\mathrm{B}}T)$$

and the partition function is

$$Q_{NVT} = \sum_{\mathbf{\Gamma}} \exp(-\mathscr{H}(\mathbf{\Gamma})/k_{\mathrm{B}}T) \tag{2.19}$$

or, in quasi-classical form, for an atomic system

$$Q_{NVT} = \frac{1}{N!}\frac{1}{h^{3N}}\int \mathrm{d}\mathbf{r}\mathrm{d}\mathbf{p}\, \exp(-\mathscr{H}(\mathbf{r},\mathbf{p})/k_{\mathrm{B}}T). \tag{2.20}$$

The appropriate thermodynamic function is the Helmholtz free energy $A$

$$A/k_{\mathrm{B}}T = -\ln Q_{NVT}. \tag{2.21}$$

In the canonical ensemble, all values of the energy are allowed, and energy fluctuations are non-zero. Thus, although $\rho_{NVT}(\mathbf{\Gamma})$ is indeed a stationary solution of the Liouville equation, the corresponding mechanical equations of motion are not a satisfactory method of sampling states in this ensemble, since they conserve energy: normal time evolution occurs on a set of independent constant-energy surfaces, each of which should be appropriately weighted, by the factor $\exp(-\mathscr{H}(\mathbf{\Gamma})/k_{\mathrm{B}}T)$. Our prescription for generating a succession of states must make provision for transitions between the energy surfaces, so that a single trajectory can probe all the accessible phase space, and yield the correct relative weighting. We shall encounter several ways of doing this in the later chapters.

Because the energy is always expressible as a sum of kinetic (**p**-dependent) and potential (**q**-dependent) contributions, the partition function factorizes into a product of kinetic (ideal gas) and potential (excess) parts.

$$\begin{aligned} Q_{NVT} &= \frac{1}{N!}\frac{1}{h^{3N}}\int \mathrm{d}\mathbf{p}\, \exp(-\mathscr{K}/k_{\mathrm{B}}T)\int \mathrm{d}\mathbf{q}\, \exp(-\mathscr{V}/k_{\mathrm{B}}T) \\ &= Q^{\mathrm{id}}_{NVT} Q^{\mathrm{ex}}_{NVT}. \end{aligned} \tag{2.22}$$

Again, for an atomic system, we see (by taking $\mathscr{V} = 0$) that

$$Q^{\mathrm{id}}_{NVT} = \frac{V^N}{N!\Lambda^{3N}} \tag{2.23}$$

$\Lambda$ being the thermal de Broglie wavelength

$$\Lambda = (h^2/2\pi m k_{\mathrm{B}}T)^{1/2}. \tag{2.24}$$

The excess part is

$$Q^{\mathrm{ex}}_{NVT} = V^{-N}\int \mathrm{d}\mathbf{r}\exp(-\mathscr{V}(\mathbf{r})/k_{\mathrm{B}}T). \tag{2.25}$$

Instead of $Q^{\mathrm{ex}}_{NVT}$, we often use the configuration integral

$$Z_{NVT} = \int \mathrm{d}\mathbf{r}\exp(-\mathscr{V}(\mathbf{r})/k_{\mathrm{B}}T). \tag{2.26}$$

Some workers include a factor $N!$ in the definition of $Z_{NVT}$. Although $Q^{\mathrm{id}}_{NVT}$ and $Q^{\mathrm{ex}}_{NVT}$ are dimensionless, the configuration integral has dimensions of $V^N$. As a consequence of the separation of $Q_{NVT}$, all the thermodynamic properties derived from $A$ can be expressed as a sum of ideal gas and configurational parts. In statistical mechanics, it is easy to evaluate ideal gas properties [Rowlinson 1963], and we may expect most attention to focus on the configurational functions. In fact, it proves possible to probe just the configurational part of phase space according to the canonical distribution, using standard Monte Carlo methods. The corresponding trajectory through phase space has essentially independent projections on the coordinate and momentum sub-spaces. The ideal gas properties are added onto the results of configuration-space Monte Carlo simulations afterwards.

The probability density for the isothermal–isobaric ensemble is proportional to

$$\exp(-(\mathscr{H} + PV)/k_{\mathrm{B}}T).$$

Note that the quantity appearing in the exponent, when averaged, gives the thermodynamic enthalpy $H = \langle \mathscr{H} \rangle + P\langle V \rangle$. Now the volume $V$ has joined the list of microscopic quantities ($\mathbf{r}$ and $\mathbf{p}$) comprising the state point $\mathbf{\Gamma}$. The appropriate partition function is

$$Q_{NPT} = \sum_{\mathbf{\Gamma}}\sum_{V}\exp(-(\mathscr{H} + PV)/k_{\mathrm{B}}T) = \sum_{V}\exp(-PV/k_{\mathrm{B}}T)\,Q_{NVT}. \tag{2.27}$$

The summation over possible volumes may also be written as an integral, in which case some basic unit of volume $V_0$ must be chosen to render $Q_{NPT}$ dimensionless. This choice is not fundamentally important [Wood 1968a]. In quasi-classical form, for an atomic system, we write:

$$Q_{NPT} = \frac{1}{N!}\frac{1}{h^{3N}}\frac{1}{V_0}\int \mathrm{d}V\int \mathrm{d}\mathbf{r}\mathrm{d}\mathbf{p}\exp(-(\mathscr{H} + PV)/k_{\mathrm{B}}T). \tag{2.28}$$

The corresponding thermodynamic function is the Gibbs free energy $G$

$$G/k_{\mathrm{B}}T = -\ln Q_{NPT}. \tag{2.29}$$

The prescription for generating state points in the constant-$NPT$ ensemble must clearly provide for changes in the sample volume as well as energy. Once more, it is possible to separate configurational properties from kinetic ones, and to devise a Monte Carlo procedure to probe configuration space only. The configuration integral in this ensemble is

$$Z_{NPT} = \int \mathrm{d}V \exp(-PV/k_{\mathrm{B}}T) \int \mathrm{d}\mathbf{r} \exp(-\mathscr{V}(\mathbf{r})/k_{\mathrm{B}}T). \tag{2.30}$$

Again, some definitions include $N!$ and $V_0$ as normalizing factors.

The density function for the grand canonical ensemble is proportional to

$$\exp(-(\mathscr{H} - \mu N)/k_{\mathrm{B}}T)$$

where $\mu$ is the specified chemical potential. Now the number of particles $N$ is a variable, along with the coordinates and momenta of those particles. The grand canonical partition function is

$$Q_{\mu VT} = \sum_{\mathbf{\Gamma}} \sum_{N} \exp(-(\mathscr{H} - \mu N)/k_{\mathrm{B}}T) = \sum_{N} \exp(\mu N/k_{\mathrm{B}}T)\, Q_{NVT}. \tag{2.31}$$

In quasi-classical form, for an atomic system,

$$Q_{\mu VT} = \sum_{N} \frac{1}{N!} \frac{1}{h^{3N}} \exp(\mu N/k_{\mathrm{B}}T) \int \mathrm{d}\mathbf{r}\mathrm{d}\mathbf{p} \exp(-\mathscr{H}/k_{\mathrm{B}}T). \tag{2.32}$$

Although it is occasionally useful to pretend that $N$ is a continuous variable, for most purposes we sum, rather than integrate, in eqns (2.31) and (2.32). The appropriate thermodynamic function is just $-PV/k_{\mathrm{B}}T$:

$$-PV/k_{\mathrm{B}}T = -\ln Q_{\mu VT}. \tag{2.33}$$

Whatever scheme we employ to generate states in the grand ensemble, clearly it must allow for addition and removal of particles. Once more, it is possible to invent a Monte Carlo method to do this, and, moreover, to probe just the configurational part of phase space; however, it turns out to be necessary to include the form of the kinetic partition function in the prescription used.

It is possible to construct many more ensembles, some of which are of interest in computer simulation. When comparing molecular dynamics with Monte Carlo, it may be convenient to add the constraint of constant (zero) total momentum, i.e. fixed centre of mass, to the constant-$NVT$ ensemble. It is also permissible to constrain certain degrees of freedom (for example, the total kinetic energy [Hoover 1983a, 1983b], or the energy in a particular chemical bond [Freasier, Jolly, Hamer, and Nordholm 1979]) while allowing others to fluctuate. Also, non-equilibrium ensembles may be set up (see Chapter 8). The possibilities are endless, the general requirements being that a phase-space density $\rho_{ens}(\Gamma)$ can be written down, and that a corresponding prescription for generating state points can be devised. The remaining questions are ones of practicality.

Not all ensembles are of interest to the computer simulator. The properties of generalized ensembles, such as the constant-$\mu PT$ ensemble, have been discussed [Hill 1956]. Here, only intensive parameters are specified: the corresponding extensive quantities show unbounded fluctuations, i.e. the system size can grow without limit. Also, $\mu$, $P$, and $T$ are related by an equation of state, so, although this equation may be unknown, they are not independently variable. For these reasons, the simulation of the constant-$\mu PT$ ensemble and related pathological examples is not a practical proposition. In all the ensembles dealt with in this book, at least one extensive parameter (usually $N$ or $V$) is fixed to act as a limit on the system size.

Finally, it is by no means guaranteed that a chosen prescription for generating phase space trajectories will correspond to any ensemble at all. It is easy to think of extreme examples of modified equations of motion for which no possible function $\rho_{ens}(\Gamma)$ is a stationary solution. In principle, some care should be taken to establish which ensemble, if any, is probed by any novel simulation technique.

## 2.3 Transforming between ensembles

Since the ensembles are essentially artificial constructs, it would be reassuring to know that they produce average properties which are consistent with one another. In the thermodynamic limit (for an infinite system size) and as long as we avoid the neighbourhood of phase transitions, this is believed to be true for the commonly used statistical ensembles [Fisher 1964]. Since we will be dealing with systems containing a finite number of particles, it is of some interest to see, in a general way, how this result comes about. The method of transformation between ensembles is standard [Hill 1956; Lebowitz, Percus, and Verlet 1967; Münster 1969; Landau and Lifshitz 1980] and we merely outline the procedure here. Nonetheless, the development is rather formal, and this section could be skipped on a first reading.

We shall be interested in transforming from an ensemble in which an extensive thermodynamic variable $F$ is fixed to one in which the intensive conjugate variable $f$ is constant. Typical conjugate pairs are $(\beta, E)$, $(\beta P, V)$, $(-\beta\mu, N)$, where $\beta = 1/k_B T$. If the old partition function and characteristic thermodynamic potential are $Q_F$ and $\Psi_F$, respectively, then the new quantities are given by:

$$Q_f = \int \mathrm{d}F' \exp(-F'f)\, Q_{F'} \tag{2.34}$$

$$\Psi_f = \Psi_F + Ff. \tag{2.35}$$

Equations (2.19)–(2.33) provide specific examples of these relations. Equation

(2.35) corresponds to the Legendre transformation of classical thermodynamics. For example, when moving from a system at constant energy to one at constant temperature (i.e. constant $\beta$), the characteristic thermodynamic potential changes from $-S/k_B$ to $-S/k_B + \beta E = \beta A$. Similarly, on going to constant temperature and pressure, the thermodynamic potential becomes $\beta A + \beta PV = \beta G$.

The average $\langle \mathscr{A} \rangle_f$ calculated in the constant-$f$ ensemble is related to the average $\langle \mathscr{A} \rangle_F$ calculated at constant-$F$ by [Lebowitz *et al.* 1967]

$$\langle \mathscr{A} \rangle_f = \exp(\Psi_f) \int \mathrm{d}F' \exp(-\Psi_{F'} - F'f) \langle \mathscr{A} \rangle_{F'}. \tag{2.36}$$

The equivalence of ensembles relies on the behaviour of the integrand of this equation for a large system: it becomes very sharply peaked around the mean value $F' = \langle F \rangle_f$. In the thermodynamic limit of infinite system size, we obtain simply

$$\langle \mathscr{A} \rangle_f = \langle \mathscr{A} \rangle_F \tag{2.37}$$

where it is understood that $F = \langle F \rangle_f$. Thus, the averages of any quantity calculated in, say, the constant-$NVE$ ensemble and the constant-$NVT$ ensemble, will be equal in the thermodynamic limit, as long as we choose $E$ and $T$ consistently so that $E = \langle E \rangle_{NVT}$. In fact, there are some restrictions on the kinds of functions $\mathscr{A}$ for which eqn (2.37) holds. $\mathscr{A}$ should be, essentially, a sum of single-particle functions,

$$\mathscr{A} = \sum_{i=1}^{N} \mathscr{A}_i \tag{2.38}$$

or, at least, a sum of independent contributions from different parts of the fluid, which may be added up in a similar way. All of the thermodynamic functions are of this short-ranged nature, insofar as they are limited by the range of intermolecular interactions. For long-ranged (e.g. dielectric) properties and long-ranged (e.g. Coulombic) forces, this becomes a more subtle point.

The situation for a finite number of particles is treated by expanding the integrand of eqn (2.36) about the mean value $\langle F \rangle_f$. If we write $F' = \langle F \rangle_f + \delta F'$, then we obtain [Lebowitz *et al.* 1967]:

$$\langle \mathscr{A} \rangle_f = \langle \mathscr{A} \rangle_{F = \langle F \rangle_f} + \frac{1}{2}\left(\frac{\partial^2}{\partial F^2} \langle \mathscr{A} \rangle_F\right)_{F = \langle F \rangle_f} \langle \delta F^2 \rangle_f + \ldots. \tag{2.39}$$

The correction term, which is proportional to the mean square fluctuations $\langle \delta F^2 \rangle$ of the quantity $F$ in the constant-$f$ ensemble, is expected to be relatively small since, as mentioned above, the distribution of $F$ values should be very sharply peaked for a many-particle system. This fluctuation term may be expressed as a straightforward thermodynamic derivative. Since $F$ and $f$ are conjugate variables, we have

$$\langle F \rangle_f = -\partial \Psi_f / \partial f \tag{2.40}$$

$$\langle \delta F^2 \rangle_f = \partial^2 \Psi_f / \partial f^2 = -\partial \langle F \rangle_f / \partial f. \tag{2.41}$$

We may write this simply as $-(\partial F/\partial f)$. Equation (2.39) is most usefully rearranged by taking the last term across to the other side, and treating it as a function of $f$ through the relation $F = \langle F \rangle_f$. Thus

$$\begin{aligned}
\langle \mathscr{A} \rangle_F &= \langle \mathscr{A} \rangle_f - \frac{1}{2} \langle \delta F^2 \rangle_f \frac{\partial^2}{\partial F^2} \langle \mathscr{A} \rangle_f \\
&= \langle \mathscr{A} \rangle_f + \frac{1}{2} \frac{\partial F}{\partial f} \frac{\partial^2}{\partial F^2} \langle \mathscr{A} \rangle_f \\
&= \langle \mathscr{A} \rangle_f + \frac{1}{2} \frac{\partial}{\partial f} \frac{\partial}{\partial F} \langle \mathscr{A} \rangle_f \\
&= \langle \mathscr{A} \rangle_f + \frac{1}{2} \frac{\partial}{\partial f} \left( \frac{\partial f}{\partial F} \right) \frac{\partial}{\partial f} \langle \mathscr{A} \rangle_f .
\end{aligned} \tag{2.42}$$

Bearing in mind that $F$ is extensive and $f$ intensive, the small relative magnitude of the correction term can be seen explicitly: it decreases as $\mathscr{O}(N^{-1})$.

Although the fluctuations are small, they are nonetheless measurable in computer simulations. They are of interest because they are related to thermodynamic derivatives (like the specific heat or the isothermal compressibility) by equations such as eqn (2.41). In general, we define the RMS deviation $\sigma(\mathscr{A})$ by the equation

$$\sigma^2(\mathscr{A}) = \langle \delta \mathscr{A}^2 \rangle_{\text{ens}} = \langle \mathscr{A}^2 \rangle_{\text{ens}} - \langle \mathscr{A} \rangle^2_{\text{ens}} \tag{2.43}$$

where

$$\delta \mathscr{A} = \mathscr{A} - \langle \mathscr{A} \rangle_{\text{ens}}. \tag{2.44}$$

It is quite important to realize that, despite the $\langle \delta \mathscr{A}^2 \rangle$ notation, we are not dealing here with the average of a mechanical quantity like $\mathscr{A}$; the best we can do is to write $\sigma^2(\mathscr{A})$ as a difference of two terms, as in eqn (2.43). Thus, the previous observations on equivalence of ensembles do not apply: fluctuations in different ensembles are not the same. As an obvious example, energy fluctuations in the constant-$NVE$ ensemble are (by definition) zero, whereas in the constant-$NVT$ ensemble, they are not. The transformation technique may be applied to obtain an equation analogous to eqn (2.42) [Lebowitz *et al.* 1967]. In the general case of the covariance of two variables $\mathscr{A}$ and $\mathscr{B}$ the result is

$$\langle \delta \mathscr{A} \delta \mathscr{B} \rangle_F = \langle \delta \mathscr{A} \delta \mathscr{B} \rangle_f + \left( \frac{\partial f}{\partial F} \right) \left( \frac{\partial}{\partial f} \langle \mathscr{A} \rangle_f \right) \left( \frac{\partial}{\partial f} \langle \mathscr{B} \rangle_f \right). \tag{2.45}$$

Now the correction term is of the same order as the fluctuations themselves. Consider, once more, energy fluctuations in the microcanonical and canonical ensembles, i.e. let $\mathscr{A} = \mathscr{B} = F = E$ and $f = \beta = 1/k_B T$. Then on the left of eqn

(2.45) we have zero, and on the right we have $\sigma^2(E)$ at constant-$NVT$ and a combination of thermodynamic derivatives which turn out to equal $(\partial E/\partial \beta) = -k_B T^2 C_V$, where $C_V$ is the constant-volume heat capacity.

## 2.4 Simple thermodynamic averages

A consequence of the equivalence of ensembles is that, provided a suitable phase function can be identified in each case, the basic thermodynamic properties of a model system may be calculated as averages in any convenient ensemble. Accordingly, we give in this section expressions for common thermodynamic quantities, omitting the subscripts which identify particular ensembles. These functions are usually derivatives of one of the characteristic thermodynamic functions $\Psi_{\text{ens}}$. Examples are $P = -(\partial A/\partial V)_{NT}$ and $\beta = (1/k_B T) = (1/k_B)(\partial S/\partial E)_{NV}$.

The kinetic, potential, and total internal energies may be calculated using the phase functions of eqns (1.1)–(1.3):

$$E = \langle \mathscr{H} \rangle = \langle \mathscr{K} \rangle + \langle \mathscr{V} \rangle . \tag{2.46}$$

The kinetic energy is a sum of contributions from individual particle momenta, while evaluation of the potential contribution involves summing over all pairs, triplets etc. of molecules, depending upon the complexity of the function as discussed in Chapter 1.

The temperature and pressure may be calculated using the virial theorem, which we write in the form of 'generalized equipartition' [Münster 1969]:

$$\langle p_k \partial \mathscr{H}/\partial p_k \rangle = k_B T \tag{2.47a}$$
$$\langle q_k \partial \mathscr{H}/\partial q_k \rangle = k_B T \tag{2.47b}$$

for any generalized coordinate $q_k$ or momentum $p_k$. These equations are examples of the general form $\langle \mathscr{A} \partial \mathscr{H}/\partial q_k \rangle = k_B T \langle \partial \mathscr{A}/\partial q_k \rangle$ which may be easily derived in the canonical ensemble. They are valid (to $\mathscr{O}(N^{-1})$) in any ensemble.

Equation (2.47a) is particularly simple when the momenta appear as squared terms in the Hamiltonian. For example, in the atomic case, we may sum up $3N$ terms of the form $p_{i\alpha}^2/m_i$, to obtain

$$\langle \sum_{i=1}^{N} |\mathbf{p}_i|^2/m_i \rangle = 2\langle \mathscr{K} \rangle = 3Nk_B T . \tag{2.48}$$

This is the familiar equipartition principle: an average energy of $k_B T/2$ per degree of freedom. It is convenient to define an instantaneous 'kinetic temperature' function

$$\mathscr{T} = 2\mathscr{K}/3Nk_B = \frac{1}{3Nk_B} \sum_{i=1}^{N} |\mathbf{p}_i|^2/m_i \tag{2.49}$$

whose average is equal to $T$. Obviously, this is not a unique definition. For a system of rigid molecules, described in terms of centre-of-mass positions and velocities together with orientational variables, the angular velocities may also appear in the definition of $\mathcal{T}$. Alternatively, it may be useful to define separate 'translational' and 'rotational' temperatures each of which, when averaged, gives $T$. In eqn (2.47a) it is assumed that the independent degrees of freedom have been identified and assigned generalized coordinates $q_k$ and momenta $p_k$. For a system of $N$ atoms, subject to internal molecular constraints, the number of degrees of freedom will be $3N - N_c$ where $N_c$ is the total number of independent internal constraints (fixed bond lengths and angles) defined in the molecular model. Then, we must replace eqn (2.49) by

$$\mathcal{T} = \frac{2\mathcal{K}}{(3N - N_c)k_B} = \frac{1}{(3N - N_c)k_B} \sum_{i=1}^{N} |\mathbf{p}_i|^2/m_i. \tag{2.50}$$

We must also include in $N_c$ any additional global constraints on the ensemble. For example, in the 'molecular dynamics' constant-$NVE\mathbf{P}$ ensemble, we must include the three extra constraints on centre of mass motion.

The pressure may be calculated via eqn (2.47b). If we choose Cartesian coordinates, and use Hamilton's equations of motion (see Chapter 3), it is easy to see that each coordinate derivative in eqn (2.47b) is the negative of a component of the force $\mathbf{f}_i$ on some molecule $i$, and we may write, summing over $N$ molecules,

$$-\tfrac{1}{3}\left\langle \sum_{i=1}^{N} \mathbf{r}_i \cdot \nabla_{\mathbf{r}_i} \mathcal{V} \right\rangle = \tfrac{1}{3}\left\langle \sum_{i=1}^{N} \mathbf{r}_i \cdot \mathbf{f}_i^{\text{tot}} \right\rangle = -Nk_BT. \tag{2.51}$$

We have used the symbol $\mathbf{f}_i^{\text{tot}}$ because this represents the sum of intermolecular forces and external forces. The latter are related to the external pressure, as can be seen by considering the effect of the container walls on the system:

$$\tfrac{1}{3}\left\langle \sum_{i=1}^{N} \mathbf{r}_i \cdot \mathbf{f}_i^{\text{ext}} \right\rangle = -PV. \tag{2.52}$$

If we define the 'internal virial' $\mathcal{W}$

$$-\tfrac{1}{3} \sum_{i=1}^{N} \mathbf{r}_i \cdot \nabla_{\mathbf{r}_i} \mathcal{V} = \tfrac{1}{3} \sum_{i=1}^{N} \mathbf{r}_i \cdot \mathbf{f}_i = \mathcal{W} \tag{2.53}$$

where now we restrict attention to intermolecular forces, then

$$PV = Nk_BT + \langle \mathcal{W} \rangle. \tag{2.54}$$

This suggests that we define an instantaneous 'pressure' function [Cheung 1977]

$$\mathcal{P} = \rho k_B \mathcal{T} + \mathcal{W}/V = \mathcal{P}^{\text{id}} + \mathcal{P}^{\text{ex}} \tag{2.55}$$

whose average is simply $P$. Again, this definition is not unique; apart from the different ways of defining $\mathcal{W}$, which we shall see below, it may be most convenient (say in a constant temperature ensemble) to use

$$\mathscr{P}' = \rho k_{\mathrm{B}}T + \mathcal{W}/V = \langle \mathscr{P}^{\mathrm{id}} \rangle + \mathscr{P}^{\mathrm{ex}} \tag{2.56}$$

instead. Both $\mathscr{P}$ and $\mathscr{P}'$ give $P$ when averaged, but their fluctuations in any ensemble will, in general, be different. It should be noted that the above derivation is not really valid for the infinite periodic systems used in computer simulations: there are no container walls and no external forces. Nonetheless, the result is the same [Erpenbeck and Wood 1977].

For pairwise interactions, $\mathcal{W}$ is more conveniently expressed in a form which is explicitly independent of the origin of coordinates. This is done by writing $\mathbf{f}_i$ as the sum of forces $\mathbf{f}_{ij}$ on atom $i$ due to atom $j$

$$\sum_i \mathbf{r}_i \cdot \mathbf{f}_i = \sum_i \sum_{j \neq i} \mathbf{r}_i \cdot \mathbf{f}_{ij} = \tfrac{1}{2} \sum_i \sum_{j \neq i} \mathbf{r}_i \cdot \mathbf{f}_{ij} + \mathbf{r}_j \cdot \mathbf{f}_{ji} \,. \tag{2.57}$$

The second equality follows because the indices $i$ and $j$ are equivalent. Newton's third law $\mathbf{f}_{ji} = -\mathbf{f}_{ij}$ is then used to switch the force indices

$$\sum_i \mathbf{r}_i \cdot \mathbf{f}_i = \tfrac{1}{2} \sum_i \sum_{j \neq i} \mathbf{r}_{ij} \cdot \mathbf{f}_{ij} = \sum_i \sum_{j > i} \mathbf{r}_{ij} \cdot \mathbf{f}_{ij} \tag{2.58}$$

where $\mathbf{r}_{ij} = \mathbf{r}_i - \mathbf{r}_j$, and the final form of the summation is usually more convenient. It is essential to use the $\mathbf{r}_{ij} \cdot \mathbf{f}_{ij}$ form in a simulation that employs periodic boundary conditions. So we have at last

$$\begin{aligned} \mathcal{W} &= \tfrac{1}{3} \sum_i \sum_{j > i} \mathbf{r}_{ij} \cdot \mathbf{f}_{ij} = -\tfrac{1}{3} \sum_i \sum_{j > i} \mathbf{r}_{ij} \cdot \nabla_{\mathbf{r}_{ij}} v(\mathbf{r}_{ij}) \\ &= -\tfrac{1}{3} \sum_i \sum_{j > i} w(r_{ij}) \end{aligned} \tag{2.59}$$

where the intermolecular pair virial function $w(r)$ is

$$w(r) = r \frac{\mathrm{d}v(r)}{\mathrm{d}r} \,. \tag{2.60}$$

Like $\mathcal{V}$, $\mathcal{W}$ is limited by the range of the interactions, and hence $\langle \mathcal{W} \rangle$ should be a well-behaved, ensemble-independent function in most cases.

For molecular fluids we may write

$$\begin{aligned} \mathcal{W} &= \tfrac{1}{3} \sum_i \sum_{j > i} \mathbf{r}_{ij} \cdot \mathbf{f}_{ij} = -\tfrac{1}{3} \sum_i \sum_{j > i} \mathbf{r}_{ij} \cdot (\nabla_{\mathbf{r}_{ij}} \mathcal{V})_{\mathbf{\Omega}_i \mathbf{\Omega}_j} \\ &= -\tfrac{1}{3} \sum_i \sum_{j > i} w(r_{ij}) \end{aligned} \tag{2.61}$$

where $\mathbf{r}_{ij}$ is the vector between the molecular centres. Here we have made it clear that the pair virial is defined as a position derivative at constant orientation of the molecules

$$w(r_{ij}) = r_{ij}\left(\frac{\partial v(r_{ij}, \mathbf{\Omega}_i, \mathbf{\Omega}_j)}{\partial r_{ij}}\right)_{\mathbf{\Omega}_i\mathbf{\Omega}_j}. \tag{2.62}$$

The pressure function $\mathscr{P}$ is defined through eqn (2.55) as before. For interaction site models, we may treat the system as a set of atoms, and use eqns (2.59), (2.60), with the summations taken over distinct pairs of sites *ia* and *jb* (compare eqn (1.12)). When doing this, however, it is important to include all intramolecular contributions (forces along the bonds for example) in the sum. Alternatively, the molecular definition, eqns (2.61), (2.62) is still valid. In this case, for computational purposes, eqn (2.62) may be rewritten in the form

$$w(r_{ij}) = \sum_a \sum_b \frac{w_{ab}(r_{ab})}{r_{ab}^2}(\mathbf{r}_{ab}\cdot\mathbf{r}_{ij}) \tag{2.63}$$

where $\mathbf{r}_{ab} = \mathbf{r}_{ia} - \mathbf{r}_{jb}$ is the vector between the sites and $w_{ab}(r_{ab})$ is the site–site pair virial function. This is equivalent to expressing $\mathbf{f}_{ij}$ in eqn (2.61) as the sum of all the site–site forces acting between the molecules. Whether the atomic or molecular definition of the virial is adopted, the ensemble average $\langle \mathscr{W} \rangle$ and hence $\langle \mathscr{P} \rangle = P$ should be unaffected.

Quantities such as $\langle N \rangle$ and $\langle V \rangle$ are easily evaluated in the simulation of ensembles in which these quantities vary, and derived functions such as the enthalpy are easily calculated from the above.

Now we turn to the question of evaluating entropy-related ('statistical') quantities such as the Gibbs and Helmholtz functions, the chemical potential $\mu$, and the entropy itself. A direct approach is to conduct a simulation of the grand canonical ensemble, in which $\mu$, or a related quantity, is specified. It must be said at the outset that there are some technical difficulties associated with grand canonical ensemble simulations, and we return to this in Chapter 4. There are also difficulties in obtaining these functions in the other common ensembles, since they are related directly to the partition function $Q$, not to its derivatives. To calculate $Q$ would mean summing over all the states of the system. It might seem that we could use the formula

$$\exp(A^{\mathrm{ex}}/k_{\mathrm{B}}T) = Q_{NVT}^{\mathrm{ex}\,-1} = \langle \exp(\mathscr{V}/k_{\mathrm{B}}T)\rangle_{NVT} \tag{2.64}$$

to estimate the excess statistical properties, but, in practice, the distribution $\rho_{NVT}$ will be very sharply peaked around the largest values of $\exp(-\mathscr{V}/k_{\mathrm{B}}T)$, i.e. where $\exp(\mathscr{V}/k_{\mathrm{B}}T)$ is comparatively small. Consequently, any simulation technique that samples according to the equilibrium distribution will be bound to give a poor estimate of $A$ by this route. Special sampling techniques have been developed to evaluate averages of this type [Valleau and Torrie 1977] and we return to this in Chapter 7. It is comparatively easy to obtain free energy differences for a given system at two different temperatures by integrating the internal energy along a line of constant density:

$$\left(\frac{A}{Nk_BT}\right)_2 - \left(\frac{A}{Nk_BT}\right)_1 = \int_{\beta_1}^{\beta_2} \left(\frac{E}{Nk_BT}\right) \frac{d\beta}{\beta} = -\int_{T_1}^{T_2} \left(\frac{E}{Nk_BT}\right) \frac{dT}{T}. \tag{2.65}$$

Alternatively, integration of the pressure along an isotherm may be used:

$$\left(\frac{A}{Nk_BT}\right)_2 - \left(\frac{A}{Nk_BT}\right)_1 = \int_{\rho_1}^{\rho_2} \left(\frac{PV}{Nk_BT}\right) \frac{d\rho}{\rho} = -\int_{V_1}^{V_2} \left(\frac{PV}{Nk_BT}\right) \frac{dV}{V}. \tag{2.66}$$

To use these expressions, it is necessary to calculate ensemble averages at state points along a reversible thermodynamic path. To calculate absolute free energies and entropies, it is necessary to extend the thermodynamic integration far enough to reach a state point whose properties can be calculated essentially exactly. In general, these calculations may be expensive, since accurate thermodynamic information is required for many closely spaced state points.

One fairly direct, and widely applicable, method for calculating $\mu$ is based on the thermodynamic identities

$$\exp(-\mu/k_BT) = Q_{N+1}/Q_N = Q_N/Q_{N-1} \tag{2.67}$$

valid at large $N$ for both the constant-$NVT$ and constant-$NPT$ ensembles. From these equations, we can obtain expressions for the chemical potential in terms of a kind of ensemble average [Widom 1963, 1982]. If we define the excess chemical potential $\mu^{ex} = \mu - \mu^{id}$ then we can write

$$\mu^{ex} = -k_BT \ln \langle \exp(-\mathscr{V}_{test}/k_BT) \rangle \tag{2.68}$$

where $\mathscr{V}_{test}$ is the potential energy change which would result from the addition of a particle (at random) to the system. This is the 'test particle insertion' method of estimating $\mu$. Eqn 2.68 also applies in the constant-$\mu VT$ ensemble [Henderson 1983]. A slightly different formula applies for constant $NVE$ because of the kinetic temperature fluctuations [Frenkel 1986]

$$\mu^{ex} = -k_B \langle \mathscr{T} \rangle \ln[\langle \mathscr{T} \rangle^{-3/2} \langle \mathscr{T}^{3/2} \exp(-\mathscr{V}_{test}/k_B\mathscr{T}) \rangle] \tag{2.69a}$$

where $\mathscr{T}$ is the instantaneous kinetic temperature. Similarly for the constant-$NPT$ ensemble, it is necessary to include the fluctuations in the volume $V$ [Shing and Chung 1987]

$$\mu^{ex} = -k_BT \ln[\langle V \rangle^{-1} \langle V \exp(-\mathscr{V}_{test}/k_BT) \rangle]. \tag{2.69b}$$

In all these cases the 'test particle', the $(N+1)$th, is not actually inserted: it is a 'ghost', i.e. the $N$ real particles are not affected by its presence. There is an alternative formula which applies to the removal of a test particle (selected at random) from the system [Powles, Evans, and Quirke 1982]. This 'test particle' is not actually removed: it is a real particle and continues to interact normally with its neighbours. In practice, this technique does not give an accurate estimate of $\mu^{ex}$, and for hard spheres (for example) it is completely unworkable [Rowlinson and Widom 1982]. We defer a detailed discussion of the applicability of these methods and more advanced techniques until Chapter 7.

## 2.5 Fluctuations

We now discuss the information that can be obtained from RMS fluctuations, calculated as indicated in eqn (2.43). The quantities of most interest are the constant-volume specific heat capacity $C_V = (\partial E/\partial T)_V$ or its constant-pressure counterpart $C_P = (\partial H/\partial T)_P$, the thermal expansion coefficient $\alpha_P = V^{-1}(\partial V/\partial T)_P$, the isothermal compressibility $\beta_T = -V^{-1}(\partial V/\partial P)_T$, the thermal pressure coefficient $\gamma_V = (\partial P/\partial T)_V$, and the adiabatic (constant-$S$) analogues of these last three. The relation $\alpha_P = \beta_T\gamma_V$ means that only two of these quantities are needed to define the third. In part, formulae for these quantities can be obtained from the standard theory of fluctuations [Landau and Lifshitz 1980], but in computer simulations we must be carful to distinguish between properly defined mechanical quantities such as the energy, or Hamiltonian, $\mathscr{H}$, the kinetic temperature $\mathscr{T}$ or the instantaneous pressure $\mathscr{P}$, and thermodynamic concepts such as $T$ and $P$, which we can only describe as ensemble averages or as parameters defining an ensemble. Thus, a standard formula such as $\sigma^2(E) = \langle \delta E^2 \rangle = k_B T^2 C_V$ can actually be used to calculate the specific heat in the canonical ensemble (provided we recognize that $E$ really means $\mathscr{H}$), whereas the analogous simple formula $\sigma^2(P) = \langle \delta P^2 \rangle = k_B T/V\beta_T$ will not be so useful (since $P$ is not the same as $\mathscr{P}$).

Fluctuations are readily computed in the canonical ensemble, and accordingly we start with this case. As just mentioned, the specific heat is given by fluctuations in the total energy

$$\langle \delta\mathscr{H}^2 \rangle_{NVT} = k_B T^2 C_V. \tag{2.70}$$

This can be divided into kinetic and potential contributions which are uncorrelated (i.e. $\langle \delta\mathscr{K}\,\delta\mathscr{V} \rangle_{NVT} = 0$):

$$\langle \delta\mathscr{H}^2 \rangle_{NVT} = \langle \delta\mathscr{V}^2 \rangle_{NVT} + \langle \delta\mathscr{K}^2 \rangle_{NVT}. \tag{2.71}$$

The kinetic part can be calculated easily, for example in the case of a system of $N$ atoms:

$$\langle \delta\mathscr{K}^2 \rangle_{NVT} = \frac{3N}{2}(k_B T)^2 = 3N/2\beta^2 \tag{2.72}$$

yielding the ideal gas part of the specific heat, $C_V^{id} = (3/2)Nk_B$. For this case, then, potential energy fluctuations are simply

$$\langle \delta\mathscr{V}^2 \rangle_{NVT} = k_B T^2 (C_V - \tfrac{3}{2}Nk_B). \tag{2.73}$$

Consideration of the cross-correlation of potential energy and virial fluctuations yields an expression for the thermal pressure coefficient $\gamma_V$ [Rowlinson 1969]:

$$\langle \delta\mathscr{V}\,\delta\mathscr{W} \rangle_{NVT} = k_B T^2 (V\gamma_V - Nk_B) \tag{2.74}$$

where $\mathscr{W}$ is defined in eqns (2.59)–(2.63). In terms of the pressure function defined in eqn (2.55) this becomes

$$\langle \delta\mathscr{V}\,\delta\mathscr{P}\rangle_{NVT} = k_{\mathrm{B}}T^2(\gamma_V - \rho k_{\mathrm{B}}) \tag{2.75}$$

once more valid for a system of $N$ atoms. Equation (2.75) also applies if $\mathscr{P}$ is replaced by $\mathscr{P}'$ or by $\mathscr{P}^{\mathrm{ex}}$ (eqn (2.56)), which is more likely to be available in a (configuration-space) constant-$NVT$ Monte Carlo calculation. Similar formulae may be derived for molecular systems. When we come to consider fluctuations of the virial itself, we must define a further 'hypervirial' function

$$\mathscr{X} = \tfrac{1}{9}\sum_i \sum_{j>i} \sum_k \sum_{l>k} (\mathbf{r}_{ij}\cdot\nabla_{\mathbf{r}_{ij}})(\mathbf{r}_{kl}\cdot\nabla_{\mathbf{r}_{kl}})\mathscr{V} \tag{2.76}$$

which becomes, for a pairwise additive potential,

$$\mathscr{X} = \tfrac{1}{9}\sum_i \sum_{j>i} x(r_{ij}) \tag{2.77}$$

where

$$x(r) = r\frac{\mathrm{d}w(r)}{\mathrm{d}r} \tag{2.78}$$

$w(r)$ being the intermolecular virial defined in eqn (2.60). It is then easy to show that

$$\langle \delta\mathscr{W}^2\rangle_{NVT} = k_{\mathrm{B}}T(Nk_{\mathrm{B}}T + \langle\mathscr{W}\rangle_{NVT} - \beta_T^{-1}V + \langle\mathscr{X}\rangle_{NVT}) \tag{2.79}$$

or

$$\langle \delta\mathscr{P}^2\rangle_{NVT} = \frac{k_{\mathrm{B}}T}{V}\left(\frac{2Nk_{\mathrm{B}}T}{3V} + \langle\mathscr{P}\rangle_{NVT} - \beta_T^{-1} + \frac{\langle\mathscr{X}\rangle_{NVT}}{V}\right). \tag{2.80}$$

The average $\langle\mathscr{X}\rangle$ is a non-thermodynamic quantity. Nonetheless, it can be calculated in a computer simulation, and so eqns (2.79) and (2.80) provide a route to the isothermal compressibility $\beta_T$. Note that Cheung [1977] uses a different definition of the hypervirial function. In terms of the fluctuations of $\mathscr{P}'$, the analogous formula is

$$\langle \delta\mathscr{P}^{\mathrm{ex}2}\rangle_{NVT} = \langle \delta\mathscr{P}'^2\rangle_{NVT} = \frac{k_{\mathrm{B}}T}{V}\left(\langle\mathscr{P}'\rangle_{NVT} - \beta_T^{-1} + \frac{\langle\mathscr{X}\rangle_{NVT}}{V}\right) \tag{2.81}$$

and this would be the formula used in most constant-$NVT$ simulations.

The desired fluctuation expressions for the microcanonical ensemble may best be derived from the above equations, by applying the transformation formula, eqn (2.45) [Lebowitz *et al.* 1967; Cheung, 1977] or directly [Ray and Graben 1981]. The equivalence of ensembles guarantees that the values of simple averages (such as $\langle\mathscr{X}\rangle$ above) are unchanged by this transformation. In the microcanonical ensemble, the energy (of course) is fixed, but the specific heat may be obtained by examining fluctuations in the separate potential and kinetic components [Lebowitz *et al.* 1967]. For $N$ atoms,

$$\langle \delta \mathscr{V}^2 \rangle_{NVE} = \langle \delta \mathscr{K}^2 \rangle_{NVE} = \tfrac{3}{2} N k_B^2 T^2 \left(1 - \frac{3Nk_B}{2C_V}\right). \tag{2.82}$$

Cross-correlations of the pressure function and (say) the kinetic energy may be used to obtain the thermal pressure coefficient:

$$\langle \delta \mathscr{P} \delta \mathscr{K} \rangle_{NVE} = \langle \delta \mathscr{P} \delta \mathscr{V} \rangle_{NVE} = \frac{N k_B^2 T^2}{V} \left(1 - \frac{3V\gamma_V}{2C_V}\right). \tag{2.83}$$

Finally, the expression for fluctuations of $\mathscr{P}$ in the microcanonical ensemble yield the isothermal compressibility, but the formula is made slightly more compact by introducing the adiabatic compressibility $\beta_S$, and using $\beta_S^{-1} = \beta_T^{-1} + TV\gamma_V^2/C_V$

$$\langle \delta \mathscr{P}^2 \rangle_{NVE} = \frac{k_B T}{V} \left( \frac{2Nk_BT}{3V} + \langle \mathscr{P} \rangle_{NVE} - \beta_S^{-1} + \frac{\langle \mathscr{X} \rangle_{NVE}}{V} \right). \tag{2.84}$$

In eqns (2.82)–(2.84) $T$ is short for $\langle \mathscr{T} \rangle_{NVE}$. All the above expressions are easily derived using the transformation technique outlined earlier, and they are all valid for systems of $N$ atoms. The same expressions (to leading order in $N$) hold in the constant-$NVE$**P** ensemble probed by molecular dynamics. Analogous formulae for molecular systems may be derived in a similar way.

Conversion from the canonical to the isothermal–isobaric ensemble is easily achieved. Most of the formulae of interest are very simple since they involve well-defined mechanical quantities. At constant $T$ and $P$, both volume and energy fluctuations may occur. The volume fluctuations are related to the isothermal compressibility

$$\langle \delta V^2 \rangle_{NPT} = V k_B T \beta_T. \tag{2.85}$$

The simplest specific heat formula may be obtained by calculating the 'instantaneous' enthalpy $\mathscr{H} + PV$, when we see

$$\langle \delta (\mathscr{H} + PV)^2 \rangle_{NPT} = k_B T^2 C_P. \tag{2.86}$$

This equation can be split into the separate terms involving $\langle \delta \mathscr{H}^2 \rangle$, $\langle \delta V^2 \rangle$ and $\langle \delta \mathscr{H} \delta V \rangle$. Finally, the thermal expansion coefficient may be calculated from the cross-correlations of 'enthalpy' and volume:

$$\langle \delta V \delta (\mathscr{H} + PV) \rangle_{NPT} = k_B T^2 V \alpha_P. \tag{2.87}$$

Other quantities may be obtained by standard thermodynamic manipulations. Finally, to reiterate, although $P$ is fixed in the above expressions, the functions $\mathscr{P}$ and $\mathscr{P}'$ defined in eqns (2.55)–(2.56) will fluctuate around the average value $P$.

In the grand canonical ensemble, energy, pressure and number fluctuations occur. The number fluctuations yield the isothermal compressibility

$$\langle \delta N^2 \rangle_{\mu VT} = k_B T (\partial N / \partial \mu)_{VT} = \frac{N^2}{V} k_B T \beta_T. \tag{2.88}$$

Expressions for the other thermodynamic derivatives are a little more complicated [Adams 1975]. The simplest formula for a specific heat is obtained by considering (by analogy with the enthalpy) a function $\mathscr{H} - \mu N$:

$$\langle \delta(\mathscr{H} - \mu N)^2 \rangle_{\mu VT} = k_B T^2 C_{\mu V} = k_B T^2 \left( \frac{\partial \langle \mathscr{H} - \mu N \rangle}{\partial T} \right)_{\mu V} \tag{2.89}$$

and the usual specific heat $C_V$ (i.e. $C_{NV}$) is obtained by thermodynamic manipulations:

$$C_V = \frac{3}{2} N k_B + \frac{1}{k_B T^2} \left( \langle \delta \mathscr{V}^2 \rangle_{\mu VT} - \frac{\langle \delta \mathscr{V} \delta N \rangle^2_{\mu VT}}{\langle \delta N^2 \rangle_{\mu VT}} \right). \tag{2.90}$$

The thermal expansion coefficient may be derived in the same way:

$$\alpha_P = \frac{P \beta_T}{T} - \frac{\langle \delta \mathscr{V} \delta N \rangle_{\mu VT}}{N k_B T^2} + \frac{\langle \mathscr{V} \rangle_{\mu VT} \langle \delta N^2 \rangle_{\mu VT}}{N^2 k_B T^2}. \tag{2.91}$$

Finally, the thermal pressure coefficient is given by

$$\gamma_V = \frac{N k_B}{V} + \frac{\langle \delta \mathscr{V} \delta N \rangle_{\mu VT}}{VT} \left( 1 - \frac{N}{\langle \delta N^2 \rangle_{\mu VT}} \right) + \frac{\langle \delta \mathscr{V} \delta \mathscr{W} \rangle_{\mu VT}}{V k_B T^2}. \tag{2.92}$$

Except within brackets $\langle \ldots \rangle$, $N$ in the above equations is understood to mean $\langle N \rangle_{\mu VT}$ and similarly $P$ means $\langle \mathscr{P} \rangle_{\mu VT}$. As emphasized by Adams [1975], when these formulae are used in a computer simulation, it is advisable to cross-check them with the thermodynamic identity $\alpha_P = \beta_T \gamma_V$.

## 2.6 Structural quantities

The structure of simple monatomic fluids is characterized by a set of distribution functions for the atomic positions, the simplest of which is the pair distribution function $g_2(\mathbf{r}_i, \mathbf{r}_j)$, or $g_2(r_{ij})$ or simply $g(r)$. This function gives the probability of finding a pair of atoms a distance $r$ apart, relative to the probability expected for a completely random distribution at the same density. To define $g(r)$, we integrate the configurational distribution function over the positions of all atoms except two, incorporating the appropriate normalization factors [McQuarrie 1976; Hansen and McDonald 1986]. In the canonical ensemble

$$g(\mathbf{r}_1, \mathbf{r}_2) = \frac{N(N-1)}{\rho^2 Z_{NVT}} \int \mathrm{d}\mathbf{r}_3 \, \mathrm{d}\mathbf{r}_4 \ldots \mathrm{d}\mathbf{r}_N \exp(-\beta \mathscr{V}(\mathbf{r}_1, \mathbf{r}_2, \ldots \mathbf{r}_N)). \tag{2.93}$$

Obviously the choice $i = 1, j = 2$ is arbitrary in a system of identical atoms. An equivalent definition takes an ensemble average over pairs

$$g(r) = \rho^{-2}\langle\sum_i\sum_{j\neq i}\delta(\mathbf{r}_i)\delta(\mathbf{r}_j-\mathbf{r})\rangle = \frac{V}{N^2}\langle\sum_i\sum_{j\neq i}\delta(\mathbf{r}-\mathbf{r}_{ij})\rangle. \quad (2.94)$$

This last form may be used in the evaluation of $g(r)$ by computer simulation; in practice, the delta function is replaced by a function which is non-zero in a small range of separations, and a histogram is compiled of all pair separations falling within each such range (see Chapter 6). Figure 2.2 shows a typical pair distribution function for the Lennard-Jones liquid close to its triple point.

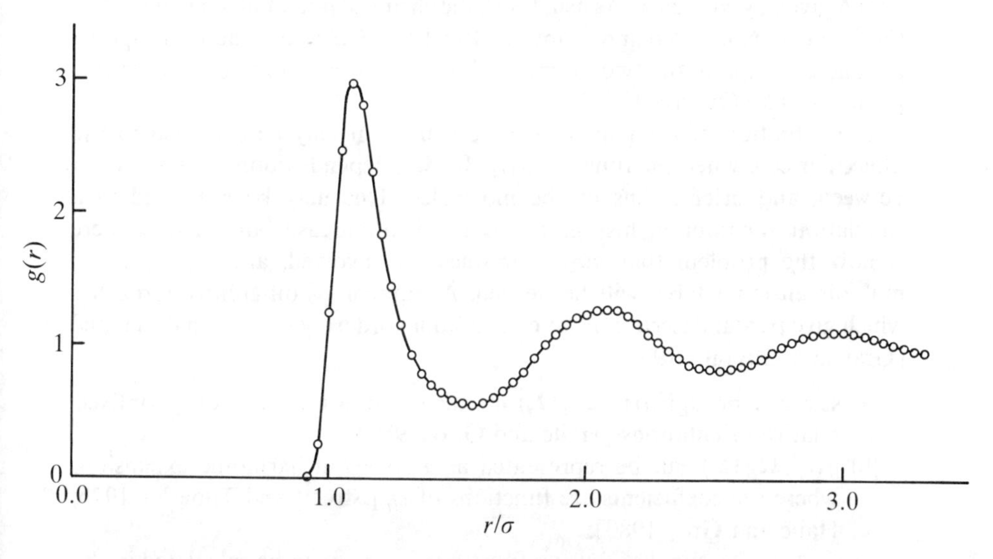


**Fig. 2.2** Pair distribution function for the Lennard-Jones fluid close to the triple point ($T^* = 0.71$, $\rho^* = 0.844$).

The pair distribution function is useful, not only because it provides insight into the liquid structure, but also because the ensemble average of any pair function may be expressed in the form

$$\langle a(\mathbf{r}_i, \mathbf{r}_j)\rangle = \frac{1}{V^2}\int \mathrm{d}\mathbf{r}_i\mathrm{d}\mathbf{r}_j\, g(\mathbf{r}_i, \mathbf{r}_j)\, a(\mathbf{r}_i, \mathbf{r}_j). \quad (2.95a)$$

or

$$\langle\mathscr{A}\rangle = \langle\sum_i\sum_{j>i} a(r_{ij})\rangle = \tfrac{1}{2}N\rho\int_0^\infty a(r)g(r)4\pi r^2\mathrm{d}r \quad (2.95b)$$

For example, we may write the energy (assuming pair additivity)

$$E = (3/2)Nk_{\mathrm{B}}T + 2\pi N\rho\int_0^\infty r^2 v(r)g(r)\,\mathrm{d}r \quad (2.96)$$

or the pressure

$$PV = Nk_{\mathrm{B}}T - (2/3)\pi N\rho \int_0^{\infty} r^2 w(r) g(r) \mathrm{d}r \tag{2.97}$$

although in practice a direct evaluation of these quantities, as discussed in Section 2.4, will usually be more accurate. Even the chemical potential may be related to $g(r)$

$$\mu = k_{\mathrm{B}}T \ln(\rho\Lambda^3) + 4\pi\rho \int_0^1 \mathrm{d}\xi \int_0^{\infty} r^2 v(r) g(r;\xi) \mathrm{d}r \tag{2.98}$$

with $\Lambda$ given by eqn (2.24). As usual with the chemical potential, there is a twist: the formula involves a pair distribution function $g(r;\xi)$ which depends upon a parameter coupling the two atoms, and it is necessary to integrate over this parameter [McQuarrie 1976].

The definition of the pair distribution function may be extended to the molecular case when the function $g(r_{ij}, \mathbf{\Omega}_i, \mathbf{\Omega}_j)$ depends upon the separation between, and orientations of the molecules. This may be evaluated in a simulation by compiling histograms, as in the atomic case, but of course there is now the problem that more variables are involved, and a very large, multidimensional table will be needed. A number of different approaches which give partial descriptions of orientational ordering have been developed [Gray and Gubbins 1984]:

(a) sections through $g(r_{ij}, \mathbf{\Omega}_i, \mathbf{\Omega}_j)$ are calculated as a function of $r_{ij}$ for fixed relative orientations [Haile and Gray 1980];
(b) $g(r_{ij}, \mathbf{\Omega}_i, \mathbf{\Omega}_j)$ can be represented as a spherical harmonic expansion, where the coefficients are functions of $r_{ij}$ [Streett and Tildesley 1976; Haile and Gray 1980];
(c) a set of site–site distribution functions $g_{ab}(r_{ab})$, can be calculated in the same way as the atomic $g(r)$ for each type of site.

The first method proceeds by compiling histograms, just as for $g(r)$, but restricting the accumulation of data to pairs of molecules which are close to a few specific relative orientations. Thus, for pairs of linear molecules, parallel configurations and T-shapes might be of interest.

The spherical harmonic expansion for a pair of linear molecules would take the form

$$g(r_{ij}, \mathbf{\Omega}_i, \mathbf{\Omega}_j) = 4\pi \sum_{l=0}^{\infty} \sum_{l'=0}^{\infty} \sum_{m} g_{ll'm}(r_{ij}) Y_{lm}(\mathbf{\Omega}_i) Y_{l'\bar{m}}(\mathbf{\Omega}_j) \tag{2.99}$$

where the functions $Y_{lm}(\mathbf{\Omega})$ are spherical harmonics and $\bar{m} = -m$. The range of the sum over $m$ values is either $(-l, l)$ or $(-l', l')$, whichever is the smaller. Note that the orientations are measured relative to the vector $\mathbf{r}_{ij}$ in each case. In a simulation, the coefficients $g_{ll'm}$ would be evaluated by averaging a product of spherical harmonics over a spherical shell around each molecule, as described in Chapter 6. The function $g_{000}(r)$ is the isotropic component, i.e. the pair distribution function for molecular centres averaged over all orientations. This approach is readily extended to non-linear molecules. The expansion can be carried out in a molecule-fixed frame [Streett and Tildesley 1976] or in a space-fixed frame [Haile and Gray 1980]. The coefficients can be recombined to give the total distribution function, but this is not profitable for elongated molecules, since many terms are required for the series to converge. Certain observable properties are related to limited numbers of the harmonic coefficients. The angular correlation parameter of rank $l$, $g_l$, may be expressed in the molecule-fixed frame

$$g_l = 1 + \frac{4\pi\rho}{2l+1} \sum_{m=-l}^{m=+l} (-1)^m \int_0^\infty g_{llm}(r)\, r^2 \mathrm{d}r \tag{2.100a}$$

$$= 1 + \frac{1}{N} \langle \sum_i \sum_{j \neq i} P_l(\cos\gamma_{ij}) \rangle \tag{2.100b}$$

where $P_l(\cos\gamma)$ is a Legendre polynomial and $\gamma_{ij}$ is the angle between the axis vectors of molecules $i$ and $j$. $g_1$ is related to the dielectric properties of polar molecules, while $g_2$ may be investigated by depolarized light scattering. Formulae analogous to eqns (2.99) and (2.100) may be written for non-linear molecules. These would involve the Wigner rotation matrices $\mathscr{D}^l_{mm'}(\boldsymbol{\Omega}_i)$ instead of the spherical harmonics [Gray and Gubbins 1984, Appendix 7].

As an alternative, a site–site description may be more appropriate. Pair distribution functions $g_{ab}(r_{ab})$ are defined for each pair of sites on different molecules, using the same definition as in the atomic case. The number of independent $g_{ab}(r_{ab})$ functions will depend on the complexity of the molecule. For example in a three-site model of OCS, the isotropic liquid is described by six independent $g_{ab}$ functions, (for OO, OC, OS, CC, CS, and SS distances) whereas for a five-site model of $CH_4$, the liquid is described by three functions (CC, CH, HH). While less information is contained in these distribution functions than in the components of $g(r_{ij}, \boldsymbol{\Omega}_i, \boldsymbol{\Omega}_j)$, they have the advantage of being directly related to the structure factor of the molecular fluid [Lowden and Chandler 1974] and hence to experimentally observable properties (for example neutron and X-ray scattering). We return to the calculation of these quantities in Chapter 6.

Finally, we turn to the definitions of quantities that depend upon wavevector rather than on position. In a simulation with periodic boundaries, we are restricted to wavevectors that are commensurate with the periodicity of the system, i.e. with the simulation box. Specifically, in a cubic box, we may examine fluctuations for which $\mathbf{k} = (2\pi/L)(k_x, k_y, k_z)$ where $L$ is the box length and $k_x, k_y, k_z$ are integers. This is a severe restriction, particularly at low $k$. One quantity of interest is the spatial Fourier transform of the number density

$$\rho(k) = \sum_{i=1}^{N} \exp(i\mathbf{k}\cdot\mathbf{r}_i). \tag{2.101}$$

Fluctuations in $\rho(k)$ are related to the structure factor $S(k)$

$$S(k) = N^{-1}\langle \rho(k)\rho(-k)\rangle \tag{2.102}$$

which may be measured by neutron or X-ray scattering experiments. Thus, $S(k)$ describes the Fourier components of density fluctuations in the liquid. It is related, through a three-dimensional Fourier transform (see Appendix D) to the pair distribution function

$$S(k) = 1 + \rho\hat{h}(k) = 1 + \rho\hat{g}(k) = 1 + 4\pi\rho\int_0^\infty r^2\frac{\sin kr}{kr}g(r)\,\mathrm{d}r \tag{2.103}$$

where we have introduced the Fourier transform of the total correlation function $h(r) = g(r) - 1$, and have ignored a delta function contribution at $k = 0$. In a similar way, $k$-dependent orientational functions may be calculated and measured routinely in computer simulations.

## 2.7 Time correlation functions and transport coefficients

Correlations between two different quantities $\mathscr{A}$ and $\mathscr{B}$ are measured in the usual statistical sense, via the correlation coefficient $c_{\mathscr{A}\mathscr{B}}$

$$c_{\mathscr{A}\mathscr{B}} = \langle \delta\mathscr{A}\,\delta\mathscr{B}\rangle/\sigma(\mathscr{A})\sigma(\mathscr{B}) \tag{2.104}$$

with $\sigma(\mathscr{A})$ and $\sigma(\mathscr{B})$ given by eqn (2.43). Schwartz inequalities guarantee that the absolute value of $c_{\mathscr{A}\mathscr{B}}$ lies between 0 and 1, with values close to 1 indicating a high degree of correlation. The idea of the correlation coefficient may be extended in a very useful way, by considering $\mathscr{A}$ and $\mathscr{B}$ to be evaluated at two different times. The resulting quantity is a function of the time difference $t$: it is a 'time correlation function' $c_{\mathscr{A}\mathscr{B}}(t)$. For identical phase functions, $c_{\mathscr{A}\mathscr{A}}(t)$ is called an autocorrelation function and its time integral (from $t = 0$ to $t = \infty$) is a correlation time $t_{\mathscr{A}}$. These functions are of great interest in computer simulation because:

(a) they give a clear picture of the dynamics in a fluid;
(b) their time integrals $t_{\mathscr{A}}$ may often be related directly to macroscopic transport coefficients;
(c) their Fourier transforms $\hat{c}_{\mathscr{A}\mathscr{A}}(\omega)$ may often be related to experimental spectra.

Good discussions of time correlation functions may be found in the standard references [Steele 1969, 1980; Berne and Harp 1970; McQuarrie 1976; Hansen and McDonald 1986]. A few comments may be relevant here. The non-normalized correlation function is defined

$$C_{\mathscr{A}\mathscr{B}}(t) = \langle \delta\mathscr{A}(t)\delta\mathscr{B}(0)\rangle_{\text{ens}} = \langle \delta\mathscr{A}(\mathbf{\Gamma}(t))\delta\mathscr{B}(\mathbf{\Gamma}(0))\rangle_{\text{ens}} \tag{2.105}$$

so that

$$c_{\mathscr{A}\mathscr{B}}(t) = C_{\mathscr{A}\mathscr{B}}(t)/\sigma(\mathscr{A})\sigma(\mathscr{B}) \tag{2.106a}$$

or

$$c_{\mathscr{A}\mathscr{A}}(t) = C_{\mathscr{A}\mathscr{A}}(t)/\sigma^2(\mathscr{A}) = C_{\mathscr{A}\mathscr{A}}(t)/C_{\mathscr{A}\mathscr{A}}(0). \tag{2.106b}$$

Just like $\langle \delta\mathscr{A}\delta\mathscr{B} \rangle$, $C_{\mathscr{A}\mathscr{B}}(t)$ is different for different ensembles, and eqn (2.45) may be used to transform from one ensemble to another. The computation of $C_{\mathscr{A}\mathscr{B}}(t)$ may be thought of as a two-stage process. First, we must select initial state points $\mathbf{\Gamma}(0)$, according to the desired distribution $\rho_{\text{ens}}(\mathbf{\Gamma})$, over which we will subsequently average. This may be done using any of the prescriptions mentioned in Section 2.1. Second, we must evaluate $\mathbf{\Gamma}(t)$. This means solving the true (Newtonian) equations of motion. By this means, time-dependent properties may be calculated in any ensemble. In practice, the mechanical equations of motion are almost always used for both purposes, i.e. we use molecular dynamics to calculate time correlation functions in the microcanonical ensemble.

Some attention must be paid to the question of ensemble equivalence, however, since the link between correlation functions and transport coefficients is made through linear response theory, which can be carried out in virtually any ensemble. This actually caused some confusion in the original derivations of expressions for transport coefficients [Zwanzig 1965]. Below, we make some general observations, and refer the reader elsewhere [McQuarrie 1976] for a fuller discussion.

Transport coefficients are defined in terms of the response of a system to a perturbation. For example, the diffusion coefficient relates the particle flux to a concentration gradient, while the shear viscosity is a measure of the shear stress induced by an applied velocity gradient. By introducing such perturbations into the Hamiltonian, or directly into the equations of motion, their effect on the distribution function $\rho_{\text{ens}}$ may be calculated. Generally, a time-dependent, non-equilibrium distribution $\rho(t) = \rho_{\text{ens}} + \delta\rho(t)$ is produced. Hence, any non-equilibrium ensemble average (in particular, the desired response) may be calculated. By retaining the linear terms in the perturbation, and comparing the equation for the response with a macroscopic transport equation, we may identify the transport coefficient. This is usually the infinite time integral of an equilibrium time correlation function of the form

$$\gamma = \int_0^\infty \mathrm{d}t \, \langle \dot{\mathscr{A}}(t)\dot{\mathscr{A}}(0) \rangle \tag{2.107}$$

where $\gamma$ is the transport coefficient, and $\mathscr{A}$ is a variable appearing in the perturbation term in the Hamiltonian. Associated with any expression of this kind, there is also an 'Einstein relation'

$$2t\gamma = \langle (\mathscr{A}(t) - \mathscr{A}(0))^2 \rangle \tag{2.108}$$

which holds at large $t$ (compared with the correlation time of $\mathscr{A}$). The connection between eqns (2.107) and (2.108) may easily be established by integration by parts. Note that only a few genuine transport coefficients exist, i.e. for only a few 'hydrodynamic' variables $\mathscr{A}$ do eqns (2.107) and (2.108) give a non-zero $\gamma$ [McQuarrie 1976].

In computer simulations, transport coefficients may be calculated from equilibrium correlation functions, using eqn (2.107), by observing Einstein relations, eqn (2.108), or indeed by going back to first principles and conducting a suitable non-equilibrium simulation. The details of calculation via eqns (2.107), (2.108) will be given in Chapter 6, and we examine non-equilibrium methods in Chapter 8. For use in equilibrium molecular dynamics, we give here the equations for calculating thermal transport coefficients in the microcanonical ensemble, for a fluid composed of $N$ identical molecules.

The diffusion coefficient $D$ is given (in three dimensions) by

$$D = \tfrac{1}{3} \int_0^\infty \mathrm{d}t \; \langle \mathbf{v}_i(t) \cdot \mathbf{v}_i(0) \rangle \tag{2.109}$$

where $\mathbf{v}_i(t)$ is the centre-of-mass velocity of a single molecule. The corresponding Einstein relation, valid at long times, is

$$2tD = \tfrac{1}{3} \langle |\mathbf{r}_i(t) - \mathbf{r}_i(0)|^2 \rangle \tag{2.110}$$

where $\mathbf{r}_i(t)$ is the molecule position. In practice, these averages would be computed for each of the $N$ particles in the simulation, the results added together, and divided by $N$, to improve statistical accuracy. Note that in the computation of the right of eqn (2.110), it is important not to switch attention from one periodic image to another, which is why it is sometimes useful to have available a set of particle coordinates which have not been subjected to periodic boundary corrections during the simulation (see Section 1.5 and Chapter 6).

The shear viscosity $\eta$ is given by

$$\eta = \frac{V}{k_\mathrm{B}T} \int_0^\infty \mathrm{d}t \; \langle \mathscr{P}_{\alpha\beta}(t) \, \mathscr{P}_{\alpha\beta}(0) \rangle \tag{2.111}$$

or

$$2t\eta = \frac{V}{k_\mathrm{B}T} \langle (\mathscr{Q}_{\alpha\beta}(t) - \mathscr{Q}_{\alpha\beta}(0))^2 \rangle . \tag{2.112}$$

Here

$$\mathscr{P}_{\alpha\beta} = \frac{1}{V} \Big( \sum_i p_{i\alpha} p_{i\beta}/m_i + \sum_i r_{i\alpha} f_{i\beta} \Big) \tag{2.113}$$

or

$$\mathscr{P}_{\alpha\beta} = \frac{1}{V} \Big( \sum_i p_{i\alpha} p_{i\beta}/m_i + \sum_i \sum_{j>i} r_{ij\alpha} f_{ij\beta} \Big) \tag{2.114}$$

is an off-diagonal ($\alpha \neq \beta$) element of the pressure tensor (compare the virial expression for the pressure function eqns (2.55) and (2.59)) and

$$\mathscr{Q}_{\alpha\beta} = \frac{1}{V} \sum_i r_{i\alpha} p_{i\beta} . \tag{2.115}$$

The negative of $\mathscr{P}_{\alpha\beta}$ is often called the stress tensor. These quantities are multi-particle properties, properties of the system as a whole, and so no additional averaging over the $N$ particles is possible. Consequently $\eta$ is subject to much greater statistical imprecision than $D$. Some improvement is possible by averaging over different components, $\alpha\beta = xy$, $yz$, $zx$, of $\mathscr{P}_{\alpha\beta}$.

The bulk viscosity is given by a similar expression:

$$\eta_V = \frac{V}{9k_BT}\sum_{\alpha\beta}\int_0^\infty \mathrm{d}t\ \langle \delta\mathscr{P}_{\alpha\alpha}(t)\,\delta\mathscr{P}_{\beta\beta}(0)\rangle$$

$$= \frac{V}{k_BT}\int_0^\infty \mathrm{d}t\ \langle \delta\mathscr{P}(t)\ \delta\mathscr{P}(0)\rangle \tag{2.116a}$$

where we sum over $\alpha, \beta = x, y, z$ and note that $\mathscr{P} = \frac{1}{3}\,\mathrm{Tr}\,\mathscr{P} = \frac{1}{3}\sum_\alpha \mathscr{P}_{\alpha\alpha}$. Rotational invariance leads to the equivalent expression

$$\eta_\mathrm{v} + \tfrac{4}{3}\eta = \frac{V}{k_BT}\int_0^\infty \mathrm{d}t\ \langle \delta\mathscr{P}_{\alpha\alpha}(t)\,\delta\mathscr{P}_{\alpha\alpha}(0)\rangle\,. \tag{2.116b}$$

Here the diagonal stresses must be evaluated with care, since a non-vanishing equilibrium average must be subtracted:

$$\delta\mathscr{P}_{\alpha\alpha}(t) = \mathscr{P}_{\alpha\alpha}(t) - \langle\mathscr{P}_{\alpha\alpha}\rangle = \mathscr{P}_{\alpha\alpha}(t) - P \tag{2.117a}$$

$$\delta\mathscr{P}(t) = \mathscr{P}(t) - \langle\mathscr{P}\rangle = \mathscr{P}(t) - P \tag{2.117b}$$

with $\mathscr{P}_{\alpha\alpha}$ given by an expression like eqn (2.114). The corresponding Einstein relation is [Alder, Gass, and Wainwright 1970]

$$2t(\eta_\mathrm{v} + \tfrac{4}{3}\eta) = \frac{V}{k_BT}\langle(\mathscr{Q}_{\alpha\alpha}(t) - \mathscr{Q}_{\alpha\alpha}(0) - Pt)^2\rangle\,. \tag{2.118}$$

The thermal conductivity $\lambda_T$ can be written [Hansen and McDonald 1986]

$$\lambda_T = \frac{V}{k_BT^2}\int_0^\infty \mathrm{d}t\ \langle j_\alpha^\varepsilon(t)\, j_\alpha^\varepsilon(0)\rangle \tag{2.119}$$

or

$$2t\lambda_\mathrm{T} = \frac{V}{k_BT^2}\langle(\delta\varepsilon_\alpha(t) - \delta\varepsilon_\alpha(0))^2\rangle\,. \tag{2.120}$$

Here, $j_\alpha^\varepsilon$ is a component of the energy current, i.e. the time derivative of

$$\delta\varepsilon_\alpha = \frac{1}{V}\sum_i r_{i\alpha}(\varepsilon_i - \langle\varepsilon_i\rangle)\,. \tag{2.121}$$

The term $\sum_i r_{i\alpha}\langle\varepsilon_i\rangle$ makes no contribution if $\sum_i r_{i\alpha} = 0$, as is the case in a normal one-component MD simulation. In calculating the energy per molecule $\varepsilon_i$, the potential energy of two molecules (assuming pairwise potentials) is taken to be divided equally between them:

$$\varepsilon_i = p_i^2/2m_i + \tfrac{1}{2}\sum_{j\neq i} v(r_{ij}) . \qquad (2.122)$$

These expressions for $\eta_V$ and $\lambda_T$ are ensemble-dependent and the above equations hold for the microcanonical case only. A fuller discussion may be found in the references [McQuarrie 1976; Zwanzig 1965].

Transport coefficients are related to the long-time behaviour of correlation functions. Short-time correlations, on the other hand, may be linked with static equilibrium ensemble averages, by expanding in a Taylor series. For example, the velocity of particle $i$ may be written

$$\mathbf{v}_i(t) = \mathbf{v}_i(0) + \dot{\mathbf{v}}_i(0)\, t + \tfrac{1}{2}\ddot{\mathbf{v}}_i(0)\, t^2 + \ldots . \qquad (2.123)$$

Multiplying by $\mathbf{v}_i(0)$ and ensemble averaging yields

$$\begin{aligned}\langle \mathbf{v}_i(t)\cdot\mathbf{v}_i(0)\rangle &= \langle v_i^2\rangle + \tfrac{1}{2}\langle \ddot{\mathbf{v}}_i\cdot\mathbf{v}_i\rangle\, t^2 + \ldots \\ &= \langle v_i^2\rangle - \tfrac{1}{2}\langle \dot{v}_i^2\rangle\, t^2 + \ldots . \end{aligned} \qquad (2.124)$$

The vanishing of the term linear in $t$, and the last step, where we set $\langle \ddot{\mathbf{v}}_i\cdot\mathbf{v}_i\rangle = -\langle \dot{\mathbf{v}}_i\cdot\dot{\mathbf{v}}_i\rangle$, follow from time reversal symmetry and stationarity [McQuarrie 1976]. Thus, the short-time velocity autocorrelation function is related to the mean square acceleration, i.e. to the mean square force. This behaviour may be used to define the Einstein frequency $\omega_{\mathrm{E}}$

$$\langle \mathbf{v}_i(t)\cdot\mathbf{v}_i(0)\rangle = \langle v_i^2\rangle\,(1 - \tfrac{1}{2}\omega_{\mathrm{E}}^2 t^2 + \ldots) . \qquad (2.125)$$

The analogy with the Einstein model, of an atom vibrating in the mean force field of its neighbours, with frequency $\omega_{\mathrm{E}}$ in the harmonic approximation, becomes clear when we replace the mean square force by the average potential curvature using

$$\begin{aligned}\langle f_{i\alpha}^2\rangle &= -\langle f_{i\alpha}\,\partial\mathscr{V}/\partial r_{i\alpha}\rangle = -k_{\mathrm{B}}T\,\langle \partial f_{i\alpha}/\partial r_{i\alpha}\rangle \\ &= k_{\mathrm{B}}T\,\langle \partial^2\mathscr{V}/\partial r_{i\alpha}^2\rangle \end{aligned} \qquad (2.126)$$

(another application of $\langle \mathscr{A}\,\partial\mathscr{H}/\partial q_k\rangle = k_{\mathrm{B}}T\langle \partial\mathscr{A}/\partial q_k\rangle$). The result is

$$\omega_{\mathrm{E}}^2 = \frac{\langle f_i^2\rangle}{m_i^2\,\langle v_i^2\rangle} = \frac{1}{3m_i}\,\langle \nabla_{\mathbf{r}_i}^2\mathscr{V}\rangle . \qquad (2.127)$$

This may be easily evaluated for, say, a pairwise additive potential. Short-time expansions of other time correlation functions may be obtained using similar techniques. The temporal Fourier transform (see Appendix D) of the velocity autocorrelation function is proportional to the density of normal modes in a purely harmonic system, and is often loosely referred to as the 'density of states' in solids and liquids. The velocity autocorrelation function and its Fourier transform for the Lennard-Jones liquid near the triple point are illustrated in Fig. 2.3.

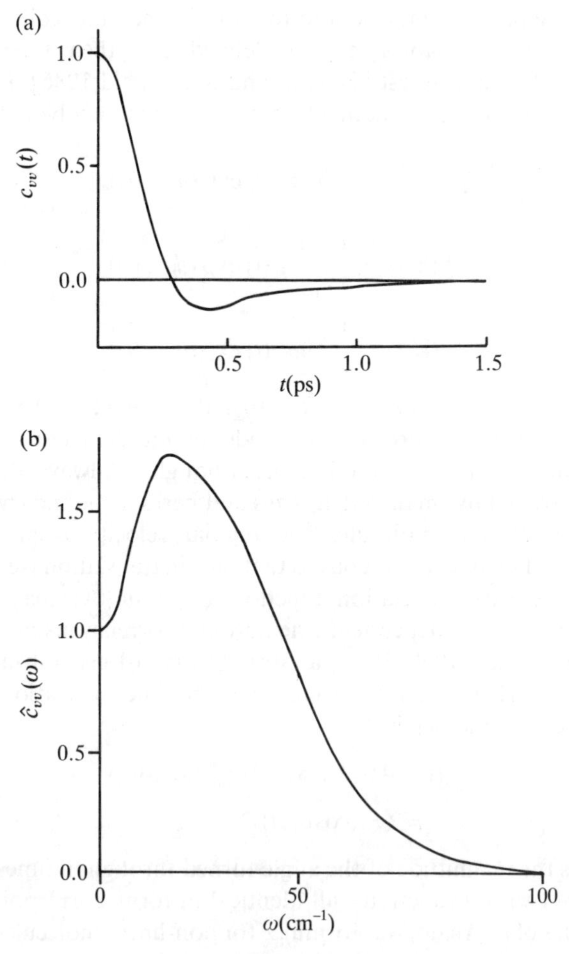


**Fig. 2.3** (a) The velocity autocorrelation function and (b) its Fourier transform, for the Lennard-Jones liquid near the triple point ($\rho^* = 0.85$, $T^* = 0.76$).

We can only mention briefly some other correlation functions of interest in computer simulations. The generalization of eqn (2.102) to the time domain yields the intermediate scattering function $I(k, t)$

$$I(k, t) = N^{-1} \langle \rho(k, t) \rho(-k, 0) \rangle \tag{2.128}$$

with $\rho(k, t)$ defined by eqn (2.101). The temporal Fourier transform of this, the dynamic structure factor $S(k, \omega)$, in principle, may be measured by inelastic neutron scattering. Spatially Fourier-transforming $I(k, t)$ yields the van Hove function $G(r, t)$, a generalization of $g(r)$ which measures the probability of finding a particle at position $r$ at time $t$, given that a particle was at the origin of coordinates at time 0. All of these functions may be divided into parts due to 'self' (i.e. single-particle) motion and due to 'distinct' (i.e. collective) effects. Other $k$-dependent variables may be defined, and their time correlation functions are of great interest [Hansen and McDonald 1986]. For example, longitudinal and transverse momentum components may be defined

$$p^{\parallel}(\mathbf{k}, t) = \frac{1}{V}\sum_i p_{ix}(t) \exp(ikr_{ix}(t)) \tag{2.129}$$

$$p_1^{\perp}(\mathbf{k}, t) = \frac{1}{V}\sum_i p_{iy}(t) \exp(ikr_{ix}(t)) \tag{2.130a}$$

$$p_2^{\perp}(\mathbf{k}, t) = \frac{1}{V}\sum_i p_{iz}(t) \exp(ikr_{ix}(t)) \tag{2.130b}$$

where for convenience we take $\mathbf{k} = (k, 0, 0)$ in the $x$ direction. These quantities are useful in discussing hydrodynamic modes in liquids. These functions may all be computed routinely in simulations, although, as always, the allowed $k$-values are restricted by small system sizes and periodic boundary conditions.

For systems of rigid molecules, the angular velocity $\boldsymbol{\omega}_i$ plays a role in reorientational dynamics analogous to that of $\mathbf{v}_i$ in translation (see Chapter 3). The angular velocity correlation function $\langle \boldsymbol{\omega}_i(t) . \boldsymbol{\omega}_i(0) \rangle$ may be used to describe rotation. Time-dependent orientational correlations may be defined [Gordon 1968; Steele 1969, 1980] as straightforward generalizations of the quantities seen earlier. For a linear molecule, the time correlation function of rank-$l$ spherical harmonics is

$$\begin{aligned} c_l(t) &= 4\pi \langle Y_{lm}(\boldsymbol{\Omega}_i(t))\, Y_{lm}^*(\boldsymbol{\Omega}_i(0)) \rangle \\ &= \langle P_l(\cos \delta\gamma(t)) \rangle \end{aligned} \tag{2.131}$$

where $\delta\gamma(t)$ is the magnitude of the angle turned through in time $t$. Note that there are $2l + 1$ rank-$l$ functions, all identical in form, corresponding to the different values of $m$. Analogous formulae for non-linear molecules involve the Wigner rotation matrices

$$\begin{aligned} c_{lmm'}(t) &= (2l+1) \langle \mathscr{D}^l_{nm}(\boldsymbol{\Omega}_i(t))\, \mathscr{D}^{l*}_{nm'}(\boldsymbol{\Omega}_i(0)) \rangle \\ &= \langle \mathscr{D}^l_{mm'}(\delta\gamma(t)) \rangle . \end{aligned} \tag{2.132}$$

These quantities are experimentally accessible and the relationships between them are of great theoretical interest (see Chapter 11). For example, first-rank autocorrelation functions may be related to infra-red absorption, and second-rank functions to light scattering. Functions of all ranks contribute to inelastic neutron scattering spectra from molecular liquids.

## 2.8 Long-range corrections

As explained in Section 1.5, computer simulations frequently use a pair potential with a spherical cutoff at a distance $r_c$. It becomes useful to correct the results of simulations to compensate for the missing long-range part of the potential. Contributions to the energy, pressure etc. for $r > r_c$ are frequently estimated by assuming that $g(r) \approx 1$ in this region, and using eqns (2.96)–(2.98)

$$E_{\text{full}} \approx E_c + E_{\text{LRC}} = E_c + 2\pi N\rho \int_{r_c}^{\infty} r^2\, v(r)\, \mathrm{d}r \tag{2.133}$$

$$(PV)_{\text{full}} \approx (PV)_c + (PV)_{\text{LRC}} = (PV)_c - (2/3)\pi N\rho \int_{r_c}^{\infty} r^2\, w(r)\, \mathrm{d}r \tag{2.134}$$

$$\mu_{\text{full}} \approx \mu_c + \mu_{\text{LRC}} = \mu_c + 4\pi\rho \int_{r_c}^{\infty} r^2\, v(r)\, \mathrm{d}r \tag{2.135}$$

where $E_{\text{full}}$, $(PV)_{\text{full}}$, $\mu_{\text{full}}$ are the desired values for a liquid with the full potential, and $E_c$, $(PV)_c$, $\mu_c$ are the values actually determined from a simulation using a potential with a cutoff. For the Lennard-Jones potential, eqn (1.6), these equations become

$$E^*_{\text{LRC}} = (8/9)\pi N\, \rho^*\, r_c^{*\,-9} - (8/3)\pi N\, \rho^*\, r_c^{*\,-3} \tag{2.136}$$

$$P^*_{\text{LRC}} = (32/9)\pi\, \rho^{*2}\, r_c^{*\,-9} - (16/3)\pi\, \rho^{*2}\, r_c^{*\,-3} \tag{2.137}$$

$$\mu^*_{\text{LRC}} = (16/9)\pi\, \rho^*\, r_c^{*\,-9} - (16/3)\pi\, \rho^*\, r_c^{*\,-3} \tag{2.138}$$

where we use Lennard-Jones reduced units (see Appendix B). In the case of the constant-$NVE$ and constant-$NVT$ ensembles, these corrections can be applied to the results after a simulation has run. However, if the volume or the number of particles is allowed to fluctuate (for example, in a constant-$NPT$ or constant-$\mu VT$ simulation) it is important to apply the corrections to the calculated instantaneous energies, pressures etc. during the course of a simulation, since they will change as the density fluctuates: it is far more tricky to attempt to do this when the simulation is over.

## 2.9 Quantum corrections

Most of this book will deal with the computer simulation of systems within the classical approximation, although we turn in Chapter 10 to the attempts which have been made to incorporate quantum effects in simulations. Quantum effects in thermodynamics may be measured experimentally via the isotope separation factor, while tunnelling, superfluidity etc. are clear manifestations of quantum mechanics.

Even within the limitations of a classical simulation, it is still possible to estimate quantum corrections of thermodynamic functions. This is achieved by expanding the partition function in powers of Planck's constant, $\hbar = h/2\pi$ [Wigner 1932; Kirkwood 1933]. For a system of $N$ atoms we have

$$Q_{NVT} = \frac{1}{\Lambda^{3N} N!} \int \mathrm{d}\mathbf{r} \left(1 - \frac{\beta\hbar^2}{24m} \sum_{i=1}^{N} (\nabla_{\mathbf{r}_i} \beta\mathscr{V}(\mathbf{r}))^2 \right) \exp(-\beta\mathscr{V}(\mathbf{r})) \tag{2.139}$$

where $\Lambda$ is defined in eqn (2.24). The expansion accounts for the leading quantum-mechanical diffraction effects; other effects, such as exchange, are small for most cases of interest. Additional details may be found elsewhere [Landau and Lifshitz 1980; McQuarrie 1976]. This leads to the following correction (quantum-classical) to the Helmholtz free energy, $\Delta A = A^{qu} - A^{cl}$

$$\Delta A_{\text{trans}} = \tfrac{1}{24} N \hbar^2 \beta^2 \langle f_i^2 \rangle / m . \tag{2.140}$$

Here, as in Section 2.7, $\langle f_i^2 \rangle$ is the mean square force on one atom in the simulation. Obviously, better statistics are obtained by averaging over all $N$ atoms. An equivalent expression is

$$\begin{aligned} \Delta A_{\text{trans}} &= \frac{N\Lambda^2 \rho}{48\pi} \int \mathrm{d}\mathbf{r}\, g(r)\, \nabla^2 v(r) \\ &= \frac{N\Lambda^2 \rho}{12} \int_0^\infty r^2\, g(r) \left( \frac{\mathrm{d}^2 v(r)}{\mathrm{d}r^2} + \frac{2}{r}\frac{\mathrm{d}v(r)}{\mathrm{d}r} \right) \mathrm{d}r \end{aligned} \tag{2.141}$$

assuming pairwise interactions. Additional corrections of order $\hbar^4$ can be estimated if the three-body distribution function $g_3$ can be calculated in a simulation [Hansen and Weis 1969]. Note that for hard systems, the leading quantum correction is of order $\hbar$: for hard spheres it amounts to replacing the hard sphere diameter $\sigma$ by $\sigma + \Lambda/\sqrt{8}$ [Hemmer 1968; Jancovici 1969]. By differentiating the above equations, quantum corrections to the energy, pressure, etc. can easily be obtained.

Equation (2.140) is also the translational correction for a system of $N$ molecules, where it is understood that $m$ now stands for the molecular mass and $\langle f_i^2 \rangle$ is the mean square force acting on the molecular centre of mass. Additional corrections must be applied for a molecular system, to take account of rotational motion [St Pierre and Steele 1969; Powles and Rickayzen 1979]. For linear molecules, with moment of inertia $I$, the additional term is [Gray and Gubbins 1984]

$$\Delta A_{\text{rot}} = \tfrac{1}{24} N \hbar^2 \beta^2 \langle \tau_i^2 \rangle / I - N\hbar^2/6I \tag{2.142}$$

where $\langle \tau_i^2 \rangle$ is the mean square torque acting on a molecule. The correction for the general asymmetric top, with three different moments of inertia $I_{xx}$, $I_{yy}$, and $I_{zz}$, is rather more complicated:

$$\begin{aligned} \Delta A_{\text{rot}} &= \frac{1}{24} N \hbar^2 \beta^2 \left( \frac{\langle \tau_{ix}^2 \rangle}{I_{xx}} + \frac{\langle \tau_{iy}^2 \rangle}{I_{yy}} + \frac{\langle \tau_{iz}^2 \rangle}{I_{zz}} \right) \\ &\quad - \left[ \frac{N\hbar^2}{24} \sum_{\text{cyclic}} \frac{2}{I_{xx}} - \frac{I_{xx}}{I_{yy} I_{zz}} \right] \end{aligned} \tag{2.143}$$

where the sum is over the three cyclic permutations of $x$, $y$, and $z$. These results are independent of ensemble, and from them the quantum corrections to any other thermodynamic quantity can easily be calculated. Moreover, it is easy to compute the mean square force and mean square torque in a simulation.

Recently, another, possibly more accurate, way of estimating quantum corrections has been proposed [Berens, Mackay, White, and Wilson 1983]. In this approach, the velocity autocorrelation function is calculated and is Fourier transformed, to obtain a spectrum, or density of states,

$$\hat{c}_{vv}(\omega) = \int_{-\infty}^{\infty} \mathrm{d}t \exp(i\omega t) \langle \mathbf{v}_i(t)\cdot\mathbf{v}_i(0)\rangle / \langle v_i^2 \rangle$$
$$= \frac{m}{3k_\mathrm{B}T}\int_{-\infty}^{\infty} \mathrm{d}t \exp(i\omega t) \langle \mathbf{v}_i(t)\cdot\mathbf{v}_i(0)\rangle . \qquad (2.144)$$

Then, quantum corrections are applied to any thermodynamic quantities of interest, using the approximation that the system behaves as a set of harmonic oscillators, whose frequency distribution is dictated by the measured velocity spectrum. For each thermodynamic function a correction function which would apply to a harmonic oscillator of frequency $\omega$, may be defined. The total correction is then obtained by integrating over all frequencies. For the Helmholtz free energy, the correction is given by

$$\Delta A = 3Nk_\mathrm{B}T \int_{-\infty}^{\infty} \frac{\mathrm{d}\omega}{2\pi}\, \hat{c}_{vv}(\omega) \ln\left(\frac{\exp(\frac{1}{2}\hbar\omega/k_\mathrm{B}T) - \exp(-\frac{1}{2}\hbar\omega/k_\mathrm{B}T)}{\hbar\omega/k_\mathrm{B}T}\right) \qquad (2.145)$$

which agrees with eqn (2.140) to $\mathcal{O}(\hbar^2)$. The rationale here is that the harmonic approximation is most accurate for the high-frequency motions that contribute the largest quantum corrections, whereas the anharmonic motions are mainly of low frequency, and thus their quantum corrections are less important. It is not clear whether this is the case for most liquids.

Quantum corrections may also be applied to structural quantities such as $g(r)$. The formulae are rather complex, and will not be given here, but they are based on the same formula eqn (2.139) for the partition function [Gibson 1974]. Again, the result is different for hard systems [Gibson 1975a, b].

When it comes to time-dependent properties, there is one quantum correction which is essential to bring the results of classical simulations into line with experiment. Quantum-mechanical autocorrelation functions obey the detailed balance condition

$$\hat{C}_{\mathscr{A}\mathscr{A}}(\omega) = \exp(\beta\hbar\omega)\hat{C}_{\mathscr{A}\mathscr{A}}(-\omega) \qquad (2.146)$$

whereas of course classical autocorrelation functions are even in frequency [Berne and Harp 1970]. The effects of detailed balance are clearly visible in experimental spectra, for example in inelastic neutron scattering, which probes $S(k,\omega)$; in fact experimental results are often converted to the symmetrized form $\exp(\frac{1}{2}\hbar\beta\omega)\, S(k,\omega)$ for comparison with classical theories. Simple empirical measures have been advocated to convert classical time correlation functions into approximate quantum-mechanical ones. Both the 'complex-time' substitutions

$$C_{\mathscr{A}\mathscr{A}}(t) \to C_{\mathscr{A}\mathscr{A}}(t - \tfrac{1}{2}i\hbar\beta) \tag{2.147}$$

[Schofield 1960] and

$$C_{\mathscr{A}\mathscr{A}}(t) \to C_{\mathscr{A}\mathscr{A}}((t^2 - i\hbar\beta t)^{1/2}) \tag{2.148}$$

[Egelstaff 1961] result in functions which satisfy detailed balance. The former is somewhat easier to apply, since it equates the symmetrized experimental spectrum with the classical simulated one, while the latter satisfies some additional frequency integral relations.

## 2.10 Constraints

In modelling large molecules such as proteins it may be necessary to include constraints in the potential model, as discussed in Section 1.3.3. This introduces some subtleties into the statistical mechanical description. The system of constrained molecules moves on a well-defined hypersurface in phase space. The generalized coordinates corresponding to the constraints and their conjugate momenta are removed from the Hamiltonian. The system is not equivalent to a fluid where the constrained degrees of freedom are replaced by harmonic springs, even in the limit of infinitely strong force constants [Fixman 1974; Pear and Weiner 1979; Chandler and Berne 1979].

To explore this difference more formally we consider a set of $N$ atoms grouped into molecules in some arbitrary way by harmonic springs. The Cartesian coordinates of the atoms are the $3N$ values ($\mathbf{r} = r_{i\alpha}$, $i = 1, 2, \ldots N$, $\alpha = x, y, z$). The system can be described by $3N$ generalized coordinates $\mathbf{q}$ (i.e. the positions of the centre of mass of each molecule, their orientations, and vibrational coordinates). The potential energy of the system can be separated into a part, $\mathscr{V}_{\mathrm{s}}$, associated with the 'soft' coordinates (the translations, rotations and internal conversions) and a part $\mathscr{V}_{\mathrm{h}}$ associated with the 'hard' coordinates (bond stretching and possibly bond angle vibrations)

$$\mathscr{V}(\mathbf{q}) = \mathscr{V}_{\mathrm{s}}(\mathbf{q}^{\mathrm{s}}) + \mathscr{V}_{\mathrm{h}}(\mathbf{q}^{\mathrm{h}}). \tag{2.149}$$

If the force constants of the hard modes are independent of $\mathbf{q}^{\mathrm{s}}$, then the canonical ensemble average of some configurational property $\mathscr{A}(\mathbf{q}^{\mathrm{s}})$ is [Berendsen and van Gunsteren 1984]

$$\langle \mathscr{A} \rangle_{NVT} = \frac{\int \mathscr{A}(\mathbf{q}^{\mathrm{s}})\, |\mathbf{G}|^{1/2} \exp(-\beta \mathscr{V}_{\mathrm{s}}(\mathbf{q}^{\mathrm{s}}))\, \mathrm{d}\mathbf{q}^{\mathrm{s}}}{\int |\mathbf{G}|^{1/2} \exp(-\beta \mathscr{V}_{\mathrm{s}}(\mathbf{q}^{\mathrm{s}}))\, \mathrm{d}\mathbf{q}^{\mathrm{s}}} \tag{2.150}$$

where $|\mathbf{G}|$ is the determinant of the mass-weighted metric tensor $\mathbf{G}$, which is associated with the transformation from Cartesian to generalized coordinates.

$$G_{kl} = \sum_{i=1}^{N} \sum_{\alpha} m_i \frac{\partial r_{i\alpha}}{\partial q_k} \frac{\partial r_{i\alpha}}{\partial q_l}. \tag{2.151}$$

$\mathbf{G}$ involves all the generalized coordinates and is a $3N \times 3N$ matrix. If the hard variables are actually constrained they are removed from the matrix $\mathbf{G}$.

$$\langle \mathscr{A} \rangle^{\mathrm{s}}_{NVT} = \frac{\int \mathscr{A}(\mathbf{q}^{\mathrm{s}})\,|\mathbf{G}^{\mathrm{s}}|^{1/2} \exp(-\beta \mathscr{V}_{\mathrm{s}}(\mathbf{q}^{\mathrm{s}}))\,\mathrm{d}\mathbf{q}^{\mathrm{s}}}{\int |\mathbf{G}^{\mathrm{s}}|^{1/2} \exp(-\beta \mathscr{V}_{\mathrm{s}}(\mathbf{q}^{\mathrm{s}}))\,\mathrm{d}\mathbf{q}^{\mathrm{s}}} \tag{2.152}$$

where

$$G^{\mathrm{s}}_{kl} = \sum_{i=1}^{N} \sum_{\alpha} m_i \frac{\partial r_{i\alpha}}{\partial q^{\mathrm{s}}_k} \frac{\partial r_{i\alpha}}{\partial q^{\mathrm{s}}_l}. \tag{2.153}$$

$\mathbf{G}^{\mathrm{s}}$ is a sub-matrix of $\mathbf{G}$ and has the dimension of the number of soft degrees of freedom. The simulation of a constrained system does not yield the same average as the simulation of the unconstrained system unless $|\mathbf{G}|/|\mathbf{G}^{\mathrm{s}}|$ is independent of the soft modes. In the simulation of large flexible molecules, it may be necessary to constrain some of the internal degrees of freedom, and in this case we would probably require an estimate of $\langle \mathscr{A} \rangle_{NVT}$ rather than $\langle \mathscr{A} \rangle^{\mathrm{s}}_{NVT}$ Fixman [1974] has suggested a solution to the problem of obtaining $\langle \mathscr{A} \rangle_{NVT}$ in a simulation employing constrained variables. A term,

$$\mathscr{V}_{\mathrm{c}} = \tfrac{1}{2} k_{\mathrm{B}} T \ln |\mathbf{H}| \tag{2.154}$$

is added to the potential $\mathscr{V}_{\mathrm{s}}$. $|\mathbf{H}|$ is given by

$$|\mathbf{H}| = |\mathbf{G}|/|\mathbf{G}^{\mathrm{s}}|. \tag{2.155}$$

Substituting $\mathscr{V}_{\mathrm{s}} + \mathscr{V}_{\mathrm{c}}$ as the potential in eqn (2.152) we recover the unconstrained average of eqn (2.150). The separate calculation of $\mathbf{G}$ and $\mathbf{G}^{\mathrm{s}}$ to obtain their determinants is difficult. However, $|\mathbf{H}|$, is the determinant of a simpler matrix

$$H_{kl} = \sum_{i=1}^{N} \sum_{\alpha} m_i \frac{\partial q^{\mathrm{h}}_k}{\partial r_{i\alpha}} \frac{\partial q^{\mathrm{h}}_l}{\partial r_{i\alpha}} \tag{2.156}$$

which has the dimensions of the number of constrained (hard) degrees of freedom.

As a simple example of the use of eqn (2.156) consider the case of a butane molecule (see Fig. 1.8). In our simplified butane, the four united atoms have the same mass $m$, the bond angles and torsional angle are free to change but the three bond lengths, $C_1$–$C_2$, $C_2$–$C_3$, and $C_3$–$C_4$ are fixed. The $3 \times 3$ matrix $\mathbf{H}$ is

$$\begin{pmatrix} 2m & -m\cos\theta & 0 \\ -m\cos\theta & 2m & -m\cos\theta' \\ 0 & -m\cos\theta' & 2m \end{pmatrix}$$

and

$$|\mathbf{H}| \propto (2 + \sin^2\theta + \sin^2\theta'). \tag{2.157}$$

Since $\theta$ and $\theta'$ can change, $\mathbf{H}$ should be included through eqn (2.154). However, it is possible to use a harmonic bond-angle potential, which keeps the bond-angles very close to their equilibrium values. In this case **H** is approximately constant and might be neglected without seriously affecting $\langle \mathscr{A} \rangle_{NVT}$. If we had also constrained the bond angles in our model of butane, then $|\mathbf{H}|$ would have been a function of the torsional angle $\phi$ as well as the $\theta$ angles. Thus **H** can change significantly when the molecules convert from the *trans* to the *gauche* state and $\mathscr{V}_{\mathrm{c}}$ must be included in the potential [van Gunsteren 1980]. In the case of a completely rigid molecule, $|\mathbf{H}|$ is a constant and need not be included. We shall discuss the consequences of constraining degrees of freedom at the appropriate points in Chapters 3 and 4.

# 3
# MOLECULAR DYNAMICS

## 3.1 Equations of motion for atomic systems

In this chapter, we deal with the techniques used to solve the classical equations of motion for a system of $N$ molecules interacting via a potential $\mathscr{V}$ as in eqn (1.4). These equations may be written down in various ways [Goldstein 1980]. Perhaps the most fundamental form is the Lagrangian equation of motion

$$\frac{\mathrm{d}}{\mathrm{d}t}(\partial \mathscr{L}/\partial \dot{q}_k) - (\partial \mathscr{L}/\partial q_k) = 0 \tag{3.1}$$

where the Lagrangian function $\mathscr{L}(\mathbf{q}, \dot{\mathbf{q}})$ is defined in terms of kinetic and potential energies

$$\mathscr{L} = \mathscr{K} - \mathscr{V} \tag{3.2}$$

and is considered to be a function of the generalized coordinates $q_k$ and their time derivatives $\dot{q}_k$. If we consider a system of atoms, with Cartesian coordinates $\mathbf{r}_i$ and the usual definitions of $\mathscr{K}$ and $\mathscr{V}$ (eqns (1.3) and (1.4)) then eqn (3.1) becomes

$$m_i \ddot{\mathbf{r}}_i = \mathbf{f}_i \tag{3.3}$$

where $m_i$ is the mass of atom $i$ and

$$\mathbf{f}_i = \nabla_{\mathbf{r}_i} \mathscr{L} = -\nabla_{\mathbf{r}_i} \mathscr{V} \tag{3.4}$$

is the force on that atom. These equations also apply to the centre of mass motion of a molecule, with $\mathbf{f}_i$ representing the total force on molecule $i$; the equations for rotational motion may also be expressed in the form of eqn (3.1), and will be dealt with in Sections 3.3 and 3.4.

The generalized momentum $p_k$ conjugate to $q_k$ is defined as

$$p_k = \partial \mathscr{L}/\partial \dot{q}_k . \tag{3.5}$$

The momenta feature in the Hamiltonian form of the equations of motion

$$\dot{q}_k = \partial \mathscr{H}/\partial p_k \tag{3.6a}$$

$$\dot{p}_k = -\partial \mathscr{H}/\partial q_k . \tag{3.6b}$$

The Hamiltonian is strictly defined by the equation

$$\mathscr{H}(\mathbf{p}, \mathbf{q}) = \sum_k \dot{q}_k p_k - \mathscr{L}(\mathbf{q}, \dot{\mathbf{q}}) \tag{3.7}$$

where it is assumed that we can write $\dot{q}_k$ on the right as some function of the momenta $\mathbf{p}$. For our immediate purposes (involving a potential $\mathscr{V}$ which is independent of velocities and time) this reduces to eqn (1.2), and $\mathscr{H}$ is automatically equal to the energy [Goldstein 1980, chapter 8]. For Cartesian coordinates, Hamilton's equations become

$$\dot{\mathbf{r}}_i = \mathbf{p}_i/m_i \tag{3.8a}$$

$$\dot{\mathbf{p}}_i = -\nabla_{\mathbf{r}_i}\mathscr{V} = \mathbf{f}_i\,. \tag{3.8b}$$

Computing centre of mass trajectories, then, involves solving either a system of $3N$ second-order differential equations, eqn (3.3), or an equivalent set of $6N$ first-order differential equations, eqns (3.8a), (3.8b). Before considering how to do this, we can make some very general remarks regarding the equations themselves.

A consequence of eqn (3.6b), or equivalently eqns (3.5) and (3.1), is that in certain circumstances a particular generalized momentum $p_k$ may be conserved, i.e. $\dot{p}_k = 0$. The requirement is that $\mathscr{L}$, and hence $\mathscr{H}$ in this case, shall be independent of the corresponding generalized coordinate $q_k$. For any set of particles, it is possible to choose six generalized coordinates, changes in which correspond to translations of the centre of mass, and rotations about the centre of mass, for the system as a whole (changes in the remaining $3N-6$ coordinates involving motion of the particles relative to one another). If the potential function $\mathscr{V}$ depends only on the magnitudes of particle separations (as is usual) and there is no external field applied (i.e. the term $v_1$ in eqn (1.4) is absent) then $\mathscr{V}$, $\mathscr{H}$, and $\mathscr{L}$ are manifestly independent of these six generalized coordinates. The corresponding conjugate momenta, in Cartesian coordinates, are the total linear momentum

$$\mathbf{P} = \sum_i \mathbf{p}_i \tag{3.9}$$

and the total angular momentum

$$\mathbf{L} = \sum_i \mathbf{r}_i \times \mathbf{p}_i = \sum_i m_i \mathbf{r}_i \times \dot{\mathbf{r}}_i \tag{3.10}$$

where we take the origin at the centre of mass of the system. Thus, these are conserved quantities for a completely isolated set of interacting molecules. In practice, we rarely consider completely isolated systems. A more general criterion for the existence of these conservation laws is provided by symmetry considerations [Goldstein 1980, Chapter 8]. If the system (i.e $\mathscr{H}$) is invariant to translation in a particular direction, then the corresponding momentum component is conserved. If the system is invariant to rotation about an axis, then the corresponding angular momentum component is conserved. Thus, we occasionally encounter systems enclosed in a spherical box, and so a spherically symmetrical $v_1$ term appears in eqn (1.4); all three components of total angular momentum about the centre of symmetry will be conserved, but total translational momentum will not be. If the surrounding walls formed a cubical box, none of these quantities would be conserved. In the case of the periodic boundary conditions described in Chapter 1, it is easy to see that translational invariance is preserved, and hence total linear momentum is conserved. Several different box geometries were considered in Chapter 1, but none of them were spherically symmetrical; in fact it is impossible (in Euclidean space) to construct a spherically symmetric periodic system. Hence, despite the fact that there may be no $v_1$ term in eqn (1.4), total angular momentum is not conserved in most molecular dynamics simulations. In the case of the spherical boundary conditions discussed in Section 1.5.5, a kind of angular momentum conservation law does apply. When we embed a two-dimensional system in the surface of a sphere the three-dimensional spherical symmetry is preserved. Similarly, for a three-dimensional system, there should be a four-dimensional conserved 'hyper-angular momentum'.

We have left until last the most important conservation law. It is easy to show, assuming $\mathscr{V}$ and $\mathscr{K}$ do not depend explicitly on time (so that $\partial\mathscr{H}/\partial t = 0$), that the form of the equations of motion guarantees that the total derivative $\dot{\mathscr{H}} = \mathrm{d}\mathscr{H}/\mathrm{d}t$ is zero, i.e. the Hamiltonian is a constant of the motion. This energy conservation law applies whether or not an external potential exists: the essential condition is that no explicitly time-dependent (or velocity-dependent) forces shall act on the system.

The second point concerning the equations of motion is that they are reversible in time. By changing the signs of all the velocities or momenta, we will cause the molecules to retrace their trajectories. If the equations of motion are solved correctly, the computer-generated trajectories will also have this property.

Our final observation concerning eqns (3.3), (3.4), and (3.6) is that the spatial derivative of the potential appears. This leads to a qualitative difference in the form of the motion, and the way in which the equations are solved, depending upon whether or not $\mathscr{V}$ is a continuous function of particle positions. To use the finite time-step method of solution to be described in the next section, it is essential that the particle positions vary smoothly with time: a Taylor expansion of $\mathbf{r}(t)$ about time $t$ may be necessary, for example. Whenever the potential varies sharply (as in the hard sphere and square well cases) impulsive 'collisions' between particles occur at which the velocities (typically) change discontinuously. The particle dynamics at the moment of each collision must be treated explicitly, and separately from the smooth inter-collisional motion. The identification of successive collisions is the key feature of a molecular dynamics program for such systems, and we shall discuss this in Section 3.6.

## 3.2 Finite difference methods

A standard method for solution of ordinary differential equations such as eqns (3.3) and (3.8) is the finite difference approach. The general idea is as follows. Given the molecular positions, velocities, and other dynamic information at time $t$, we attempt to obtain the positions, velocities etc. at a later time $t+\delta t$, to a sufficient degree of accuracy. The equations are solved on a step-by-step basis; the choice of the time interval $\delta t$ will depend somewhat on the method of solution, but $\delta t$ will be significantly smaller than the typical time taken for a molecule to travel its own length. Many different algorithms fall into the general finite difference pattern, and several have been reviewed [Gear 1966, 1971; van Gunsteren and Berendsen 1977; Berendsen and van Gunsteren 1986]. To illustrate the principles of the method, we will choose one (a predictor–corrector algorithm) and then proceed to discuss the technical details which affect the choice in practice.

If the classical trajectory is continuous, then an estimate of the positions, velocities etc. at time $t+\delta t$ may be obtained by Taylor expansion about time $t$:

$$\begin{aligned}
\mathbf{r}^{\mathrm{p}}(t+\delta t) &= \mathbf{r}(t)+\delta t\,\mathbf{v}(t)+\tfrac{1}{2}\delta t^2\,\mathbf{a}(t)+\tfrac{1}{6}\delta t^3\,\mathbf{b}(t)+\ \ldots\\
\mathbf{v}^{\mathrm{p}}(t+\delta t) &= \mathbf{v}(t)+\delta t\,\mathbf{a}(t)+\tfrac{1}{2}\delta t^2\,\mathbf{b}(t)+\ \ldots\\
\mathbf{a}^{\mathrm{p}}(t+\delta t) &= \mathbf{a}(t)+\delta t\,\mathbf{b}(t)+\ \ldots\\
\mathbf{b}^{\mathrm{p}}(t+\delta t) &= \mathbf{b}(t)+\ \ldots\,.
\end{aligned} \tag{3.11}$$

The superscript marks these as 'predicted' values: we shall be 'correcting' them shortly. Just as $\mathbf{r}$ and $\mathbf{v}$ stand for the complete set of positions and velocities, so $\mathbf{a}$ is short for all the accelerations, and $\mathbf{b}$ denotes all the third time derivatives of $\mathbf{r}$. If we truncate the expansion, retaining (for example) just the terms given explicitly in eqn (3.11), then we seem to have achieved our aim of (approximately) advancing the values of the stored coordinates and derivatives from one time step to the next. In this example, we would store four 'vectors' $\mathbf{r}$, $\mathbf{v}$, $\mathbf{a}$, and $\mathbf{b}$. Equivalent alternatives would be to base the prediction on $\mathbf{r}$, $\mathbf{v}$ and 'old' values of the velocities $\mathbf{v}(t-\delta t)$, $\mathbf{v}(t-2\delta t)$, or on $\mathbf{r}$, $\mathbf{v}$, $\mathbf{a}$ and 'old' accelerations $\mathbf{a}(t-\delta t)$, using slightly different predictor equations. However, there is a snag. An equation like eqn (3.11) will not generate correct trajectories as time advances, because we have not introduced the equations of motion. These enter through the correction step. We may calculate, from the new positions $\mathbf{r}^{\mathrm{p}}$, the forces at time $t+\delta t$, and hence the correct accelerations $\mathbf{a}^{\mathrm{c}}(t+\delta t)$. These can be compared with the predicted accelerations from eqn (3.11), to estimate the size of the error in the prediction step:

$$\Delta\mathbf{a}(t+\delta t) = \mathbf{a}^{\mathrm{c}}(t+\delta t)-\mathbf{a}^{\mathrm{p}}(t+\delta t). \tag{3.12}$$

This error, and the results of the predictor step, are fed into the corrector step, which reads, typically,

$$\begin{aligned}
\mathbf{r}^{\mathrm{c}}(t+\delta t) &= \mathbf{r}^{\mathrm{p}}(t+\delta t)+c_0\,\Delta\mathbf{a}(t+\delta t)\\
\mathbf{v}^{\mathrm{c}}(t+\delta t) &= \mathbf{v}^{\mathrm{p}}(t+\delta t)+c_1\,\Delta\mathbf{a}(t+\delta t)\\
\mathbf{a}^{\mathrm{c}}(t+\delta t) &= \mathbf{a}^{\mathrm{p}}(t+\delta t)+c_2\,\Delta\mathbf{a}(t+\delta t)\\
\mathbf{b}^{\mathrm{c}}(t+\delta t) &= \mathbf{b}^{\mathrm{p}}(t+\delta t)+c_3\,\Delta\mathbf{a}(t+\delta t).
\end{aligned} \tag{3.13}$$

The idea is that $\mathbf{r}^c(t+\delta t)$ etc. are now better approximations to the true positions, velocities etc. Gear [1966, 1971] has discussed the 'best' choice for the coefficients $c_0, c_1, c_2, c_3 \ldots$ (i.e. the choice leading to optimum stability and accuracy of the trajectories). Different values of the coefficients are required if we include more (or fewer) position derivatives in our scheme, and the coefficients also depend upon the order of the differential equation being solved (here it is second order, since the double time derivative of position is being compared with the accelerations computed from $\mathbf{r}$). Tables of the coefficients proposed by Gear [1966, 1971] are given in Appendix E, and the way in which eqns (3.11)–(3.13) are used in a program is illustrated in F.2.

The corrector step may be iterated: new 'correct' accelerations are calculated from the positions $\mathbf{r}^c$ and compared with the current values of $\mathbf{a}^c$, so as to further refine the positions, velocities etc. through an equation like eqn (3.13). In many applications this iteration is the key to obtaining an accurate solution. The predictor provides an initial guess at the solution, which in principle does not have to be a very good one, since the successive corrector iterations should then converge rapidly onto the correct answer. In molecular dynamics, however, the evaluation of accelerations (i.e. forces) from particle positions is the most time-consuming part of a simulation, and since this is implicit in each corrector step, a large number of corrector iterations would be very expensive. Normally, just one (occasionally two) corrector steps are carried out, and so an accurate predictor stage (such as the Taylor series of eqn (3.11)) is essential. The general scheme of a stepwise MD simulation, based on a predictor–corrector algorithm, may be summarized as follows:

(a) predict the positions, velocities, accelerations etc., at a time $t+\delta t$, using the current values of these quantities;
(b) evaluate the forces, and hence accelerations $\mathbf{a}_i = \mathbf{f}_i/\mathbf{m}_i$, from the new positions;
(c) correct the predicted positions, velocities, accelerations etc., using the new accelerations;
(d) calculate any variables of interest, such as the energy, virial, order parameters, ready for the accumulation of time averages, before returning to (a) for the next step.

The simple predictor–corrector algorithm just described is only one of many possibilities. Which is the best algorithm to use in MD? The choice is not so wide as it may seem at first. Firstly, a large class of algorithms may be interconverted by simple matrix transformation [Gear 1966, 1971; van Gunsteren and Berendsen 1977; Berendsen and van Gunsteren 1986] and hence are essentially equivalent, although there may be small differences due to round-off errors. Secondly, we have a choice in that we may treat the equations of motion as first-order differential equations, or we may integrate them directly in the form $\ddot{\mathbf{r}} = \mathbf{f}/m$. It is usually possible to relate algorithms expressed in these two different ways.

A shortlist of desirable qualities for a successful simulation algorithm might be as follows.

(a) It should be fast, and require little memory.
(b) It should permit the use of a long time step $\delta t$.
(c) It should duplicate the classical trajectory as closely as possible.
(d) It should satisfy the known conservation laws for energy and momentum, and be time-reversible.
(e) It should be simple in form and easy to program.

For molecular dynamics, not all of the above points are very important. Compared with the time-consuming force calculation, which is carried out at every time step, the raw speed of the integration algorithm is not crucial. It is far more important to be able to employ a long time step $\delta t$: in this way, a given period of 'simulation' time can be covered in a modest number of integration steps, i.e. in an acceptable amount of computer time. Clearly, the larger $\delta t$, the less accurately will our solution follow the correct classical trajectory. How important are points (c) and (d) above?

The accuracy and stability of a simulation algorithm are measured by its local and global truncation errors, and algorithms may be tested on a simple model, such as the harmonic oscillator [Gear 1966, 1971; Hockney and Eastwood 1981]. The algorithms to be discussed below have been chosen with these criteria in mind, but it is unreasonable to expect that any approximate method of solution will dutifully follow the exact classical trajectory indefinitely. Any two classical trajectories which are initially very close will eventually diverge from one another exponentially with time. In the same way, any small perturbation, even the tiny error associated with finite precision arithmetic, will tend to cause a computer-generated trajectory to diverge from the true classical trajectory with which it is initially coincident. We illustrate the effect in Fig. 3.1: using one simulation as a reference, we show that a small perturbation applied at time $t = 0$ causes the trajectories in the perturbed simulation to diverge from the reference trajectories, and become statistically uncorrelated, within a few hundred time steps [see also Stoddard and Ford 1973, and Erpenbeck and Wood 1977]. In this example, we show the growing average 'distance in configuration space', defined as $|\Delta\mathbf{r}|$ where $|\Delta\mathbf{r}|^2 = (1/N)\sum|\mathbf{r}_i(t) - \mathbf{r}_i^0(t)|^2$, $\mathbf{r}_i^0(t)$ being the position of molecule $i$ at time $t$ in a reference simulation, and $\mathbf{r}_i(t)$ being the position of the same molecule at the same time in the perturbed simulation. In the three cases illustrated here, all the molecules in the perturbed runs are initially displaced in random directions from their reference positions at $t = 0$, by $10^{-3}\,\sigma$, $10^{-6}\,\sigma$, and $10^{-9}\,\sigma$, respectively, where $\sigma$ is the molecular diameter. In all other respects, the runs are identical; in particular, each corresponds to essentially the same total energy. As the runs proceed, however, other mechanical quantities eventually become statistically uncorrelated. In Fig. 3.1, we show the percentage difference in kinetic energies between perturbed and reference simulations. On the scale of the figure, the

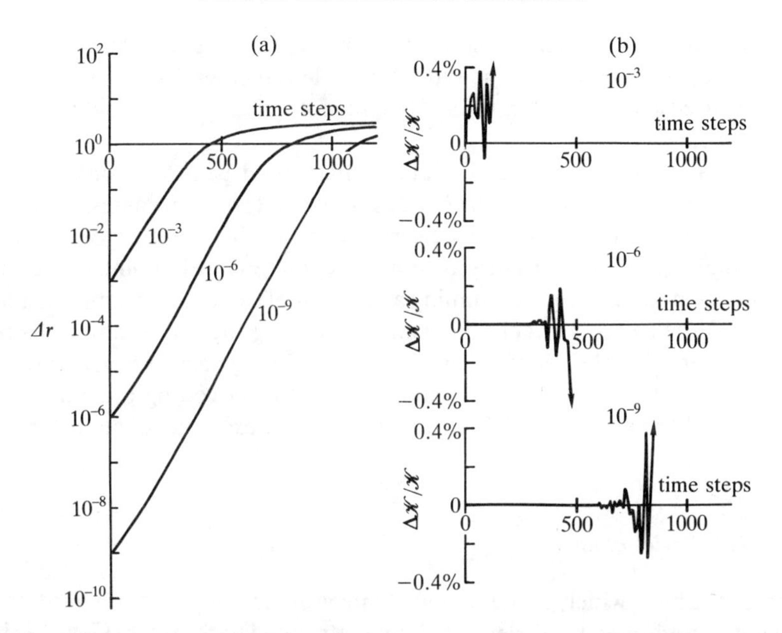


**Fig. 3.1** The divergence of trajectories in molecular dynamics. Atoms interacting through the potential $v^{\mathrm{RLJ}}(r)$, eqn (1.10a), were used, and a dense fluid state was simulated ($\rho^* = 0.6$, $T^* = 1.05$, $\delta t^* = 0.005$). The curves are labelled with the initial displacement in units of $\sigma$. (a) $\Delta r$ is the phase space separation between perturbed and reference trajectories. (b) $\Delta\mathscr{K}/\mathscr{K}$ is the percentage difference in kinetic energies.

kinetic energies remain very close for a period whose length depends on the size of the initial perturbation; after this point the differences become noticeable very rapidly. Presumably, both the reference trajectory and the perturbed trajectory are diverging from the true solution of Newton's equations.

Clearly, no integration algorithm will provide an essentially exact solution for a very long time. Fortunately, we do not need to do this. Remember that molecular dynamics serves two roles. Firstly, we need essentially exact solutions of the equations of motion for times comparable with the correlation times of interest, so that we may accurately calculate time correlation functions. Secondly, we use the method to generate states sampled from the microcanonical ensemble. We do not need exact classical trajectories to do this, but must lay great emphasis on energy conservation as being of primary importance for this reason. Momentum conservation is also important, but this can usually be easily arranged. The point is that the particle trajectories must stay on the appropriate constant-energy hypersurface in phase space, otherwise correct ensemble averages will not be generated. Energy conservation is degraded as the time step is increased, and so all simulations involve a trade-off between economy and accuracy: a good algorithm permits a large time step to be used while preserving acceptable energy conservation. Other factors dictating the energy-conserving properties are the shape of the potential energy curves and the typical particle velocities. Thus, shorter time steps are used at high temperatures, for light molecules, and for rapidly varying potential functions.

The final quality an integration algorithm should possess is simplicity. A simple algorithm will involve the storage of only a few coordinates, velocities etc., and will be easy to program. Bearing in mind that solution of ordinary differential equations is a fairly routine task, there is little point in wasting valuable man-hours on programming a very complicated algorithm, when the time might be better spent checking and optimizing the calculation of forces (see Chapter 5). Little computer time is to be gained by increases in algorithm speed, and the consequences of making a mistake in coding a complicated scheme might be significant. We now turn to specific examples of algorithms in common use.

### *3.2.1 The Verlet algorithm*

Perhaps the most widely used method of integrating the equations of motion is that initially adopted by Verlet [1967] and attributed to Störmer [Gear 1971]. This method is a direct solution of the second-order equations (3.3). The method is based on positions $\mathbf{r}(t)$, accelerations $\mathbf{a}(t)$, and the positions $\mathbf{r}(t-\delta t)$ from the previous step. The equation for advancing the positions reads as follows:

$$\mathbf{r}(t+\delta t) = 2\mathbf{r}(t) - \mathbf{r}(t-\delta t) + \delta t^2 \mathbf{a}(t) . \tag{3.14}$$

There are several points to note about eqn (3.14). It will be seen that the velocities do not appear at all. They have been eliminated by addition of the equations obtained by Taylor expansion about $\mathbf{r}(t)$:

$$\begin{aligned} \mathbf{r}(t+\delta t) &= \mathbf{r}(t) + \delta t\, \mathbf{v}(t) + (1/2)\delta t^2\, \mathbf{a}(t) + \ \ldots \\ \mathbf{r}(t-\delta t) &= \mathbf{r}(t) - \delta t\, \mathbf{v}(t) + (1/2)\delta t^2 \mathbf{a}(t) - \ \ldots . \end{aligned} \tag{3.15}$$

The velocities are not needed to compute the trajectories, but they are useful for estimating the kinetic energy (and hence the total energy). They may be obtained from the formula

$$\mathbf{v}(t) = \frac{\mathbf{r}(t+\delta t) - \mathbf{r}(t-\delta t)}{2\delta t} . \tag{3.16}$$

Whereas eqn (3.14) is correct except for errors of order $\delta t^4$ (the local error) the velocities from eqn (3.16) are subject to errors of order $\delta t^2$. More accurate estimates of $\mathbf{v}(t)$ can be made, if more variables are stored, but this adds to the inconvenience already implicit in eqn (3.16), namely that $\mathbf{v}(t)$ can only be computed once $\mathbf{r}(t+\delta t)$ is known. A second observation regarding the Verlet algorithm is that it is properly centred (i.e. $\mathbf{r}(t-\delta t)$ and $\mathbf{r}(t+\delta t)$ play symmetrical roles in eqn (3.14)), making it time-reversible. Thirdly, the advancement of positions takes place all in one go, rather than in two stages as in the predictor–corrector example. This means that the algorithm is coded differently from the standard predictor–corrector (given on microfiche F.2). Assume that we have available the current and old positions. The current accelerations are evaluated in the force loop as usual. Then, the coordinates are advanced in the following way.

```
        SUMVSQ = 0.0
        SUMVX  = 0.0
        SUMVY  = 0.0
        SUMVZ  = 0.0

        DO 100 I = 1, N

           RXNEWI    = 2.0 * RX(I) - RXOLD(I) + DTSQ * AX(I)
           RYNEWI    = 2.0 * RY(I) - RYOLD(I) + DTSQ * AY(I)
           RZNEWI    = 2.0 * RZ(I) - RZOLD(I) + DTSQ * AZ(I)
           VXI       = ( RXNEWI - RXOLD(I) ) / DT2
           VYI       = ( RYNEWI - RYOLD(I) ) / DT2
           VZI       = ( RZNEWI - RZOLD(I) ) / DT2
           SUMVSQ    = SUMVSQ + VXI ** 2 + VYI ** 2 + VZI ** 2
           SUMVX     = SUMVX + VXI
           SUMVY     = SUMVY + VYI
           SUMVZ     = SUMVZ + VZI
           RXOLD(I) = RX(I)
           RYOLD(I) = RY(I)
           RZOLD(I) = RZ(I)
           RX(I)     = RXNEWI
           RY(I)     = RYNEWI
           RZ(I)     = RZNEWI

100     CONTINUE
```

The variables DTSQ and DT2 store, respectively, $\delta t^2$ and $2\delta t$. Note the use of temporary variables RXNEWI, RYNEWI and RZNEWI to store the new positions within the loop. This is necessary because the current values must be transferred over to the 'old' position variables before being overwritten with the new values. This shuffling operation takes place in the last six statements within the loop. Note also that the calculation of kinetic energy (from SUMVSQ) and total linear momentum (from SUMVX, SUMVY and SUMVZ), is included in the loop, since this is the only moment at which both $\mathbf{r}(t+\delta t)$ and $\mathbf{r}(t-\delta t)$ are available to compute velocities. Following the particle move, we are ready to evaluate the forces for the next step. The overall scheme is illustrated in Fig. 3.2.

As we can see, the Verlet method requires essentially $9N$ words of storage, making it very compact, and it is simple to program. The algorithm is exactly reversible in time and, given conservative forces, is guaranteed to conserve linear momentum. The method has been shown to have excellent energy-conserving properties even with long time steps. As an example, for simulations of liquid argon near the triple point, RMS energy fluctuations of the order 0.01 per cent of the potential well depth are observed using

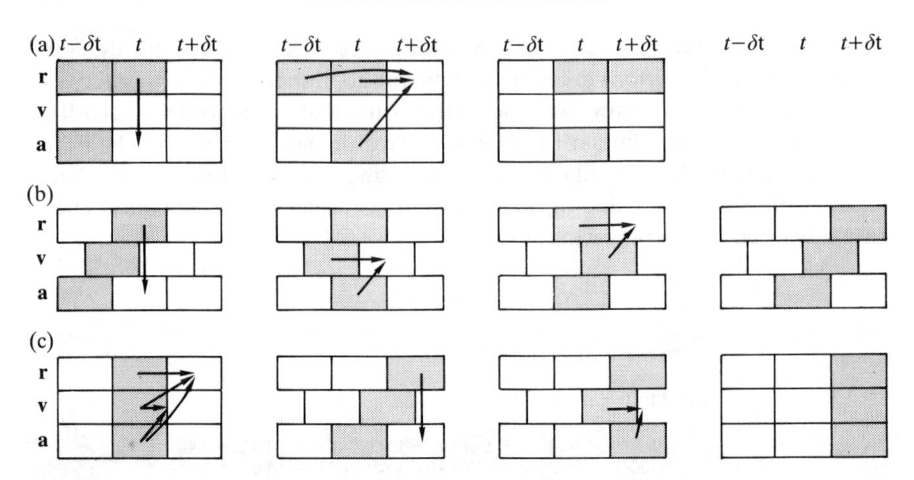


**Fig. 3.2** Various forms of the Verlet algorithm. (a) Verlet's original method. (b) The leap-frog form. (c) The velocity form. We show successive steps in the implementation of each algorithm. In each case, the stored variables are in grey boxes.

$\delta t \approx 10^{-14}$ s, and these increase to 0.2 per cent for $\delta t \approx 4 \times 10^{-14}$ s [Verlet 1967; Fincham and Heyes 1982; Heyes and Singer 1982]. Against the Verlet algorithm, we must say that the handling of velocities is rather awkward, and that the form of the algorithm may needlessly introduce some numerical imprecision [Dahlquist and Björk 1974]. This arises because, in eqn (3.14), a small term ($\mathcal{O}(\delta t^2)$) is added to a difference of large terms ($\mathcal{O}(\delta t^0)$), in order to generate the trajectory.

Modifications to the basic Verlet scheme have been proposed to tackle these deficiencies. One of these is a so-called half-step 'leap-frog' scheme [Hockney 1970; Potter 1972, Chapter 5]. The origin of the name becomes apparent when we write the algorithm down:

$$\mathbf{r}(t+\delta t) = \mathbf{r}(t) + \delta t\, \mathbf{v}(t+\tfrac{1}{2}\delta t) \tag{3.17a}$$

$$\mathbf{v}(t+\tfrac{1}{2}\delta t) = \mathbf{v}(t-\tfrac{1}{2}\delta t) + \delta t\, \mathbf{a}(t). \tag{3.17b}$$

The stored quantities are the current positions $\mathbf{r}(t)$ and accelerations $\mathbf{a}(t)$ together with the mid-step velocities $\mathbf{v}(t-1/2\delta t)$. The velocity equation (3.17b) is implemented first, and the velocities leap over the coordinates to give the next mid-step values $\mathbf{v}(t+1/2\delta t)$. During this step, the current velocities may be calculated

$$\mathbf{v}(t) = \tfrac{1}{2}(\mathbf{v}(t+\tfrac{1}{2}\delta t) + \mathbf{v}(t-\tfrac{1}{2}\delta t)). \tag{3.18}$$

This is necessary so that the energy ($\mathcal{H} = \mathcal{K} + \mathcal{V}$) at time $t$ can be calculated, as well as any other quantities that require positions and velocities at the same instant. Following this, eqn (3.17a) is used to propel the positions once more ahead of the velocities. After this, the new accelerations may be evaluated ready for the next step. This is illustrated in Fig. 3.2. Elimination of the velocities from these equations shows that the method is algebraically equivalent to Verlet's algorithm. There are some advantages in programming eqns (3.17)–(3.18), however, since the velocities (admittedly not at time $t$) appear explicitly [Fincham and Heyes 1982]; for example, adjusting the simulation energy is usually achieved by appropriately scaling the velocities. Numerical benefits derive from the fact that at no stage do we take the difference of two large quantities to obtain a small one; this minimizes loss of precision on a computer. If there is a desperate need to conserve storage space, the accelerations may be directly accumulated onto the velocities, thus making the overall requirements of order $6N$ words [Fincham and Heyes 1982]. The cost is that eqn (3.18) may no longer be used, and it becomes necessary to estimate the kinetic energy at time $t$ from the known mid-step values. An example of the leap-frog technique in use in a low-storage program coded in FORTRAN and in BASIC (for a microcomputer) is given in F.3. Finally, we note that the leap-frog approach may be applied to other algorithms as well as Verlet's [Fincham and Heyes 1982].

As eqn (3.18) shows, leap-frog methods still do not handle the velocities in a completely satisfactory manner. A Verlet-equivalent algorithm which does store positions, velocities, and accelerations all at the same time $t$, and which minimizes round-off error, has recently been proposed [Swope, Andersen, Berens, and Wilson 1982]. This 'velocity Verlet' algorithm takes the form

$$\mathbf{r}(t+\delta t) = \mathbf{r}(t) + \delta t\,\mathbf{v}(t) + \tfrac{1}{2}\delta t^2\,\mathbf{a}(t) \tag{3.19a}$$

$$\mathbf{v}(t+\delta t) = \mathbf{v}(t) + \tfrac{1}{2}\delta t\,[\mathbf{a}(t) + \mathbf{a}(t+\delta t)]\,. \tag{3.19b}$$

Again, the Verlet algorithm may be recovered by eliminating the velocities. In this form, the method resembles a three-value predictor–corrector algorithm (see Appendix E), where the position corrector coefficient is zero [van Gunsteren and Berendsen 1977]. The algorithm only requires storage of $\mathbf{r}$, $\mathbf{v}$, and $\mathbf{a}$. Although it is not implemented in exactly the form of a Gear predictor–corrector, it does involve two stages, with a force evaluation in between. Firstly, the new positions at time $t+\delta t$ are calculated using eqn (3.19a), and the velocities at mid-step are computed using

$$\mathbf{v}(t+\tfrac{1}{2}\delta t) = \mathbf{v}(t) + \tfrac{1}{2}\delta t\,\mathbf{a}(t)\,. \tag{3.20}$$

The forces and accelerations at time $t+\delta t$ are then computed, and the velocity move completed

$$\mathbf{v}(t+\delta t) = \mathbf{v}(t+\tfrac{1}{2}\delta t) + \tfrac{1}{2}\,\delta t\,\mathbf{a}(t+\delta t). \tag{3.21}$$

At this point, the kinetic energy at time $t+\delta t$ is available. The potential energy at this time will have been evaluated in the force loop. The whole process is shown in Fig. 3.2. The method once more uses $9N$ words of storage, and its numerical stability, convenience, and simplicity make it perhaps the most attractive proposed to date. The code for the velocity version of Verlet's method is a straightforward transcription of eqns (3.19)–(3.21) (see program F.4).

Before we leave Verlet, we should mention the investigation by Beeman [1976] of several algorithms, one of which reduces to eqn (3.14) when the velocities are eliminated [Sangster and Dixon 1976; Hockney and Eastwood 1981]. The algorithm is

$$\mathbf{r}(t+\delta t) = \mathbf{r}(t) + \delta t\,\mathbf{v}(t) + \tfrac{2}{3}\delta t^2\,\mathbf{a}(t) - \tfrac{1}{6}\delta t^2\,\mathbf{a}(t-\delta t) \tag{3.22a}$$

$$\mathbf{v}(t+\delta t) = \mathbf{v}(t) + \tfrac{1}{3}\delta t\,\mathbf{a}(t+\delta t) + \tfrac{5}{6}\delta t\,\mathbf{a}(t) - \tfrac{1}{6}\delta t\,\mathbf{a}(t-\delta t)\,. \tag{3.22b}$$

The method stores $\mathbf{r}(t)$, $\mathbf{v}(t)$, $\mathbf{a}(t)$, and $\mathbf{a}(t-\delta t)$. Offsetting the complexity of these formulae, and the need to store the 'old' accelerations, is a more accurate equation for the velocities than eqn (3.16), and consequently an apparent improvement in energy conservation. However, once again, all the methods described in this section are essentially equivalent in that they have identical global errors and in fact generate identical position trajectories.

### *3.2.2 The Gear predictor–corrector*

We have already discussed the basic predictor–corrector algorithm with Gear's set of corrector coefficients. How does it compare with the various forms of the Verlet algorithm? In the form described here, a four-value Gear algorithm (as described earlier) requires $15N$ words of storage ($\mathbf{r}$, $\mathbf{v}$, $\mathbf{a}$, $\mathbf{b}$, and the new accelerations/forces); a more accurate five-value method would need $18N$ words. This is a large requirement compared with Verlet, and this may be an important factor. Unfortunately, increasing the order of a Gear method does not result in a great improvement in accuracy for molecular dynamics. It is instructive to consider the reasons for this. In a liquid, the forces on a molecule, and hence its motion during a time step, are dictated by the motion of its neighbours, particularly the close neighbours, which move into and out of a small region of strong interaction with the molecule. This makes any Taylor series predictor, which takes no account of the motion of the neighbours, unreliable, and a high-order predictor is no significant improvement over a low-order one. Since the success of the predictor–corrector method relies on the accuracy of the predictor, especially if we perform only one corrector iteration, the higher-order Gear algorithms have little to offer over the simpler low-order Verlet type methods. If we measure performance by calculating the root-mean-square energy fluctuations $\langle \delta\mathcal{H}^2 \rangle^{1/2}$ for a short run using a particular time step $\delta t$, then the results typically take the form of Fig. 3.3 [see also van Gunsteren and Berendsen 1977; Fincham and Heyes 1982; Berendsen and van Gunsteren 1986]. At short $\delta t$, the higher-order Gear methods are

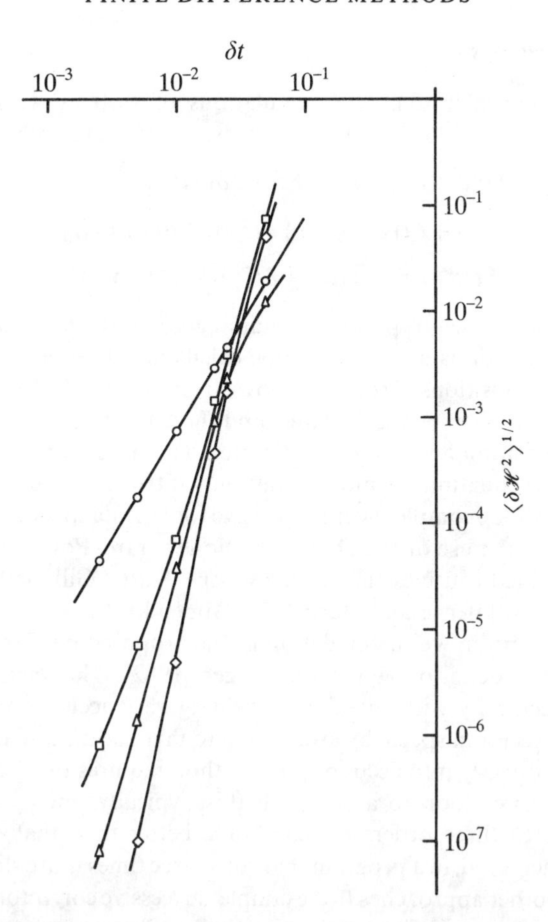


**Fig. 3.3** Energy conservation of various algorithms. The system studied is as for Fig. 3.1. We calculate RMS energy fluctuations $\langle \delta \mathscr{H}^2 \rangle^{1/2}$ for various runs starting from the same initial conditions, and proceeding for the same total simulation time $t_{\text{run}}$, but using different time steps $\delta t$ and corresponding numbers of steps $\tau_{\text{run}} = t_{\text{run}}/\delta t$. The plot uses log-log scales. The curves correspond to velocity Verlet (circles), Gear fourth-order (squares), Gear fifth-order (triangles), and Gear sixth-order (diamonds) algorithms.

more accurate, but in most simulations we are interested in making $\delta t$ as high as possible. With a longer time step, the Verlet algorithm is more attractive. In fact, $\langle \delta \mathscr{H}^2 \rangle^{1/2}$ is closely proportional to $\delta t^2$ for Verlet-equivalent algorithms, while energy conservation for the higher-order methods worsens much more rapidly with increasing $\delta t$. However, in many of the applications to be described in later chapters, it is very convenient to use a standard Gear method, which may be easily adapted to handle modified first- and second-order equations of motion. For an extensive discussion of the merits of various MD algorithms see Berendsen and van Gunsteren [1986].

### *3.2.3 Other methods*

The earliest molecular dynamics simulations of systems with continuous potentials [Rahman 1964] utilized a predictor–corrector algorithm of the form

$$\mathbf{r}^{\mathrm{p}}(t+\delta t) = \mathbf{r}(t-\delta t) + 2\delta t\,\mathbf{v}(t) \tag{3.23a}$$

$$\mathbf{v}(t+\delta t) = \mathbf{v}(t) + \tfrac{1}{2}\delta t\,[\mathbf{a}(t+\delta t) + \mathbf{a}(t)] \tag{3.23b}$$

$$\mathbf{r}(t+\delta t) = \mathbf{r}(t) + \tfrac{1}{2}\delta t[\mathbf{v}(t+\delta t) + \mathbf{v}(t)]\,. \tag{3.23c}$$

Equation (3.23a) is used to provide an initial guess at the new positions, from which the accelerations $\mathbf{a}(t+\delta t)$ may be calculated. The new velocities, and then the new positions proper, follow from eqns (3.23b), (3.23c); the accelerations may then be recalculated, and these last two equations iterated, to refine the position and velocity estimates. The method provides accurate solutions of the equations of motion, but only if two or three passes through eqns (3.23b), (3.23c), complete with expensive force evaluation, are carried out. For this reason, the use of the above equations is rare. For the same reason, standard packaged routines such as the Runge–Kutta–Gill method, are only occasionally used [Berne and Harp 1970, Appendix A].

The final algorithm we shall mention is one proposed by Toxvaerd [1982] and examined in detail by Heyes and Singer [1982]. This method has been developed specifically with the aim of yielding very accurate trajectories at some cost in execution speed, by attempting to take the motion of neighbouring molecules directly into account. The method requires two passes through the expensive force loop to accomplish this; typically, energy conservation (which is two to three orders of magnitude better than that of the Verlet algorithm) is achieved, in a program that runs three times more slowly. It is not clear whether other approaches (for example, successive corrector iterations or simply a reduced time step) would be equally effective in generating more accurate trajectories.

## 3.3 Molecular dynamics of rigid non-spherical bodies

Molecular systems, of course, are not rigid bodies in any sense: they consist of fundamental particles interacting via intra- and intermolecular forces. In principle, we should not distinguish between these forces, but as a practical definition we take the forces acting within molecules to be at least an order of magnitude greater than those acting between molecules. If treated classically, as in the earliest molecular simulations [Harp and Berne 1968, 1970; Berne and Harp 1970] molecular bond vibrations would occur so rapidly that an extremely short time step would be required to solve the equations of motion. In any case, the classical approach is highly questionable for bond vibrations.

A common solution to these problems is to take the intramolecular bonds to be of fixed length. This is not an inconsequential step, but seems reasonable if, as is commonly true at normal temperatures, the amplitude of vibration (classical or quantum) is small compared with molecular dimensions. For polyatomic molecules, we must also consider whether all bond angles should be assumed to be fixed. This is less reasonable in molecules with low-frequency torsional degrees of freedom, or indeed where conformer interconversion is of interest. In this section, we consider the molecular dynamics of molecules in which all bond lengths and internal angles are taken to be constant, i.e. in which the molecule is a single rigid unit. In Section 3.4 we discuss the simulation of flexible polyatomic molecules.

In classical mechanics, it is natural to divide molecular motion into translation of the centre of mass and rotation about the centre of mass [Goldstein 1980]. The former motion is handled by the methods of the previous sections: we simply interpret the force $\mathbf{f}_i$ in the equation $m\ddot{\mathbf{r}}_i = \mathbf{f}_i$ as being the vector sum of all the forces acting on molecule $i$ at the centre of mass $\mathbf{r}_i$. The rotational motion is governed by the torque $\boldsymbol{\tau}_i$ about the centre of mass. When the interactions have the form of forces $\mathbf{f}_{ia}$ acting on sites $\mathbf{r}_{ia}$ in the molecule, the torque is simply defined

$$\boldsymbol{\tau}_i = \sum_a (\mathbf{r}_{ia} - \mathbf{r}_i) \times \mathbf{f}_{ia} = \sum_a \mathbf{d}_{ia} \times \mathbf{f}_{ia} . \tag{3.24}$$

The positions of atoms relative to the molecular centre of mass are written $\mathbf{d}_{ia}$ here. When multipolar terms appear in the potential, the expression for the torque is more complicated [Price *et al.* 1984], but it may still be calculated from the molecular positions and orientations (see Appendix C). The torque enters the rotational equations of motion in the same way that the force enters the translational equations; the nature of orientation space, however, guarantees that the equations of reorientational motion will not be as simple as the translational equations. In this section, we consider the motion of a non-linear molecule under the influence of external forces, taking our origin of coordinates to lie at the centre of mass. We then consider the special case of linear diatomic and polyatomic molecules. To simplify the notation, we drop the suffix $i$ in this section, understanding $\mathbf{d}_a$ and $\mathbf{f}_a$ etc. to refer to the atoms in a single molecule.

### 3.3.1 *Non-linear molecules*

The orientation of a rigid body specifies the relation between an axis system fixed in space and one fixed with respect to the body, usually the 'principal' body-fixed system in which the inertia tensor is diagonal. Any unit vector $\mathbf{e}$ may be expressed in terms of components in the body-fixed or space-fixed frames: we use the notation $\mathbf{e}^{\mathrm{b}}$ and $\mathbf{e}^{\mathrm{s}}$, respectively. These components are related by the rotation matrix $\mathbf{A}$

$$\mathbf{e}^{\mathrm{b}} = \mathbf{A} \cdot \mathbf{e}^{\mathrm{s}} . \tag{3.25}$$

The nine components of the rotation matrix are the direction cosines of the body-fixed axis vectors in the space-fixed frame, and they completely define the molecular orientation. In fact there is substantial redundancy in this formula: only three independent quantities (generalized coordinates) are needed to define **A**. These are generally taken to be the Euler angles $\phi\theta\psi$ in a suitable convention (see Fig. 3.4 and Goldstein [1980]).

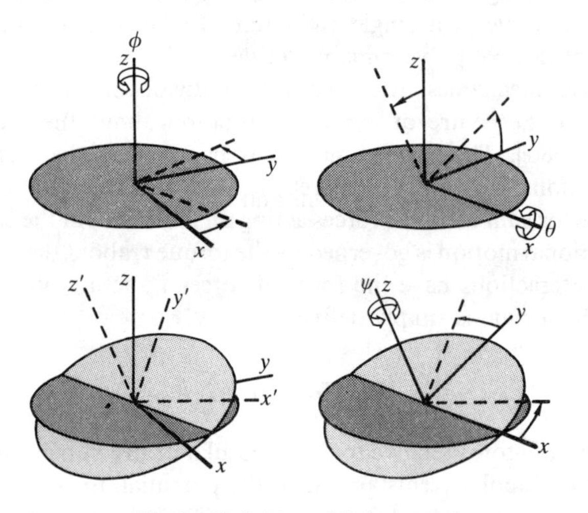


**Fig. 3.4** Definition of Euler angles.

$$\mathbf{A} = \begin{pmatrix} \cos\phi\cos\psi - \sin\phi\cos\theta\sin\psi & \sin\phi\cos\psi + \cos\phi\cos\theta\sin\psi & \sin\theta\sin\psi \\ -\cos\phi\sin\psi - \sin\phi\cos\theta\cos\psi & -\sin\phi\sin\psi + \cos\phi\cos\theta\cos\psi & \sin\theta\cos\psi \\ \sin\phi\sin\theta & -\cos\phi\sin\theta & \cos\theta. \end{pmatrix} \tag{3.26}$$

Clearly, if **e** is a vector fixed in the molecular frame (e.g. a bond vector) then $\mathbf{e}^b$ will not change with time. In space-fixed coordinates, though, the components of $\mathbf{e}^s$ will vary. This is a specific case of the general equation linking time derivatives in the two systems

$$\dot{\mathbf{e}}^s = \dot{\mathbf{e}}^b + \boldsymbol{\omega}^s \times \mathbf{e}^s = \boldsymbol{\omega}^s \times \mathbf{e}^s . \tag{3.27}$$

The time derivative of the angular velocity vector $\boldsymbol{\omega}$ is dictated by the torque $\boldsymbol{\tau}$ acting on the molecule. Although the torque is most easily evaluated in space-fixed axes, it is most convenient to make the connection with $\boldsymbol{\omega}$, via the angular momentum **l**, in the body-fixed principal axis system where the inertia tensor **I** is diagonal. Thus we may write

$$\dot{\mathbf{l}}^s = \boldsymbol{\tau}^s \tag{3.28a}$$

$$\dot{\mathbf{l}}^b + \boldsymbol{\omega}^b \times \mathbf{l}^b = \boldsymbol{\tau}^b. \tag{3.28b}$$

Note that we have taken the components of $\boldsymbol{\omega} \times \mathbf{l}$ in the body-fixed frame, since $\dot{\mathbf{l}}^{\mathrm{b}}$ and $\boldsymbol{\tau}^{\mathrm{b}}$ are expressed in that frame. The resulting equations for the components of $\boldsymbol{\omega}$ in the body-fixed frame are

$$\dot{\omega}_x^{\mathrm{b}} = \frac{\tau_x^{\mathrm{b}}}{I_{xx}} + \left(\frac{I_{yy} - I_{zz}}{I_{xx}}\right)\omega_y^{\mathrm{b}}\omega_z^{\mathrm{b}} \tag{3.29a}$$

$$\dot{\omega}_y^{\mathrm{b}} = \frac{\tau_y^{\mathrm{b}}}{I_{yy}} + \left(\frac{I_{zz} - I_{xx}}{I_{yy}}\right)\omega_z^{\mathrm{b}}\omega_x^{\mathrm{b}} \tag{3.29b}$$

$$\dot{\omega}_z^{\mathrm{b}} = \frac{\tau_z^{\mathrm{b}}}{I_{zz}} + \left(\frac{I_{xx} - I_{yy}}{I_{zz}}\right)\omega_x^{\mathrm{b}}\omega_y^{\mathrm{b}} \tag{3.29c}$$

where $I_{xx}$, $I_{yy}$, and $I_{zz}$ are the three principal moments of inertia. Conversion from space-fixed to body-fixed systems and back is handled by the analogues of eqn (3.25), i.e.

$$\boldsymbol{\tau}^{\mathrm{b}} = \mathbf{A} \cdot \boldsymbol{\tau}^{\mathrm{s}} \tag{3.30}$$

$$\boldsymbol{\omega}^{\mathrm{s}} = \mathbf{A}^{-1} \cdot \boldsymbol{\omega}^{\mathrm{b}} = \mathbf{A}^{\mathrm{T}} \cdot \boldsymbol{\omega}^{\mathrm{b}} \tag{3.31}$$

since the inverse of $\mathbf{A}$ is its transpose. To complete the picture, we need an equation of motion for the molecular orientation itself, i.e. for $\mathbf{A}$. This may take the form of three equations like eqn (3.27) for the basic vectors of the molecular frame, or we may write down the equations of motion of the Euler angles themselves

$$\begin{aligned}
\dot{\phi} &= -\omega_x^{\mathrm{s}} \frac{\sin\phi\cos\theta}{\sin\theta} + \omega_y^{\mathrm{s}} \frac{\cos\phi\cos\theta}{\sin\theta} + \omega_z^{\mathrm{s}} \\
\dot{\theta} &= \omega_x^{\mathrm{s}}\cos\phi + \omega_y^{\mathrm{s}}\sin\phi \\
\dot{\psi} &= \omega_x^{\mathrm{s}} \frac{\sin\phi}{\sin\theta} - \omega_y^{\mathrm{s}} \frac{\cos\phi}{\sin\theta} .
\end{aligned} \tag{3.32}$$

Once more, we emphasize that these equations would apply to each molecule separately. In principle, eqns (3.26) with (3.29)–(3.32), may be solved in a step-by-step fashion, just as we deal with the translational equations of motion. However, they suffer from a serious drawback. The presence of the $\sin\theta$ terms in eqn (3.32) means that a divergence occurs whenever $\theta$ approaches 0 or $\pi$. The molecular motion is unaffected when this occurs (of course!) but, because of our choice of axes, the angles $\phi$ and $\psi$ become degenerate (see Fig. 3.4). Thus, the equations of motion are unsatisfactory when written in this form. One way to cope with this would be to reduce the time step, so as to deal with the more rapidly varying quantities, whenever any molecule approached the critical values $\theta \approx 0$ or $\pi$. This would be very expensive and awkward, and a slightly more satisfactory solution was employed by Barojas, Levesque, and Quentrec [1973]: two alternative sets of space-fixed axes for each molecule were used, and whenever the angle $\theta$ in one system approached 0 or $\pi$, a switchover to the other set was made.

In recent years, a much more elegant and straightforward solution to the problem of divergence in the orientational equations of motion has been proposed. Recognizing that singularity-free equations could not be obtained in terms of three independent variables, Evans [1977] suggested the use of four quaternion parameters as generalized coordinates. Quaternions fulfil the requirements of having well-behaved equations of motion. The four quaternions are linked by one algebraic equation, so there is just one 'redundant' variable. The basic simulation algorithm has been described by Evans and Murad [1977].

A quaternion **Q** is a set of four scalar quantities

$$\mathbf{Q} = (q_0, q_1, q_2, q_3) \tag{3.33}$$

and it is often useful to think of the last three elements $(q_1, q_2, q_3)$ as constituting a vector. The quaternions of interest here satisfy the constraint

$$q_0^2 + q_1^2 + q_2^2 + q_3^2 = 1 \tag{3.34}$$

and the way in which such a quaternion may represent the orientation of a rigid body is discussed by Goldstein [1980]. In the Euler angle convention of Fig. 3.4 and eqn (3.26), it is most convenient to define

$$\begin{aligned} q_0 &= \cos\tfrac{1}{2}\theta\cos\tfrac{1}{2}(\phi+\psi) \\ q_1 &= \sin\tfrac{1}{2}\theta\cos\tfrac{1}{2}(\phi-\psi) \\ q_2 &= \sin\tfrac{1}{2}\theta\sin\tfrac{1}{2}(\phi-\psi) \\ q_3 &= \cos\tfrac{1}{2}\theta\sin\tfrac{1}{2}(\phi+\psi) \end{aligned} \tag{3.35}$$

when the rotation matrix becomes

$$\mathbf{A} = \begin{pmatrix} q_0^2+q_1^2-q_2^2-q_3^2 & 2(q_1q_2+q_0q_3) & 2(q_1q_3-q_0q_2) \\ 2(q_1q_2-q_0q_3) & q_0^2-q_1^2+q_2^2-q_3^2 & 2(q_2q_3+q_0q_1) \\ 2(q_1q_3+q_0q_2) & 2(q_2q_3-q_0q_1) & q_0^2-q_1^2-q_2^2+q_3^2 \end{pmatrix}. \tag{3.36}$$

The quaternions for each molecule satisfy the equations of motion

$$\begin{pmatrix} \dot{q}_0 \\ \dot{q}_1 \\ \dot{q}_2 \\ \dot{q}_3 \end{pmatrix} = \frac{1}{2}\begin{pmatrix} q_0 & -q_1 & -q_2 & -q_3 \\ q_1 & q_0 & -q_3 & q_2 \\ q_2 & q_3 & q_0 & -q_1 \\ q_3 & -q_2 & q_1 & q_0 \end{pmatrix}\begin{pmatrix} 0 \\ \omega_x^{\mathrm{b}} \\ \omega_y^{\mathrm{b}} \\ \omega_z^{\mathrm{b}} \end{pmatrix}. \tag{3.37}$$

The equations of motion, eqn (3.37) with (3.29), using the matrix of eqn (3.36) to transform between space-fixed and body-fixed coordinates, contain no unpleasant singularities. They are a system of first-order differential equations which may be solved by the Gear predictor–corrector method, with appropriate values for the corrector coefficients as described in Appendix E. The scheme involves storing all four quaternions, together with their first four (say) time derivatives, and also the angular velocities and four time derivatives, for each molecule. The general scheme of Section 3.2 is followed, with the slight complication that prediction and correction are applied to the angular velocities (based on eqn (3.29)) and then to the quaternions (eqn (3.37)). This is illustrated in program F.5.

It is possible to go further, and eliminate the angular velocities from the equations of motion altogether, obtaining second-order differential equations for the quaternions [Powles, Evans, McGrath, Gubbins, and Murad 1979; Allen 1984a]. An advantage of this approach is that, for multipolar potentials, the torques may be expressed very easily as derivatives of the potential with respect to the quaternions themselves, thus simplifying the algebra.

As in the translational case, an alternative to the Gear predictor–corrector algorithm seems preferable. A leap-frog formulation for quaternions has been proposed [Fincham 1981]. The simple leap-frog described in section 3.2.1 cannot be used directly, because the quaternion derivatives $\dot{\mathbf{Q}}$ appearing in eqn (3.37) depend not only upon the angular velocity but also on $\mathbf{Q}$ itself. However, a modified leap-frog [Potter 1972] may be applied in this case. The method is based on stored values of $\mathbf{l}^s(t-\frac{1}{2}\delta t)$, $\mathbf{Q}(t)$ and the torques $\boldsymbol{\tau}^s(t)$ just computed from positions and orientations at time $t$. The first step is to bring all the angular momenta up to date

$$\mathbf{l}^s(t) = \mathbf{l}^s(t-\tfrac{1}{2}\delta t) + \tfrac{1}{2}\delta t\boldsymbol{\tau}^s(t)\,. \tag{3.38}$$

These quantities are used to form the body-fixed angular velocity at time $t$, which in turn gives the time derivative of the quaternions $\dot{\mathbf{Q}}(t)$ through eqn (3.37). Then, a guess at $\mathbf{Q}(t+\frac{1}{2}\delta t)$ is made

$$\mathbf{Q}(t+\tfrac{1}{2}\delta t) = \mathbf{Q}(t) + \tfrac{1}{2}\delta t\dot{\mathbf{Q}}(t)\,. \tag{3.39}$$

None of these values need to be stored away permanently; the sole aim of the auxiliary equations (3.38) and (3.39) is to obtain an estimate of $\mathbf{Q}(t+\frac{1}{2}\delta t)$ so that transformations from space-fixed to body-fixed angular momentum, and the calculation of $\dot{\mathbf{Q}}$, can be implemented at the half-step time. The main algorithm equations are

$$\mathbf{l}^s(t+\tfrac{1}{2}\delta t) = \mathbf{l}^s(t-\tfrac{1}{2}\delta t) + \delta t\boldsymbol{\tau}^s(t) \tag{3.40a}$$

and

$$\mathbf{Q}(t+\delta t) = \mathbf{Q}(t) + \delta t\dot{\mathbf{Q}}(t+\tfrac{1}{2}\delta t) \tag{3.40b}$$

where eqn (3.40b) is implemented after we have converted $\mathbf{l}^s(t+\frac{1}{2}\delta t)$ into $\dot{\mathbf{Q}}(t+\frac{1}{2}\delta t)$ using the results of the auxiliary equations. The torques and forces may now be evaluated (assuming that the centre of mass cordimates have also been advanced to time $t+\delta t$) and the whole process repeated. The algorithm seems to be stable and accurate with a moderately large time step [Fincham 1981] and the associated code appears on microfiche F.6.

Whichever method is used to integrate forward the quaternion equations of motion, in principle the constraint $q_0^2+q_1^2+q_2^2+q_3^2=1$ should be preserved: it is easy to show that, according to eqn (3.37), the time derivative of the sum of squares is zero. In practice, of course, the fact that the equations are only being solved approximately means that small errors may build up over a period of time. To avoid this, it is common practice to 'renormalize' the quaternions, so as to guarantee that the sum of squares for each molecule is unity, at frequent intervals (e.g. every time step).

### 3.3.2 *Linear molecules*

For a linear molecule, one of the principal moments of inertia vanishes, and the other two inertia components become equal. It is possible to treat the motion of linear molecules by adaptations of the algorithms discussed in the previous section. For example, a quaternion parameter algorithm can be used, where one of the body-fixed angular velocity components (corresponding to rotation about the axis) is always kept at a zero value. This approach however is a little clumsy, and contains much redundant information: to specify the orientation of a linear molecule, we only need to know the components of a unit vector pointing along the axis. A simulation algorithm which takes advantage of the special properties of linear molecules, and uses a Gear predictor–corrector technique, was proposed by Cheung and Powles [1975]. The way in which the predictor–corrector method is used is very much like the quaternion parameter approach described in the last section: the equations for the angular velocity (plus derivatives) and for the bond vector (plus derivatives) are treated successively. The approach we shall describe here, however, uses the leap-frog method.

For a linear molecule, the angular velocity and the torque must be perpendicular to the molecular axis at all times. If $\mathbf{e}^s$ is the unit vector along the axis, this means that the torque on a molecule can always be written as

$$\boldsymbol{\tau}^s = \mathbf{e}^s \times \mathbf{g}^s \tag{3.41}$$

where $\mathbf{g}^s$ can be determined from the intermolecular forces. In the particular case of an interaction site model, the position of each site relative to the centre of mass may be written

$$\mathbf{d}_a{}^s = d_a \mathbf{e}^s \tag{3.42}$$

so we may then write

$$\mathbf{g}^s = \sum_a d_a \mathbf{f}_a{}^s \tag{3.43}$$

(compare eqn (3.24)). The vector $\mathbf{g}^s$ can always be replaced by its component perpendicular to the molecular axis without affecting eqn (3.41), i.e. we may write

$$\boldsymbol{\tau}^s = \mathbf{e}^s \times \mathbf{g}^\perp \tag{3.44}$$

with

$$\mathbf{g}^\perp = \mathbf{g}^s - (\mathbf{g}^s \cdot \mathbf{e}^s)\mathbf{e}^s. \tag{3.45}$$

The vector $\mathbf{g}^\perp$ has the useful property that $|\mathbf{g}^\perp|^2 = |\boldsymbol{\tau}^s|^2$, so that it can be used to calculate the mean squared torque on a given molecule.

The equations of rotational motion can now be written as two first-order differential equations [Singer *et al.* 1977]

$$\dot{\mathbf{e}}^s = \mathbf{u}^s \tag{3.46a}$$

$$\dot{\mathbf{u}}^s = \mathbf{g}^\perp / I + \lambda \mathbf{e}^s \tag{3.46b}$$

where $I$ is the moment of inertia. The first equation simply defines $\mathbf{u}^s$ as the time derivative of $\mathbf{e}^s$. The second equation looks unusual, but, on time-differentiating the relation $\mathbf{e}^s \times \mathbf{u}^s = \boldsymbol{\omega}^s$ (see eqn 3.27), eqn (3.46b) reduces to the more familiar eqn (3.28a) with $\dot{\mathbf{l}}^s = I\dot{\boldsymbol{\omega}}^s$. Physically, the two terms in eqn (3.46b) correspond to the force $\mathbf{g}^\perp$ responsible for rotation of the molecule, and the force $\lambda\mathbf{e}^s$ along the bond which constrains the bond length to be a constant of the motion. The quantity $\lambda$ can be thought of as a Lagrange multiplier, and we shall meet further examples of this approach to bond constraints in the next section. Fincham [1984] has proposed a solution to eqn (3.46) using a leap-frog algorithm. An expression for $\lambda$ is obtained by considering the advancement of coordinates over half a time step

$$\mathbf{u}^s(t) = \mathbf{u}^s(t - \tfrac{1}{2}\delta t) + \tfrac{1}{2}\delta t[\mathbf{g}^\perp(t)/I + \lambda(t)\mathbf{e}^s(t)]\,. \tag{3.47}$$

Taking the scalar product of both sides with the vector $\mathbf{e}^s(t)$, and using $\mathbf{e}^s(t)\cdot\mathbf{u}^s(t) = 0$ and $\mathbf{e}^s(t)\cdot\mathbf{g}^\perp(t) = 0$ gives

$$\lambda(t)\delta t = -2\mathbf{u}^s(t - \tfrac{1}{2}\delta t)\cdot\mathbf{e}^s(t) \tag{3.48}$$

and so

$$\delta t\dot{\mathbf{u}}^s(t) = \delta t\mathbf{g}^\perp(t)/I - 2[\mathbf{u}^s(t - \tfrac{1}{2}\delta t)\cdot\mathbf{e}^s(t)]\mathbf{e}^s(t)\,. \tag{3.49}$$

This equation is used to advance a full step in the integration algorithm

$$\mathbf{u}^s(t + \tfrac{1}{2}\delta t) = \mathbf{u}^s(t - \tfrac{1}{2}\delta t) + \delta t\dot{\mathbf{u}}^s(t)\,. \tag{3.50}$$

The step is completed using

$$\mathbf{e}^s(t + \delta t) = \mathbf{e}^s(t) + \delta t\,\mathbf{u}^s(t + \tfrac{1}{2}\delta t)\,. \tag{3.51}$$

The above equations are applied to each molecule. This algorithm seems to produce stable and accurate trajectories, requires the storage of $\mathbf{e}^s(t)$, $\mathbf{u}^s(t - \tfrac{1}{2}\delta t)$ and $\mathbf{g}^\perp(t)$, and is simple to program (see program F.6).

## 3.4 Constraint dynamics

In polyatomic systems, it becomes necessary to consider not only the stretching of interatomic bonds, but also bending motions, which change the angle between bonds, and twisting motions, which alter torsional angles (see Fig. 1.8). These torsional motions are, typically, of much lower frequency than bond vibrations, and are very important in long-chain organic molecules: they lead to conformational interconversion and have a direct influence on polymer dynamics. Clearly, these effects must be treated properly in molecular dynamics, within the classical approximation. It would be quite unrealistic to assume total rigidity of such a molecule, although bond lengths can be thought of as fixed and a case might be made out for a similar approximation in the case of bond bending angles.

Of course, for any system with such holonomic constraints applied (i.e. a set of algebraic equations connecting the coordinates) it is possible to construct a set of generalized coordinates obeying constraint-free equations of motion (i.e. ones in which the constraints appear implicitly). For any molecule of moderate complexity, such an approach would be very complicated, although it was used in the first simulations of butane [Ryckaert and Bellemans 1975]. The equations of motion in such a case are derived from first principles, starting with the Lagrangian (eqns (3.1) and (3.2)).

A special technique has been developed to handle the dynamics of a molecular system in which certain arbitrarily selected degrees of freedom (such as bond lengths) are constrained, while others remain free to evolve under the influence of intermolecular and intramolecular forces. This constraint dynamics approach [Ryckaert, Ciccotti, and Berendsen 1977] in effect uses a set of undetermined multipliers to represent the magnitudes of forces directed along the bonds, which are required to keep the bond lengths constant. The technique is to solve the equations of motion for one time step in the absence of the constraint forces, and subsequently determine their magnitudes and correct the atomic positions. The method can be applied equally well to totally rigid and non-rigid molecules. Its great appeal is that it reduces even a complex polyatomic liquid simulation to the level of difficulty of an atomic calculation plus a constraint package based on molecular geometry. The method is described in detail by Ryckaert *et al.* [1977], but we will illustrate its application with a simple example. In this section, as in the last one, $\mathbf{r}_a$ will represent the position of atom $a$ in a specific molecule.

Consider a bent triatomic molecule such as water, in which we wish to constrain two of the bonds to be of fixed length, but allow the remaining bond, and hence the inter-bond angle, to vary under the influence of the intramolecular potential. Numbering the central (oxygen) atom 2, and the two outer (hydrogen) atoms 1 and 3, we write the equations of motion in the form

$$m_1 \ddot{\mathbf{r}}_1 = \mathbf{f}_1 + \mathbf{g}_1 \tag{3.52a}$$

$$m_2\ddot{\mathbf{r}}_2 = \mathbf{f}_2 + \mathbf{g}_2 \tag{3.52b}$$

$$m_3\ddot{\mathbf{r}}_3 = \mathbf{f}_3 + \mathbf{g}_3\,. \tag{3.52c}$$

Here $\mathbf{f}_1$, $\mathbf{f}_2$, and $\mathbf{f}_3$ are the forces due to intermolecular interactions and to those intramolecular effects that are explicitly included in the potential. The remaining terms $\mathbf{g}_1$ etc. are the constraint forces: their role is solely to keep the desired bond lengths constant, that is to ensure that the equations

$$\chi_{12} = r_{12}^2(t) - d_{12}^2 = 0 \tag{3.53a}$$

$$\chi_{23} = r_{23}^2(t) - d_{23}^2 = 0 \tag{3.53b}$$

(where $d_{12}$ and $d_{23}$ are the bond lengths, and $r_{12} = |\mathbf{r}_1 - \mathbf{r}_2|$etc.) are satisfied at all times. The Lagrangian equations of motion are derived from these constraints [Bradbury 1968, Chapter 11]; they are eqns (3.52) with

$$\mathbf{g}_a = \tfrac{1}{2}\lambda_{12}\nabla_{\mathbf{r}_a}\chi_{12} + \tfrac{1}{2}\lambda_{23}\nabla_{\mathbf{r}_a}\chi_{23} \tag{3.54}$$

and $\lambda_{12}$ and $\lambda_{23}$ are undetermined (Lagrangian) multipliers. The factors of (1/2) are introduced so that this definition of the multipliers agrees with later equations. So far, we have made no approximations, and, in principle, could solve for the constraint forces [Orban and Ryckaert 1974]. However, because we are bound to solve the equations of motion approximately, using finite difference methods, in practice this will lead to bond lengths that steadily diverge from the desired values.

Instead, Ryckaert *et al.* [1977] suggested an approach in which the constraint forces are calculated so as to guarantee that the constraints are satisfied at each time step; by implication, the constraint forces themselves are only correct to the same order of accuracy as the integration algorithm. Thus, we write

$$m_a\ddot{\mathbf{r}}_a = \mathbf{f}_a + \mathbf{g}_a \approx \mathbf{f}_a + \mathbf{g}_a^{(r)} \tag{3.55}$$

where $\mathbf{g}_a^{(r)}$ is an approximation (the form of which will be given below) to the true forces of constraint, $\mathbf{g}_a$ acting on each atom $a$. By considering the way in which these forces enter into the Verlet algorithm, eqn (3.14), we can write

$$\mathbf{r}_a(t+\delta t) = \mathbf{r}'_a(t+\delta t) + (\delta t^2/m_a)\mathbf{g}_a^{(r)}(t) \tag{3.56}$$

where $\mathbf{r}'_a(t+\delta t)$ is the position which would have been reached in the absence of any constraints. Returning to our example of water, and recognizing that the constraint forces must be directed along the bonds and must conform to Newton's third law, we see that

$$\mathbf{g}_1^{(r)} = \lambda_{12}\mathbf{r}_{12} \tag{3.57a}$$

$$\mathbf{g}_2^{(r)} = \lambda_{23}\mathbf{r}_{23} - \lambda_{12}\mathbf{r}_{12} \tag{3.57b}$$

$$\mathbf{g}_3^{(r)} = -\lambda_{23}\mathbf{r}_{23} \tag{3.57c}$$

where $\lambda_{12}$ and $\lambda_{23}$ are the undetermined multipliers. These may be calculated if we write out eqn (3.56) explicitly:

$$\mathbf{r}_1(t+\delta t) = \mathbf{r}'_1(t+\delta t) + (\delta t^2/m_1)\lambda_{12}\mathbf{r}_{12}(t) \tag{3.58a}$$

$$\mathbf{r}_2(t+\delta t) = \mathbf{r}'_2(t+\delta t) + (\delta t^2/m_2)\lambda_{23}\mathbf{r}_{23}(t) - (\delta t^2/m_2)\lambda_{12}\mathbf{r}_{12}(t) \tag{3.58b}$$

$$\mathbf{r}_3(t+\delta t) = \mathbf{r}'_3(t+\delta t) - (\delta t^2/m_3)\lambda_{23}\mathbf{r}_{23}(t). \tag{3.58c}$$

Thus

$$\mathbf{r}_{12}(t+\delta t) = \mathbf{r}'_{12}(t+\delta t) + \delta t^2(m_1^{-1}+m_2^{-1})\lambda_{12}\mathbf{r}_{12}(t) - \delta t^2 m_2^{-1}\lambda_{23}\mathbf{r}_{23}(t) \tag{3.59a}$$

$$\mathbf{r}_{23}(t+\delta t) = \mathbf{r}'_{23}(t+\delta t) - \delta t^2 m_2^{-1}\lambda_{12}\mathbf{r}_{12}(t) + \delta t^2(m_2^{-1}+m_3^{-1})\lambda_{23}\mathbf{r}_{23}(t). \tag{3.59b}$$

Now we can take the square modulus of both sides, and apply our desired constraints: $|\mathbf{r}_{12}(t+\delta t)|^2 = |\mathbf{r}_{12}(t)|^2 = d_{12}^2$ and similarly for $\mathbf{r}_{23}$. The result is a pair of quadratic equations in $\lambda_{12}$ and $\lambda_{23}$, the coefficients in which are all known (given that we already have the 'unconstrained' positions $\mathbf{r}'_a$) and which can be solved for the undetermined multipliers. In practice, since terms linear in $\lambda_{12}, \lambda_{23}$ are proportional to $\delta t^2$, while the second order terms are proportional to $\delta t^4$, these equations are solved in an iterative fashion. The quadratic terms are dropped and the remaining linear equations solved for $\lambda_{12}$ and $\lambda_{23}$; these values are substituted into the quadratic terms to give new linear equations, which yield improved estimates of $\lambda_{12}$ and $\lambda_{23}$, and so on. Finally, these values are used in eqn (3.58). The way in which the above equations translate into code is shown in program F.7.

We have examined this case in some detail so as to bring out the important features in a more general scheme. Bond angle (as opposed to bond length) constraints present no fundamental difficulty, and may be handled by introducing additional length constraints. For example, the H–O–H bond angle in water may be fixed by constraining the H–H distance, in addition to the O–H bond lengths. Instead of eqn (3.58) we would then have

$$\mathbf{r}_1(t+\delta t) = \mathbf{r}'_1(t+\delta t) + (\delta t^2/m_1)\lambda_{12}\mathbf{r}_{12}(t) - (\delta t^2/m_1)\lambda_{31}\mathbf{r}_{31}(t) \tag{3.60a}$$

$$\mathbf{r}_2(t+\delta t) = \mathbf{r}'_2(t+\delta t) + (\delta t^2/m_2)\lambda_{23}\mathbf{r}_{23}(t) - (\delta t^2/m_2)\lambda_{12}\mathbf{r}_{12}(t) \tag{3.60b}$$

$$\mathbf{r}_3(t+\delta t) = \mathbf{r}'_3(t+\delta t) + (\delta t^2/m_3)\lambda_{31}\mathbf{r}_{31}(t) - (\delta t^2/m_3)\lambda_{23}\mathbf{r}_{23}(t) \tag{3.60c}$$

and eqn (3.59) would be replaced by

$$\mathbf{r}_{12}(t+\delta t) = \mathbf{r}'_{12}(t+\delta t) + \delta t^2(m_1^{-1}+m_2^{-1})\lambda_{12}\mathbf{r}_{12}(t) - \delta t^2 m_2^{-1}\lambda_{23}\mathbf{r}_{23}(t) - \delta t^2 m_1^{-1}\lambda_{31}\mathbf{r}_{31}(t) \tag{3.61a}$$

$$\mathbf{r}_{23}(t+\delta t) = \mathbf{r}'_{23}(t+\delta t) - \delta t^2 m_3^{-1}\lambda_{31}\mathbf{r}_{31}(t) + \delta t^2(m_2^{-1}+m_3^{-1})\lambda_{23}\mathbf{r}_{23}(t) - \delta t^2 m_2^{-1}\lambda_{12}\mathbf{r}_{12}(t) \tag{3.61b}$$

$$\mathbf{r}_{31}(t+\delta t) = \mathbf{r}'_{31}(t+\delta t) - \delta t^2 m_1^{-1}\lambda_{12}\mathbf{r}_{12}(t) - \delta t^2 m_3^{-1}\lambda_{23}\mathbf{r}_{23}(t) + \delta t^2(m_3^{-1}+m_1^{-1})\lambda_{31}\mathbf{r}_{31}(t). \tag{3.61c}$$

This process of 'triangulating' the molecule by introducing fictitious bonds is straightforwardly applied to more complex systems. Figure 1.8 shows bond length constraints applied to the carbon units in a recent model of butane, which leaves just one internal parameter (the torsion angle $\phi$) free to evolve under the influence of the potential. The extension to n-alkanes is discussed by Ryckaert *et al.* [1977] and an application to the case of n-decane has been described [Ryckaert and Bellemans 1978].

For very small molecules, as in the example above, the (linearized) constraint equations may be solved by straightforward algebra. For a larger polyatomic molecule, with $n_c$ constraints, the solution of these equations essentially requires inversion of an $n_c \times n_c$ matrix at each time step. This could become time-consuming for very large molecules, such as proteins. Assuming, however, that only near-neighbour atoms and bonds are related by constraint equations, the constraint matrix will be sparse, and special inversion techniques might be applicable. An alternative procedure is to go through the constraints one by one, cyclically, adjusting the coordinates so as to satisfy each in turn. The procedure may be iterated until all the constraints are satisfied to within a given tolerance. This approach has been called SHAKE [Ryckaert *et al.* 1977] and is most useful when large molecules are involved. An example of the SHAKE algorithm for a chain molecule is given in program F.8.

Problems may arise in the construction of a constraint scheme for certain molecules. Consider the linear molecule $CS_2$: it has three atoms and five degrees of freedom (two rotational and three translational) so we require $n_c = 3 \times 3 - 5 = 4$ constraints. This is impossible with only three bond lengths available to be specified. A more subtle example is that of benzene, modelled as six united CH atoms in a hexagon. For six degrees of freedom (three rotational and three translational) we require $n_c = 3 \times 6 - 6 = 12$ constraints, and this number may indeed be accommodated. However, the constraint matrix is then found to be singular, i.e. its determinant vanishes. Physically, the problem is that all the constraints act in the plane of the molecule, and none of them act to preserve planarity. The solution to both these problems is to choose a subset of atoms sufficient to define the molecular geometry, apply constraints to those atoms, and express the coordinates of the remaining atoms as linear combinations of those of the primary 'core' [Ciccotti, Ferrario, and Ryckaert 1982]. In computing the dynamics of the core, there is a simple prescription for transferring the forces acting on the 'secondary' atoms to the core atoms, so as to generate the correct linear and angular accelerations. Recently, the SHAKE method has been extended to handle more general geometrical constraints [Ryckaert 1985] needed to specify (for example) the arrangement of side-chains or substituent atoms in flexible hydrocarbons. A review of these techniques has recently appeared [Ciccotti and Ryckaert 1986].

SHAKE is most easily applied to the Verlet algorithm, in which only positions and accelerations appear, although van Gunsteren and Berendsen [1977] have described how SHAKE could be fitted into a higher order predictor–corrector method. The aim of such a constraint package in an algorithm which includes velocities (and higher derivatives of position) is to ensure that not only do we satisfy the constraint equations but also we satisfy the derivatives of those equations: if $r_{ab}^2 = \mathbf{r}_{ab} \cdot \mathbf{r}_{ab} =$ constant, then we should ensure that $\mathrm{d}(r_{ab}^2)/\mathrm{d}t = 2\mathbf{r}_{ab} \cdot \dot{\mathbf{r}}_{ab} = 0$. Recently, a modification of the method of constraints, built around the velocity version of Verlet's algorithm (Section 3.2.1) has been proposed [Andersen 1983]. Again, approximations to the true constraint forces are needed in the equations of motion to guarantee that the constraints are satisfied at all time steps. The velocity Verlet algorithm is a two-stage process, with each stage involving the forces, including the forces of constraint. Accordingly, at each stage an approximation to $\mathbf{g}$ is made, so as to ensure that the constraints are satisfied. Referring to eqns (3.19)–(3.21), we see that the constraint forces enter into the algorithm as follows. In the first stage we have

$$\mathbf{r}_a(t+\delta t) = \mathbf{r}'_a(t+\delta t) + \tfrac{1}{2}(\delta t^2/m_a)\mathbf{g}_a^{(r)}(t) \tag{3.62}$$

and

$$\mathbf{v}_a(t+\tfrac{1}{2}\delta t) = \mathbf{v}'_a(t+\tfrac{1}{2}\delta t) + \tfrac{1}{2}(\delta t/m_a)\mathbf{g}_a^{(r)}(t)\,. \tag{3.63}$$

The constraint forces, $\mathbf{g}_a^{(r)}$, for this stage are directed along the bond vectors $\mathbf{r}_{ab}(t)$. They are determined by solving eqn (3.62) by matrix inversion or iteratively as in SHAKE. At the same time, the velocities $\mathbf{v}'_a$ at time $t+\frac{1}{2}\delta t$, obtained using eqn (3.20), are adjusted according to eqn (3.63). The second part of the algorithm follows evaluation of the non-constraint forces $\mathbf{f}_a(t+\delta t)$, which are used in eqn (3.21) to give $\mathbf{v}'_a(t+\delta t)$. The second stage is:

$$\mathbf{v}_a(t+\delta t) = \mathbf{v}'_a(t+\delta t) + \tfrac{1}{2}(\delta t/m_a)\mathbf{g}_a^{(v)}(t+\delta t)\,. \tag{3.64}$$

These constraint forces $\mathbf{g}_a^{(v)}(t+\delta t)$, are directed along the bonds $\mathbf{r}_{ab}(t+\delta t)$, and are chosen so that the velocities satisfy the constraints exactly at time $t+\delta t$. Note that, in the next integration step, a different approximation to these same constraint forces, namely $\mathbf{g}_a^{(r)}(t+\delta t)$, will be used. This step follows immediately. By analogy with SHAKE, Andersen has termed the iterative solution of these equations 'RATTLE'. Code for the RATTLE algorithm in a simulation of a chain molecule is given in program F.9 (compare with SHAKE in F.8).

This is a good point at which to reflect on the relative merits of the constraint methods and quaternion parameter methods for the solution of rigid body equations of motion. Constraint dynamics provides a neat and simple approach, readily adapted to different molecular geometries. Quaternions are an equally elegant and general solution. At first sight, it would seem that for large 'rigid' structures (such as benzene and adamantane) the number of constraints required would be excessive for an essentially simple problem in rigid body motion, and that quaternions would have a clear advantage. However, Ciccotti *et al.* [1982] have substantially rationalized this situation, reducing the constraint equations to dealing with a rigid 'core' of atoms (two for a linear molecule, three for a planar molecule, and four for a non-planar one) for any rigid molecule geometry. Comparison with the quaternion predictor–corrector method used in simulations of liquid $CS_2$ [Tildesley and Madden 1981] suggests that constraint dynamics permits a much longer time step, while being, in other respects, of comparable efficiency [Ciccotti *et al.* 1982]. However, it is now clear [Fincham 1981] that low order quaternion algorithms, such as the modified leap-frog discussed in Section 3.3.1, are superior to the predictor–corrector approach, and are at least as accurate at long time steps as the method of constraints. Possibly the only case where constraint dynamics is simpler, as far as rigid bodies are concerned, is that of the diatomic or other linear molecules, in which special techniques, as discussed in Section 3.3.2, can be used.

On the other hand, as soon as any non-rigidity is introduced into the molecular model, constraint dynamics as typified by SHAKE and RATTLE provide the only realistic option, with outstanding advantages of generality and convenience over the alternative methods based on generalized coordinates. For flexible molecules, we have ample choice as to where to apply constraints, and it is generally believed that, while constraining bond lengths is worthwhile, it is best to leave bond angles (and certainly torsion angles) free to evolve under the influence of appropriate terms in the potential energy. This is partly on the grounds of program efficiency: the SHAKE algorithm iterations converge very slowly when rigid 'triangulated' molecular units are involved, often necessitating a reduced time step, which might as well be used in a proper integration of the bond 'wagging' motions instead [van Gunsteren and Berendsen 1977; van Gunsteren 1980]. The other reason is that the relatively low frequencies of these motions makes the constraint approximation less valid. As discussed in Section 2.10, a model with a strong harmonic potential is different from one in which the potential is replaced by a rigid constraint. This point has been recognized for some time in the field of polymer dynamics, and has been tested by computer simulation [Fixman 1974, 1978a, b; Go and Scheraga 1976; Helfand 1979; Pear and Weiner 1979]. In practical terms, for a model of a protein molecule, van Gunsteren and Karplus [1982] have shown that the introduction of bond length constraints into a model based otherwise on realistic intramolecular potential functions has little effect on the structure and dynamics, but the further introduction of constraints on bond angles seriously affects the torsion angle distributions and the all-important conformational interconversion rates. This effect can be countered by adding the additional constraint potential of eqn (2.154), which involves the calculation of the metric determinant $|\mathbf{H}|$. This is time-consuming and algebraically complicated for all but the simplest flexible molecules, and the lesson seems to be that, for realistic molecular dynamics simulations, bond length constraints are permissible, but bond angle constraints should not be introduced without examining their effects.

Two final points should be made, in relation to the calculation of thermodynamic properties of model systems incorporating constraints. The calculation of the total kinetic energy of such a system is a simple matter of summing the individual atomic contributions in the usual way. When using this quantity to estimate the temperature, according to eqn (2.50), we must divide by the number of degrees of freedom. It should be clear from the specification of the molecular model how many independent constraints have been applied, and hence what the number of degrees of freedom is. Secondly, in molecular systems quantities such as the pressure may be calculated in several ways, the two most important of which focus on the component atoms, and on the molecular centres of mass, respectively. Consider the evaluation of the virial function (eqn (2.59)) interpreting the sum as being taken over all atom–atom separations $\mathbf{r}_{ab}$ and forces $\mathbf{f}_{ab}$. In this case, all intramolecular contributions to $\mathcal{W}$ including the constraint forces should be taken into account. Now consider the alternative interpretation of eqn (2.61), in which we take the $\mathbf{f}_{ij}$ to represent the sum of all the forces acting on a molecule $i$, due to its interactions with molecule $j$, and take each such force to act at the centre of mass. In this case, all the intramolecular forces, including the constraint forces, cancel out and can be omitted from the sum. It is easy to show that, at equilibrium, the average pressure computed by either route is the same.

## 3.5 Checks on accuracy

Is it working properly? This is the first question that must be asked when a simulation is run for the first time, and the answer is frequently in the negative. Here, we discuss the tell-tale signs of a non-functioning MD program.

The first check must be that the conservation laws are properly obeyed, and in particular that the energy should be 'constant'. In fact small changes in the energy will occur (see Fig. 3.3). For a simple Lennard-Jones system, fluctuations of order 1 part in $10^4$ are generally considered to be acceptable, although some workers are less demanding, and some more so. No systematic investigation of this point has been carried out [Fincham 1985]. Energy fluctuations may be reduced by decreasing the time step. If one of the Verlet algorithms is being used, then a suggestion due to Andersen [see Berens *et al.* 1983] may be useful. Several short runs should be undertaken, each starting from the same initial configuration and covering the same total time $t_{\text{run}}$: each run should employ a different time step $\delta t$, and hence consist of a different number of steps $\tau_{\text{run}} = t_{\text{run}}/\delta t$. The RMS energy fluctuations for each run should be calculated. If the program is functioning correctly, and other sources of energy fluctuations (such as potential truncation) have been eliminated, then the Verlet algorithm should give RMS energy fluctuations which are accurately proportional to $\delta t^2$ (see Fig. 3.3). A good initial estimate of $\delta t$ is that it should be roughly an order of magnitude less than the Einstein period $t_E = 2\pi/\omega_E$, where the Einstein frequency $\omega_E$ is given by eqn. (2.127). If a typical liquid starting configuration is available, then $\omega_E$ may be obtained by averaging over all the molecules in the system. Otherwise, a guess may be made by considering a hypothetical solid phase of the same density as the system of interest.

A slow upward drift of energy may also be due to a time step that is too long, to potential truncation effects (see Section 5.2.4), or might indicate a program error. Effects with a 'physical' origin and those due to time step problems can be distinguished by the procedure outlined above, i.e. duplicating a short run but using a larger number of smaller time steps. If the drift as a function of simulation time is unchanged, then it is presumably connected with the system under study, whereas if it is substantially reduced, the method used to solve the equations of motion (possibly the size of the time step) is responsible. In the category of program error, we should mention the possibility that the wrong quantity is being calculated. If the total energy varies significantly but the simulation is 'stable' in the sense that no inexorable climb in energy occurs, then the way in which the energy is calculated should be examined. Are potential and kinetic contributions added together correctly? Is the pairwise force (appearing in the double loop) in fact correctly derived from the potential? This last possibility may be tested by including statements that calculate the force on a given particle numerically, from the potential energy, by displacing it slightly in each of the three coordinate directions. The result may be compared with the analytical formula encoded in the program. As emphasized earlier, although small fluctuations are permissible, it is essential to eliminate any traces of a drift in the total energy over periods of thousands of time steps, if the simulation is to probe the microcanonical ensemble correctly.

Rather than a slow drift, a very rapid, even catastrophic, increase in energy may occur within the first few time steps. There are two possibilities here: either a starting configuration with particle overlaps has been chosen (so that the intermolecular forces are unusually large) or there is a serious program error. The starting configuration may be tested simply by printing out the initial coordinates and inspecting the numbers. Alternatively, particularly when the number of particles is large, statements may temporarily be incorporated into the force loop so as to test each of the pair separations and print out particle coordinates and identifiers whenever a very close pair is detected.

Tracking down a serious program error may be a difficult task. It is a favourite mistake, particularly when reading in the potential parameters in real (e.g. SI) units to make a small, but disastrous, error in unit conversion. There is much to be said for testing out a program on a small number of particles before tackling the full-size system, but beware! Is the potential cutoff distance still smaller than half of the box length? Frequent program errors involve mismatching of number, length or type of variables passed between routines in COMMON blocks or in argument lists. Simple typographical errors, while hard to spot, may have far-reaching effects. It is hard to overemphasize how useful modern software development tools can be in locating and eliminating mistakes of this kind. A good editor may be used to check the source code much more efficiently than simple visual inspection. Many FORTRAN compilers produce compilation listings which include a summary of the types and lengths of all variables used in each routine. Examining these listings is a good way to spot misspelt variables. On modern computers, excellent interactive FORTRAN debugging facilities exist, which allow the program to be run under user control, with constant monitoring of the program flow and the values of variables of interest. Needless to say, a program written in a simple, logical, and modular fashion will be easier to debug (and will contain fewer errors!) than one which has not been planned in this way [Balfour and Marwick 1979; Ledgard and Chmura 1978]. Some programming considerations appear in Appendix A.

For molecular simulations, errors may creep into the program more easily than in the simple atomic case. Energy should be conserved just as for atomic simulations, although, for small molecules, a rather short time step may be needed to achieve this, since rotational motion occurs so rapidly. If non-conservation is a problem, several points may need checking. Incorrectly differentiating the potential on the way to the torques may be a source of error: this is more complicated for potentials incorporating multipolar terms [see Appendix C]. Again, this may be tested numerically, by subjecting selected molecules to small rotations, and observing the change in potential energy. If the angular part of the motion is suspect, the rest of the program may be tested by 'freezing out' the rotation. This is accomplished by disengaging the rotational algorithm; physically this corresponds to giving the molecules an infinite moment of inertia and zero angular velocity. Energy should still be conserved under these conditions. Conversely, the angular motion may be tested out by omitting, temporarily, the translational algorithm, thus fixing the molecular centres at their initial positions.

Two final points should be made. When the program appears to be running correctly, the user should check that the monitored quantities are in fact evolving in time. Even conserved variables will fluctuate a little if only due to round-off errors, and any quantity that appears to be constant to 10 significant figures should be regarded with suspicion: it is probably not being updated at all. Excellent conservation, but no science, will result from a program that does not, in fact, move the particles (due to some error associated with the predictor and corrector routines, for example). A time step that is too small (or that has been accidentally set to zero) will be very wasteful of computer time, and the extent to which $\delta t$ can be increased without prejudicing the stability of the simulation should be investigated. Finally, the problems discussed above are all 'mechanical' rather than 'thermodynamic', i.e. they are associated with the correct solution of the equations of motion. The quite separate question of attaining thermodynamic equilibrium will be discussed in Chapter 5. If a well-known system is being simulated (e.g. Lennard-Jones, soft-sphere potentials etc.) then it is obviously sensible to compare the simulation output, when equilibrium has been attained, with the known thermodynamic properties.

## 3.6 Molecular dynamics of hard systems

The molecular dynamics of molecules interacting via hard potentials (i.e. discontinuous functions of distance) must be solved in a way which is qualitatively different from the molecular dynamics of soft bodies. Whenever the distance between two particles becomes equal to a point of discontinuity in the potential, then a 'collision' (in a broad sense) occurs: the particle velocities will change suddenly, in a specified manner, depending upon the particular model under study. Thus, the primary aim of a simulation program here is to locate the time, collision partners, and all impact parameters, for every collision occurring in the system, in chronological order. Instead of a regular, step-by-step, approach, as for soft potentials, hard potential programs evolve on a collision-by-collision basis, computing the collision dynamics and then searching for the next collision. The general scheme may be summarized as follows:

(a) locate next collision;
(b) move all particles forward until collision occurs;
(c) implement collision dynamics for the colliding pair;
(d) calculate any properties of interest, ready for averaging, before returning to (a).

Because of the need to locate accurately future collision times, simulations have been restricted in the main to systems in which force-free motion occurs between collisions, and in which the molecular geometry is spherical. In these simple cases, which include hard spheres [Alder and Wainwright 1959, 1960], rough, and otherwise-modified, hard spheres [O'Dell and Berne 1975; Berne 1977], and square-well molecules [Alder and Wainwright 1959], location of the time of collision between any two particles requires the solution of a quadratic equation. We examine this in detail in the next section. The computational problems become quite daunting when we consider solving the highly non-linear equations that result from models in which the hard cores are supplemented with long-range soft potentials. An example is the primitive model of electrolytes, consisting of hard spheres plus Coulomb interactions. By contrast, such systems may be handled easily using Monte Carlo simulation (see Chapter 4). Recent developments suggest that it may be possible to treat these 'hybrid' hard + soft systems by returning to an approximate 'step-by-step' approach [Stratt, Holmgren, and Chandler 1981; McNeill and Madden 1982]. We consider this briefly in Section 3.6.2.

### 3.6.1 *Hard spheres*

A program to solve hard-sphere molecular dynamics has two functions to perform: the calculation of collision times and the implementation of collision dynamics. The collision time calculation is the expensive part of the program, since, in principle, all possible collisions between distinct pairs must be considered.

Consider two spheres, $i$ and $j$, of diameter $\sigma$, whose positions at time $t$ are $\mathbf{r}_i$ and $\mathbf{r}_j$, and whose velocities are $\mathbf{v}_i$ and $\mathbf{v}_j$. If these particles are to collide at time $t + t_{ij}$ then the following equation will be satisfied:

$$|\mathbf{r}_{ij}(t+t_{ij})| = |\mathbf{r}_{ij} + \mathbf{v}_{ij}t_{ij}| = \sigma \tag{3.65}$$

where $\mathbf{r}_{ij} = \mathbf{r}_i - \mathbf{r}_j$ and $\mathbf{v}_{ij} = \mathbf{v}_i - \mathbf{v}_j$. If we define $b_{ij} = \mathbf{r}_{ij} \cdot \mathbf{v}_{ij}$, then this equation becomes

$$v_{ij}^2 t_{ij}^2 + 2b_{ij}t_{ij} + r_{ij}^2 - \sigma^2 = 0\,. \tag{3.66}$$

This is a quadratic equation in $t_{ij}$. If $b_{ij} > 0$, then the molecules are going away from each other and they will not collide. If $b_{ij} < 0$, it may still be true that $b_{ij}^2 - v_{ij}^2(r_{ij}^2 - \sigma^2) < 0$, in which case eqn (3.66) has complex roots and again no collision occurs. Otherwise (assuming that the spheres are not already overlapping) two positive roots arise, the smaller of which corresponds to impact

$$t_{ij} = \frac{-b_{ij} - (b_{ij}^2 - v_{ij}^2(r_{ij}^2 - \sigma^2))^{1/2}}{v_{ij}^2}\,. \tag{3.67}$$

A simple piece of code to locate the next possible collision for each particle is given below. As will become clear, it is useful to store away all the collision times (in an array COLTIM) and the collision partners (PARTNR) as we find them. Also, in the following, TIMBIG just stores a long time (i.e. a large number, say $10^{10}$) and SIGSQ is the square of the particle diameter SIGMA. Finally, in applying the minimum image convention for periodic boundaries (taking the unit cube as the simulation box) we are assuming that we only need to examine the nearest images of any two particles in order to pick out the collision between them. Relaxing this assumption makes the simulation program a little more complicated; it should only break down at very low densities.

```
        DO 100 I = 1, N

           COLTIM(I) = TIMBIG

100     CONTINUE

        DO 200 I = 1, N-1

           DO 199 J = I + 1, N

              RXIJ = RX(I) - RX(J)
              RYIJ = RY(I) - RY(J)
              RZIJ = RZ(I) - RZ(J)
              RXIJ = RXIJ - ANINT ( RXIJ )
              RYIJ = RYIJ - ANINT ( RYIJ )
              RZIJ = RZIJ - ANINT ( RZIJ )
              VXIJ = VX(I) - VX(J)
              VYIJ = VY(I) - VY(J)
              VZIJ = VZ(I) - VZ(J)
              BIJ  = RXIJ * VXIJ + RYIJ * VYIJ + RZIJ * VZIJ

              IF ( BIJ .LT. 0.0 ) THEN

                 RIJSQ = RXIJ ** 2 + RYIJ ** 2 + RZIJ ** 2
                 VIJSQ = VXIJ ** 2 + VYIJ ** 2 + VZIJ ** 2
                 DISCR = BIJ ** 2 - VIJSQ * ( RIJSQ - SIGSQ )

                 IF ( DISCR .GT. 0.0 ) THEN

                    TIJ = ( - BIJ - SQRT ( DISCR ) ) / VIJSQ

                    IF ( TIJ .LT. COLTIM(I) ) THEN

                       COLTIM(I) = TIJ
                       PARTNR(I) = J

                    ENDIF

                    IF ( TIJ .LT. COLTIM(J) ) THEN

                       COLTIM(J) = TIJ
                       PARTNR(J) = I

                    ENDIF

                 ENDIF

              ENDIF

199        CONTINUE

200     CONTINUE
```

In the interests of clarity, we have not optimized this code; a more efficient version appears on microfiche F.10. Notice how the collision times are initially all set to be very large, and are only reduced to reasonable values when all the requirements for a collision are met. The next stage of the program is to locate the earliest collision and the colliding pair *i* and *j*.

```
          TIJ = TIMBIG

          DO 300 K = 1, N

             IF ( COLTIM(K) .LT. TIJ ) THEN

                TIJ = COLTIM(K)
                I   = K

             ENDIF

300       CONTINUE

          J = PARTNR(I)
```

All molecules are moved forward by the time $t_{ij}$, the periodic boundary conditions are applied, and the table of future collision times is adjusted accordingly:

```
        DO 400 K = 1, N

           COLTIM(K) = COLTIM(K) - TIJ
           RX(K)     = RX(K) + VX(K) * TIJ
           RY(K)     = RY(K) + VY(K) * TIJ
           RZ(K)     = RZ(K) + VZ(K) * TIJ
           RX(K)     = RX(K) - ANINT ( RX(K) )
           RY(K)     = RY(K) - ANINT ( RY(K) )
           RZ(K)     = RZ(K) - ANINT ( RZ(K) )

400     CONTINUE
```

Now we are ready to carry through the second part of the calculation, namely the collision dynamics themselves. The changes in velocities of the colliding pair are completely dictated by the requirements that energy and linear momentum be conserved and (for smooth hard spheres) that the impulse acts along the line of centres, as shown in Fig. 3.5.

Using conservation of total linear momentum and (kinetic) energy, and assuming equal masses, the velocity change $\delta \mathbf{v}_i$, such that

$$\mathbf{v}_i \text{ (after)} = \mathbf{v}_i \text{ (before)} + \delta \mathbf{v}_i \tag{3.68a}$$

$$\mathbf{v}_j \text{ (after)} = \mathbf{v}_j \text{ (before)} - \delta \mathbf{v}_i \tag{3.68b}$$

is given by

$$\delta \mathbf{v}_i = -(b_{ij}/\sigma^2)\mathbf{r}_{ij} = -\mathbf{v}_{ij}{}^{\parallel} \tag{3.69}$$

with $b_{ij} = \mathbf{r}_{ij} \cdot \mathbf{v}_{ij}$ evaluated now at the moment of impact (it is still a negative number). Thus, $\delta \mathbf{v}_i$ is simply the negative of the projection of $\mathbf{v}_{ij}$ along the $\mathbf{r}_{ij}$ direction, which we denote $\mathbf{v}_{ij}{}^{\parallel}$ (see Fig. 3.5). The code for the collision dynamics is simply a transcription of eqn (3.69) followed by eqns (3.68a) and (3.68b).

Now we could return to the initial loop and recalculate all collision times afresh. In fact, there is no need to carry out this calculation in entirety, since many of the details in COLTIM and PARTNR will have been unaffected by the collision between $i$ and $j$. Obviously, we must look for the next collision partners of $i$ and $j$; also, we have to discover the fate of any other molecules which *were* due to collide with $i$ and $j$, had these two not met each other first. Apart from these, the information in our collision lists is still quite valid. The 'update' procedure can take the following form:

```
           II = I
           JJ = J

           DO 500 I = 1, N

              IF ( ( I .EQ. II ) .OR. ( PARTNR(I) .EQ. II ) .OR.
     :              ( I .EQ. JJ ) .OR. ( PARTNR(I) .EQ. JJ )      ) THEN

                 COLTIM(I) = TIMBIG

                 DO 499 J = 1, N

                    IF ( J .NE. I ) THEN

                       ... usual calculation for IJ collision ...

                       IF ( TIJ .LT. COLTIM(I) ) THEN

                          COLTIM(I) = TIJ
                          PARTNR(I) = J

                       ENDIF

                       IF ( TIJ .LT. COLTIM(J) ) THEN

                          COLTIM(J) = TIJ
                          PARTNR(J) = I

                       ENDIF

                    ENDIF

499              CONTINUE

              ENDIF

500        CONTINUE
```

Following this, the smallest time in COLTIM is located, the particles are moved on, and the whole procedure is repeated. A complete (and fairly efficient) hard-sphere program is given in F.10.

The generalization of this program to the case of the square-well potential is straightforward. Now, for each pair, there are two distances at which 'collisions' occur, so the algorithm for determining collision times is slightly more involved. Collisions at the inner sphere obey normal hard-sphere dynamics; at the outer boundary, the change in momentum is determined by the usual conservation laws. For molecules approaching each other, the

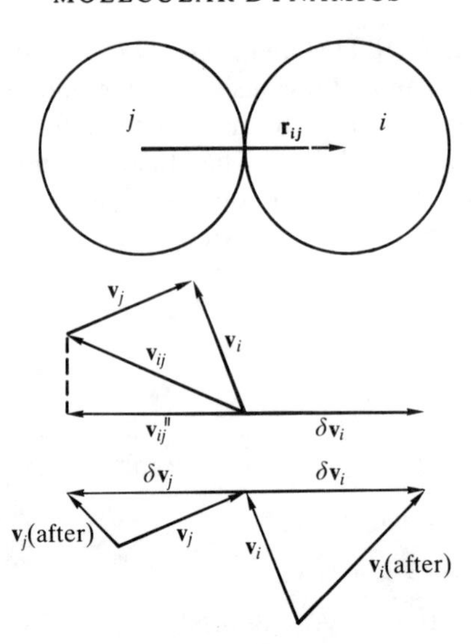


**Fig. 3.5** A smooth hard-sphere collision. The vector construction gives the change in velocities for each particle. For illustrative purposes, we have taken all the vectors to be coplanar in this example.

potential energy drops on crossing the boundary, and so the kinetic energy shows a corresponding increase. If the molecules are separating, two possibilities arise. If the total kinetic energy is sufficient, the molecules cross the boundary with a loss in $\mathscr{K}$ to compensate the rise in $\mathscr{V}$. Alternatively, if $\mathscr{K}$ is insufficient, reflection at the outer boundary occurs and the particles remain 'bound'.

More complicated potentials involving several 'steps' can be treated in the same way; a quite realistic potential can be constructed from a large number of vertical and horizontal segments, but of course the simulation becomes more expensive as more 'collisions' have to be dealt with per unit time [Chapela, Martinez-Casas, and Alejandre 1984].

The other main modification to the hard-sphere model which preserves spherical symmetry is the introduction of roughness. Rough spheres [Subramanian and Davis 1975; O'Dell and Berne 1975], differ from simple hard spheres only in their collision dynamics: the free flight dynamics between collisions, and hence the techniques used to locate future collisions, are identical. Rough spheres are characterized by a diameter $\sigma$, a mass $m$, and a moment of inertia $I$ or, alternatively, a parameter $\kappa = 4I/m\sigma^2$. They have translational velocities $\mathbf{v}$ and angular or 'spin' velocities $\boldsymbol{\omega}$. Rough sphere collision dynamics are subject to the usual conservation laws: a collision between two molecules will preserve total energy (rotational plus translational) total linear momentum, and total angular momentum defined by

$$\mathbf{J} = \mathbf{L} + \mathbf{S} = \sum_i m\mathbf{r}_i \times \mathbf{v}_i + \sum_i I\boldsymbol{\omega}_i . \tag{3.70}$$

The difference between rough and smooth hard spheres may be viewed in the following way (see Fig. 3.6). Consider the two points on the sphere surfaces that come together at the moment of impact. The relative velocity vector of these points is

$$\mathbf{v}_{ij}^{\mathrm{imp}} = \mathbf{v}_i^{\mathrm{imp}} - \mathbf{v}_j^{\mathrm{imp}} = (\mathbf{v}_i - \mathbf{v}_j) - \tfrac{1}{2}(\boldsymbol{\omega}_i + \boldsymbol{\omega}_j) \times \mathbf{r}_{ij}. \tag{3.71}$$

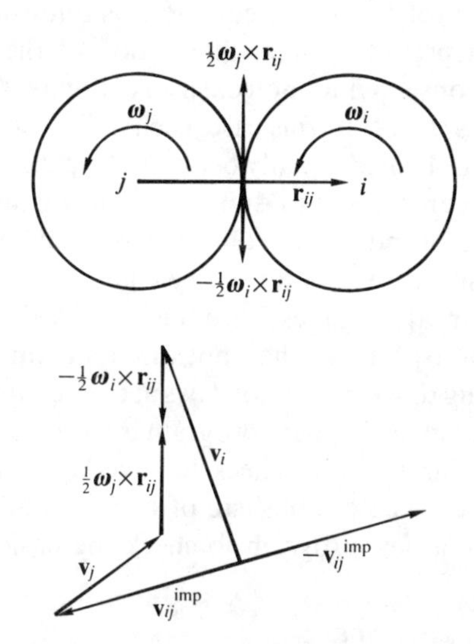


**Fig. 3.6** A rough hard-sphere collision. We show the reversal of the relative velocity vector of the impact points. For illustrative purposes, all vectors are taken to be coplanar, except for the angular velocity vectors which point upwards, normal to the plane.

We may resolve this vector into components parallel to the line of the centres of the colliding pair, and perpendicular to this line

$$\mathbf{v}_{ij}^{\mathrm{imp}} = \mathbf{v}_{ij}^{\mathrm{imp}\parallel} + \mathbf{v}_{ij}^{\mathrm{imp}\perp}. \tag{3.72}$$

Because the spheres are hard, the parallel component of $\mathbf{v}_{ij}^{\mathrm{imp}}$ is reversed on impact. For smooth hard spheres, the perpendicular part is not altered, and so there is no change in the angular velocities: these become redundant. For rough hard spheres, both parts of the relative velocity vector $\mathbf{v}_{ij}^{\mathrm{imp}}$ are reversed on impact. Hence

$$\delta\mathbf{v}_{ij}^{\mathrm{imp}} = \mathbf{v}_{ij}^{\mathrm{imp}}\,(\text{after}) - \mathbf{v}_{ij}^{\mathrm{imp}}\,(\text{before}) = -2\mathbf{v}_{ij}^{\mathrm{imp}}\,(\text{before}). \tag{3.73}$$

Then the conservation laws lead to an expression for the impulse

$$\delta\mathbf{p}_i = \tfrac{1}{2}m\left(\delta\mathbf{v}_{ij}^{\mathrm{imp}\parallel} + \frac{\kappa}{1+\kappa}\,\delta\mathbf{v}_{ij}^{\mathrm{imp}\perp}\right) \tag{3.74}$$

in terms of which the changes in velocities and angular velocities become

$$m\mathbf{v}_i \text{ (after)} = m\mathbf{v}_i \text{ (before)} + \delta\mathbf{p}_i$$
$$m\mathbf{v}_j \text{ (after)} = m\mathbf{v}_j \text{ (before)} - \delta\mathbf{p}_i$$
$$I\boldsymbol{\omega}_i \text{ (after)} = I\boldsymbol{\omega}_i \text{ (before)} - \tfrac{1}{2}\mathbf{r}_{ij} \times \delta\mathbf{p}_i$$
$$I\boldsymbol{\omega}_j \text{ (after)} = I\boldsymbol{\omega}_j \text{ (before)} - \tfrac{1}{2}\mathbf{r}_{ij} \times \delta\mathbf{p}_i\,. \tag{3.75}$$

The computer code for the above equations is quite straightforward. The rough sphere system provides a nice illustration of the fact that, although $\mathbf{J} = \mathbf{L} + \mathbf{S}$ is indeed conserved in molecular encounters, the intrinsic or 'spin' angular momentum $\mathbf{S}$ is not. In this case, each collision creates (or destroys) spin, and the change is divided equally between the partners. Consequently, $\mathbf{L}$ is also not separately conserved. For quite different reasons, having nothing to do with the dynamics of individual collisions, periodic boundary conditions will destroy the conservation law for $\mathbf{J}$ (see Chapter 1). Attempts have been made to introduce 'partial roughness' into the basic hard-sphere model [Berne 1977; Lyklema 1979a, b] but we shall not discuss them here.

Apart from verifying the conservation laws at each collision, a simple test for the proper working of a hard-sphere program is to examine configurations at intervals during the simulation to check for overlaps. The properties of the hard sphere system are well-established, of course, and could be compared with the output from a new program to check the basic method.

### *3.6.2 Hard non-spherical bodies*

For any non-spherical rigid body model, calculating the collision point for two molecules, even in the case of free flight between collisions, becomes a taxing numerical problem. In general, a highly non-linear equation for the time at which the contact condition is met must be solved. This has only been attempted for systems of hard spherocylinders [Rebertus and Sando 1977], for fused dumb-bell diatomics [Bellemans, Orban, and van Belle 1980; Allen and Imbierski, 1987], and for hard lines and related models [Frenkel and Maguire 1983; Allen and Cunningham 1986]. Since the collision equation must be solved for each possible colliding pair, any technique that speeds up the numerical solution is of great benefit.

Such a method might simply be a way of rapidly establishing upper and lower bounds on the possible roots. In some circumstances, all we need to know is whether or not a root exists within a finite time interval, i.e. a time step. For many hard systems, the result of not detecting a collision between two molecules will be that overlap occurs. By checking the overlap condition at regular intervals $\delta t$, we can detect collisions and locate the collision time within that interval. The only errors involved here will be if two particles enter and then leave the overlap region within the course of one time step, i.e. if multiple roots of the collision equation exist in that interval. This error can be minimized if $\delta t$ is chosen to be sufficiently small. Within a very small distance of a known root, solution of the collision equation becomes a straightforward procedure.

The above method was essentially that employed by Rebertus and Sando [1977] in their simulations of hard spherocylinders. The way in which we can deal 'retrospectively' with the collisions occurring in a time step has been investigated in some detail by Stratt *et al.* [1981]. A bonus associated with the return to a step-by-step approach is that we are no longer restricted to free flight between collisions: it is just as easy to use any of the algorithms described earlier in this chapter to integrate the equations of motion forward in time. Thus, it is now possible to treat 'hybrid' systems of hard cores plus soft attractive (or repulsive) potentials [McNeill and Madden 1982].

As an example of a rigid body model that cannot be treated in the above fashion, we should mention the study by Frenkel and Maguire [1983] of a system of infinitely thin hard lines of finite length. In a system of this kind, in three dimensions, overlaps can never occur, so it is less easy to detect collisions retrospectively. In this situation, a brute force solution of the collision equations is unavoidable, although it is possible to optimize the root-searching procedure. These comments apply equally to more complicated molecular models based on the hard line unit [Allen and Cunningham 1986].

The extension of the step-by-step algorithm to the case of flexible polyatomic molecules with hard and soft potentials should present no additional problems (a simple example is the use of rigid, angle-constraining 'windows' in a simulation of a butane-like model [Stratt *et al.* 1981]). We should mention one elegant approach to the model of a flexible chain of hard spheres [Rapaport 1978, 1979; Bellemans *et al.* 1980] which once more reduces the complexity of a polyatomic simulation to the level of a simple atomic simulation. In the Rapaport model, the length of the bond between two adjacent atoms in the chain is not fixed, but is constrained to lie between two values $\sigma$ and $\sigma + \delta\sigma$. Interactions between non-bonded atoms, and between atoms on different polymer molecules, are of the usual hard sphere form. The spherical atoms undergo free flight between collisions that are of the usual kind: in fact the 'bonds' are no more than extreme examples of the square-well potential (with infinite walls on both sides of the well). By choosing $\delta\sigma$ to be small, the bond lengths may be constrained as closely as desired, at the expense (of course) of there being more 'bond collisions' per unit time. The model can be extended so that we can construct nearly-rigid, as well as more complicated flexible, molecules from the basic building blocks [Chapela *et al.* 1984].

Checks on the working of a program which simulates hard molecular systems must include tests of the basic conservation laws on collision, and periodic examination of the configuration for unphysical overlaps. It is also sensible to conduct preliminary runs for any special cases of the molecular model (for example, hard spheres or lines) whose properties are well-known.

# 4
# MONTE CARLO METHODS

## 4.1 Introduction

The Monte Carlo method was developed by von Neumann, Ulam, and Metropolis at the end of the Second World War to study the diffusion of neutrons in fissionable material. The name 'Monte Carlo', chosen because of the extensive use of random numbers in the calculation, was coined by Metropolis in 1947 and used in the title of a paper describing the early work at Los Alamos [Metropolis and Ulam 1949].

Statisticians had used model sampling experiments to investigate problems long before this time. The English statistician W. S. Gossett ('Student') [1908] estimated the correlation coefficients in his '$t$' distribution with the help of a sampling experiment, and Lord Kelvin's assistant generated 5000 random trajectories to study the elastic collisions of particles with shaped walls [Kelvin 1901]. The novel contribution of von Neumann and Ulam [1945] was to realize that determinate mathematical problems could be treated by finding a probabilistic analogue which is then solved by a stochastic sampling experiment.

These sampling experiments involve the generation of random numbers followed by a limited number of arithmetic and logical operations, which are often the same at each step. These are tasks that are well suited to a computer and the growth in the importance of the method can be linked to the rapid development of these machines. The arrival of the MANIAC computer at Los Alamos in 1952 prompted the study of the many-body problem by Metropolis *et al.* [1953] and the development of the Metropolis Monte Carlo method [Wood 1986], which is the subject of this chapter.

As always, there are those who cannot wait for technology. Buffon, the eminent eighteenth-century French naturalist, discovered a beautiful theorem in geometrical probability. If a needle of length $l$ is thrown at random onto a set of equally spaced parallel lines, $d$ apart (where $d > l$), the probability of the needle crossing a line is $2l/\pi d$. In 1901, the Italian mathematician Lazzerini performed a simulation by spinning round and dropping a needle 3407 times. He estimated $\pi$ to be 3.1415929 [Pedoe 1958]. We shall use this as an example of a simple Monte Carlo integration in the next section. From this exhausting beginning the method has grown to the point where it is, arguably, 'the most powerful and commonly used technique for analysing complex problems' [Rubinstein 1981].

As outlined in Chapter 2, the Metropolis Monte Carlo method aims to generate a trajectory in phase space which samples from a chosen statistical ensemble. There are several difficulties involved in devising such a prescription and making it work for a system of molecules in a liquid. So we take care to introduce the Monte Carlo method through some simple examples in the following sections.

## 4.2 Monte Carlo integration

### *4.2.1 Hit and miss*

We can illustrate the use of the MC technique as a method of integration by returning to the evaluation of $\pi$. This can be done by finding the area of a circle of unit radius. The circle, centred at the origin and inscribed in a square, is shown in Fig. 4.1.

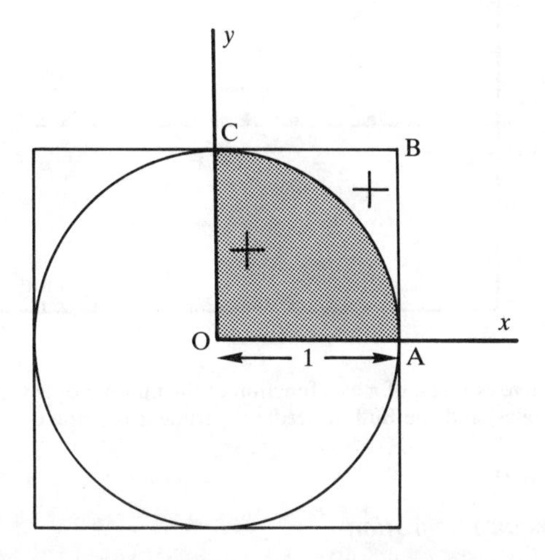


**Fig. 4.1** The geometry for the hit and miss integration to find the area of the circle.

A number of trial shots are generated in the square OABC. At each trial two independent random numbers are chosen from a uniform distribution on (0, 1). These numbers are used as the coordinates of a point, (examples are marked as crosses in the figure). The distance from the random point to the origin is calculated. If this distance is less than or equal to one, the shot has landed in the shaded region and a hit is scored. If a total of $\tau_{\text{shot}}$ shots are fired and $\tau_{\text{hit}}$ hits scored then

$$\pi \approx \frac{4 \times \text{Area under the curve CA}}{\text{Area of the square OABC}} = \frac{4\tau_{\text{hit}}}{\tau_{\text{shot}}}. \tag{4.1}$$

The key to this method is the generation of $2\tau_{\text{shot}}$ random numbers from a uniform distribution. Random number generators are simple programs and their construction and performance are discussed in Appendix G.

The estimate of this area will depend on the numbers of trials; in fact the error in the estimate is $\mathcal{O}(\tau_{\text{shot}}^{-1/2})$. The results from a hit and miss experiment are shown in Fig. 4.2; the correct value for the area of the circle is, of course, $\pi$ and after $10^7$ shots the MC estimate is 3.14173 correct to four figures. To calculate another decimal place would require an order of magnitude increase in the number of shots. It is straightforward to devise a similar hit-and-miss experiment to simulate Buffon's needle. In Fig. 4.2 an estimate of $\pi$ obtained in this way has been included. After $10^7$ shots the result is 3.140472 (only accurate to three figures), confirming that Lazzerini had a lucky afternoon.

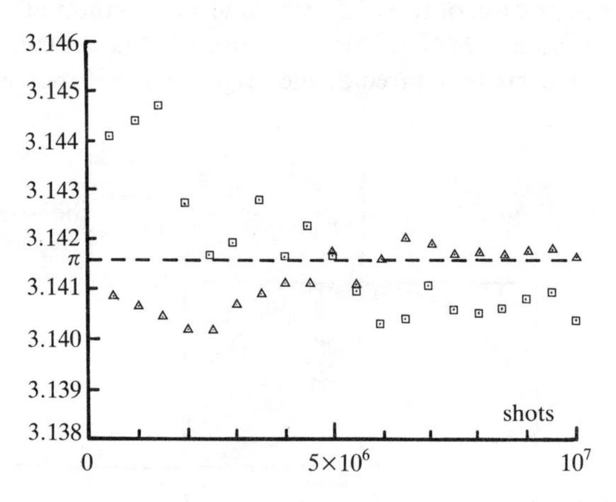


**Fig. 4.2** The cumulative estimate of $\pi$ as a function of the number of MC shots by hit-and-miss area of a circle (triangles) and the Buffon needle experiment (squares).

### 4.2.2 *Sample mean integration*

Hit and miss integration is conceptually easy to understand but the sample mean method is more generally applicable and offers a more accurate estimate for most integrals [Hammersley and Handscomb 1964; Rubinstein 1981]. In this case the integral of interest

$$F = \int_{x_1}^{x_2} \mathrm{d}x\, f(x) \tag{4.2}$$

is rewritten as

$$F = \int_{x_1}^{x_2} \mathrm{d}x \left(\frac{f(x)}{\rho(x)}\right) \rho(x) \tag{4.3}$$

where $\rho(x)$ is an arbitrary probability density function. Consider performing a number of trials $\tau$, each consisting of choosing a random number $\zeta_\tau$ from the distribution $\rho(x)$ in the range $(x_1, x_2)$. Then

$$F = \left\langle \frac{f(\zeta_\tau)}{\rho(\zeta_\tau)} \right\rangle_{\text{trials}} \tag{4.4}$$

where the brackets represent an average over all trials. A simple application would be to choose $\rho(x)$ to be uniform, i.e.

$$\rho(x) = \frac{1}{(x_2 - x_1)} \qquad x_1 \leqslant x \leqslant x_2 \tag{4.5}$$

and then the integral $F$ can be estimated as

$$F \approx \frac{(x_2 - x_1)}{\tau_{\max}} \sum_{\tau=1}^{\tau_{\max}} f(\zeta_\tau). \tag{4.6}$$

To apply this approach to the estimation of $\pi$ we consider the equation for the circle in the first quadrant, $f(x) = (1 - x^2)^{-1/2}$, with $x$ between $x_1 = 0$ and $x_2 = 1$. In a typical experiment the estimate of $\pi$ after $10^7$ trials using eqn (4.6) is 3.14169.

For the simple one-dimensional integration, eqn (4.2), the MC technique is not competitive with straightforward numerical methods such as Simpson's rule (the Simpson's rule estimate of $\pi$ with only $10^4$ function evaluations is 3.141593). However, for the multidimensional integrals of statistical mechanics, the sample mean method, with a suitable choice of $\rho(x)$, is the only sensible solution. To understand this, we consider the evaluation of the configurational integral $Z_{NVT} = \int d\mathbf{r} \exp(-\beta\mathscr{V})$, (eqn (2.26)), for a system of, say, $N = 100$ molecules in a cube of side $L$. Even a crude Simpson's rule integration might require 10 function evaluations for each of the 300 coordinates, so as to span the range $(-\frac{1}{2}L, \frac{1}{2}L)$. This total of $10^{300}$ function evaluations is quite infeasible. Moreover, the overwhelming proportion of these would give a zero result since the Boltzmann factor is extremely small (zero for hard spheres) whenever molecules overlap significantly. The sample mean approach to this integral, using a uniform distribution, might proceed as follows. A trial $\tau$ is carried out:

(a) pick a point at random in the 300-dimensional configuration space, by generating 300 random numbers, on $(-\frac{1}{2}L, \frac{1}{2}L)$, which, taken in triplets, specify the coordinates of each molecule;
(b) calculate the potential energy, $\mathscr{V}(\tau)$, and hence the Boltzmann factor for this configuration.

This procedure is repeated for many trials and the configurational integral is estimated using

$$Z_{NVT} \approx \frac{V^N}{\tau_{\max}} \sum_{\tau=1}^{\tau_{\max}} \exp(-\beta\mathscr{V}(\tau)). \tag{4.7}$$

In principle, the number of trials $\tau_{\max}$ may be increased until $Z_{NVT}$ is estimated to the desired accuracy. We would not expect to have to conduct $10^{300}$ function evaluations, as for Simpson's rule, but again a large number of the trials would give a very small contribution to the average. An accurate estimation of $Z_{NVT}$ for a dense liquid using a uniform sample mean method is beyond the capabilities of current computers, although methods of this type have been used to examine the structural properties of the hard sphere fluid at low densities [Alder, Frankel, and Lewinson 1955]. The difficulties in the calculation of $Z_{NVT}$ apply equally to the calculation of ensemble averages such as

$$\langle \mathcal{A} \rangle_{NVT} = \frac{\int \mathrm{d}\mathbf{r} \mathcal{A} \exp(-\beta \mathcal{V})}{\int \mathrm{d}\mathbf{r} \exp(-\beta \mathcal{V})} \approx \frac{\sum_{\tau=1}^{\tau_{\max}} \mathcal{A}(\tau) \exp(-\beta \mathcal{V}(\tau))}{\sum_{\tau=1}^{\tau_{\max}} \exp(-\beta \mathcal{V}(\tau))}, \tag{4.8}$$

if we attempt to estimate the numerator and denominator separately by using the uniform sample mean method. However, at realistic liquid densities the problem can be solved using a sample mean integration where the random coordinates are chosen from a non-uniform distribution. This method of 'importance sampling' is discussed in the next section.

## 4.3 Importance sampling

Importance sampling techniques choose random numbers from a distribution $\rho(x)$, which allows the function evaluation to be concentrated in the regions of space that make important contributions to the integral. Consider the canonical ensemble. In this case the desired integral is

$$\langle \mathcal{A} \rangle_{NVT} = \int \mathrm{d}\boldsymbol{\Gamma} \rho_{NVT}(\boldsymbol{\Gamma}) \mathcal{A}(\boldsymbol{\Gamma})$$

i.e. the integrand is $f = \rho_{NVT} \mathcal{A}$. By sampling configurations at random, from a chosen distribution $\rho$ we can estimate the integral as

$$\langle \mathcal{A} \rangle_{NVT} = \langle \mathcal{A} \rho_{NVT} / \rho \rangle_{\text{trials}}. \tag{4.9}$$

For most functions $\mathcal{A}(\boldsymbol{\Gamma})$, the integrand will be significant where $\rho_{NVT}$ is significant. In these cases choosing $\rho = \rho_{NVT}$ should give a good estimate of the integral. In this case

$$\langle \mathcal{A} \rangle_{NVT} = \langle \mathcal{A} \rangle_{\text{trials}}. \tag{4.10}$$

(This is not always true and sometimes we choose alternative distributions $\rho(\boldsymbol{\Gamma})$ (see Section 7.2.2).)

Such a method, with $\rho = \rho_{NVT}$, was originally developed by Metropolis *et al.* [1953]. The problem is not solved, simply rephrased. The difficult job is finding a method of generating a sequence of random states so that by the end of the simulation each state has occurred with the appropriate probability. It turns out that it is possible to do this without ever calculating the normalizing factor for $\rho_{NVT}$, i.e. the partition function (see eqns (2.11)–(2.13)).

The solution is to set up a Markov chain of states of the liquid, which is constructed so that it has a limiting distribution of $\rho_{NVT}$. A Markov chain is a sequence of trials that satisfies two conditions:

(a) The outcome of each trial belongs to a finite set of outcomes, $\{\mathbf{\Gamma}_1, \mathbf{\Gamma}_2, \ldots \mathbf{\Gamma}_m, \mathbf{\Gamma}_n, \ldots\}$, called the state space.
(b) The outcome of each trial depends only on the outcome of the trial that immediately precedes it.

Two states $\mathbf{\Gamma}_m$ and $\mathbf{\Gamma}_n$ are linked by a transition probability $\pi_{mn}$, which is the probability of going from state $m$ to state $n$. The properties of a Markov chain are best illustrated with a simple example. Suppose the reliability of your mainframe computer follows a certain pattern. If it is up and running on one day it has a 60 per cent chance of running correctly on the next. If, however, it is down, it has a 70 per cent chance of also being down the next day. The state space has two components, up (↑) and down (↓), and the transition matrix has the form

$$\pi = \begin{matrix} \\ \uparrow \\ \downarrow \end{matrix} \begin{matrix} \uparrow & \downarrow \\ \end{matrix} \begin{pmatrix} 0.6 & 0.4 \\ 0.3 & 0.7 \end{pmatrix}. \tag{4.11}$$

If the computer is equally likely to be up or down to begin with, then the initial probability can be represented as a vector, which has the dimensions of the state space

$$\boldsymbol{\rho}^{(1)} = \overset{\uparrow}{(0.5} \quad \overset{\downarrow}{0.5)}\,. \tag{4.12}$$

The probability that the computer is up on the second day is given by the matrix equation

$$\boldsymbol{\rho}^{(2)} = \boldsymbol{\rho}^{(1)}\boldsymbol{\pi} = (0.45,\ 0.55) \tag{4.13}$$

i.e. there is a 45 per cent chance of running a program. The next day would give

$$\boldsymbol{\rho}^{(3)} = \boldsymbol{\rho}^{(2)}\boldsymbol{\pi} = \boldsymbol{\rho}^{(1)}\boldsymbol{\pi}\boldsymbol{\pi} = \boldsymbol{\rho}^{(1)}\boldsymbol{\pi}^2 = (0.435,\ 0.565), \tag{4.14}$$

and a 43.5 per cent chance of success. If you are anxious to calculate your chances in the long run, then the limiting distribution is given by

$$\boldsymbol{\rho} = \lim_{\tau \to \infty} \boldsymbol{\rho}^{(1)}\boldsymbol{\pi}^{\tau}\,. \tag{4.15}$$

A few applications of eqn (4.15) show that the result converges to $\boldsymbol{\rho} = (0.4286,\ 0.5714)$. It is clear from eqn (4.15) that the limiting distribution, $\boldsymbol{\rho}$, must satisfy the eigenvalue equation

$$\boldsymbol{\rho}\boldsymbol{\pi} = \boldsymbol{\rho} \tag{4.16a}$$

$$\sum_m \rho_m \pi_{mn} = \rho_n \tag{4.16b}$$

with eigenvalue unity. $\boldsymbol{\pi}$ is termed a stochastic matrix since its rows add to one

$$\sum_n \pi_{mn} = 1\,. \tag{4.17}$$

It is the transition matrix for an irreducible Markov chain. (An irreducible or ergodic chain is one where every state can eventually be reached from another state.) More formally, we note that the Perron–Frobenius theorem [Chung 1960; Feller 1957] states that an irreducible stochastic matrix has one left eigenvalue which equals unity, and the corresponding eigenvector is the limiting distribution of the chain. The other eigenvalues are less than unity and they govern the rate of convergence of the Markov chain. The limiting distribution, $\boldsymbol{\rho}$ implied by the chain is quite independent of the initial condition $\boldsymbol{\rho}^{(1)}$ (so don't worry if your machine is likely to be down today).

In the case of a liquid, we must construct a much larger transition matrix, which is stochastic and ergodic (see Chapter 2). In contrast to the previous problem, the elements of the transition matrix are unknown, but the limiting distribution of the chain is the vector with elements $\rho_m = \rho_{NVT}(\boldsymbol{\Gamma}_m)$ for each point $\boldsymbol{\Gamma}_m$ in phase space. It is possible to determine elements of $\boldsymbol{\pi}$ which satisfy eqns (4.16) and (4.17) and thereby generate a phase space trajectory in the canonical ensemble. We have considerable freedom in finding an appropriate transition matrix, with the crucial constraint that the elements of the matrix should be independent of $Q_{NVT}$. A useful trick in searching for a solution of eqn (4.16) is to replace it by the unnecessarily strong condition of 'microscopic reversibility':

$$\rho_m \pi_{mn} = \rho_n \pi_{nm} . \tag{4.18}$$

Summing over all states $m$ and making use of eqn (4.17) we regain eqn (4.16)

$$\sum_m \rho_m \pi_{mn} = \sum_m \rho_n \pi_{nm} = \rho_n \sum_m \pi_{nm} = \rho_n . \tag{4.19}$$

A suitable scheme for constructing a phase space trajectory in the canonical ensemble involves choosing a transition matrix which satisfies eqns (4.17) and (4.18). The first such scheme was suggested by Metropolis *et al.* [1953] and is often known as the asymmetrical solution. If the states $m$ and $n$ are distinct, this solution considers two cases

$$\pi_{mn} = \alpha_{mn} \qquad \rho_n \geqslant \rho_m \qquad m \neq n \tag{4.20a}$$

$$\pi_{mn} = \alpha_{mn}(\rho_n/\rho_m) \qquad \rho_n < \rho_m \qquad m \neq n . \tag{4.20b}$$

It is also important to allow for the possibility that the liquid remains in the same state,

$$\pi_{mm} = 1 - \sum_{n \neq m} \pi_{mn} . \tag{4.20c}$$

In this solution $\boldsymbol{\alpha}$ is a symmetrical stochastic matrix, ($\alpha_{mn} = \alpha_{nm}$), often called the underlying matrix of the Markov chain. The symmetric properties of $\boldsymbol{\alpha}$ can be used to show that for the three cases ($\rho_m = \rho_n$, $\rho_m < \rho_n$, and $\rho_m > \rho_n$) the transition matrix defined in eqn (4.20) satisfies eqns (4.17) and (4.18). It is worth stressing that it is the symmetric property of $\alpha$ that is essential in satisfying microscopic reversibility in this case. Non-symmetrical $\alpha$ matrices which satisfy microscopic reversibility or just the weaker condition, eqn (4.16), can be constructed but these are not part of the basic Metropolis recipe [Owicki and Scheraga 1977a]. Finally, this solution only involves the ratio $\rho_n/\rho_m$ and is therefore independent of $Q_{NVT}$.

There are other solutions to eqns (4.17) and (4.18). The symmetrical solution [Wood and Jacobson 1959; Flinn and McManus 1961; Barker 1965], is often referred to as Barker sampling:

$$\pi_{mn} = \alpha_{mn}\rho_n/(\rho_n + \rho_m) \quad m \neq n \tag{4.21a}$$

$$\pi_{mm} = 1 - \sum_{n \neq m} \pi_{mn} . \tag{4.21b}$$

Equation (4.21) also satisfies the condition of microscopic reversibility.

If states of the fluid are generated using transition matrices such as eqns (4.20) and (4.21), then a particular property, $\langle \mathscr{A} \rangle_{\text{run}}$, obtained by averaging over the $\tau_{\text{run}}$ trials in the Markov chain is related to the average in the canonical ensemble [Chung 1960, p. 99; Wood 1968a]

$$\langle \mathscr{A} \rangle_{NVT} = \langle \mathscr{A} \rangle_{\text{run}} + \mathscr{O}(\tau_{\text{run}}^{-1/2}) \tag{4.22}$$

As mentioned in Chapter 2, we usually restrict simulations to the configurational part of phase space, calculate average configurational properties of the fluid, and add the ideal gas parts after the simulation.

Since there are a number of suitable transition matrices, it is useful to choose a particular solution which minimizes the variance in the estimate of $\langle \mathscr{A} \rangle_{\text{run}}$. Suitable prescriptions for defining the variance in the mean, $\sigma^2(\langle \mathscr{A} \rangle_{\text{run}})$, are discussed in Chapter 7. In particular the 'statistical inefficiency' $s$ (Section 6.4.1)

$$s = \lim_{\tau_{\text{run}} \to \infty} \tau_{\text{run}} \sigma^2 (\langle \mathscr{A} \rangle_{\text{run}})/\sigma^2(\mathscr{A}) \tag{4.23}$$

measures how slowly a run converges to its limiting value. Peskun [1973] has shown that it is reasonable to order two transition matrices,

$$\boldsymbol{\pi}_1 \leqslant \boldsymbol{\pi}_2 \tag{4.24}$$

if each off-diagonal element of $\boldsymbol{\pi}_1$ is less than the corresponding element in $\boldsymbol{\pi}_2$. If this is the case, then

$$s(\langle \mathscr{A} \rangle, \boldsymbol{\pi}_1) \geqslant s(\langle \mathscr{A} \rangle, \boldsymbol{\pi}_2) \tag{4.25}$$

for any property $\mathscr{A}$. If the off-diagonal elements of $\boldsymbol{\pi}$ are large then the probability of remaining in the same state is small and the sampling of phase space will be improved. With the restriction that $\rho_m$ and $\rho_n$ are positive, eqns (4.20) and (4.21) show that the Metropolis solution leads to a lower statistical inefficiency of the mean than the Barker solution.

Valleau and Whittington [1977a] stress that a low statistical inefficiency is not the only criterion for choosing a particular $\pi$. Since the simulations are of finite length, it is essential that the Markov chain samples a representative portion of phase space in a reasonable number of moves. All the results derived in this section depend on the ergodicity of the chain (i.e. that there is some non-zero multi-step transition probability of moving between any two allowed states of the fluid). If these allowed states are not connected the MC run may produce a low $s$ but in addition a poor estimate of the canonical average. When the path between two allowed regions of phase space is difficult to find, the situation is described as a bottleneck (see Fig. 2.1). These bottlenecks are always a worry in MC simulations but are particularly troublesome in the simulation of two-phase coexistence [Lee *et al.* 1974], in the simulation of phase transitions [Evans, Tildesley, and Sluckin 1984], and in simulations of ordinary liquids at unusually high density.

Where a comparison has been made between the two common solutions to the transition matrix, eqns (4.20) and (4.21), the Metropolis solution appears to lead to a faster convergence of the chain [Valleau and Whittington 1977b]. The Metropolis method becomes more favourable as the number of available states at a given step increases and as the energy difference between the states increases. (For two-state problems such as the Ising model the symmetric algorithm may be favourable [Cunningham and Meijer 1976]). In the next section we describe the implementation of the asymmetric solution.

## 4.4 The Metropolis method

To implement the Metropolis solution to the transition matrix, it is necessary to specify the underlying stochastic matrix $\boldsymbol{\alpha}$. This matrix is designed to take the system from state $m$ into any one of its neighbouring states $n$ with equal probability. There is considerable freedom in choosing $\boldsymbol{\alpha}$ and the only constraint is that $\alpha_{mn} = \alpha_{nm}$. A useful but arbitrary definition of a neighbouring state is illustrated in Fig. 4.3. This diagram shows six atoms in a state $m$; to construct a neighbouring state $n$ one atom ($i$) is chosen at random and displaced from its position $\mathbf{r}_i^m$ with equal probability to any point $\mathbf{r}_i^n$ inside the square $\mathscr{R}$. This square is of side $2\delta r_{\max}$ and is centred at $\mathbf{r}_i^m$. In a three-dimensional example, $\mathscr{R}$ would be a small cube. On the computer there are a large but finite number of new positions, $N_{\mathscr{R}}$, for the atom $i$ and in this case $\alpha_{mn}$ can be simply defined as

$$\begin{aligned} \alpha_{mn} &= 1/N_{\mathscr{R}} \qquad && \mathbf{r}_i^n \in \mathscr{R} \\ \alpha_{mn} &= 0 && \mathbf{r}_i^n \notin \mathscr{R}. \end{aligned} \tag{4.26}$$

With this choice of $\boldsymbol{\alpha}$, eqn (4.20) is readily implemented. At the beginning of an MC move an atom is picked at random and given a uniform random displacement along each of the coordinate directions. The maximum displace-

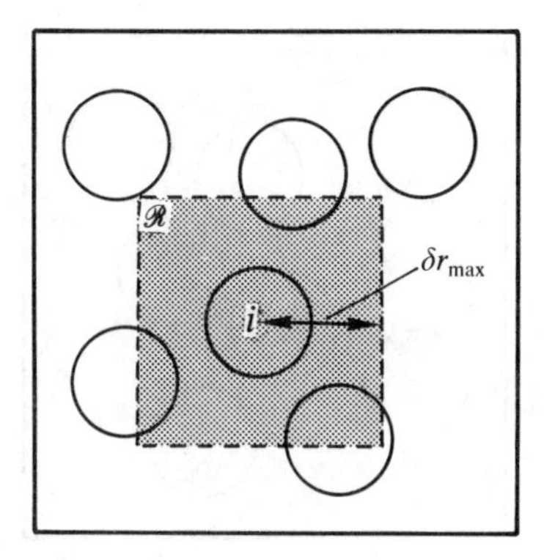


**Fig. 4.3** State $n$ is obtained from state $m$ by moving atom $i$ with a uniform probability to any point in the shaded region $\mathscr{R}$.

ment, $\delta r_{\max}$, is an adjustable parameter that governs the size of the region $\mathscr{R}$ and controls the convergence of the Markov chain. The new position is obtained with the following code. RANF(DUMMY) is a library function for generating a uniform random number on (0, 1); a dummy argument is required by FORTRAN-77 syntax. DRMAX is the maximum displacement $\delta r_{\max}$.

```
RXINEW = RX(I) + ( 2.0 * RANF ( DUMMY ) - 1.0 ) * DRMAX
RYINEW = RY(I) + ( 2.0 * RANF ( DUMMY ) - 1.0 ) * DRMAX
RZINEW = RZ(I) + ( 2.0 * RANF ( DUMMY ) - 1.0 ) * DRMAX
```

The appropriate element of the transition matrix depends on the relative probabilities of the initial state $m$ and the final state $n$. There are two cases to consider. If $\delta \mathscr{V}_{nm} = \mathscr{V}_n - \mathscr{V}_m \leqslant 0$ then $\rho_n \geqslant \rho_m$ and eqn (4.20a) applies. If $\delta \mathscr{V}_{nm} > 0$ then $\rho_n < \rho_m$ and eqn (4.20b) applies. (The symbol $\mathscr{V}_m$ is used as a shorthand for $\mathscr{V}(\mathbf{\Gamma}_m)$.) The next step in an MC move is to determine $\delta \mathscr{V}_{nm}$. The determination of $\delta \mathscr{V}_{nm}$ does not require a complete recalculation of the configurational energy of the $m$th state, just the changes associated with the moving atom. For example (see Fig. 4.4) the change in potential energy is calculated by computing the energy of atom $i$ with all the other atoms before and after the move

$$\delta \mathscr{V}_{nm} = \left( \sum_{j=1}^{N} v(r_{ij}^{n}) - \sum_{j=1}^{N} v(r_{ij}^{m}) \right) \tag{4.27}$$

where the sum over the atoms excludes atom $i$. In calculating the change of energy, the explicit interaction of atom $i$ with all its neighbours out to a cutoff distance $r_c$ is considered. The contribution from atoms beyond the cutoff could be estimated using a mean field correction (see Section 2.8), but in fact the correction for atom $i$ in the old and new positions is exactly the same and does not need to be included explicitly in the calculation of $\delta \mathscr{V}_{nm}$.

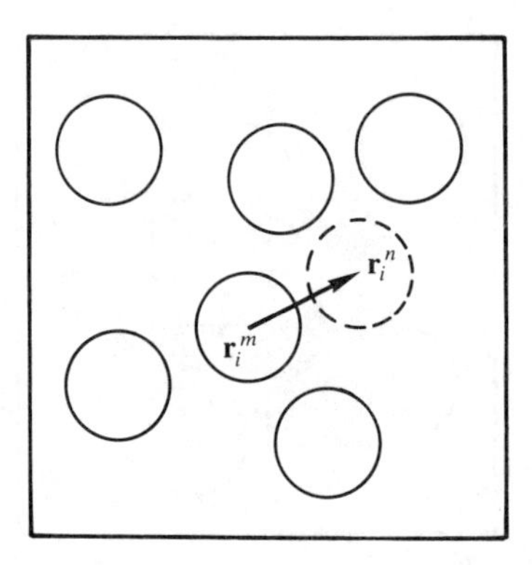


**Fig. 4.4** State $n$ is generated from state $m$ by displacing atom $i$ from $\mathbf{r}_i^m$ to $\mathbf{r}_i^n$.

If the move is downhill in energy ($\delta\mathcal{V}_{nm} \leqslant 0$), then the probability of state $n$ is greater than state $m$ and the new configuration is accepted. The method of choosing trial moves ensures that the transition probability $\pi_{mn} = \alpha_{mn}$, the value required by eqn (4.20a).

If the move is uphill in energy ($\delta\mathcal{V}_{nm} > 0$), then the move is accepted with a probability $\rho_n/\rho_m$ according to eqn (4.20b). Again the factor $\alpha_{mn}$ is automatically included in making the move. This ratio can be readily expressed as the Boltzmann factor of the energy difference:

$$\frac{\rho_n}{\rho_m} = \frac{Z_{NVT}^{-1}\exp(-\beta\mathcal{V}_n)}{Z_{NVT}^{-1}\exp(-\beta\mathcal{V}_m)} = \frac{\exp(-\beta\mathcal{V}_n)\exp(-\beta\delta\mathcal{V}_{nm})}{\exp(-\beta\mathcal{V}_n)} = \exp(-\beta\delta\mathcal{V}_{nm}). \tag{4.28}$$

To accept a move with a probability of $\exp(-\beta\delta\mathcal{V}_{nm})$, a random number $\xi$ is generated uniformly on $(0, 1)$. The random number is compared with $\exp(-\beta\delta\mathcal{V}_{nm})$. If it is less than $\exp(-\beta\delta\mathcal{V}_{nm})$ the move is accepted. This procedure is illustrated in Fig. 4.5. During the run, suppose that a particular uphill move,

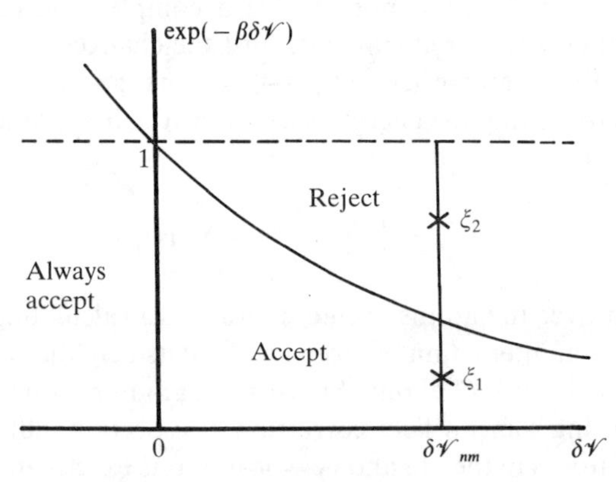


**Fig. 4.5** Accepting uphill moves in the MC simulation.

$\delta\mathscr{V}_{nm}$, is attempted. If at that point a random number $\xi_1$ is chosen (see Fig. 4.5), the move is accepted. If $\xi_2$ is chosen the move is rejected. Over the course of the run the net result is that energy changes such as $\delta\mathscr{V}_{nm}$ are accepted with a probability $\exp(-\beta\delta\mathscr{V}_{nm})$. If the uphill move is rejected, the system remains in state $m$ in accord with the finite probability $\pi_{mm}$ of eqn (4.20c). In this case, the atom is retained at its old position and the old configuration is recounted as a new state in the chain. This procedure can be summarized by noting that we accept any move (uphill or downhill) with probability $\min(1, \exp(-\beta\delta\mathscr{V}_{nm}))$, where min has the same meaning as the FORTRAN MIN function.

A complete MC program for a fluid of Lennard-Jones atoms is given in F.11. Here, we show the typical code for the heart of the program, the acceptance and rejection of moves. In this code, DELTV is the energy difference $\delta\mathscr{V}_{nm}$ between the states. One point to note is that we must guard against a trial move which results in significant molecular overlap, since a very large value of $\delta\mathscr{V}_{nm}$ might cause underflow problems in the computation of $\exp(-\beta\delta\mathscr{V}_{nm})$. We do this by testing $\beta\delta\mathscr{V}_{nm}$ (DELTVB below). If it is too large (say $> 75$) then the move is immediately rejected. This also results in a saving of time, since exponentiation is usually an expensive operation.

```
DELTV  = VNEW - VOLD
DELTVB = BETA * DELTV

IF ( DELTVB .LT. 75.0 ) THEN

   IF ( DELTVB .LE. 0.0 ) THEN

      V      = V + DELTV
      RX(I)  = RXINEW
      RY(I)  = RYINEW
      RZ(I)  = RZINEW
      NACCPT = NACCPT + 1

   ELSEIF ( EXP ( - DELTVB ) .GT. RANF ( DUMMY ) ) THEN

      V      = V + DELTV
      RX(I)  = RXINEW
      RY(I)  = RYINEW
      RZ(I)  = RZINEW
      NACCPT = NACCPT + 1

   ENDIF

ENDIF

NTRIAL = NTRIAL + 1

... accumulate averages ...
```

So far we have said little about the maximum allowed displacement of the atom, $\delta r_{max}$, which governs the size of the trial move. If this parameter is too small then a large fraction of moves are accepted but the phase space of the liquid is explored slowly, i.e. consecutive states are highly correlated. If $\delta r_{max}$ is too large then nearly all the trial moves are rejected and again there is little movement through phase space. In fact $\delta r_{max}$ is often adjusted during the simulation so that about half the trial moves are rejected. This adjustment can be handled automatically using the following code, which adjusts the maximum displacement every NADJST trial moves.

```
IF ( MOD ( NTRIAL, NADJST ) .EQ. 0 ) THEN

   RATIO = REAL ( NACCPT ) / REAL ( NADJST )

   IF ( RATIO .GT. 0.5 )  THEN

      DRMAX = DRMAX * 1.05

   ELSE

      DRMAX = DRMAX * 0.95

   ENDIF

   NACCPT = 0

ENDIF
```

It is not clear that an acceptance ratio of 0.5 is optimum. A reported study of the parameter $\delta r_{max}$ [Wood and Jacobson 1959] suggests that an acceptance ratio of only 0.1 maximizes the root mean square displacement of atoms as a function of computer time. The root mean square displacement is one possible measure of the movement through phase space and the work suggests that a small number of large moves is most cost effective. Few simulators would have the courage to reject nine out of ten moves on this limited evidence and an acceptance ratio of 0.5 is still common. This issue highlights a difficulty in assessing particular simulation methods. The work of Wood and Jacobson was performed on 32 hard spheres, at a particular packing fraction, on a first generation computer. There is no reason to believe that their results would be the same for a different potential, at a different state point on a different machine. The MC technique is time-consuming and since most researchers are more interested in new results rather than methodology there has been little work on the optimization of parameters such as $\delta r_{max}$ and the choice of transition matrix.

In the original Metropolis method one randomly chosen atom is moved to generate a new state. The underlying stochastic matrix can be changed so that several or all of the atoms are moved simultaneously [Ree 1970; Ceperley, Chester, and Kalos 1977]. $\delta \mathscr{V}_{nm}$ is calculated using a straightforward extension of eqn (4.27) and the move is accepted or rejected using the normal criteria. Chapman and Quirke [1985] have performed a simulation of 32 Lennard-Jones atoms at a typical liquid density and temperature. In this study, all 32 atoms were moved simultaneously, and an acceptance ratio of $\approx$ 30 per cent was obtained using $\delta r_{max} \approx 0.3\sigma$. Chapman and Quirke found that equilibration (see Chapter 5) was achieved more rapidly by employing multi-particle moves rather than single-particle moves. The relative efficiency of multi-particle and single-particle moves, as measured by their ability to sample phase space in a given amount of computer time, has not been subjected to systematic study.

A common practice in MC simulation is to select the atoms to move sequentially (i.e. in order of atom index) rather than randomly. This cuts down on the amount of random number generation and is an equally valid method of generating the correctly weighted states [Hastings 1970]. The length of a MC simulation is conveniently measured in 'cycles', i.e. $N$ trial moves whether selected sequentially or randomly. The computer time involved in a MC cycle is comparable (although obviously not equivalent) to that in a MD time step.

The simulation of hard spheres is particularly easy using the MC method. The same Metropolis procedure is used, except that, in this case, the overlap of two spheres results in an infinite positive energy change and $\exp(-\beta\delta\mathscr{V}_{nm}) = 0$. All trial moves involving an overlap are immediately rejected since $\exp(-\beta\delta\mathscr{V}_{nm})$ would be smaller than any random number generated on $(0, 1)$. Equally all moves that do not involve overlap are immediately accepted. As before in the case of a rejection the old configuration is recounted in the average.

The importance sampling technique only generates states that make a substantial contribution to ensemble averages such as the energy. In practice we cannot sum over all the possible states of the fluid and so cannot calculate $Z_{NVT}$. Consequently, this is not a direct route to the 'statistical' properties of the fluid such as $A$, $S$, and $\mu$. In the canonical ensemble there are a number of ways around this problem, such as thermodynamic integration and the particle insertion methods (see Section 2.4). It is also possible to use umbrella sampling to calculate free energy differences (see Chapter 7). Alternatively the problem can be tackled at root by conducting simulations in the grand canonical ensemble (Section 4.6).

## 4.5 Isothermal–isobaric Monte Carlo

An advantage of the MC method is that it can be readily adapted to the calculation of averages in any ensemble. Wood [1968a, b; 1970] first showed that the method could be extended to the isothermal–isobaric ensemble. This ensemble was introduced in Section 2.2, and in designing a simulation method we should recall that the number of molecules, the temperature, and the pressure are fixed while the volume of the simulation box is allowed to fluctuate. The original constant-$NPT$ simulations were performed on hard spheres and disks, but McDonald [1969, 1972] extended the technique to cover continuous potentials in his study of Lennard-Jones mixtures. This ensemble was thought to be particularly appropriate for simulating mixtures since experimental measurements of excess properties are recorded at constant pressure and theories of mixing are often formulated with this assumption. The method has also been used in the simulation of single-component fluids [Vorontstov-Vel'yaminov, El'y-Ashevich, Morgenshtern, and Chakovskikh 1970] and in the study of phase transitions [Abraham 1982]. It is worth recalling that at constant $N$, $P$, $T$ we should not see two phases coexisting in the same simulation cell, a problem which bedevils the simulation of phase transitions in the canonical ensemble.

In the constant-$NPT$ ensemble the configurational average of a property $\mathcal{A}$ is given by

$$\langle \mathcal{A} \rangle_{NPT} = \frac{\int_0^{\infty} \mathrm{d}V \exp(-\beta PV)\, V^N \int \mathrm{d}\mathbf{s}\, \mathcal{A}(\mathbf{s}) \exp(-\beta \mathcal{V}(\mathbf{s}))}{Z_{NPT}}. \tag{4.29}$$

In eqn (4.29), $Z_{NPT}$ is the appropriate configurational integral eqn. (2.30) and $V$ is the volume of the fluid. Note that in this equation we use a set of scaled coordinates $\mathbf{s} = (\mathbf{s}_1, \mathbf{s}_2, \ldots, \mathbf{s}_N)$ where

$$\mathbf{s} = L^{-1}\mathbf{r}. \tag{4.30}$$

In this case the configurational integral in eqn (4.29) is over the unit cube and the additional factor of $V^N$ comes from the volume element d$\mathbf{r}$. (In this section the simulation box is assumed to be a cube of side $L = V^{1/3}$; the arguments can be easily extended to non-cubic boxes.)

The Metropolis scheme is implemented by generating a Markov chain of states which has a limiting distribution proportional to

$$\exp(-\beta(PV + \mathcal{V}(\mathbf{s})) + N \ln V)$$

and the method used is a direct extension of the ideas discussed in Section 4.4.

A new state is generated by displacing a molecule randomly and/or making a random volume change from $V_m$ to $V_n$

$$\begin{aligned} \mathbf{s}_i^n &= \mathbf{s}_i^m + \delta s_{\max} (2\boldsymbol{\xi} - \mathbf{1}) \\ V_n &= V_m + \delta V_{\max} (2\xi - 1). \end{aligned} \tag{4.31}$$

Here, as usual, $\xi$ is a random number generated uniformly on (0, 1), while $\boldsymbol{\xi}$ is a vector whose components are also uniform random numbers on (0, 1) and $\mathbf{1}$ is the vector (1, 1, 1). $\delta s_{\max}$ and $\delta V_{\max}$ govern the maximum changes in the scaled coordinates of the particles, and in the volume of the simulation box, respectively. Their precise values will depend on the state point studied and they are chosen to produce an acceptance ratio of 35–50 per cent [McDonald 1972]. These values are initial guesses and can be automatically adjusted by the program, although in this case there are two independent maximum displacements and many different combinations will produce a given acceptance ratio.

Once the new state $n$ has been produced the quantity $\delta H$ is calculated,

$$\delta H_{nm} = \delta \mathcal{V}_{nm} + P(V_n - V_m) - N\beta^{-1} \ln(V_n/V_m). \tag{4.32}$$

$\delta H_{nm}$ is closely related to the enthalpy change in moving from state $m$ to state $n$. Moves are accepted with a probability equal to $\min(1, \exp(-\beta\delta H_{nm}))$ using the techniques discussed in Section 4.4. A move may proceed with a change in particle position or a change in volume or a combination of both.

Eppenga and Frenkel [1984] have pointed out that it may be more convenient to make random changes in $\ln V$ rather than in $V$ itself. A random number $\delta(\ln V)$ is chosen uniformly in some range $(-\delta(\ln V)_{\max}, \delta(\ln V)_{\max})$, the volume multiplied by $\exp(\delta(\ln V))$ and the molecular positions scaled accordingly. The only change to the acceptance/rejection procedure is that the factor $N$ in eqn (4.32) is replaced by $N+1$.

One important difference between this ensemble and the canonical ensemble is that when a move involves a change in volume the density of the liquid changes. In this case the long-range corrections to the energy in states $m$ and $n$ are different and must be included directly in the calculation of $\delta\mathcal{V}_{nm}$ (see Section 2.8).

In the general case, changing the volume is computationally more expensive than displacing a molecule. For a molecule displacement there are at most $2(N-1)$ calculations of the pair potential in calculating $\delta\mathcal{V}_{nm}$. In general, a volume change in a pair-additive fluid requires the recalculation of all the $\frac{1}{2}N(N-1)$ interactions. Fortunately, for the simplest potentials, the change in $\mathcal{V}$ with volume can be calculated by scaling. As an example, consider the configurational energy of a Lennard-Jones fluid in state $m$:

$$\mathcal{V}_m = 4\varepsilon\sum_i \sum_{j>i}\left(\frac{\sigma}{L_m s_{ij}^m}\right)^{12} - 4\varepsilon\sum_i \sum_{j>i}\left(\frac{\sigma}{L_m s_{ij}^m}\right)^{6}$$

$$= \mathcal{V}_m^{(12)} + \mathcal{V}_m^{(6)}. \tag{4.33}$$

Here we have divided up the potential into its separate twelfth-power and sixth-power components. If the only change between the states $m$ and $n$ is the length of the box then the energy of the new state is

$$\mathcal{V}_n = \mathcal{V}_m^{(12)}\left(\frac{L_m}{L_n}\right)^{12} + \mathcal{V}_m^{(6)}\left(\frac{L_m}{L_n}\right)^{6}$$

and

$$\delta\mathcal{V}_{nm} = \delta\mathcal{V}_{nm}^{\text{vol}} = \mathcal{V}_m^{(12)}\left[\left(\frac{L_m}{L_n}\right)^{12} - 1\right] + \mathcal{V}_m^{(6)}\left[\left(\frac{L_m}{L_n}\right)^{6} - 1\right]. \tag{4.34}$$

This calculation is extremely rapid and only requires that the two components of the potential energy, $\mathcal{V}^{(12)}$ and $\mathcal{V}^{(6)}$, be stored separately. If the potential cutoff is taken to scale with the box length (i.e. $r_c = s_c L$ with $s_c$ constant) then the separate terms $\mathcal{V}_{\text{LRC}}^{(12)}$ and $\mathcal{V}_{\text{LRC}}^{(6)}$ scale just like $\mathcal{V}^{(12)}$ and $\mathcal{V}^{(6)}$ respectively. If in addition to a box-length change a molecule is simultaneously displaced, then there are two contributions

$$\delta\mathcal{V}_{nm} = \delta\mathcal{V}_{nm}^{\text{dis}} + \delta\mathcal{V}_{nm}^{\text{vol}} \tag{4.35}$$

where $\delta\mathscr{V}_{nm}^{\text{vol}}$ is given by eqn (4.34) and

$$\delta\mathscr{V}_{nm}^{\text{dis}} = \mathscr{V}_n(L_n) - \mathscr{V}_m(L_n). \tag{4.36}$$

Thus the energy change on displacement is obtained using the new box-length $L_n$ (think of scaling the box, followed by moving the molecule).

This simple prescription for the calculation of $\delta\mathscr{V}_{nm}$ relies on there being just one characteristic length in the potential function. This may not be the case for some complicated pair potentials, and it is also not true for most molecular models, where intramolecular bond lengths as well as site–site potentials appear. For an interaction site model, simple scaling would imply a non-physical change in the molecular shape. For these cases the calculation of $\delta\mathscr{V}_{nm}^{\text{vol}}$ is expensive and so volume changes must be carried out much less frequently than the displacement of a particle [Owicki and Scheraga 1977b]. In a constant-$NPT$ simulation of 125 $H_2O$ molecules, Jorgensen [1982] attempted to change the volume of the box once every sixth cycle. The range for a possible volume move was $\sim \pm 50$ Å$^3$. The code for a constant-$NPT$ simulation is given in program F.12.

By averaging over the states in the Markov chain it is possible to calculate mechanical properties such as the volume and the enthalpy, and various properties related to their fluctuations. In common with the constant-$NVT$ simulation, this method only samples important regions of phase space and it is not possible to calculate the 'statistical' properties such as the Gibbs free energy. During the course of a particular run the virial can be calculated in the usual manner to produce an estimate of the pressure. This calculated pressure (including the long-range correction) should be equal to the input pressure, $P$, used in eqn (4.32) to generate the Markov chain. This test is a useful check of a properly coded constant-$NPT$ program.

From the limited evidence available, it appears that the fluctuations of averages calculated in a constant-$NPT$ MC simulation are greater than those associated with the averages in a constant-$NVT$ simulation. However, the error involved in calculating excess properties of mixtures in the two ensembles is comparable, since they can be arrived at more directly in a constant-$NPT$ calculation [McDonald 1972].

Finally, constant-pressure simulations of hard disks and spheres, [Wood 1968b, 1970], can be readily performed using the methods described in this section. Wood [1968b] has also developed an elegant method for hard-core systems where the integral over $\exp(-\beta PV)$ in eqn (4.29) is used to define a Laplace transform. The simulation is performed by generating a Markov chain in the transform space using a suitably defined pseudo-potential. This method avoids direct scaling of the box; details can be found in the original paper.

## 4.6 Grand canonical Monte Carlo

In grand canonical ensemble MC (GCMC) the chemical potential is fixed while the number of molecules fluctuates. The simulations are carried out at constant $\mu$, $V$, and $T$, and the average of some property $\mathscr{A}$ is given by

$$\langle \mathscr{A} \rangle_{\mu VT} = \frac{\sum_{N=0}^{\infty} (N!)^{-1} V^N z^N \int \mathrm{d}\mathbf{s}\, \mathscr{A}(\mathbf{s}) \exp(-\beta \mathscr{V}(\mathbf{s}))}{Q_{\mu VT}} \tag{4.37}$$

where $z = \exp(\beta\mu)/\Lambda^3$ is the activity, $\Lambda$ is defined in eqn (2.24) and $Q_{\mu VT}$ in eqn (2.32). Again it is convenient to use a set of scaled coordinates $\mathbf{s} = (\mathbf{s}_1, \mathbf{s}_2, \ldots, \mathbf{s}_N)$ defined as in eqn (4.30) for each particular value of $N$. In common with the other ensembles discussed in this chapter only the configurational properties are calculated during the simulation and the ideal gas contributions are added at the end. A minor complication is that these contributions will depend on $\langle N \rangle_{\mu VT}$, which must be calculated during the run. $N$ is not a continuous variable (the minimum change in $N$ is one), and the sum in eqn (4.37) will not be replaced by an integral.

In GCMC the Markov chain is constructed so that the limiting distribution is proportional to

$$\exp(-\beta(\mathscr{V}(\mathbf{s}) - N\mu) - \ln N! - 3N \ln \Lambda + N \ln V). \tag{4.38}$$

A number of methods of generating this chain have been proposed. A method applied in early studies of lattice systems [Salsburg, Jacobson, Fickett, and Wood 1959; Chesnut 1963], uses a set of variables $(c_1, c_2 \ldots)$, each taking the value 0 (unoccupied) or 1 (occupied), to define a configuration. In the simplest approach a trial move attempts to turn either a 'ghost' site ($c_i = 0$) into a real site ($c_i = 1$) or vice versa.

This method has been extended to continuous fluids by Rowley, Nicholson, and Parsonage [1975] and used more recently by Yao, Greenkorn, and Chao [1982]. In this application real and ghost molecules are moved throughout the system using the normal Metropolis method for displacement. This means that 'ghost' moves are always accepted because no interactions are involved. In addition there are frequent conversion attempts between 'ghost' and real molecules. Unfortunately a 'ghost' molecule tends to remain close to the position at which its real precursor was destroyed, and is likely to re-materialize, at some later step in the simulation, in this same 'hole' in the liquid. This memory effect does not lead to incorrect results [Barker and Henderson 1976], but may result in a slow convergence of the chain. The total number of real and ghost molecules, $M$, must be chosen so that if all the molecules became real $\mathscr{V}$ would be very high for all possible configurations. In this case the sum in eqn (4.37) can be truncated at $M$. This analysis makes it clear that, in GCMC simulations, we are essentially transferring molecules between our system of interest and an ideal gas system, each of which is limited to a maximum of $M$ molecules. Thus the system properties are measured relative to those of this restricted ideal gas; if $M$ is sufficiently large this should not matter.

Most workers now adopt the original method of Norman and Filinov [1969]. In this technique there are three different types of move:

(a) a molecule is displaced;
(b) a molecule is destroyed (no record of its position is kept);
(c) a molecule is created at a random position in the fluid.

Displacement is handled using the normal Metropolis method. If a molecule is destroyed the ratio of the probabilities of the two states is ($N$ is the number of molecules initially in state $m$)

$$\frac{\rho_n}{\rho_m} = \exp(-\beta\delta\mathscr{V}_{nm})\exp(-\beta\mu)\frac{N\Lambda^3}{V}, \tag{4.39}$$

which in terms of the activity is

$$\frac{\rho_n}{\rho_m} = \exp(-\beta\delta\mathscr{V}_{nm} + \ln(N/zV)) = \exp(-\beta\delta D_{nm}). \tag{4.40}$$

Here we have defined the 'destruction function' $\delta D_{nm}$. A destruction move is accepted with probability $\min(1, \exp(-\beta\delta D_{nm}))$ using the methods of Section 4.4. Finally, in a creation step, similar arguments give

$$\frac{\rho_n}{\rho_m} = \exp(-\beta\delta\mathscr{V}_{nm} + \ln(zV/N+1)) = \exp(-\beta\delta C_{nm}) \tag{4.41}$$

(defining the 'creation function' $\delta C_{nm}$) and the move is accepted or rejected using the same criteria.

In this scheme there is the danger of using an underlying stochastic matrix which is unsymmetric with respect to creation/destruction. The condition of microscopic reversibility can be satisfied by making the probability of an attempted creation, $\alpha^{c}$, equal to the probability of an attempted destruction, $\alpha^{d}$ [Nicholson and Parsonage 1982, p. 154]. The method outlined allows for the destruction or creation of only one molecule at a time. Except at low densities, moves which involve the addition or removal of more than one molecule would be highly improbable and such changes are not cost effective [Norman and Filinov 1969].

Although $\alpha^{d}$ must equal $\alpha^{c}$ there is some freedom in chosing between creation/destruction and a simple displacement, $\alpha^{m}$. Again Norman and Filinov [1969] varied $\alpha^{m}$ and found that $\alpha^{m} = \alpha^{d} = \alpha^{c} = 1/3$ gave the fastest convergence of the chain, and these are the values commonly employed. Thus moves, destructions, and creations are selected at random, with equal probability.

Typically, the configurational energy, pressure, and density are calculated as ensemble averages during the course of the GCMC simulations. The beauty of this type of simulation is that the free energy can be calculated directly,

$$A/N = \mu - \langle\mathscr{P}\rangle_{\mu VT}V/\langle N\rangle_{\mu VT} \tag{4.42}$$

and using eqn (4.42) it is possible to determine all the 'statistical' properties of the liquid.

Variations on the method described in this section have been employed by a number of workers. The Metropolis method for creation and destruction can be replaced by a symmetrical algorithm. In this case the decisions for creation and destruction are respectively

$$\text{create if} \quad \left(1+\frac{N+1}{zV}\exp\left(\beta\delta\mathcal{V}_{nm}\right)\right)^{-1} \geqslant \xi$$

and

$$\text{destroy if} \quad \left(1+\frac{zV}{N}\exp\left(\beta\delta\mathcal{V}_{nm}\right)\right)^{-1} \geqslant \xi$$

with $\xi$ generated uniformly on (0, 1).

Adams [1974, 1975] has also suggested an alternative formulation which splits the chemical potential into the ideal gas and excess parts:

$$\begin{aligned} \mu &= \mu^{\text{ex}} + \mu^{\text{id}} \\ &= (\mu^{\text{ex}} + kT\ln\langle N\rangle_{\mu VT}) + kT\ln(\Lambda^3/V) \\ &= kTB + kT\ln(\Lambda^3/V). \end{aligned} \tag{4.43}$$

Adams performed the MC simulation at constant $B$, $V$, and $T$, where $B$ is defined by eqn (4.43). $\mu$ can be obtained by calculating $\langle N\rangle_{\mu VT}$ during the run and using it in eqn (4.43). The technique is completely equivalent to the normal method at constant $z$, $V$, and $T$.

There are a number of technical points to be considered in performing GCMC. In common with the constant-$NPT$ ensemble, the density is not constant during the run. In these cases the long-range corrections must be included directly in the calculation of $\delta\mathcal{V}_{nm}$. The corrections should also be applied during the run to other configurational properties such as the virial. If this is not done, difficulties may arise in correcting the pressure at the end of the simulation: this can affect the calculation of the free energy through eqn (4.42) [Barker and Henderson 1976; Rowley, Nicholson, and Parsonage 1978].

A problem which is peculiar to GCMC is that, when molecules are created or destroyed, the array indices which identify the molecule need to be reordered. This problem can be handled neatly using the following technique [Nicholson 1984]. In this simple illustration, we consider a simulation which begins with six molecules and where we expect a maximum of ten. An array LOCATE is the key to which molecules are 'alive' at the current step of the simulation. At the first step LOCATE looks like

| I | 1 | 2 | 3 | 4 | 5 | 6 | 7 | 8 | 9 | 10 |
|---|---|---|---|---|---|---|---|---|---|---|
| LOCATE(I) | 1 | 2 | 3 | 4 | 5 | 6 | 0 | 0 | 0 | 0 |

↑ $N = 6$ (under I = 6)

(a)

If molecule 3 is destroyed the array is updated and 3 is moved to the 'dead' area of the array.

| I | 1 | 2 | 3 | 4 | 5 | 6 | 7 | 8 | 9 | 10 |
|---|---|---|---|---|---|---|---|---|---|---|
| LOCATE(I) | 1 | 2 | 4 | 5 | 6 | 3 | 0 | 0 | 0 | 0 |

↑ $N = 5$ (under I = 5)

(b)

If a new molecule is created it is given the index LOCATE($N + 1$) i.e. the array would remain unchanged but $N$ would be increased by one.

| I | 1 | 2 | 3 | 4 | 5 | 6 | 7 | 8 | 9 | 10 |
|---|---|---|---|---|---|---|---|---|---|---|
| LOCATE(I) | 1 | 2 | 4 | 5 | 6 | 3 | 0 | 0 | 0 | 0 |

↑ $N = 6$ (under I = 6)

(c)

Suppose a second new molecule is created. In this case LOCATE($N + 1$) = 0 so the new molecule index is set to $N + 1$ and $N$ is then increased by one.

| I | 1 | 2 | 3 | 4 | 5 | 6 | 7 | 8 | 9 | 10 |
|---|---|---|---|---|---|---|---|---|---|---|
| LOCATE(I) | 1 | 2 | 4 | 5 | 6 | 3 | 7 | 0 | 0 | 0 |

↑ $N = 7$ (under I = 7)

(d)

At any stage in the program it is easy to search over all the molecules actually present by running over a loop with upper index $N$ as follows:

```
      DO 10 I = 1, N

         IATOM = LOCATE(I)
         RXI   = RX(IATOM)
         RYI   = RY(IATOM)
         RZI   = RZ(IATOM)

         ... calculate energy etc. ...

10    CONTINUE
```

In the Norman and Filinov method, the new molecule 3, which is in LOCATE(6) at the end of step (c), has coordinates RX(3), RY(3), RZ(3), which are chosen at random and which are not related to the original coordinates of molecule 3 at step (a). The code for creation and destruction attempts in GCMC is given in program F.13 and the code for updating and tidying the array LOCATE is given in program F.14.

Grand canonical simulations are more complicated to program than those in the canonical ensemble. The advantage of the method is that it provides a direct route to the 'statistical' properties of the fluid. For example, by determining the free energy of two different solid structures in two independent GCMC simulations we can say which of the two structures is thermodynamically stable at a particular $\mu$ and $T$. GCMC is particularly useful for studying inhomogeneous systems such as monolayer and multilayer adsorption near a surface [Whitehouse, Nicholson, and Parsonage 1983] or the electrical double-layer [Carnie and Torrie 1984; Guldbrand, Jönsson, Wennerström, and Linse 1984]. In these systems the surface often attracts the molecules strongly so that when a molecule diffuses into the vicinity of the surface it may tend to remain there throughout the simulation. GCMC additionally destroys particles in the dense region near the surface and creates them in the dilute region away from the surface. In this way it should encourage efficient sampling of some less likely but allowed regions of phase space as well as helping to break up metastable structures near the surface.

GCMC simulations of fluids have not been used widely. The problem is that as the density of the fluid is increased the probability of successful creation or destruction steps becomes small. Creation attempts fail because of the high risk of overlap. Destruction attempts fail because the removal of a particle without the subsequent relaxation of the liquid structure results in the loss of attractive interactions. Clearly this means that destructions in the vicinity of a surface may be infrequent and this somewhat offsets the advantage of GCMC in the simulation of adsorption [Nicholson 1984]. To address these problems, Mezei [1980] has extended the basic method to search for cavities in the fluid which are of an appropriate size to support a creation. Once these cavities are located, creation attempts are made more frequently in the region of the cavity. In the Lennard-Jones fluid at $T^* = 2.0$, the highest density at which the system could be successfully studied was increased from $\rho^* = 0.65$ (conventional GCMC) to $\rho^* = 0.85$ (extended GCMC). The techniques for preferential sampling close to a molecule or a cavity are discussed in Section 7.3.

## 4.7 Molecular liquids

### *4.7.1 Rigid molecules*

In the MC simulation of a molecular liquid the underlying matrix of the Markov chain is altered to allow moves which usually consist of a combined translation and rotation of one molecule. Chains involving a number of purely translational and purely rotational steps are perfectly proper but are not usually exploited in the simulation of molecular liquids. (There have been a number of simulations of idealized models of liquid crystals and plastic crystals where the centres of the molecules are fixed to a three-dimensional lattice. These simulations consist of purely rotational moves [see e.g. Luckhurst and Simpson 1982; O'Shea 1978].)

The translational part of the move is carried out by randomly displacing the centre of mass of a molecule along each of the space-fixed axes. As before the maximum displacement is governed by the adjustable parameter $\delta r_{\max}$. The orientation of a molecule is often described in terms of the Euler angles defined in Section 3.3.1. A change in orientation can be achieved by taking small random displacements in each of the Euler angles of molecule $i$.

$$\phi_i^n = \phi_i^m + (2\xi_1 - 1)\delta\phi_{\max} \tag{4.44a}$$

$$\theta_i^n = \theta_i^m + (2\xi_2 - 1)\delta\theta_{\max} \tag{4.44b}$$

$$\psi_i^n = \psi_i^m + (2\xi_3 - 1)\delta\psi_{\max} \tag{4.44c}$$

where $\delta\phi_{\max}$, $\delta\theta_{\max}$, and $\delta\psi_{\max}$ are the maximum displacements in the Euler angles.

In an MC step the ratio of the probabilities of the two states is given by

$$\frac{\rho_n}{\rho_m} = \frac{\exp(-\beta(\mathcal{V}_m + \delta\mathcal{V}_{nm}))\mathrm{d}\mathbf{r}^n \mathrm{d}\mathbf{\Omega}^n}{\exp(-\beta\mathcal{V}_m)\mathrm{d}\mathbf{r}^m \mathrm{d}\mathbf{\Omega}^m}. \tag{4.45}$$

The appropriate volume elements have been included to convert the probability densities into probabilities. $\mathrm{d}\mathbf{\Omega}^m = \prod_{i=1}^{N} \mathrm{d}\mathbf{\Omega}_i^m$ and $\mathrm{d}\mathbf{\Omega}_i^m = \sin\theta_i^m \mathrm{d}\theta_i^m \mathrm{d}\psi_i^m \mathrm{d}\phi_i^m/\Omega$ for molecule $i$ in state $m$. $\Omega$ is a constant which is $8\pi^2$ for non-linear molecules. In the case of linear molecules, the angle $\psi$ is not required to define the orientation, and $\Omega = 4\pi$. The volume elements for states $m$ and $n$ have not previously been included in the ratio $\rho_n/\rho_m$ (see eqn (4.28)), for the simple reason that they are the same in both states for a translational move, and cancel. For a move which only involves one molecule $i$

$$\frac{\rho_n}{\rho_m} = \exp(-\beta\delta\mathcal{V}_{nm})\frac{\sin\theta_i^n}{\sin\theta_i^m}. \tag{4.46}$$

The ratio of the sines must appear in the transition matrix $\pi_{mn}$ either in the acceptance/rejection criterion or in the underlying matrix element $\alpha_{mn}$. This last approach is most convenient. It amounts to choosing random displacements in $\cos\theta_i$ rather than in $\theta_i$:

$$\cos\theta_i^n = \cos\theta_i^m + (2\xi_2 - 1)\delta(\cos\theta)_{\max} \tag{4.47}$$

and adopting the usual Metropolis recipe of accepting or rejecting with a probability of min $(1, \exp(-\beta\delta\mathcal{V}_{nm}))$. Including the $\sin\theta$ factor in the underlying chain avoids difficulties with $\theta_i^m = 0$ analogous to the problems mentioned in Section 3.3.1. Equations (4.44a), (4.44c), and (4.47) move a molecule from one orientational state into any one of its neighbouring orientational states with equal probability and fulfil the condition of microscopic reversibility.

It is useful to keep the angles which describe the orientation of a particular molecule in the appropriate range $(-\pi, \pi)$ for $\psi$ and $\phi$, and $(0, \pi)$ for $\theta$. This is not essential, but avoids unnecessary work and possible overflow in the subsequent evaluation of any trigonometric functions. This can be done by a piece of code which is rather like that used to implement periodic boundary conditions. If DPHIMX is the maximum change in $\phi$, and TWOPI stores the value $2\pi$,

```
PHINEW = PHIOLD + ( 2.0 * RANF ( DUMMY ) - 1.0 ) * DPHIMX
PHINEW = PHINEW - ANINT ( PHINEW / TWOPI ) * TWOPI
```

with similar code for $\psi$. In the case of eqn (4.47), it is necessary to keep $\cos\theta$ in the range $(-1, 1)$:

```
COSNEW = COSOLD + ( 2.0 * RANF ( DUMMY ) - 1.0 ) * DCOSMX
COSNEW = COSNEW - ANINT ( COSNEW / 2.0 ) * 2.0
```

Note that when the ANINT function is not zero the molecule is rotated by $\pi$.

An alternative method for rotating the molecules was originally proposed by Barker and Watts [1969] in their MC simulation of water. It involves selecting a molecule and rotating it by a random amount $\delta\gamma$ (selected uniformly in the usual way) about one of the three space-fixed axes chosen at random. For example we consider a fluid of linear molecules (these ideas can be readily extended to non-linear molecules). In this case it is more convenient to represent the molecular orientation by a unit vector $\mathbf{e}$ fixed in the molecule. The orientation of molecule $i$ is represented by a vector with components

$$\begin{aligned} e_{ix} &= \cos\phi_i \sin\theta_i \\ e_{iy} &= \sin\phi_i \sin\theta_i \\ e_{iz} &= \cos\theta_i \,. \end{aligned} \tag{4.48}$$

A new configuration is generated using

$$\mathbf{e}_i^n = \mathbf{A}_x \mathbf{e}_i^m \tag{4.49}$$

where

$$\mathbf{A}_x = \begin{pmatrix} 1 & 0 & 0 \\ 0 & \cos\delta\gamma & \sin\delta\gamma \\ 0 & -\sin\delta\gamma & \cos\delta\gamma \end{pmatrix} \tag{4.50}$$

and the equation corresponds to a rotation of $\delta\gamma$ about the space-fixed $x$ axis. There are similar equations for rotations about the $y$ and $z$ axes,

$$\mathbf{A}_y = \begin{pmatrix} \cos\delta\gamma & 0 & -\sin\delta\gamma \\ 0 & 1 & 0 \\ \sin\delta\gamma & 0 & \cos\delta\gamma \end{pmatrix} \tag{4.51}$$

and

$$\mathbf{A}_z = \begin{pmatrix} \cos\delta\gamma & \sin\delta\gamma & 0 \\ -\sin\delta\gamma & \cos\delta\gamma & 0 \\ 0 & 0 & 1 \end{pmatrix}. \tag{4.52}$$

The advantage of this method is that the orientation of the molecule can be stored as one or more vectors. For interaction site model fluids, this means that the expensive evaluation of trigonometric functions can be avoided completely. This is not the case with methods that involve changes in the Euler angles. Similar formulae apply when quaternions are used to represent molecular orientations [Vesely 1982].

A third method for changing the orientation of a molecule has been suggested by Jansoone [1974]. The new trial orientation, $\mathbf{e}_i^n$, is chosen randomly and uniformly on a region of the surface of a sphere with the constraint that

$$1 - \mathbf{e}_i^n \cdot \mathbf{e}_i^m < d \ll 1. \tag{4.53}$$

$d$ controls the size of the maximum displacement and a sensible first guess for this parameter is 0.2. There are a number of methods for generating a random vector on the surface of a sphere (see Appendix G.3). These can be easily combined with the method for generating randomly and uniformly in a restricted region (see Appendix G.4) to produce a simple algorithm which will generate orientations with the constraint eqn (4.53). Examples of code for these three methods of generating a new orientation are given in program F.15.

One difficulty with MC methods for molecular fluids is that there are usually a number of parameters governing the maximum translational and orientational displacement of a molecule during a move. As usual these parameters can be adjusted automatically to give an acceptance rate of $\approx 0.5$, but there is not a unique set of maximum displacement parameters which will achieve this. A sensible set of values is best obtained by trial and error for the particular simulation in hand.

The MC method is particularly useful for simulating hard-core molecules. The complicated MD schemes mentioned in Section 3.6.2 can be avoided and the program consists simply of choosing one of the above schemes for moving a molecule and an algorithm for checking for overlap. The heart of a simple MC program for hard dumb-bells is given in F.16.

The MC method has been used successfully in the canonical ensemble for simulating hard-core molecules [Streett and Tildesley 1978; Wojcik and Gubbins 1983] and more realistic linear and non-linear molecules [Barker and Watts 1969; Romano and Singer 1979]. Simulations of molecular fluids have also been attempted in the isothermal–isobaric ensemble [Owicki and Scheraga 1977b; Eppenga and Frenkel 1984]. To our knowledge there have been no simulations of molecular liquids in the grand canonical ensemble.

*4.7.2 Non-rigid molecules*

Non-rigidity introduces new difficulties into the MC technique. The problem in this case is to find a suitable set of generalized coordinates to describe the positions and momenta of the molecules. Once the generalized momentum coordinates have been established, the integrations over the momenta can be performed analytically which will leave just the configurational part of the ensemble average. However, the integration over momenta will produce complicated Jacobians in the configurational integral, one for each molecule (see Section 2.10). The Jacobian will be some function of the generalized orientational variables, $\theta$, $\phi$ which describe the overall orientation of the molecule and the bond bending and torsion angles which describe the internal configuration. A simple example of this type of term is the $\sin\theta_i$ in the configurational integral for rigid molecules, which comes from the integration over the momenta $(p_\phi)_i$. As we have already seen in Section 4.7.1, these Jacobians are important in calculating the ratio $\rho_n/\rho_m$ used in generating the Markov chain in the Metropolis method or, correspondingly, in designing the correct underlying stochastic matrix. For non-rigid molecules, correctly handling the Jacobian terms is more difficult.

This problem can be solved satisfactorily for the class of non-rigid molecules where the overall moment of inertia is independent of the coordinates of internal rotation (e.g. *iso*-butane, acetone) [Pitzer and Gwinn 1942]. Generalized coordinates have also been developed for a non-rigid model of butane, which does not fall into this simple class [Ryckaert and Bellemans 1975; Pear and Weiner 1979], but the expressions are complicated and become increasingly so for longer molecules.

One way of working with generalized coordinates is as follows. In butane (see Section 1.3), it is possible to constrain bond lengths and bond bending angles, while allowing the torsional angle to change according to its potential function. The movement of the molecule in the simulation is achieved by random movements of randomly chosen atoms subject to the required constraints [Curro 1974]. An example of such a technique is shown for butane in Fig. 4.6.

A typical MC sequence might be (assuming that each move is accepted):

(a) atom 1 is moved by rotating around the 2–3 bond;
(b) atoms 1 and 2 are moved simultaneously by rotating around the 3–4 bond;
(c) atom 4 is moved by rotating around the 2–3 bond.

Moves (a) and (c) involve a random displacement of the torsional angle $\phi$, in the range $(-\pi, \pi)$. The entire molecule is translated and rotated through space by making random rotations of atoms around randomly chosen bonds. We can also include an explicit translation of the whole molecule, and an overall rotation about one of the space-fixed axes. The disadvantage of this simple

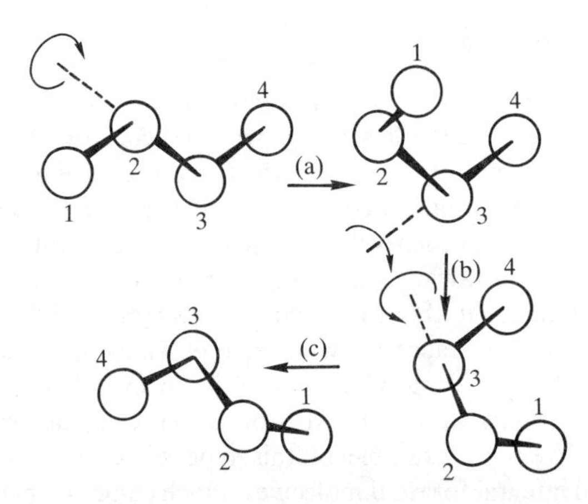


**Fig. 4.6** A possible method for moving a chain molecule (butane), subject to bond length and angle constraints, in an MC simulation.

approach at high density is that a small rotation around the 1–2 bond can cause a substantial movement of atom 4, which is likely to result in overlap and a high rejection rate for new configurations.

If we consider the case of the simplified butane molecule introduced in Sections 1.3.3 and 2.10, then a trial MC move might consist of a translation and rotation of the whole molecule and a change in the internal configuration made by choosing a random increment in $d(\cos\theta)$, $d(\cos\theta')$, and $d\phi$ (see Fig. 1.8). To avoid the artefacts associated with the constraint approximation, the Markov chain should be generated with a limiting distribution proportional to

$$\exp(-\beta(\mathscr{V}+\mathscr{V}_c)) = \exp(-\beta(\mathscr{V}+\tfrac{1}{2}k_BT\ln[2+\sin^2\theta+\sin^2\theta'])) . \tag{4.54}$$

If $\theta$ and $\theta'$ stay close to their equilibrium values throughout, it might be possible to introduce only a small error by neglecting the constraint potential $\mathscr{V}_c$ in eqn (4.54). The constraint term becomes more complicated and important in the case of bond-angle constraints. For this reason there have been few Metropolis MC simulations of long-chain flexible molecules. The technique of choice is constraint dynamics, using quadratic bond-angle potentials to avoid the metric term in the potential (see Section 3.4).

There have been a considerable number of studies of polymer systems using the MC method [Binder 1984]. Single chains can be simulated using crude MC methods. In this technique a polymer chain of specified length is built up randomly in space [Lal and Spencer 1971] or on a lattice [Suzuki and Nakata 1970]. A chain is abandoned if a substantial overlap is introduced during its construction. When a large number $N$ of chains of the required length have been produced, the average of a property (such as the end-to-end distance) is calculated from

$$\langle \mathcal{A} \rangle = \frac{\sum_{i=1}^{N} \mathcal{A}_i \exp(-\beta \mathcal{V}_i)}{\sum_{i=1}^{N} \exp(-\beta \mathcal{V}_i)} \tag{4.55}$$

where the sums range over all the $N$ polymer chains. The approach is inapplicable for a dense fluid of chains. A more conventional MC method, which avoids this problem, was suggested by Wall and Mandel [1975]. In a real fluid a chain is likely to move in a slithering fashion: the head of the chain moves to a new position and the rest of the chain follows like a snake or lizard. This type of motion is termed 'reptation' [de Gennes 1971]. A successful MC algorithm would mimic this motion. The 'slithering snake' model was originally applied to a polymer on a two-dimensional lattice and a simple example is shown in Fig. 4.7.

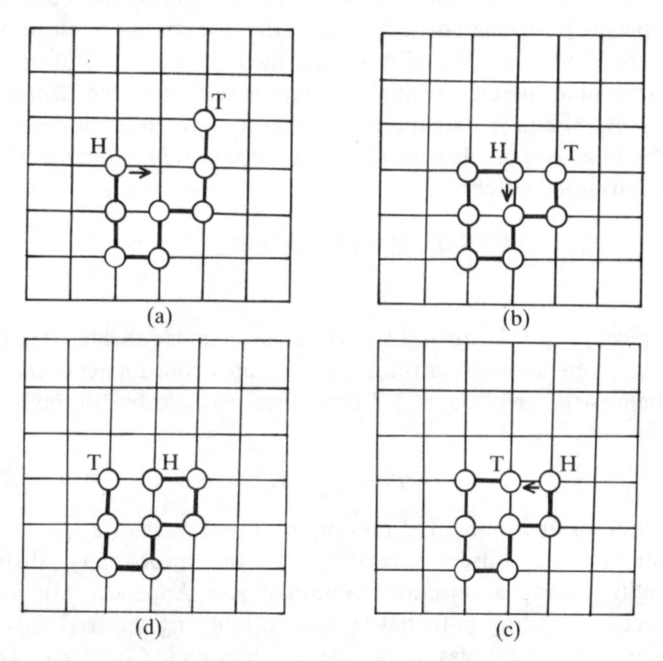


**Fig. 4.7** The slithering snake polymer on a two-dimentional lattice. The configurations are generated in order (a), (b), (c), and (d).

A polymer of eight segments is simulated. One end of the molecule is chosen at random to be the head (H) while the other is the tail (T). The head is moved to a new position on the lattice, all the other atoms move one site along the chain and the tail position becomes vacant. In Fig. 4.7(b) the proposed head move is rejected, since the chains are not allowed to overlap. The resulting configuration, which is identical to the previous one, is included in the averaging, and the simulation proceeds. We can see that random selection of the head and tail is important, since otherwise the system might become locked with the head unable to move. In Fig. 4.7(c), head and tail have been interchanged, and the proposed move is accepted, since the tail position of the chain will be empty when the move is complete. Although this example illustrates the method for a single polymer chain, it is easily extended to a dense fluid of chains, since only one atom moves in generating each new configuration.

Such 'reptation MC' algorithms have been applied to chains on a three-dimensional lattice [Wall, Chin, and Mandel 1977] and to continuum fluids [Brender and Lax 1983]. Bishop, Ceperley, Frisch, and Kalos [1980] have developed a reptation algorithm which is suitable for a chain with arbitrary intermolecular and intramolecular potentials in a continuum fluid. The method exploits the Metropolis solution to the transition matrix to asymptotically sample the Boltzmann distribution. In the case studied by Bishop and co-workers, the model consists of $N$ chains each containing $n_a$ atoms. All the atoms in the fluid interact through the repulsive part of the Lennard-Jones potential, $v^{\mathrm{RLJ}}(r)$, eqn (1.10a); this interaction controls the excluded volume of the chains. In addition, adjacent atoms in the same chain interact through a modified harmonic potential,

$$v^{\mathrm{H}}(r) = \begin{cases} -0.5\,k\,\sigma_1^2 \ln[1-(r/\sigma_1)^2] & 0 \leqslant r \leqslant \sigma_1 \\ \infty & r > \sigma_1 \end{cases} \tag{4.56}$$

where, typically, $\sigma_1 = 1.95\sigma$ and $k = 20$. Each chain is considered in turn and one end is chosen randomly as the head. The initial coordinates of the atoms in the $i$th chain are $(\mathbf{r}_{i1}, \mathbf{r}_{i2}, \ldots, \mathbf{r}_{in_a})$. A new position is selected for the head, atom $n_a$,

$$\mathbf{r}' = \mathbf{r}_{in_a} + \delta\mathbf{r}\,. \tag{4.57}$$

The direction of $\delta\mathbf{r}$ is chosen at random on the surface of a sphere, and the magnitude $\delta r$ is chosen according to the probability distribution $\exp(-\beta v^{\mathrm{H}}(r))$ using a rejection technique (see Appendix G). Thus, the intramolecular bonding potential is used in selecting the trial move (other examples of introducing bias in this way will be seen in Chapter 7). The chain has a new trial configuration $(\mathbf{r}_{i2}, \mathbf{r}_{i3}, \ldots, \mathbf{r}_{in_a}, \mathbf{r}')$. The change in non-bonded interactions in creating a new configuration is calculated by summing over all the atoms

$$\begin{aligned} \delta\mathcal{V}_{nm} = {} & \sum_{a=2}^{n_a} v^{\mathrm{RLJ}}(|\mathbf{r}_{ia}-\mathbf{r}'|) - v^{\mathrm{RLJ}}(|\mathbf{r}_{ia}-\mathbf{r}_{i1}|) \\ & + \sum_{j\neq i}\sum_{a=1}^{n_a} v^{\mathrm{RLJ}}(|\mathbf{r}_{ja}-\mathbf{r}'|) - v^{\mathrm{RLJ}}(|\mathbf{r}_{ja}-\mathbf{r}_{i1}|)\,. \end{aligned} \tag{4.58}$$

Non-bonded interactions from chain $i$ and from all other chains $j$ are included here. $\delta\mathscr{V}_{nm}$ is used to decide whether the move should be accepted or rejected according to the Metropolis criteria. As usual, rejected moves are recounted. The approach works well; there are no geometrical constraints to take into account in this example, all the atoms being free to move under the influence of the potentials.

# 5
# SOME TRICKS OF THE TRADE

## 5.1 Introduction

The purpose of this chapter is to put flesh on the bones of the techniques that have been outlined in Chapters 3 and 4. There is a considerable gulf between understanding the ideas behind the MC and MD methods, and writing and running efficient programs. In this chapter, we describe some of the programming techniques commonly used in the simulation of fluids. There are a number of similarities in the structure of MC and MD programs. They involve a start-up from an initial configuration of molecules, the generation of new configurations in a particular ensemble, and the calculation of observable properties by averaging over a finite number of configurations. Because of the similarities, most of the ideas developed in this chapter are applicable to both techniques, and we shall proceed with this in mind, pointing out any specific exceptions. The first part of this chapter describes the methods used to speed up the evaluation of the interactions between molecules, which are at the heart of a simulation program. The second part describes the overall structure of a typical program and gives details of running a simulation.

## 5.2 The heart of the matter

In Chapter 1, we gave an example of the calculation of the potential energy for a system of particles interacting via the pairwise Lennard-Jones potential. At that point, we paid little attention to the efficiency of that calculation, although we have mentioned points such as the need to avoid the square root function, and the relative speeds of arithmetic operations (see Chapter 1 and Appendix A). The calculation of the potential energy of a particular configuration (and, in the case of MD, the forces acting on all molecules), is the heart of a simulation program, and is executed many millions of times. Great care must be taken to make this particular section of code as efficient as possible. In this section we return to the force/energy routine with the following questions in mind. Is it possible to avoid expensive function evaluations when we calculate the forces on a molecule? What happens when the form of the potential makes taking a square root inevitable? What can we do with much more complicated forms of pair potential?

### *5.2.1 Efficient calculation of forces, energies, and pressures*

Consider, initially, an atomic system with pairwise potentials. Assume that we have identified a pair of atoms $i$ and $j$. Using the minimum image separations, the squared interatomic distance is readily calculated. The force on atom $i$ due to $j$ is

$$\mathbf{f}_{ij} = -\nabla_{\mathbf{r}_i} v(r_{ij}) = -\nabla_{\mathbf{r}_{ij}} v(r_{ij}). \tag{5.1}$$

This force is directed along the interatomic vector $\mathbf{r}_{ij} = \mathbf{r}_i - \mathbf{r}_j$ (see Fig. 5.1) and it is easy to show that

$$\mathbf{f}_{ij} = -\frac{1}{r_{ij}}\left(\frac{\mathrm{d}v(r_{ij})}{\mathrm{d}r_{ij}}\right)\mathbf{r}_{ij} = -\frac{w(r_{ij})}{r_{ij}^2}\mathbf{r}_{ij}. \tag{5.2}$$

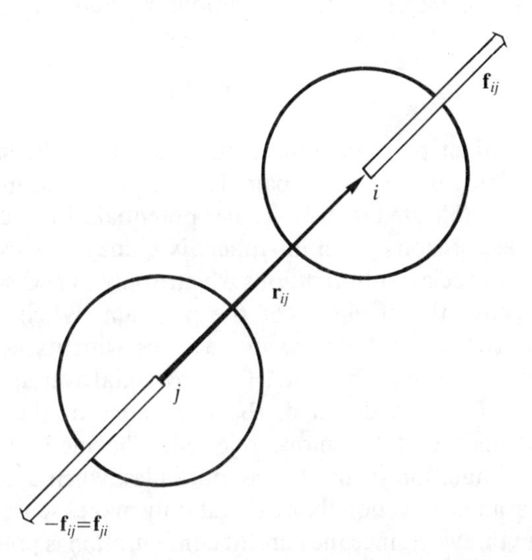


**Fig. 5.1** The separation vector and force between two molecules.

This equation makes it clear that if $v(r_{ij})$ is an even function of $r_{ij}$, then the force vector can be calculated without ever working out the absolute magnitude of $\mathbf{r}_{ij}$: $r_{ij}^2$ will do. The function $w(r_{ij})$ is the pair virial function introduced in eqns (2.59)–(2.63). If $v(r_{ij})$ is even in $r_{ij}$, then so is $w(r_{ij})$. Taking the Lennard-Jones potential, eqn (1.6), as our example, we have

$$\mathbf{f}_{ij} = \frac{24\varepsilon}{r_{ij}^2}\left[2(\sigma/r_{ij})^{12} - (\sigma/r_{ij})^6\right]\mathbf{r}_{ij}. \tag{5.3}$$

In an MD simulation, $v(r_{ij})$, $w(r_{ij})$, and $\mathbf{f}_{ij}$ are calculated within a double loop over all pairs $i$ and $j$ as outlined in Chapter 1. The force on particle $j$ is calculated from the force on $i$ by exploiting Newton's third law. There are one or two elementary steps that can be taken to make this calculation efficient, and these appear in program F.17. In a MC calculation, $v(r_{ij})$ and $w(r_{ij})$ will typically be calculated in a loop over $j$, with $i$ (the particle being given a trial move) specified. This is illustrated in program F.11.

The calculation of the configurational energy and the force can be readily extended to molecular fluids in the interaction site formalism. In this case the potential energy is given by eqn (1.12) and the virial by (for example) eqn (2.63). If required, the forces are calculated in a straightforward way. In this case, it may be simplest to calculate the virial by using the definitions (compare eqns (2.59), (2.61))

$$w(r_{ab}) = -\mathbf{r}_{ab} \cdot \mathbf{f}_{ab} \tag{5.4}$$

summed over all distinct site–site separations $\mathbf{r}_{ab}$ and forces $\mathbf{f}_{ab}$ (including intramolecular ones) or

$$w(r_{ij}) = -\mathbf{r}_{ij} \cdot \mathbf{f}_{ij} \tag{5.5}$$

summed over distinct pairs of molecules, where $\mathbf{f}_{ij}$ is the sum of site–site interactions $\mathbf{f}_{ab}$ acting between each pair. These equations translate easily into code. For more complicated intermolecular potentials, for example involving multipoles, the expressions given in Appendix C may be used.

There are some special considerations which apply to MC simulations, and which may improve the efficiency of the program. When a molecule $i$ is subjected to a trial move, the new interactions with its neighbours $j$ are calculated. It is possible to keep a watch for substantial overlap energies during this calculation: if one is detected, the remainder of the loop over $j$ is immediately skipped and the move rejected. The method is particularly effective in the simulation of hard-core molecules, when a single overlap is sufficient to guarantee rejection. Note that it only makes sense to test the trial configuration in this way, since the current configuration is presumably free of substantial overlaps. For soft-core potentials care should be taken not to set the overlap criterion at too low an energy: occasional significant overlaps may make an important contribution to some ensemble averages.

If no big overlaps are found, the result of the loops over $j$ is a change in potential energy which is used in the MC acceptance/rejection test. If the move is accepted, this number can be used to update the current potential energy (as seen in Section 4.4): there is no need to recalculate the energy from scratch. It may be worth considering a similar approach when calculating the virial, i.e. compute the change in this function which accompanies each trial move, and update $\mathcal{W}$ if it is accepted. Whether this is cost-effective compared with a less frequent complete recalculation of $\mathcal{W}$ depends on the acceptance ratio: it would not be worthwhile if a large fraction of moves were rejected. In any case, a complete recalculation of the energy and the virial should be carried out at the end of the simulation as a check that all is well.

The calculation of the pressure in systems of hard molecules (whether in MC or in MD) is carried out in a slightly different fashion, not generally within the innermost loop of the program, and we return to this in Section 5.6.

### 5.2.2 *Avoiding the square root*

When the potential has an odd power of $r_{ij}$ (ion–ion, dipole–dipole for example), there is no method of calculating the potential function without evaluating the square root of the intermolecular separation. To avoid the use of the SQRT function, Singer [1983] suggests the following algorithm for the case where $r_{ij}^2$ lies between 0.1 and 1.0. This should be true in most cases if we elect to use the $(-1, +1)$ or $(-\frac{1}{2}, +\frac{1}{2})$ simulation box (Chapter 1) so long as the number of molecules is not too large (when scaled intermolecular distances might fall below the lower limit). The square root is expanded as a least-squares polynomial. Let $s = r_{ij}^2$. The equation

$$r_{ij} = s^{1/2} \tag{5.6a}$$

is replaced by

$$r_{ij} \approx c_0 + c_1 s + c_2 s^2 + c_3 s^3 \tag{5.6b}$$

where the coefficients are [Powles 1984a] $c_0 = 0.188030699$, $c_1 = 1.48359853$, $c_2 = -1.0979059$, and $c_3 = 0.430357353$. Refinement cycles are used to make improvements on the first guess:

$$r_{ij} \rightarrow r_{ij}' = r_{ij} + \delta s / 2 r_{ij} \tag{5.7}$$

where $\delta s = s - r_{ij}^2$. Three refinement steps gives an accuracy of six figures. This algorithm is approximately 20 times faster than the intrinsic SQRT function on the CDC 7600, when efficiently programmed: see F.18. However, this performance varies dramatically from machine to machine, and there may be no improvement at all on some computers.

### 5.2.3 *Table look-up and spline fit potentials*

As the potentials used in simulations become more complicated, a direct evaluation of the potential and forces can be avoided by using a prepared table. This technique has been used in the simulation of the Barker–Fisher–Watts potential for argon [Barker *et al.* 1971], which contains 11 adjustable parameters and an exponential. The table is constructed once, at the beginning of the simulation program, and the potential and force are calculated as functions of $s = r_{ij}^2$. In the following code, as an example, we set up a table to calculate the exponential-6 potential,

$$v^{E6}(r) = -A/r^6 + B \exp(-Cr) \tag{5.8}$$

where $A$, $B$, and $C$ are parameters. The table interval $\delta s$ is stored in DS, and KMAX is the number of table entries. A variable RLOW is used to prevent arithmetic overflow resulting from attempting to calculate the potential at very short distances; typically RLOW would be set to one-quarter of the potential minimum distance.

```
      DO 100 K = 1, KMAX

         S = REAL ( K ) * DS
         RIJ = SQRT ( S )

         IF ( RIJ .GT. RLOW ) THEN

            VTAB(K) = - A / ( RIJ ** 6 ) + B * EXP ( - C * RIJ )

         ENDIF

100   CONTINUE
```

During the course of the run, values of $s = r_{ij}^2$ are calculated for a particular pair of molecules, and the potential is interpolated from the table. There are a variety of interpolation algorithms, and as an example we use the Newton–Gregory forward difference method [Booth 1972]. Suppose we have tabulated a function $v(s)$, i.e. we have a set of values $v_1 = v(s_1)$, $v_2 = v(s_2)$, etc. at equal intervals $\delta s$. Define the first differences $\delta v_k = v_{k+1} - v_k$ and the second differences $\delta^2 v_k = \delta v_{k+1} - \delta v_k$. If we have a value $s$ lying between the table values $s_k$ and $s_{k+1}$, then $v(s)$ may be interpolated from the values $v_k$, $\delta v_k$, and $\delta^2 v_k$ (see Fig. 5.2). Setting $\xi = (s - s_k)/\delta s$, we have

$$v(s) \approx v_k + \xi \delta v_k + \tfrac{1}{2}\xi(\xi - 1)\delta^2 v_k. \tag{5.9}$$

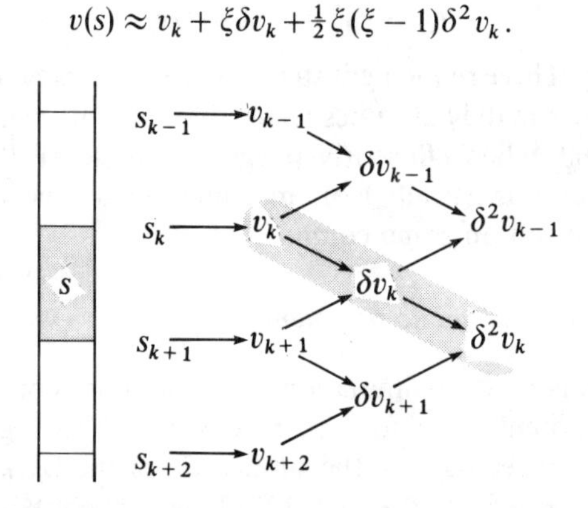


**Fig. 5.2** The Newton–Gregory forward difference interpolation.

This interpolation can be programmed efficiently using the following code. Assume that $r_{ij}^2$ has been calculated in the usual way, and stored in the variable RIJSQ. SDS is $r_{ij}^2/\delta r_{ij}^2$, i.e. $s/\delta s$.

```
      SDS = RIJSQ / DS
      K   = INT ( SDS )
      XI  = SDS - REAL ( K )
      VK  = VTAB(K)
      VK1 = VTAB(K+1)
      VK2 = VTAB(K+2)
      T1  = VK  + ( VK1 - VK  ) *   XI
      T2  = VK1 + ( VK2 - VK1 ) * ( XI - 1.0 )
      VIJ = T1 + ( T2 - T1 ) * XI * 0.5
```

In a molecular dynamics simulation, of course, we also need to evaluate the forces. We may compute these by constructing a separate table of values of the function $w(r_{ij})/r_{ij}^2$, which enters into the force calculation through eqn (5.2). It is simpler, however, to note that this function is given by

$$\frac{w(r_{ij}^2)}{r_{ij}^2} = \frac{w(s)}{s} = \frac{2\mathrm{d}v}{\mathrm{d}s} \tag{5.10}$$

and obtain it by differentiation of eqn (5.9). The success of the interpolation method depends on the careful choice of the table-spacing. Typically, we find that $\delta s = \delta r_{ij}^2 = 0.01 r_m^2$, where $r_m$ is the position of the potential minimum, produces a sufficiently fine grid for use in MD and MC simulations.

An improvement to this method has been suggested [Andrea, Swope, and Andersen 1983] which cuts down the storage requirements for the tables. This is particularly important if the fluid is characterized by many different site–site potentials. The function $v(s)$ (where $s = r_{ij}^2$) is again divided into a number of regions by grid points or knots $s_k$. In each interval $(s_k, s_{k+1})$ the function is approximated by a fifth-order polynomial

$$v(s) \approx c_0 + c_1 \delta s + c_2 \delta s^2 + c_3 \delta s^3 + c_4 \delta s^4 + c_5 \delta s^5 \tag{5.11}$$

where now $\delta s = s - s_k$. The coefficients $c_0 \ldots c_5$ are uniquely determined by the exact values of $v(s)$, $\mathrm{d}v(s)/\mathrm{d}s$, and $\mathrm{d}^2 v(s)/\mathrm{d}s^2$ at the two ends of the interval [Andrea *et al.* 1983, Appendix]. Thus, we need to store the grid points $s_k$ (which need not be evenly spaced) and six coefficients for each interval. In their simulation of water, Andrea *et al.* represented the O–O, O–H, and H–H potentials using 14, 16, and 26 intervals respectively. For MD, the forces are easily obtained by differentiating eqn (5.11) and using eqn (5.10), as before.

### 5.2.4 *Shifted and shifted-force potentials*

The truncation of the intermolecular potential at a cutoff introduces some difficulties in defining a consistent potential and force for use in the MD method. The function $v(r_{ij})$ used in a simulation contains a discontinuity at $r_{ij} = r_c$: whenever a pair of molecules crosses this boundary, the total energy will not be conserved. We can avoid this by shifting the potential function by an amount $v_c = v(r_c)$, i.e. using instead the function

$$v^{\mathrm{S}}(r_{ij}) = \begin{cases} v(r_{ij}) - v_c & r_{ij} \leqslant r_c \\ 0 & r_{ij} > r_c \end{cases}. \tag{5.12}$$

The small additional term is constant for any pair interaction, and does not affect the forces, and hence the equations of motion of the system. However, its contribution to the total energy varies from time step to time step, since the total number of pairs within cutoff range varies. This term should certainly be included in the calculation of the total energy, so as to check the conservation law. However, there is a further problem. The force between a pair of molecules is still discontinuous at $r_{ij} = r_c$. For example, in the Lennard-Jones case, the force is given by eqn (5.3) for $r_{ij} \leqslant r_c$, but is zero for $r_{ij} > r_c$. The magnitude of the discontinuity is $0.039\varepsilon\sigma^{-1}$ for $r_c = 2.5\sigma$. It can cause instability in the numerical solution of the differential equations. To avoid this difficulty, a number of workers have used a 'shifted-force potential' [Stoddard and Ford 1973; Streett, Tildesley, and Saville 1978a; Nicolas *et al.* 1979; Powles *et al.* 1982]. A small linear term is added to the potential, so that its derivative is zero at the cutoff distance

$$v^{\mathrm{SF}}(r_{ij}) = \begin{cases} v(r_{ij}) - v_c - \left(\dfrac{\mathrm{d}v(r_{ij})}{\mathrm{d}r_{ij}}\right)_{r_{ij}=r_c}(r_{ij}-r_c) & r_{ij} \leqslant r_c \\ 0 & r_{ij} > r_c \end{cases} \tag{5.13}$$

The discontinuity now appears in the gradient of the force, not in the force itself. The shifted-force potential for the Lennard-Jones case is shown in Fig. 5.3. The force goes smoothly to zero at the cutoff $r_c$, removing problems in energy conservation and any numerical instability in the equations of motion. Making the additional term quadratic [Stoddard and Ford 1973] avoids taking a square root. Of course, the difference between the shifted-force potential and the original potential means that the simulation no longer corresponds to the desired model liquid. However, the thermodynamic properties of a fluid of particles interacting with the unshifted potential can be recovered from the shifted-force potential simulation results, using a simple perturbation scheme [Nicolas *et al.* 1979; Powles 1984b].

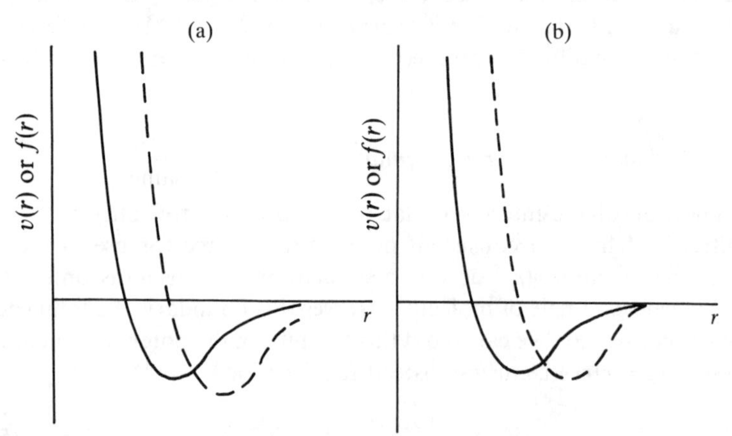


**Fig. 5.3** Magnitude of the pair potential (solid line) and force (dashed line) for (a) the Lennard-Jones potential and (b) its shifted-force modification.

## 5.3 Neighbour lists

In the inner loops of the MD and MC programs, we consider a molecule $i$ and loop over all molecules $j$ to calculate the minimum image separations. If molecules are separated by distances greater than the potential cutoff, the program skips to the end of the inner loop, avoiding expensive calculations, and considers the next neighbour. In this method, the time to examine all pair separations is proportional to $N^2$. Verlet [1967] suggested a technique for improving the speed of a program by maintaining a list of the neighbours of a particular molecule, which is updated at intervals. Between updates of the neighbour list, the program does not check through all the $j$ molecules, but just those appearing on the list. The number of pair separations explicitly considered is reduced. This saves time in looping through $j$, minimum imaging, calculating $r_{ij}^2$, and checking against the cutoff, for all those particles not on the list. Obviously, there is no change in the time actually spent calculating the energy and forces arising from neighbours within the potential cutoff. In this section, we describe some useful time-saving neighbour list methods. These methods are equally applicable to MC and MD simulations, and for convenience we use the MD method to illustrate them. There are differences concerning the relative sizes of the neighbour lists required in MC and MD and we return to this point at the end of the next section. Related techniques may be used to speed up MD of hard systems [Erpenbeck and Wood 1977]. In this case, the aim is to construct and maintain, as efficiently as possible, a table of future collisions between pairs of molecules. The entire question of the scheduling of molecular collisions has been discussed by Rapaport [1980].

### *5.3.1 The Verlet neighbour list*

In the original Verlet method, the potential cutoff sphere, of radius $r_c$, around a particular molecule is surrounded by a 'skin', to give a larger sphere of radius $r_l$, as shown in Fig. 5.4. At the first step in a simulation, a list is constructed of all the neighbours of each molecule, for which the pair separation is within $r_l$. These neighbours are stored in a large array, called, shall we say, LIST. LIST is quite large, of dimension roughly $4\pi r_l^3 \rho N/6$. At the same time a second indexing array, POINT, of size $N$, is constructed. POINT(I) points to the position in the array LIST where the first neighbour of molecule I can be found. Since POINT(I + 1) points to the first neighbour of molecule I + 1, then POINT(I + 1) − 1 points to the last neighbour of molecule I. Thus, using POINT, we can readily identify the part of the large LIST array which contains neighbours of I. The code for setting up the arrays LIST and POINT is given in program F.19.

Over the next few time steps, the list is used in the force/energy evaluation routine. For each molecule I, the program identifies the neighbours J, by running over LIST from POINT(I) to POINT(I + 1) − 1. It is essential to check that POINT(I + 1) is actually greater than POINT(I): if this is not the case, then molecule I has no neighbours, and can be skipped. This is certainly possible in dilute systems. A sample force routine using the Verlet list is given in program F.19. From time to time, the neighbour list is reconstructed, and

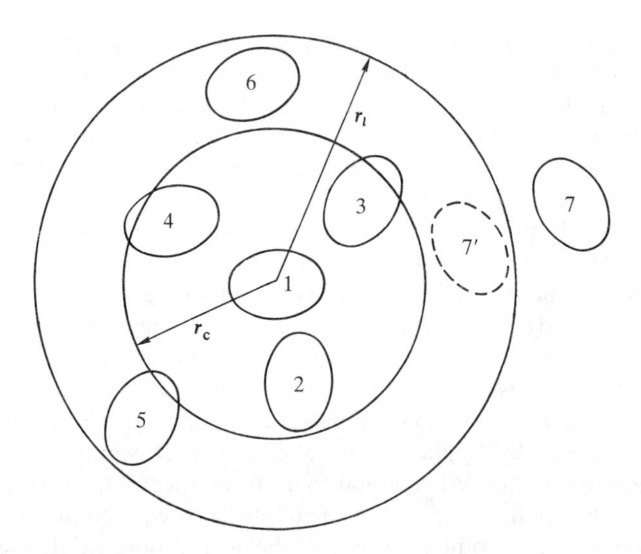


**Fig. 5.4** The cutoff sphere, and its skin, around a molecule 1. Molecules 2, 3, 4, 5, and 6 are on the list of molecule 1; molecule 7 is not. Only molecules 2, 3, and 4 are within the range of the potential at the time the list is constructed.

the cycle is repeated. The algorithm is successful because the skin around $r_c$ is chosen to be thick enough so that between reconstructions a molecule, such as 7 in Fig. 5.4, which is not on the list of molecule 1, cannot penetrate through the skin into the important $r_c$ sphere. Molecules such as 3 and 4 can move in and out of this sphere, but since they are on the list of molecule 1, they are always considered regardless, until the list is next updated.

The interval between updates of the table is often fixed at the beginning of the program, and intervals of 10–20 steps are quite common. An important refinement allows the program to update the neighbour list automatically. When the list is constructed, a vector for each molecule is set to zero. At subsequent steps, the vector is incremented with the displacement of each molecule. Thus it stores the total displacement for each molecule since the last update. When the sum of the magnitudes of the two largest displacements exceeds $r_l - r_c$, the neighbour list should be updated again [Fincham and Ralston 1981; Thompson 1983]. The code for automatic updating of the neighbour list is given on microfiche F.19.

The list sphere radius, $r_l$, is a parameter that we are free to choose. As $r_l$ is increased, the frequency of updates of the neighbour list will decrease. However, with a large list, the efficiency of the non-update steps will decrease. This balance has been examined by Thompson [1983] for MD simulations of 256 and 500 Lennard-Jones atoms. Simulations at $\rho^* = 0.8$, $T^* = 0.76$, were run for 1000 time steps; $r_c$ was fixed at $2.5\sigma$, and $r_l$ varied. The results are given in Table 5.1. A list cutoff of about $2.7\sigma$ would provide substantial speed

**Table 5.1** *Time saving using a Verlet neighbour list* [*Thompson* 1983].

| List Radius | Update[a] interval | Time[b] | |
|---|---|---|---|
| | | $N = 256$ | $N = 500$ |
| no list | — | 3.33 | 10.00 |
| 2.60 | 5.78 | 2.24 | 4.93 |
| 2.70 | 12.50 | 2.17 | 4.55 |
| 2.90 | 26.32 | 2.28 | 4.51 |
| 3.10 | 43.48 | 2.47 | 4.79 |
| 3.43 | 83.33 | 2.89 | — |
| 3.50 | 100.00 | — | 5.86 |

[a]Update interval is the average number of steps between updates. It is essentially independent of system size.
[b]Time is CPU time per step, in seconds. The runs were performed on a PDP 11/70.

increases for both systems, but for systems of size $N \approx 500$ and larger, the improvement is dramatic. As the size of the system becomes larger, the size of the LIST array grows, approximately $\propto N$. If storage is a priority, then a binary representation of the list can be employed [O'Shea 1983]. Each bit in a two-dimensional array represents a pair of molecules $i$ and $j$. This bit is set to 1 if the molecules are neighbours, and zero otherwise. The array is used to check for neighbours at subsequent steps, and is revised at suitable intervals. For a system of 256 molecules, a conventional neighbour list would require approximately $64 \times 256$ words; on a 16-bit machine, the binary array reduces this to $8 \times 256$ words.

In the MC method, the array POINT has a size $N + 1$ rather than $N$, since the index I runs over all $N$ atoms rather than $N - 1$ as in MD. In addition, the array LIST is roughly twice as large in MC as in a corresponding MD program. In the MD technique, the list for a particular molecule $i$ contains only the molecules $j$ with an index greater than $i$, since in this method we use Newton's third law to calculate the force on $j$ from $i$ at the same time as the force on $i$ from $j$. In the MC method particles $i$ and $j$ are moved independently and the list must contain separately the information that $i$ is a neighbour of $j$ and $j$ a neighbour of $i$. In this case the binary representation discussed by O'Shea [1983] is particularly useful.

### *5.3.2 Cell structures and linked lists*

As the size of the system increases towards 1000 molecules, the conventional neighbour list becomes too large to store easily, and the logical testing of every pair in the system is inefficient. An alternative method of keeping track of neighbours for large systems is the cell index method [Quentrec and Brot 1975; Hockney and Eastwood 1981]. The cubic simulation box (extension to non-cubic cases is possible) is divided into a regular lattice of M × M × M cells. A two-dimensional representation of this is shown in Fig. 5.5. These cells are chosen so that the side of the cell $l = L/M$ is greater than the cutoff distance for the forces, $r_c$. For the two-dimensional example of Fig. 5.5, the neighbours of any molecule in cell 13 are to be found in the cells 7, 8, 9, 12, 13, 14, 17, 18, 19. If there is a separate list of molecules in each of those cells, then searching through the neighbours is a rapid process. For the two-dimensional system illustrated, there are approximately $N_c = N/M^2$ molecules in each cell; the analogous result in three dimensions would be $N_c = N/M^3$. Using the cell

(a)

| | | | | |
|---|---|---|---|---|
| 21 | 22 | 23 | 24 | 25 |
| 16 | 17 | 18 | 19 | 20 |
| 11 | 12 | 13 | 14 | 15 |
| 6 | 7 | 8 | 9 | 10 |
| 1 | 2 | 3 | 4 | 5 |

(b)

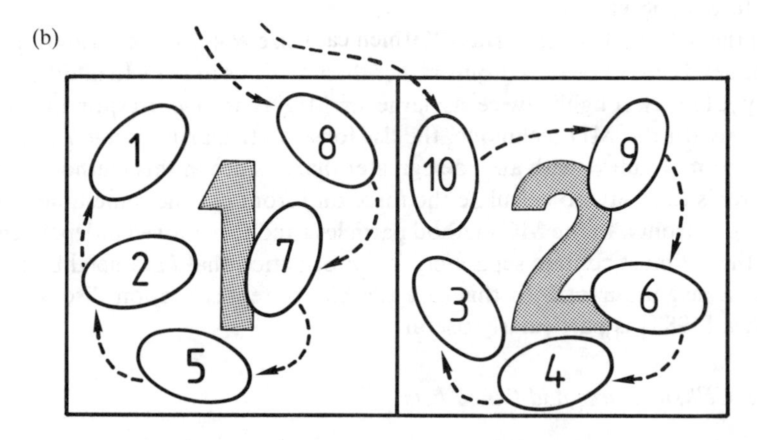


**Fig. 5.5** The cell method in two dimensions. (a) The central box is divided into $M \times M$ cells ($M = 5$). (b) A close-up of cells 1 and 2, showing the molecules and the link-list structure.

structure in two dimensions we need only examine $9NN_c$ pairs (or just $4.5NN_c$ if we take advantage of the third law in the MD method). This contrasts with $N^2$ (or $\frac{1}{2}N(N-1)$) for the brute force approach. When the cell structure is used in three dimensions, then we compute $27NN_c$ interactions ($13.5NN_c$ for MD) as compared with $N^2$ (or $\frac{1}{2}N(N-1)$).

The cell structure may be set up and used by the method of linked lists [Knuth 1973, Chapter 2; Hockney and Eastwood 1981, Chapter 8]. The first part of the method involves sorting all the molecules into their appropriate cells. This sorting is rapid, and may be performed every step. Two arrays are created during the sorting process. The 'head-of-chain' array (HEAD) has one element for each cell. This element contains the identification number of one of the molecules sorted into that cell. This number is used to address the element of a linked-list array (LIST), which contains the number of the next molecule in that cell. In turn, the LIST array element for that molecule is the index of the next molecule in the cell, and so on. If we follow the trail of link-list references, we will eventually reach an element of LIST which is zero. This indicates that there are no more molecules in that cell, and we move on to the head-of-chain molecule for the next cell. To illustrate this, imagine a simulation of particles in two cells: 1, 2, 5, 7, and 8 in cell one and 3, 4, 6, 9, and 10 in cell two (see Fig. 5.5). The HEAD and LIST arrays are

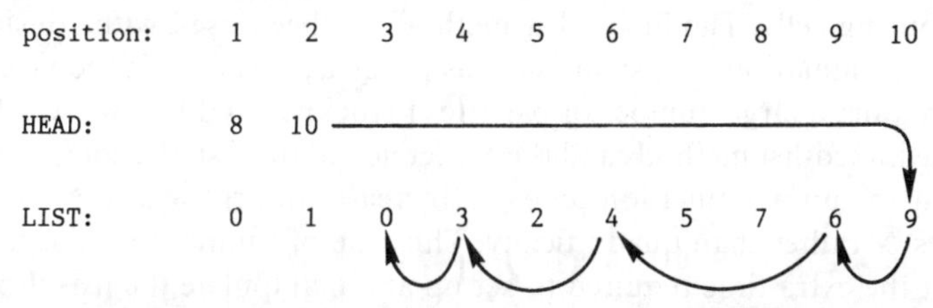


For cell two, HEAD(2) = 10, and the route through the linked list is arrowed. The construction of HEAD and LIST is straightforward. We take the usual unit cube simulation box, in which case the cell length is just $1/M$. The inverse of this quantity is stored in CELLI.

```
        DO 100 ICELL = 1, NCELL
           HEAD(ICELL) = 0
100     CONTINUE
        CELLI = REAL ( M )
        DO 200 I = 1, N
           ICELL = 1 + INT ( ( RX(I) + 0.5 ) * CELLI )
    :                + INT ( ( RY(I) + 0.5 ) * CELLI ) * M
    :                + INT ( ( RZ(I) + 0.5 ) * CELLI ) * M * M
           LIST(I) = HEAD(ICELL)
           HEAD(ICELL) = I
200     CONTINUE
```

In MD, the calculation of the forces is performed by looping over all cells. For a given cell, the program sorts through the linked list. A particular molecule on the list may interact with all molecules in the same cell that are further down the list. This avoids counting the *ij* interaction twice. A particular molecule also may interact with the molecules in the neighbouring cells. To avoid double counting of these forces, only a limited number of the neighbouring cells are considered. This idea is most easily explained with reference to our two-dimensional example shown in Fig. 5.5. A molecule in cell 13 interacts with eight neighbouring cells, but the program only checks the chains in cells 9, 14, 18, and 19. Interactions between cells 13 and 12 are checked when cell 12 is the focus of attention, and so on. In this way, we can make full use of Newton's third law in calculating the forces. We note that, for a cell at the edge of the basic simulation box, (15 for example in Fig. 5.5) it is necessary to consider the periodic cells (6, 11, and 16). An example of the code for constructing and searching through cells is given in F.20. The cell structure may be used somewhat more efficiently if the molecules in each cell are sorted into order of increasing (say) $x$-coordinate [Hockney and Eastwood 1981].

The linked-list method can also be used with the MC technique. In this case, a molecule move involves checking molecules in the same cell, and in all the neighbouring cells. The linked-list method has been used with considerable success in simulation of systems such as plasmas, galaxies, and ionic crystals, which require a large number of particles [Hockney and Eastwood 1981]. In both the linked-list methods and the Verlet neighbour list, the computing time required to run a simulation tends to increase linearly with the number of particles $N$ rather than quadratically. This rule of thumb does not take into account the extra time required to set up and manipulate the lists. For small systems ($N \approx 100$) the overheads involved make the use of lists unprofitable.

A modification of the cell structure employs cells that are sufficiently small that at most one particle can occupy each cell. In this case, a linked-list structure as described above is not required: the program simply loops over all cells, and conducts an inner loop over the cells that are within a distance $r_c$ of the cell of interest. Advantages of this method are that the list of cells 'within range' of each given cell may be computed at the start of the program, remaining unaltered throughout, and that a simple decision (is the cell occupied or not?) is required at each stage.

## 5.4 Multiple time step methods

In a typical MD simulation, as much as 95 per cent of the computing time is spent in examining the complete set of $\frac{1}{2}N(N-1)$ pairs, identifying those pairs separated by less than the cutoff distance $r_c$, and computing the forces for this subset. The remaining pair interactions do not influence the dynamics of the system. For each molecule, the set of neighbours within the cutoff changes with time; the task of identifying and rejecting the others is time-consuming, but the book-keeping methods discussed in the previous sections reduce this time considerably. Any further increase in speed can only be achieved by cutting down the time spent actually evaluating forces between pairs within cutoff range.

The multiple time-step method [Streett, Tildesley, and Saville 1978a, b] is designed to do this. Figure 5.6 shows the close neighbours of an atom $i$, which are split into two groups: 'primary' neighbours, which lie within a distance $r_p$ of atom $i$, and 'secondary' neighbours, which are at a distance between $r_p$ and $r_c$ from $i$. In this way, the total force $\mathbf{f}_i$ on $i$ is separated into a primary component $\mathbf{f}_i^p$ due to the close neighbours, and a secondary component $\mathbf{f}_i^s$ from the more remote neighbours. Typically, for a Lennard-Jones fluid, $r_p$ would be chosen to lie between $\sigma$ and $1.5\sigma$. The motion of atom $i$ is dominated by a rapidly changing primary force resulting from collisions with the nearest neighbour 'cage'. The secondary force is smaller, and changes more slowly with time. The multiple time step method takes advantage of this separation, and aims to calculate the secondary pair interactions less frequently than the more important primary ones. The method is most conveniently used in conjunction with the Gear predictor–corrector algorithm (Section 3.2 and Appendix E).

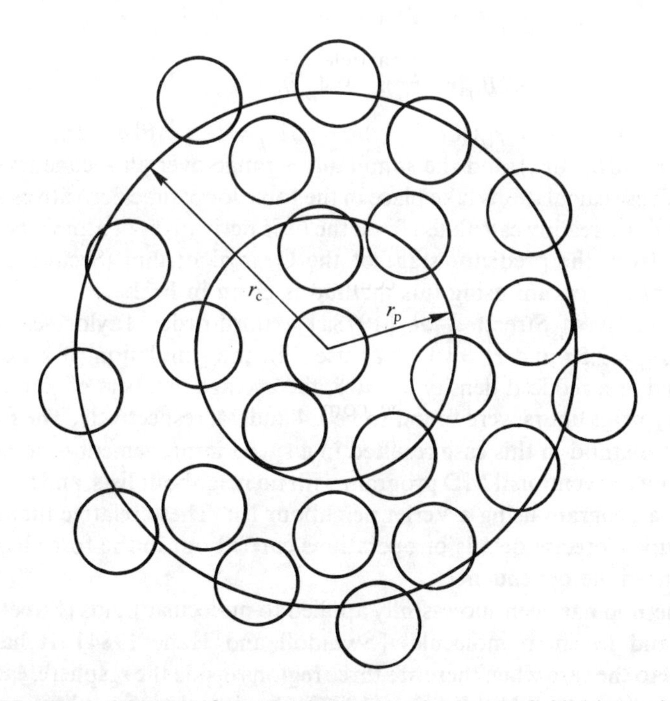


**Fig. 5.6** The close neighbours of a molecule. These molecules are divided into primary and secondary neighbours.

At time $t$ the primary and secondary forces on each atom $i$ are calculated in the usual way. At the same time, the time derivatives of the secondary forces, $\dot{\mathbf{f}}_i^{\mathrm{s}}(t)$, $\ddot{\mathbf{f}}_i^{\mathrm{s}}(t) \ldots$, up to order $m$ say, are calculated, and a list of primary neighbours is compiled. At each of the next $\tau_{\mathrm{mts}} - 1$ steps, the primary force is computed explicitly using the list. The secondary force is estimated using a Taylor series of order $m$:

$$\mathbf{f}_i^{\mathrm{s}}(t+\tau\delta t) \approx \mathbf{f}_i^{\mathrm{s}}(t) + (\tau\delta t)\dot{\mathbf{f}}_i^{\mathrm{s}}(t) + \tfrac{1}{2}(\tau\delta t)^2\ddot{\mathbf{f}}_i^{\mathrm{s}} + \ldots \quad (5.14)$$

Following these $\tau_{\mathrm{mts}}$ steps (one involving a full calculation of forces and the secondary time derivatives, and $\tau_{\mathrm{mts}} - 1$ using eqn (5.14) for the secondary forces) the entire process is repeated, starting with a recalculation of the primary list. Thus, the method effectively uses two time steps: $\delta t$ for the primary interactions and $\tau_{\mathrm{mts}}\delta t$ for the secondary ones.

Convenient expressions for the force derivatives can be readily obtained in terms of the time derivatives of particle positions and spatial derivatives of the potential. For example,

$$\dot{\mathbf{f}}_i^{\mathrm{s}} = \sum_j A_{ij}\dot{\mathbf{r}}_{ij} + B_{ij}(\mathbf{r}_{ij}\cdot\dot{\mathbf{r}}_{ij})\mathbf{r}_{ij} \quad (5.15)$$

and

$$\ddot{\mathbf{f}}_i^{\mathrm{s}} = \sum_j [B_{ij}(\mathbf{r}_{ij}\cdot\ddot{\mathbf{r}}_{ij} + \dot{\mathbf{r}}_{ij}\cdot\dot{\mathbf{r}}_{ij}) + C_{ij}(\mathbf{r}_{ij}\cdot\dot{\mathbf{r}}_{ij})^2]\mathbf{r}_{ij} + 2B_{ij}(\mathbf{r}_{ij}\cdot\dot{\mathbf{r}}_{ij})\dot{\mathbf{r}}_{ij} + A_{ij}\ddot{\mathbf{r}}_{ij} \quad (5.16)$$

where $A_{ij} = -(1/r_{ij})(\mathrm{d}v_{ij}/\mathrm{d}r_{ij})$, $B_{ij} = (1/r_{ij})(\mathrm{d}A_{ij}/\mathrm{d}r_{ij})$, and $C_{ij} = (1/r_{ij})(\mathrm{d}B_{ij}/\mathrm{d}r_{ij})$, and the summations range over all secondary neighbours $j$. These calculations take place in the force loop; time derivatives such as $\ddot{\mathbf{r}}_{ij} = \ddot{\mathbf{r}}_i - \ddot{\mathbf{r}}_j$ are readily calculated from the time derivatives of atomic positions available from the predictor stage of the Gear algorithm (Section 3.2). An example of a program using this method is given in F.21.

In the study of Streett *et al.* [1978a] a third-order Taylor series, with $\tau_{\mathrm{mts}} = 10$, $r_{\mathrm{c}} = 2.5\sigma$ and $r_{\mathrm{p}} = 1.1\sigma$ was used. In the simulation of a Lennard-Jones fluid at a reduced density $\rho^* = 0.8$, the average numbers of primary and secondary neighbours were found to be 1.4 and 24, respectively. The multiple time-step method in this case resulted in a speed improvement of a factor of 7–10 over a conventional MD program with no neighbour lists, and a factor of 3–5 over a program using a Verlet neighbour list. These relative timings will depend upon precise details of operations carried out in the force loop, and may be machine dependent.

The method has been successfully applied to molecular fluids [Streett, *et al.* 1978b] and to chain molecules [Swindoll and Haile 1984]. It has been extended to the case when there are three regions inside the $r_{\mathrm{c}}$ sphere, each with a different time step [Nicolas *et al.* 1979]. One interesting application of the method is in the simulation of fluids with three-body forces [Haile 1978]. For the Axilrod–Teller potential (see Appendix C and Maitland *et al.* [1981]), the three-body contribution to the force varies slowly with time, and it is possible to treat the entire three-body component as a secondary interaction. In this way, the prohibitively expensive summation over triplets may be avoided for most of the steps in the simulation.

There are still some difficulties with the method. Firstly, there is a considerable programming effort required to incorporate the relevant code into a conventional program. Secondly, the method has only been used with a time step $\delta t$ rather smaller than that used in conventional simulations (e.g. $5 \times 10^{-15}$ s [Streett *et al.* 1978a], as opposed to the more usual $10^{-14}$ [Verlet 1967]). It is found that the use of a larger time step requires an increase in $r_p$ and a decrease in the update interval $\tau_{mts}$, thus offsetting some of the advantages gained by using multiple time steps. The efficiency of the method is clearly very sensitive to the choice of $r_p$; perhaps the best way to test this method is to fix the desired time step, decide upon a satisfactory level of energy conservation, and adjust $r_p$ and the update interval $\tau_{mts}$, to maximize the efficiency of the simulation subject to these constraints. For molecular fluids, where a shorter time step is often required in any case, these difficulties are less evident.

## 5.5 How to handle long-range forces

### *5.5.1 Introduction*

In the previous sections, we have discussed the core of the program when the forces are short ranged. In this section, we turn our attention to the handling of long-range forces in simulations.

A long-range force is often defined as one in which the spatial interaction falls off no faster than $r^{-d}$ where $d$ is the dimensionality of the system. In this category are the charge–charge interaction between ions ($v^{zz}(r) \sim r^{-1}$) and the dipole–dipole interaction between molecules ($v^{\mu\mu}(r) \sim r^{-3}$). These forces are a serious problem for the computer simulator, since their range is greater than half the box length for a typical simulation of $\approx 500$ molecules. The brute force solution to this problem would be to increase the size of the central box $L$ to hundreds of nanometres so that the screening by neighbours would diminish the effective range of the potentials. Even with modern computers, this solution is impracticable, since the time required to run such a simulation is approximately proportional to $N^2$, i.e. $L^6$.

How can such potentials be handled? The problem is particularly acute for $v^{zz}(r)$. Straightforward spherical truncation of the potential can be ruled out. The resulting sphere around a given ion could be charged, since the number of anions and cations need not balance at any instant. The tendency of ions to migrate back and forth across the spherical surface would create artificial effects at $r = r_c$. This can be countered by distributing a new charge over the surface of the sphere, equal in magnitude and opposite in sign to the net charge of the sphere, so as to guarantee local electroneutrality. This is rather like shifting the potential as described in Section 5.2.4. Adams [1983b] shows that the results from this approach are system size-dependent but that for a system of 512 ions, they compare well with those obtained from the Ewald sum (see the next section). However, some undesirable structural effects are inevitable.

In contrast, the basic minimum image method corresponds to cutting off the potential at the surface of a cube surrounding the ion in question (see Fig. 1.12). This cube will be electrically neutral. However, the drawback is that similarly charged ions will tend to occupy positions in opposite corners of the cube: the periodic image structure will be imposed directly on what should be an isotropic liquid, and this results in a distortion of the liquid structure. This effect might be reduced in the non-cubic boxes of Fig. 1.10 [Adams 1980]. Similar, if less dramatic, effects would be seen by applying spherical cutoff or minimum image to a polar system ($v^{\mu\mu}(r)$).

In Sections 5.5.2 and 5.5.3, we concentrate on two methods which can be used to tackle the problem of long-range forces. The lattice methods, such as the Ewald sum, include the interaction of an ion or molecule with all its periodic images. These methods will tend to overemphasize the periodic nature of the model fluid. The reaction field methods assume that the interaction from molecules beyond a cutoff distance can be handled in an average way, using macroscopic electrostatics. These methods will tend to overemphasize the continuum nature of a polar fluid and require an *a priori* estimate of the relative permittivity. Both methods use well-known ideas from the theory of electrostatics [see e.g. Fröhlich 1949]. In particular, a charge distribution within a spherical cavity polarizes the surrounding medium. This polarization, which depends upon the relative permittivity of the medium, has an effect on the charge distribution in the cavity.

### *5.5.2 The Ewald sum*

The Ewald sum is a technique for efficiently summing the interaction between an ion and all its periodic images. It was originally developed in the study of ionic crystals [Ewald 1921; Madelung 1918]. In Fig. 1.9, ion 1 interacts with ions 2, $2_A$, $2_B$, and all the other images of 2. The potential energy can be written as

$$\mathscr{V}^{zz} = \tfrac{1}{2}\sum_{\mathbf{n}}{}'\left(\sum_{i=1}^{N}\sum_{j=1}^{N} z_i z_j |\mathbf{r}_{ij} + \mathbf{n}|^{-1}\right) \tag{5.17}$$

where $z_i$, $z_j$, are the charges. Remember that, for simplicity of notation, we are omitting all factors of $4\pi\varepsilon_0$: this corresponds to adopting a non-SI unit of charge (see Appendix B). The sum over $\mathbf{n}$ is the sum over all simple cubic lattice points, $\mathbf{n} = (n_x L, n_y L, n_z L)$ with $n_x$, $n_y$, $n_z$ integers. This vector reflects the shape of the basic box. The prime indicates that we omit $i = j$ for $\mathbf{n} = 0$. For long-range potentials, this sum is conditionally convergent, i.e. the result depends on the order in which we add up the terms. A natural choice is to take boxes in order of their proximity to the central box. The unit cells are added in sequence: the first term has $|\mathbf{n}| = 0$, i.e. $\mathbf{n} = (0, 0, 0)$; the second term, $|\mathbf{n}| = L$, comprises the six boxes centred at $\mathbf{n} = (\pm L, 0, 0), (0, \pm L, 0), (0, 0, \pm L)$; etc. As we add further terms to the sum, we are building up our infinite system in roughly spherical layers (see Fig. 5.7). When we adopt this approach, we must specify the nature of the medium surrounding the sphere, in particular its relative permittivity (dielectric constant) $\varepsilon_s$. The results for a sphere surrounded by a good conductor such as a metal ($\varepsilon_s = \infty$) and for a sphere surrounded by vacuum ($\varepsilon_s = 1$) are different [de Leeuw, Perram, and Smith 1980].

$$\mathscr{V}^{zz}(\varepsilon_s = \infty) = \mathscr{V}^{zz}(\varepsilon_s = 1) - \frac{2\pi}{3L^3}\left|\sum_i z_i \mathbf{r}_i\right|^2. \tag{5.18}$$

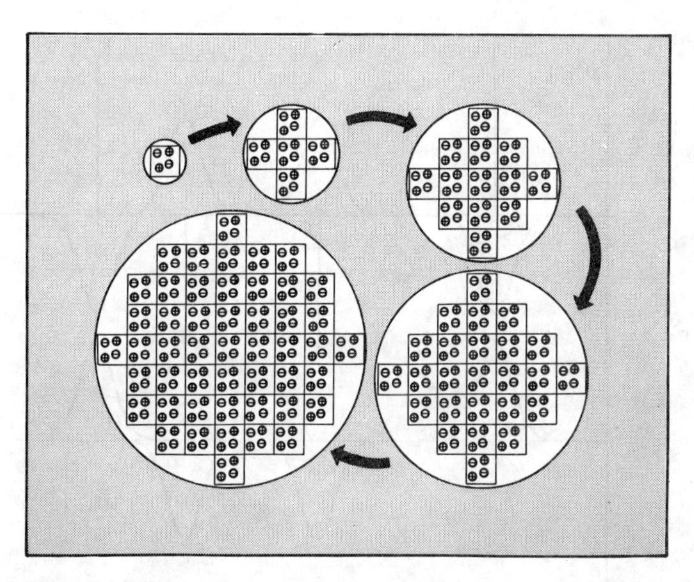

**Fig. 5.7** Building up the sphere of simulation boxes. We illustrate a very small system of two ion pairs for simplicity. The shaded region represents the external dielectric continuum of relative permittivity $\varepsilon_s$.

This equation applies in the limit of a very large sphere of boxes. In the vacuum, the sphere has a dipolar layer on its surface: the last term in eqn (5.18) cancels this. For the sphere in a conductor there is no such layer. The Ewald method is a way of efficiently calculating $\mathscr{V}^{zz}(\varepsilon_s = \infty)$. Equation (5.18) enables us to use the Ewald sum in a simulation where the large sphere is in a vacuum, if this is more convenient. The mathematical details of the method are given by de Leeuw *et al.* [1980] and Heyes [1981]. Here we concentrate on the physical ideas. At any point during the simulation, the distribution of charges in the central cell constitutes the unit cell for a neutral lattice which extends throughout space. In the Ewald method, each point charge is surrounded by a charge distribution of equal magnitude and opposite sign, which spreads out radially from the charge. This distribution is conveniently taken to be Gaussian

$$\rho_i^z(\mathbf{r}) = z_i \kappa^3 \exp(-\kappa^2 r^2)/\pi^{3/2} \tag{5.19}$$

where the arbitrary parameter $\kappa$ determines the width of the distribution, and $\mathbf{r}$ is the position relative to the centre of the distribution. This extra distribution acts like an ionic atmosphere, to screen the interaction between neighbouring charges. The screened interactions are now short-ranged, and the total screened potential is calculated by summing over all the molecules in the central cube and all their images in the real space lattice of image boxes. This is illustrated in Fig. 5.8(a).

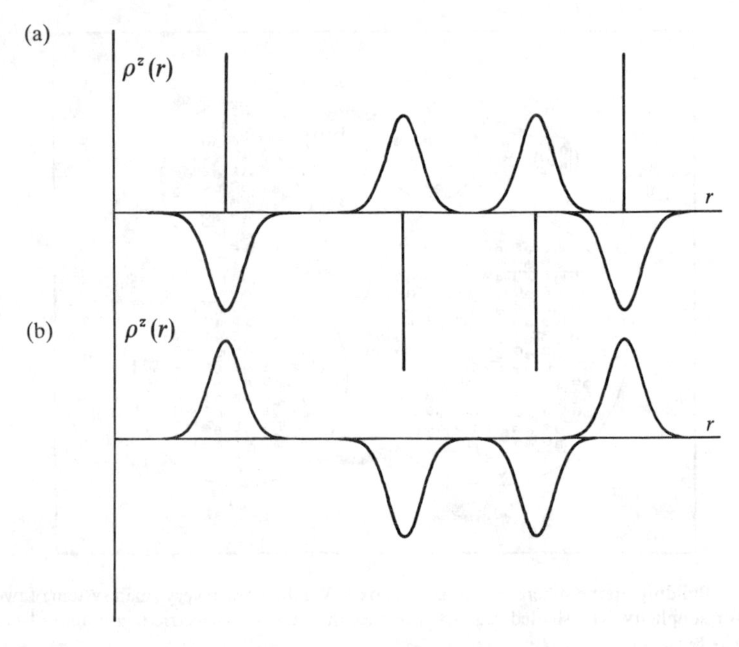


**Fig. 5.8** Charge distribution in the Ewald sum. (a) Original point charges plus screening distribution. (b) Cancelling distribution.

A charge distribution of the same sign as the original charge, and the same shape as the distribution $\rho_i^z(\mathbf{r})$ is also added (see Fig. 5.8(b)). This cancelling distribution reduces the overall potential to that due to the original set of charges. The cancelling distribution is summed in reciprocal space. In other words, the Fourier transforms of the cancelling distributions (one for each original charge) are added, and the total transformed back into real space. (Fourier transforms are discussed in Appendix D.) There is an important correction: the recipe includes the interaction of the cancelling distribution centred at $\mathbf{r}_i$ with itself, and this self term must be subtracted from the total. Thus, the final potential energy will contain a real space sum plus a reciprocal space sum minus a self-term plus the surface term already discussed. The final result is

$$\mathscr{V}^{zz}(\varepsilon_s = 1) = \tfrac{1}{2} \sum_{i=1}^{N} \sum_{j=1}^{N} \left( \sum_{|\mathbf{n}|=0}^{\infty}{}' z_i z_j \frac{\operatorname{erfc}(\kappa|\mathbf{r}_{ij} + \mathbf{n}|)}{|\mathbf{r}_{ij} + \mathbf{n}|} + (1/\pi L^3) \sum_{\mathbf{k} \neq \mathbf{0}} z_i z_j (4\pi^2/k^2) \exp(-k^2/4\kappa^2) \cos(\mathbf{k} \cdot \mathbf{r}_{ij}) \right) - (\kappa/\pi^{1/2}) \sum_{i=1}^{N} z_i^2 + (2\pi/3L^3) \left| \sum_{i=1}^{N} z_i \mathbf{r}_i \right|^2 . \tag{5.20}$$

Here erfc $(x)$ is the complementary error function (erfc $(x) = (2/\pi^{1/2}) \times \int_x^\infty \exp(-t^2)\mathrm{d}t$) which falls to zero with increasing $x$. Thus, if $\kappa$ is chosen to be large enough, the only term which contributes to the sum in real space is that with $n = 0$, and so the first term reduces to the normal minimum image convention. The second term is a sum over reciprocal vectors $\mathbf{k} = 2\pi\mathbf{n}/L^2$. A large value of $\kappa$ corresponds to a sharp distribution of charge, so that we need to include many terms in the $k$-space summation to model it. In a simulation, the aim is to choose a value of $\kappa$ and a sufficient number of $k$-vectors, so that eqn (5.20) (with the real space sum truncated at $\mathbf{n} = 0$) and eqn (5.18) give the same energy for typical liquid configurations. In practice, $\kappa$ is typically set to $5/L$ and 100–200 wave vectors are used in the $k$-space sum [Woodcock and Singer 1971]. We stress that checks should be carried out on the reliability of eqn (5.20) for each individual system which is simulated before beginning the run.

The modified charge–charge interaction is calculated in the normal way in the main loop of the simulation program; erfc $(x)$ is available as an intrinsic function in many extensions of FORTRAN-77, and in some mathematical function libraries. The sum over $k$ vectors is normally carried out in a separate subroutine; using complex arithmetic, it is possible to replace the triple sum over $|\mathbf{k}|$, $i$, and $j$, by a double sum over $|\mathbf{k}|$ and $i$. A version of this subroutine is given in program F.22. This part of the program can be quite expensive on conventional computers, but it may be efficiently vectorized, and so is well-suited for pipeline processing. Before leaving the Ewald sum, we note that there is nothing unique about the Gaussian form of the charge distribution. This general method is due to Berthaut [1952], and Heyes [1981] has given the appropriate functional forms for nine different charge distributions.

The original method of Ewald can be readily extended to dipolar systems. In the derivation of eqn (5.20), $z_i$ is simply replaced by $\boldsymbol{\mu}_i . \boldsymbol{\nabla}_{\mathbf{r}_i}$, where $\boldsymbol{\mu}_i$ is the molecular dipole. The resulting expression is [Kornfeld 1924; Adams and McDonald 1976; de Leeuw *et al.* 1980]

$$\mathscr{V}^{\mu\mu}(\varepsilon_s = 1) = \tfrac{1}{2} \sum_{i=1}^{N} \sum_{j=1}^{N} \Bigg( \sum_{|\mathbf{n}|=0}^{\infty}{}' (\boldsymbol{\mu}_i \cdot \boldsymbol{\mu}_j) B(\mathbf{r}_{ij} + \mathbf{n}) - (\boldsymbol{\mu}_i \cdot \mathbf{r}_{ij})(\boldsymbol{\mu}_j \cdot \mathbf{r}_{ij}) C(\mathbf{r}_{ij} + \mathbf{n})$$

$$+ \sum_{\mathbf{k} \neq \mathbf{0}} (1/\pi L^3)(\boldsymbol{\mu}_i \cdot \mathbf{k})(\boldsymbol{\mu}_j \cdot \mathbf{k})(4\pi^2/k^2) \exp(-k^2/4\kappa^2) \cos(\mathbf{k} \cdot \mathbf{r}_{ij}) \Bigg)$$

$$- \sum_{i=1}^{N} 2\kappa^3 \mu_i^2 / 3\pi^{1/2} + \tfrac{1}{2} \sum_{i=1}^{N} \sum_{j=1}^{N} (4\pi/3L^3) \boldsymbol{\mu}_i \cdot \boldsymbol{\mu}_j \qquad (5.21)$$

where again factors of $4\pi\varepsilon_0$ are omitted. In this equation, the sums over $i$ and $j$ are for dipoles in the central box and

$$B(r) = \operatorname{erfc}(\kappa r)/r^3 + (2\kappa/\pi^{1/2}) \exp(-\kappa^2 r^2)/r^2 \qquad (5.22)$$

$$C(r) = 3 \operatorname{erfc}(\kappa r)/r^5 + (2\kappa/\pi^{1/2})(2\kappa^2 + 3/r^2) \exp(-\kappa^2 r^2)/r^2. \qquad (5.23)$$

This expression can be used in the same way as the Ewald sum, with the real space sum truncated at $|\mathbf{n}| = 0$ and a separate subroutine to calculate the $k$-vector sum. Smith [1982a] has given an elegant formulation of the extension of the Ewald method to dipoles and quadrupoles; his article contains explicit expressions for forces and torques which will be of use in MD simulations.

A conceptually simple method for modelling dipoles and higher moments is to represent them as partial charges within the core of a molecule. In this case, the Ewald method may be applied directly to each partial charge at a particular site. The only complication in this case is in the self term. In a normal ionic simulation, we subtract the spurious term in the $k$-space summation that arises from the interaction of a charge at $\mathbf{r}_i$ with the distributed charge also centred at $\mathbf{r}_i$. In the simulation of partial charges within a molecule, it is necessary to subtract the terms that arise from the interaction of the charge at $\mathbf{r}_{ia}$ with the distributed charges centred at all the other sites within the same molecule ($\mathbf{r}_{ia}$, $\mathbf{r}_{ib}$, etc.) [Heyes 1983a]. This gives rise to a self-energy of

$$\mathscr{V}^{\text{self}} = \tfrac{1}{2} \sum_i \sum_{a=1}^{n_s} z_{ia} \Bigg( 2\kappa z_{ia}/\pi^{1/2} + \sum_{b \neq a}^{n_s} z_{ib} \operatorname{erf}(\kappa d_{ab})/d_{ab} \Bigg)$$

$$= \sum_i \Bigg( \sum_{a=1}^{n_s} \kappa z_{ia}^2/\pi^{1/2} + \tfrac{1}{2} \sum_{a=1}^{n_s} \sum_{b \neq a}^{n_s} z_{ia} z_{ib} \operatorname{erf}(\kappa d_{ab})/d_{ab} \Bigg) \qquad (5.24)$$

where there are $n_s$ sites on molecule $i$ and the intramolecular separation of sites

$a$ and $b$ is $d_{ab}$. This term must be subtracted from the potential energy.

Detailed theoretical studies [de Leeuw *et al.*, 1980; Felderhof, 1980; Neumann and Steinhauser, 1983a, b; Neumann, Steinhauser and Pawley, 1984] have revealed the precise nature of the simulation when we employ a lattice sum. In this short section we present a simplified picture. For a review of the underlying ideas in dielectric theory, the reader is referred to Madden and Kivelson [1984] and for a more detailed account of their implementation in computer simulations to McDonald [1986].

The simulations which use the potential energy of eqn (5.20) are for a very large sphere of periodic replications of the central box *in vacuo*. As $\varepsilon_s$, the relative permittivity of the surroundings is changed, the potential energy function is altered. For a dipolar system, de Leeuw *et al.* [1980] have shown

$$\mathscr{V}^{\mu\mu}(\varepsilon_s) = \mathscr{V}^{\mu\mu}(\varepsilon_s = 1) - \frac{3k_B T}{N\mu^2} y \frac{(\varepsilon_s - 1)}{(2\varepsilon_s + 1)} \sum_{i=1}^{N} \sum_{j=1}^{N} \boldsymbol{\mu}_i \cdot \boldsymbol{\mu}_j \tag{5.25}$$

where $y = 4\pi\rho\mu^2/9k_B T$. If, instead of vacuum, the sphere is considered to be surrounded by metal ($\varepsilon_s \rightarrow \infty$), the last term in this equation exactly cancels the surface term in eqn (5.21). The potential functions of eqns (5.21) and (5.20) are often used without the final surface terms [Woodcock and Singer 1971; Adams and McDonald 1976] corresponding to a sphere surrounded by a metal. That the sum of the first three terms in these equations corresponds to the case $\varepsilon_s = \infty$ can be traced to the neglect of the term $k = 0$ in the reciprocal space summations.

The relative permittivity $\varepsilon$ of the system of interest is, in general, not the same as that of the surrounding medium. The appropriate formula for calculating $\varepsilon$ in a particular simulation does, however, depend on $\varepsilon_s$ in the following way

$$\frac{1}{\varepsilon - 1} = \frac{1}{3yg(\varepsilon_s)} - \frac{1}{(2\varepsilon_s + 1)} \tag{5.26}$$

where $g$ is related to the fluctuation in the total dipole moment of the central simulation box

$$g = g(\varepsilon_s) = \frac{\left\langle \left| \sum_{i=1}^{N} \boldsymbol{\mu}_i \right|^2 \right\rangle - \left\langle \left| \sum_{i=1}^{N} \boldsymbol{\mu}_i \right| \right\rangle^2}{N\mu^2}. \tag{5.27}$$

Note that the calculated value of $g$ depends upon $\varepsilon_s$ through the simulation Hamiltonian. For $\varepsilon_s = 1$, eqn (5.26) reduces to the Clausius–Mosotti result

$$\frac{\varepsilon - 1}{\varepsilon + 2} = yg(1) \tag{5.28}$$

and for $\varepsilon_s \rightarrow \infty$

$$\varepsilon = 1 + 3yg(\infty). \tag{5.29}$$

A sensible way of calculating $\varepsilon$ is to run the simulation using the potential energy of eqn (5.21) without the surface term, and to substitute the calculated value of $g$ into eqn (5.29). The error magnification in using eqn (5.28) is substantial, and this route to $\varepsilon$ should be avoided. In summary, the thermodynamic properties ($E$ and $P$) and the permittivity are independent of $\varepsilon_s$, whereas the appropriate Hamiltonian and $g$-factor are not. There may also be a small effect on the structure. The best choice for $\varepsilon_s$ would be $\varepsilon$, in which case eqn (5.26) reduces to the Kirkwood formula (Fröhlich 1949]

$$\frac{(2\varepsilon+1)(\varepsilon-1)}{9\varepsilon} = yg(\varepsilon). \tag{5.30}$$

However, of course, we do not know $\varepsilon$ in advance.

### *5.5.3 The reaction field method*

In the reaction field method, the field on a dipole in the simulation consists of two parts: the first is a short-range contribution from molecules situated within a cutoff sphere or 'cavity' $\mathscr{R}$, and the second arises from molecules outside $\mathscr{R}$ which are considered to form a dielectric continuum ($\varepsilon_s$) producing a reaction field within the cavity [Onsager 1936] (see Fig. 5.9). The size of the reaction field acting on molecule $i$ is proportional to the moment of the cavity surrounding $i$,

$$\mathscr{E}_i = \frac{2(\varepsilon_s - 1)}{2\varepsilon_s + 1} \frac{1}{r_c^3} \sum_{j \in \mathscr{R}} \boldsymbol{\mu}_j \tag{5.31}$$

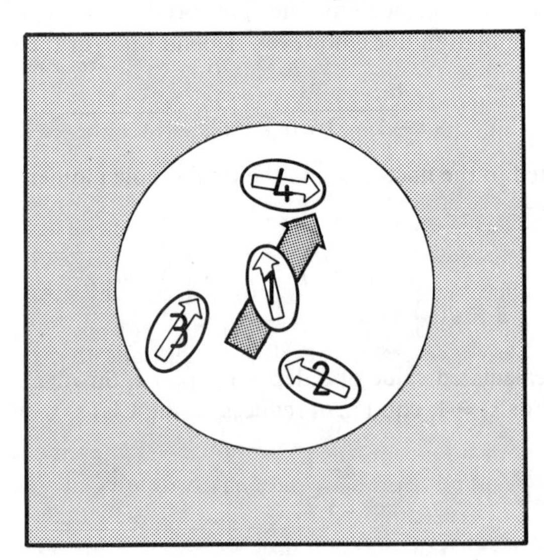


**Fig. 5.9** A cavity and reaction field. Molecules 2, 3, and 4 interact directly with molecule 1. The continuum polarized by the molecules in the cavity produces a reaction field at 1 (shaded).

where the summation extends over the molecules in the cavity, including $i$, and $r_c$ is the radius of the cavity. The contribution to the energy from the reaction field is $-\frac{1}{2}\boldsymbol{\mu}_i \cdot \boldsymbol{\mathscr{E}}_i$. The torque on molecule $i$ from the reaction field is $\boldsymbol{\mu}_i \times \boldsymbol{\mathscr{E}}_i$. Barker and Watts [1973] first used the reaction field in a simulation of water, and there are useful discussions by Friedman [1975] and Barker [1980].

Whenever a molecule enters or leaves the cavity surrounding another, a discontinuous jump occurs in the energy due to direct interactions within the cavity and in the reaction field contribution. These changes do not exactly cancel, and the result is poor energy conservation in MD. In addition, spurious features appear in the radial distribution function at $r = r_c$. These problems may be avoided by tapering the interactions at the cavity surface [Adams, Adams, and Hills 1979]: the explicit interactions between molecules $i$ and $j$ are weighted by a factor $f(r_{ij})$, which approaches zero continuously at $r_{ij} = r_c$. For example, linear tapering may be used:

$$f(r_{ij}) = \begin{cases} 1.0 & r_{ij} < r_t \\ (r_c - r_{ij})/(r_c - r_t) & r_t \leqslant r_{ij} \leqslant r_c \\ 0.0 & r_c < r_{ij} \end{cases} \tag{5.32}$$

where an appropriate value of $r_t$ is $r_t = 0.95\, r_c$. The contribution of the molecular dipoles to the cavity dipole, and hence the reaction field, are correspondingly weighted. Recent investigations of tapering methods [Adams *et al.* 1979; Andrea *et al.* 1983; Berens *et al.* 1983] have suggested that it may be rewarding to adopt more sophisticated formulae than the linear one given above. When partial charges (or multipoles) are used to represent a molecular system, it is also best to define the cavity $\mathscr{R}$ using the centre of a molecule as the origin for the cutoff $r_c$, rather than attempting to apply it site-by-site.

Neumann and Steinhauser [1980], and Neumann [1983] have considered the nature of the simulations when a reaction field is applied. The appropriate formula for calculating the dielectric constant is eqn (5.26) where $g(\varepsilon_s)$ is calculated from the fluctuation in the total moment of the complete central simulation box [Patey, Levesque, and Weiss 1982] as in eqn (5.27).

The static reaction field is straightforward to calculate in a conventional MD or MC simulation, and it involves only a modest increase in execution time. A potential difficulty with the reaction field method is the need for an *a priori* knowledge of the external dielectric constant ($\varepsilon_s$). Fortunately, the thermodynamic properties of a dipolar fluid are reasonably insensitive to the choice of $\varepsilon_s$, and the dielectric constant can be calculated using eqn (5.26).

A possible modification is to take account of the finite time required for the reaction field to respond to changes in the cavity dipole. This 'delayed reaction field' method has been employed in simulations of water [van Gunsteren, Berendsen, and Rullmann 1978]. A disadvantage of this method is that the reaction field does work on the cavity dipole, and so energy is not conserved in a constant-$NVE$ MD simulation.

### 5.5.4 *Other methods*

Our survey of the inclusion of long-range forces in fluid simulation has been necessarily brief. There are two other important techniques, which we can only outline.

The first is the particle–particle and particle–mesh (PPPM) algorithm for ionic systems [Eastwood, Hockney, and Lawrence 1980]. In common with the Ewald method, this algorithm separates the total force on ion $i$ into a long-range and short-range part. The short-range part of the potential is handled normally (see Sections 5.2 and 5.3). The total long-range part of the force on $i$ is calculated using the particle–mesh technique. There are three distinct steps.

(a) The charge density in the fluid is approximated by assigning charges to a finely-spaced mesh in the simulation box.
(b) The fast Fourier transform technique is used to solve Poisson's equation for the electrostatic potential due to the charge distribution on the mesh. This gives the potential at each mesh point.
(c) The field at each mesh point is calculated by numerically differentiating the potential, and then the force on a particular particle $i$ is calculated from the mesh field by interpolation.

This method has been employed in the simulation of the melting of ionic crystals and is described fully by Hockney and Eastwood [1981]. It has the advantage over the Ewald method of taking a time $\mathcal{O}(N)$ at large $N$ rather than $\mathcal{O}(N^2)$, which is particularly useful in the study of large systems.

The second method is a technique due to Ladd [1977, 1978] for studying dipolar systems. The interaction of a dipole $\mu_i$ with all its minimum image neighbours is considered explicitly. As usual, the minimum image 'box' surrounding the particle of interest is considered at the centre of a periodic array of its own replicas. Ladd expands the electric field due to the particles in each of the neighbouring boxes in a multipole expansion around the centre of the simulation box. The energy of a molecular dipole at $\mathbf{r}_i$ is a sum of the interactions with these neighbouring box multipoles plus the explicit minimum image neighbour interaction.

### 5.5.5 *Summary*

Finally, we address the question of when we need to use these complicated schemes for handling the long-range forces. In the case of an ionic liquid, where we use the unscreened Coulomb potential, there is no real choice but to implement the Ewald method, or the particle–mesh method, for accurate calculations of thermodynamic and structural properties. Valleau [1980] and Valleau and Whittington [1977a] have criticized the use of the Ewald sum, because of the unrealistic way an instantaneous dipolar fluctuation of charge in the simulation box is duplicated (rather than being damped out) in the infinite replica system. This objection may be particularly important when behaviour near an electrode is being simulated (see Chapter 11). The evidence so far indicates that structure in a homogeneous fluid, simulated with the Ewald method, is remarkably isotropic, as long as a sufficient number of $k$-vectors are used [Adams 1983b].

In the case of the dipole–dipole interaction, the problem becomes less clear cut. Table 5.2 contains the simple thermodynamic properties for a Lennard-Jones fluid, where the atoms have additional point dipoles at the centre (the Stockmayer potential) [Adams *et al.* 1979]. The long-range forces are handled in a variety of different ways. Apart from the minimum image method, all the techniques give essentially the same result for $E$ and $P$, considering the slight differences in temperature.

**Table 5.2** *Thermodynamic properties of a dipolar fluid [after Adams et al. 1979].* $\rho^* = 0.80$, *and* $\mu^{*2} = 1.0$.

| Method | $-\mathscr{V}^*$ | $P^*$ | $T^*$ |
|---|---|---|---|
| Minimum image | 5.98 | 1.66 | 1.235 |
| Spherical truncation | 6.18 | 1.61 | 1.242 |
| Ewald–Kornfeld | 6.22 | 1.52 | 1.228 |
| Reaction field | 6.17 | 1.69 | 1.266 |
| Tapered spherical truncation | 6.19 | 1.59 | 1.236 |
| Tapered reaction field | 6.20 | 1.64 | 1.243 |

Energy, pressure and temperature are measured in Lennard-Jones reduced units (Appendix B). For the reaction field methods, $\varepsilon_s = 7.0$.

It appears that the long-range forces are not important for simple thermodynamic properties of dipolar fluids. This is also true for the single-particle time correlation functions.

For accurate calculation of dielectric properties, collective correlation functions, and spherical harmonic coefficients of the angular pair distribution function, it is essential to deal with the long-range forces properly. An example is the calculation of the relative permittivity for the Stockmayer fluid [Neumann *et al.* 1984]. Table 5.3 contains $\varepsilon$ calculated in four simulations of the same fluid at the same state point. Two of the simulations use the reaction field method with different values of the relative permittivity of the surrounding medium ($\varepsilon_s$). The other two simulations use the lattice sum method due to Ladd (see Section 5.5.4). In this case the sphere of periodic replications is surrounded by a material of relative permittivity $\varepsilon_s$.

There is good agreement between the results obtained using the reaction field and the lattice sum, and the results for different values of $\varepsilon_s$ are also

**Table 5.3** *The relative permittivity* ($\varepsilon$) *of the Stockmayer fluid.* $\rho^* = 0.822$, $\mu^{*2} = 3.0$ *and* $y = 2.994$ [*after Neumann et al.* 1984].

| Method | $\varepsilon_s$ | $N$ | $T^*$ | $\varepsilon$ | MD time steps |
|---|---|---|---|---|---|
| Reaction field | 70 | 512 | 1.149 | 66.6 | 100 000 |
| Reaction field | $\infty$ | 512 | 1.150 | 66.1 | 150 000 |
| Lattice sum | 70 | 512 | 1.152 | 66.5 | 100 000 |
| Lattice sum | $\infty$ | 256 | 1.149 | 67.2 | 100 000 |

consistent. Recent studies indicate that it is possible to make an accurate estimate of $\varepsilon$ for a model polar fluid using either the reaction field or the lattice sum with a system of 256 molecules.

## 5.6 When the dust has settled

At the end of the central loop of the program, a new configuration of molecules is created, and there are a number of important configurational properties that can be calculated. At this point in the program, the potential energy and the forces on particular molecules are available. The square of the configurational energy, $\mathscr{V}^2$, is calculated so that, at the end of the simulation, $(\langle \mathscr{V}^2 \rangle - \langle \mathscr{V} \rangle^2)$ can be used in eqns (2.73) or (2.82) to calculate the specific heat in the canonical or microcanonical ensemble.

Although the average force and torque on a molecule in a fluid are zero, the mean-square values of these properties can be used to calculate the quantum corrections to the free energy given by eqns (2.140)–(2.143). In a MD simulation, the force $\mathbf{f}_i$ on molecule $i$ from its neighbours is calculated to move the molecules. The forces are not required in the implementation of the Metropolis MC method, so that they must be evaluated in addition to the potential energy if the quantum corrections are to be evaluated. The alternative of calculating the corrections via $g(r)$ (eqn (2.141)) is less accurate.

This is the point in the simulation at which a direct calculation of the chemical potential can be carried out. A test particle, which is identical to the other molecules in the simulation, is inserted into the fluid at random [Widom 1963]. The particle does not disturb the phase trajectory of the fluid, but the energy of interaction with the other molecules, $\mathscr{V}_{\text{test}}$, is calculated. This operation is repeated many times, and the quantity $\exp(-\beta\mathscr{V}_{\text{test}})$ is used in eqn (2.68) to compute the chemical potential.

Another simple way of implementing the method is as follows [Powles *et al.* 1982]. A lattice of points (typically $7 \times 7 \times 7$) is set up in the basic simulation box, and a fictitious particle is placed at each of the lattice points. The grid is inserted at regular intervals (say every 20 time steps or cycles) and $\exp(-\beta\mathscr{V}_{\text{test}})$ evaluated for each particle. In the MD method, the total kinetic temperature, $\mathscr{T}$, fluctuates and it is essential to use eqn (2.68b). The insertion subroutine increases the running time by approximately 20 per cent.

The difficulty with this method is that a large number of substantial overlaps occur when particles are inserted. The exponential is then negligible, and we do not improve the statistics in the estimation of $\mu$. Special techniques may be needed in such cases and we address these in Chapter 7. This is not a severe problem for the Lennard-Jones fluid, where Powles *et al.* [1982] have calculated $\mu$ close to the triple point with runs of less than 8000 time steps. For molecular fluids, Romano and Singer [1979] calculated $\mu$ for a model of liquid chlorine up to a reduced density of $\rho\sigma^3 = 0.4$ (the triple point density is $\approx 0.52$); Fincham, Quirke, and Tildesley [1986] obtained an accurate estimate of $\mu$ by direct insertion at $\rho\sigma^3 = 0.45$ for the same model. Finally, in the case of mixtures, the chemical potential of each species can be determined by inserting a grid containing only molecules of that species. A full discussion of the particle insertion method is given by Powles *et al.* [1982] and Frenkel [1986].

For systems of hard molecules, the pressure is generally not calculated in the expensive inner loop region of the program (whether MC or MD). In fact, in Monte Carlo simulations of hard systems, the usual virial expression for $P$ cannot be used. Eppenga and Frenkel [1984] have reported a trick for estimating $P$ which is rather like the test particle determination of $\mu$. As described by them, the method is restricted to convex hard molecules, and involves calculating the probability of acceptance of a trial box size contraction, just as is used in constant-$NPT$ MC simulations. In this case, the move is a fake: it does not actually take place. For molecules that are not convex, acceptance of a fake box expansion would also need to be considered. Of course, in genuine constant-$NPT$ MC, the pressure is a parameter of the simulation, so these problems need not arise. In molecular dynamics of hard molecules, the virial expression can be recast into a form involving an average over collisions

$$\langle \mathscr{W} \rangle = \frac{1}{3t_{\mathrm{obs}}} \sum_{\mathrm{colls}} \mathbf{r}_{ij} \cdot \delta \mathbf{p}_i \tag{5.33}$$

where $i$ and $j$ represent a colliding pair, $\mathbf{r}_{ij}$ is the vector between the molecular centres at the time of collision, and $\delta\mathbf{p}_i = -\delta\mathbf{p}_j$ is the collisional impulse, i.e. the change in momentum. The sum is over all collisions occurring in time $t_{\mathrm{obs}}$. This expression may also be written in terms of the collision rate and the average of $\mathbf{r}_{ij} \cdot \delta\mathbf{p}_i$ per collision. Further details, including a discussion of the system-size dependence of these formulae may be found elsewhere [Alder and Wainwright 1960; Hoover and Alder 1967; Erpenbeck and Wood 1977].

In this book, we have assumed that calculation of most other properties of interest will be carried out after the simulation, by analysis of an output tape or disk file. This analysis will be the subject of Chapter 6. In some cases, however, it may be preferable to calculate properties such as time correlation functions and structural distributions during the simulation itself. Against this, it must be said that the simulation program may become overcomplicated if too much calculation is included in it. For example, the pair distribution function $g(r)$ involves a sum over pairs of molecules, and this is sometimes included in the inner loop of an MD program. However, $g(r)$ is generally of interest for separations $r$ much greater than the potential cutoff $r_c$, and so we need to examine many more pairs than would be required to calculate the energy and forces. It is sensible to carry out this expensive summation less frequently than once per time step. This means that a logical switch must be included in the inner loop, to indicate when the $g(r)$ calculation is to be switched on, or alternatively a completely separate routine for calculating $g(r)$ should be written, and called (say) every 10 steps or MC cycles. Since a logical switch may inhibit vectorization of the inner loop on some pipeline machines, the latter course of action seems preferable, and the $g(r)$ routine will turn out to be very similar to that used in tape analysis, as described in Chapter 6. In a similar way, time correlation functions may be calculated during an MD simulation, by methods very similar to some of those described in Chapter 6. However, this will require extra storage in the simulation program, and will make the program itself more complicated.

## 5.7 Starting up

In the remainder of this chapter, we consider the overall structure of the simulation programs. It is usual to carry out sequences of runs at different state points, each run following on from the previous one. In both MD and MC techniques, it is necessary to design a starting configuration for the first simulation. For MC, the molecular positions and orientations are specified, and for MD, in addition, the initial velocities and angular velocities must be chosen. For the first run in a series it is important to choose a configuration that can relax quickly to the structure and velocity distribution appropriate to a fluid. This period of equilibration must be monitored carefully, since the disappearance of the initial structure may be quite slow. As a series of runs progresses, the coordinates and velocities from the last configuration of the previous run can be scaled (giving a new density, energy etc.) to form the initial configuration for the next run. Again, with each change in state point, a period of equilibration must be set aside before attempting to compute proper simulation averages.

### *5.7.1 The initial configuration*

The simplest method of constructing a liquid structure is to place molecules at random inside the simulation box (see Appendix G). The difficulty with this technique is that the configuration so constructed may contain substantial overlaps. This would be totally unphysical for a hard-core system. For soft potentials, the energy for most random configurations, although high, can be calculated (provided no two molecules are centred at exactly the same point), so this type of configuration can be used to start Monte Carlo simulations, provided that the system is allowed to relax. In molecular dynamics, on the other hand, the large intermolecular potentials and the correspondingly large forces can cause difficulties in the solution of the stiff differential equations of motion.

It is more usual to start from a lattice. Almost any lattice is suitable, but historically the face-centred cubic structure, with its $4M^3$ ($M = 2, 3, 4, 5 \ldots$) lattice points has been the starting configuration for many simulations. This lattice is shown in Fig. 5.10. The lattice spacing is chosen so that the appropriate liquid state density is obtained. During the course of the simulation the lattice structure will disappear, to be replaced by a typical liquid structure. This process of 'melting' can be enhanced by giving each molecule a small random displacement from its initial lattice point along each of the space-fixed axes [Schofield 1973].

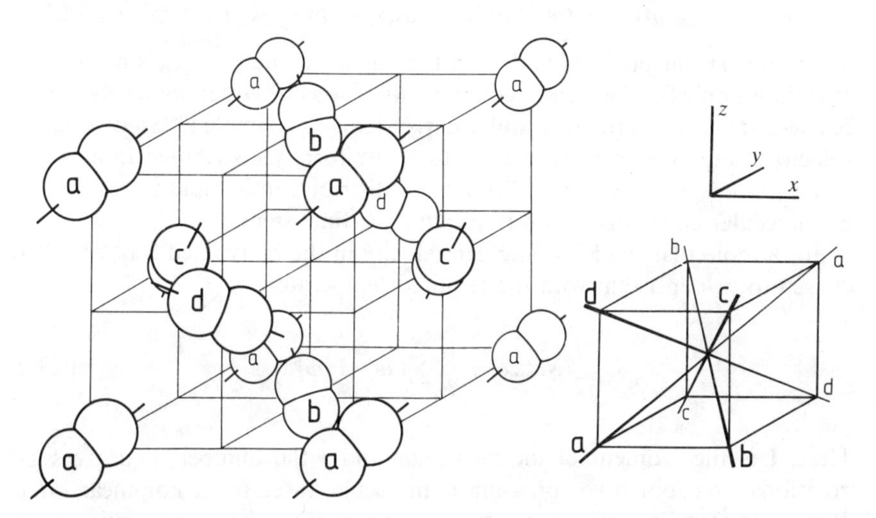


**Fig. 5.10** Unit cell of the f.c.c. structure for linear molecules. The centre-of-mass positions are as in the argon lattice. The orientations are in four sublattices: (a) (1, 1, 1); (b) (1, −1, −1); (c) (−1, 1, −1); (d) (−1, −1, 1).

In the case of a molecular fluid, it is also necessary to assign the initial orientations of the molecules. A model commonly used for linear molecules is the α-f.c.c. lattice, which is the solid structure of $CO_2$ and one of the phases of $N_2$ (see Fig. 5.10). In this structure, there are four sublattices of molecules oriented along the four diagonals of the unit cell. A code for generating the α-f.c.c. lattice is given in program F.23. For non-linear molecules, any suitable known crystal structure could be used. Small random displacements can also be applied to the lattice orientations so as to speed up melting. Some workers prefer to choose the orientations completely randomly given a centre of mass structure, although at high densities, with elongated molecules, random assignment of the directions can result in non-physical overlaps.

### *5.7.2 The initial velocities*

For a molecular dynamics simulation, the initial velocities of all the molecules must be specified. It is usual to choose random velocities, with magnitudes conforming to the required temperature, corrected so that there is no overall momentum

$$\mathbf{P} = \sum_{i=1}^{N} m_i \mathbf{v}_i = 0. \tag{5.34}$$

The velocities may be chosen randomly from a Gaussian distribution (see Appendix G). For example, in an atomic system

$$\rho(v_{ix}) = (m_i/2\pi k_B T)^{1/2} \exp(-\tfrac{1}{2} m_i v_{ix}^2 / k_B T) \tag{5.35}$$

where $\rho(v_{ix})$ is the probability density for velocity component $v_{ix}$, and similar equations apply for the $y$ and $z$ components. The same equations apply to the centre-of-mass velocities in a molecular system. As a simple alternative, each velocity component may be chosen to be uniformly distributed in a range $(-v_{max}, +v_{max})$; the Maxwell–Boltzmann distribution is rapidly established by molecular collisions within (typically) 100 time steps.

For a molecular fluid, the angular velocity in the body-fixed frame is also chosen to be consistent with the required temperature

$$\frac{f}{2} N k_B \mathcal{T} = \frac{1}{2} \sum_{i=1}^{N} \boldsymbol{\omega}_i^b \cdot \mathbf{I} \cdot \boldsymbol{\omega}_i^b . \tag{5.36}$$

Here, $\mathbf{I}$ is the moment of inertia tensor and $f$ the number of degrees of rotational freedom (two for a linear molecule, three for a nonlinear one). Because the total angular momentum is not conserved, it is not essential to set the initial value of this quantity to zero, but it is sensible to ensure that the molecular angular momenta roughly cancel each other. For linear molecules, each angular velocity $\boldsymbol{\omega}_i$ must be chosen perpendicular to the molecular axis (see Appendix G). An example of this technique is given in program F.24. One method for initializing the angular velocity for a lattice configuration involves choosing pairs of molecules with identical orientations and assigning them equal and opposite angular velocities chosen at random. An alternative method is to set the angular velocity of every molecule to zero at the start of the run, and to choose the translational kinetic temperature to be greater than required. The normal process of equilibration will then redistribute the energy amongst the different degrees of freedom. Precise adjustments to the kinetic temperature are made by scaling velocities during equilibration.

### *5.7.3 Equilibration*

If a simulation is started from a lattice, or from a disordered configuration at a different density and temperature, it is necessary to run for a period so that the system can come to equilibrium at the new state point. At the end of this equilibration period, all memory of the initial configuration should have been lost. A simple way to monitor system equilibration is to record the instantaneous values of the potential energy and pressure during this period. In the case of a lattice start, the potential energy rises from a large negative value to a value typical of a dense liquid, as shown in Fig. 5.11(a). The behaviour of the instantaneous pressure is also shown in Fig. 5.11(b). The equilibration period should be extended at least until these quantities have ceased to show a systematic drift and have started to oscillate about steady mean values.

Equilibration is especially important when the initial configuration is a lattice, and the state point of interest is in the liquid region of the phase diagram. There are a number of parameters that can be monitored to track the 'melting' of the lattice, and subsequent progress to equilibrium. The degree of translational order in the centres of mass is tested by evaluating the translational order parameter

$$\rho(k) = \frac{1}{N} \sum_{i=1}^{N} \cos(\mathbf{k} \cdot \mathbf{r}_i) \tag{5.37}$$

where $\mathbf{r}_i$ is the position vector of the centre of mass of the $i$th molecule and $\mathbf{k}$ is a reciprocal lattice vector of the initial lattice. For example, $\mathbf{k} = (2\pi/l)(-1, 1, -1) = ((2N)^{1/3}\pi/L)(-1, 1, -1)$ for f.c.c. where $l$ is the unit cell size, which may be set equal to $L/(N/4)^{1/3}$ in a cubic simulation box. It is, of course, possible to monitor several such components. For a solid, $\rho(k)$ is of order unity, whereas for a liquid it oscillates, with amplitude $\mathcal{O}(N^{-1/2})$ about zero [Verlet 1967]. The translational order parameter for a simulation starting in the f.c.c. lattice is shown in Fig. 5.11(c). It is clear that, in this instance, $\rho(k)$ is a much more sensitive indicator of the persistence of a lattice structure, and of the need to extend the equilibration period, than the 'thermodynamic' quantities shown in Fig. 5.11(a) and (b).

The rotational order parameter, as introduced by Vieillard-Baron [1972] is given, for linear molecules by

$$P_1 = \frac{1}{N} \sum_{i=1}^{N} P_1(\cos\gamma_i) = \frac{1}{N} \sum_{i} \cos\gamma_i \tag{5.38}$$

where $\gamma_i$ is the angle between the molecular axis of molecule $i$ and the original axis direction in the perfect crystal. Several other parameters of this type (for example, higher-order Legendre polynomials) can be monitored. $P_1$ is 1 for the initial configuration and fluctuates around zero with amplitude $\mathcal{O}(N^{-1/2})$ when the fluid is rotationally disordered. For non-linear molecules, several

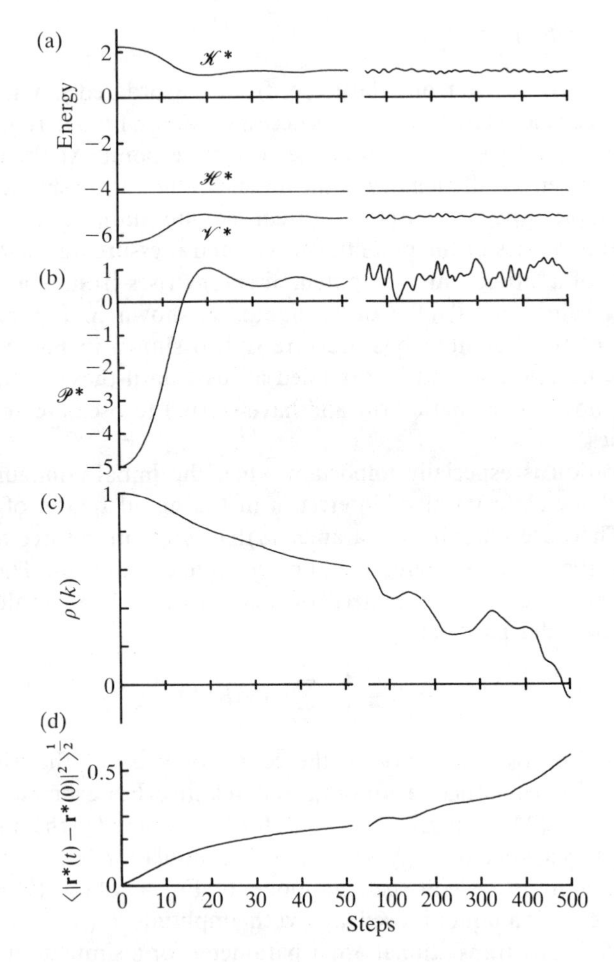


**Fig. 5.11** The equilibration phase of a MD simulation. The first 50 steps are shown in detail. The system consists of 108 atoms interacting via the Lennard-Jones pair potential, starting from an f.c.c. lattice with Maxwell–Boltzman velocity distribution. The system is near the triple point ($\rho^* = 0.8442$, $T^* = 0.722$, $\delta t^* = 0.005$, $r_c^* = 2.5$, no long-range corrections applied). (a) Potential, kinetic, and total energies; (b) instantaneous pressure; (c) translational order parameter; (d) root-mean-square displacement.

similar order parameters, based on different molecular axes, may be examined; they should all vanish simultaneously on 'melting'. An example of an order parameter evaluation subroutine is given in F.25.

An additional strategy involves monitoring the mean squared displacements of molecules from their initial lattice positions. This function increases during the course of a liquid simulation (see eqn (2.110)) but oscillates around a mean value in a solid. A useful rule of thumb is that when the root-mean-squared displacement per particle exceeds $0.5\sigma$ and is clearly increasing, then the system has 'melted' and the equilibration can be terminated. Care should be taken to exclude periodic boundary corrections in the computation of this quantity. This technique is useful for monitoring equilibration not only from a lattice but also from a disordered starting configuration, particularly when there is a danger that the system may become trapped in a glassy state rather than forming a liquid: eqn (5.37) would not be appropriate for these cases.

An additional danger during the equilibration period is that the system may enter a region of gas-liquid coexistence. If a study of a homogeneous fluid is attempted in the two-phase region, large, slow density fluctuations occur in the central simulation box. This is most clearly manifest in the radial distribution function [Jacucci and Quirke 1980a]. In the two-phase region, $g(r)$ has an unusually large first peak ($g(r) \approx 5$), it exhibits long-ranged, slow, oscillations, and does not decay to its correct long-range value of 1. The structure factor $S(k)$ diverges as $k \to 0$, indicating long-wavelength fluctuations. Monitoring these structural quantities may give a warning that the system has entered a two-phase region, in which case extremely long equilibration times will be required.

One useful trick that may be used to increase the rate of equilibration from a lattice, is to raise the kinetic temperature to a high value (e.g. $T^* = 5$ for Lennard-Jones atoms) for the initial 500 steps (for example, by scaling all the velocities). The temperature is reset to the desired value during the second part of the equilibration period. It is sometimes convenient to continually adjust the kinetic temperature of an MD simulation throughout the equilibration phase, using one of the methods described in Chapter 7.

It is difficult to say how long a run is needed to guarantee equilibration, but periods of 500–1000 time steps or MC cycles are typical (remember (Section 4.4) that for an $N$-atom system, one MC cycle is $N$ attempted moves). More time should be set aside for equilibration from an initial lattice, or whenever it is suspected that a phase transition is close; somewhat less time is required for high-temperature fluids. The golden rule is to examine carefully the parameters mentioned above, as the simulation proceeds. At the end of equilibration, they should have clearly reached the expected limiting behaviour. In an MD simulation, it is also worthwhile to check the proper partitioning of kinetic temperature for a molecular system (i.e. $\mathcal{T}_{\text{rot}} \approx \mathcal{T}_{\text{trans}}$) although it should be remembered that these instantaneous values are subject to significant fluctuations. At the end of the equilibration period, the accumulators for the ensemble averages are reset to zero, and the production phase of the simulation begins.

## 5.8 Organization of the simulation

Computer simulations are programs that require a substantial amount of CPU time. A typical MD or MC simulation of a fluid will take over an hour of time on a CDC 7600. Simulations that are run on a minicomputer may take many days or weeks of dedicated computing. For this reason, simulations should be designed so that they can be restarted with the minimum difficulty. The restart facility enables the total simulation to be broken up into manageable chunks of computing time. In the event of an unexpected machine failure, the program can be started again with a minimum loss of computing resources. It may even be possible to make the simulation self-starting, so that it can be run as a series of small jobs without human intervention. The details of job organization clearly depend on the particular computer being used; in this section, we try to illustrate one scheme which, we hope, will be fairly generally applicable and easily adapted.

### *5.8.1 Input/output and file handling*

Ideally, manipulation of files by the user should be kept to a minimum. Only a handful of parameters define the important features of a simulation: the run length, step size, desired temperature etc. These numbers can be stored in a small input file, which can easily be accessed and altered by the user. Accordingly, they should be read in by the program as formatted records—preferably free format should be used, and it seems sensible to associate this file with the FORTRAN-77 default input channel. Consequently, the basic parameters are read in by a READ(*, *) statement. Equally, the user should be able to read the essential output from the simulation: the instantaneous values of energy, pressure etc. will be needed at frequent intervals, and at the end of the run, the simulation averages of a number of quantities will be of immediate interest. This information can be directed to the FORTRAN-77 default output channel, using WRITE(*, $f$) statements, where $f$ represents a format statement number or specifier. This channel is also the destination of any run time error messages generated by the program. Instead of files on disk, of course, the default input and output channels could be associated with a card reader and line printer, or with a terminal keyboard and screen, respectively.

The remaining information required by the program need only be in machine-readable form, i.e. for reasons of economy of storage space can be accessed using unformatted READ and WRITE statements. Typically, the molecular positions, velocities, accelerations etc. (the starting configuration) can be stored in a file of this kind. It is useful to update this information, and store the current configuration together with average accumulators at regular intervals during the run. By over-writing this file frequently (perhaps every 100–500 steps or cycles) we make it easy to restart the simulation in the event of program (or machine) failure; consequently, the file is best placed on disk. We shall refer to this file as the configuration file; it is written out for the last time at the end of the simulation, and is saved for future use (i.e. as the starting point for a new run). Finally, we come to the very large file that stores molecular positions, velocities, and accelerations, taken at frequent intervals (typically 10 steps) during the run, for future analysis. Eventually, this vast amount of information must be stored on magnetic tape, although on some computers it may be possible to use an intermediate, temporary disk file during the simulation, copying the results to tape afterwards; this avoids dedicating a tape deck to the program for the entire period of the run. Since these configurations are not required for restarting the simulation, it may be possible to condense the information (e.g. using reduced precision variables) before writing it out (unformatted) so as to save space. We shall refer to this file as the tape file, or simply as the tape.

### *5.8.2 Program structure*

In the following, we will assume that all manipulation of the configuration file is performed in separate utility programs. This includes the initial generation of a configuration at the start of a series of simulations (for more details, see Section 5.7), the scaling of coordinates or velocities to generate a different state point from a previous final configuration, and the setting to zero of the ancillary information (step number, average accumulators etc.), which should be carried out at the start of a simulation at a new state point. In short, the utility programs produce a configuration file which is 'ready to go'. By separating these activities from the main simulation program, the latter is kept simple in structure: it accepts an initial configuration and, during the course of the run, maintains it in a state suitable for continuing the run. The 'configuration handling' utility programs may be run interactively or as batch jobs in between runs of the main program.

In our scheme, then, the main program begins by reading in the run parameters, with statements like the following.

```
READ (*,*) TITLE
READ (*,*) CNFILE
READ (*,*) TPFILE
READ (*,*) NSTEP, ISAVE, IPRINT, ITAPE, EQLBRT
READ (*,*) DT, DENS, TEMP

WRITE (*,'(1X,''SIMULATION RUN      '',1X,A     )') TITLE
WRITE (*,'(1X,''CONFIGURATION FILE  '',1X,A     )') CNFILE
WRITE (*,'(1X,''TAPE FILE           '',1X,A     )') TPFILE
WRITE (*,'(1X,''NUMBER OF STEPS     '',1X,I10   )') NSTEP
WRITE (*,'(1X,''SAVE INTERVAL       '',1X,I10   )') ISAVE
WRITE (*,'(1X,''PRINT INTERVAL      '',1X,I10   )') IPRINT
WRITE (*,'(1X,''TAPE WRITE INTERVAL'',1X,I10   )') ITAPE
WRITE (*,'(1X,''EQUILIBRATION FLAG  '',1X,I10   )') EQLBRT
WRITE (*,'(1X,''TIME STEP           '',1X,F15.8)') DT
WRITE (*,'(1X,''DENSITY             '',1X,F15.8)') DENS
WRITE (*,'(1X,''TEMPERATURE         '',1X,F15.8)') TEMP
```

In the first three statements, we read in a run title, and the names of the configuration and tape files, to character variables, which will have been declared in a statement of the kind

```
CHARACTER TITLE*50, CNFILE*30, TPFILE*30
```

at the start of the program. For list-directed input, as here, the data items will take the form of character constants

'Descriptive run title'
'Filename1'
'Filename2'

whereas the apostrophes may be omitted if a statement with format control such as READ (*, '(A)') CNFILE is used instead. It is very useful to identify each run using a suitably descriptive title, if only to allow easy recognition of the output at a later date. If date and time routines are available, then again it is convenient to write these out at the start of the run, for easy reference. In the same way, it is most advisable to write out the basic run parameters, exactly as they are read in, at the start of the output file, thus avoiding any ambiguity about the nature of the simulation, when the results are studied later. Here, we are reading in the number of steps in the entire run, and the intervals (in steps) between successive print-outs of thermodynamic information, between successive writes to the tape file, and between successive updates of the configuration file. We also read a flag or switch controlling (as an example) which of two algorithms to use for moving the molecules: these might be a standard molecular dynamics algorithm (EQLBRT=0) and a special equilibration algorithm (EQLBRT=1), which continually rescales the molecular velocities so as to maintain a desired temperature. Information defining the state point (for example temperature and density) and parameters such as the time step (or in Monte Carlo, the maximum size of moves) are also read in. Unless we are dealing with a very simple pair potential, we would want to include potential parameters, the length of potential cutoffs, and perhaps the relative molecular masses in this list of input parameters. Where appropriate, it will be necessary to read in parameters governing the size and precision of any potential lookup tables, and also the size of any neighbour list structures to be used in the program.

Following the initial input of parameters, it will be necessary to define subsidiary quantities. For example, we might require the value of $\sigma$ for a pair potential, in units of the box length, and this must be computed from the density and the number of molecules in the simulation. Other parameters, which may be most conveniently input in reduced units based on the potential, or even in SI units, will have to be converted into the units appropriate to the simulation program. This may be conveniently done in a separate routine, called by a statement of the form

```
CALL SETUP ( N, DENS, ......, SIGMA, ...... )
```

If we are using potential tables and neighbour lists, we will also have to initialize them. Generally, it is advisable to delegate each separate task of this kind to a separate subroutine, so as to maintain a simple and clear, modular, program structure. The next stage of the program is to open the configuration file and read it in. Typically, the code might look like this:

```
        OPEN ( UNIT = CONFIG, FILE = CNFILE,
     :         STATUS = 'UNKNOWN', FORM = 'UNFORMATTED' )

        READ ( CONFIG ) RX, RY, RZ, VX, VY, VZ
        READ ( CONFIG ) STEP, ACE, ACES, ACV, ACVS, ......

        CLOSE ( UNIT = CONFIG )
```

Here, CONFIG is the unit number, an INTEGER variable, perhaps specified in a PARAMETER statement at the start of the program. The STATUS='OLD' option should ensure that an error will be generated, and the program will crash, if the configuration file does not exist. If a 'softer' failure option is required, perhaps with an explanatory error message, the ERR = parameter in the OPEN statement may be used. The first record of the file, we have assumed, contains molecular positions, velocities etc., while the second contains the INTEGER variable STEP, which indicates how many steps had been completed when these variables were written out, and the various accumulators (for energy, squared energy, potential energy, squared potential energy etc.) which are incremented as the program proceeds. At the start of the run, these quantities will usually have been initialized with zero values by the configuration handling program. They will be non-zero if it has been decided to extend a previous run to a larger number of steps (when we simply change the value of NSTEP in the input file and let the run pick up where it left off) or when we are restarting a crashed simulation with the same input file and the configuration file, which will have been updated by the program some time before the crash. The configuration file is closed in the statements above, but it will be reopened, overwritten, and closed again many times during the simulation, for this purpose.

Next, if desired, the large tape file should be opened ready for output. Let us assume that a value of ITAPE=0 indicates that no tape output is required. In that case,

```
        IF ( ITAPE .GT. 0 ) THEN

           OPEN ( UNIT = TAPE, FILE = TPFILE,
     :            STATUS = 'UNKNOWN', FORM = 'UNFORMATTED' )

        ENDIF
```

Here, the STATUS = 'UNKNOWN' option should simply open the file if it exists already, but should create a new file otherwise. This is important, since a new file will normally be required at the start of a new run, but the existing information will be needed in the case of a restart. This information may be preserved by reading through the appropriate number of records in the tape file, and appending records as the simulation proceeds afterwards. It is easiest to do this if each tape file record includes the time step at which it was written out (TSTEP) and if we may assume that the interval between updates of the configuration file (ISAVE) is longer than the interval between writes to the tape file (ITAPE). In that case, if STEP is non-zero, we can be sure that it will be necessary to read in at least one record from the tape file, and we can use the following simple code, which compares the time-step for each tape record with the time step of the configuration file from which we are starting, or restarting, the run. The code would probably be more elegant if we could use the DO WHILE . . . ENDDO construction available in many extended FORTRAN-77 compilers, but unfortunately this is not part of the standard language.

```
      IF ( ( STEP .GT. 0 ) .AND. ( ITAPE .NE. 0 ) ) THEN
10       READ ( TAPE ) TSTEP, ... and information for this step ...
         IF ( ( TSTEP + ITAPE ) .LE. STEP ) GOTO 10
      ENDIF
```

This positions the tape file at exactly the right point for future output. Note that this may not be the end of the tape file: if tape output is more frequent than the updating of the configuration file, then all the records between the time of this last update and the (presumed) crash of the previous program are not wanted, but will be reproduced in the current simulation. This code will need modification if, for example, we ever wish to extend a run but prefer to create a new tape file rather than append to an existing one.

We are almost ready to deal with the body of the simulation. Before doing so, for a molecular dynamics simulation, we may have to call some routine to compute the initial forces on the molecules, if these were not included in the configuration file. For Monte Carlo, there will be an initial calculation of the total energy, and perhaps other properties, of the system. Thus, we should CALL FORCE or CALL ENERGY as appropriate. Finally, if any of the 'interval' parameters ISAVE, IPRINT, ITAPE etc., have been given zero values (to indicate that no output is required) then these values should be reset so that they can be used in the FORTRAN MOD ( . . . ) function without causing an overflow.

```
      IF ( ISAVE  .EQ. 0 ) ISAVE  = NSTEP + 1
      IF ( IPRINT .EQ. 0 ) IPRINT = NSTEP + 1
      IF ( ITAPE  .EQ. 0 ) ITAPE  = NSTEP + 1
```

These new values will fulfil the required function, as will be seen below. The main program loop may be written in the following way; note that, again, we are essentially programming a construct of the form DO WHILE (STEP .LT. NSTEP) . . . ENDDO.

```
100     IF ( STEP .LT. NSTEP ) THEN

           CALL MOVE ( N, DT, RX, RY, RZ, VX, VY, VZ,
     :                 AX, AY, AZ, V, K, W, ... )

           CALL CALC ( ... any other quantities for this step ... )

           STEP = STEP + 1
           ACE  = ACE  + E
           ACES = ACES + E ** 2

           ... other accumulators treated similarly ...

           IF ( MOD ( STEP, IPRINT ) .EQ. 0 ) THEN

              AVE = ACE / REAL ( STEP )

              ... calculate other running averages similarly ...
              ... write out information to output channel    ...

           ENDIF

           IF ( MOD ( STEP, ITAPE ) .EQ. 0 ) THEN

              ... compute data needed for tape file ...

              WRITE ( TAPE ) STEP, ... and desired information ...

           ENDIF

           IF ( MOD ( STEP, ISAVE ) .EQ. 0 ) THEN

              OPEN ( UNIT = CONFIG, FILE = CNFILE,
     :               STATUS = 'OLD', FORM = 'UNFORMATTED' )

              WRITE ( CONFIG ) RX, RY, RZ, VX, VY, VZ
              WRITE ( CONFIG ) STEP, ACE, ACES, ...

              CLOSE ( UNIT = CONFIG )

           ENDIF

           GOTO 100

        ENDIF

        IF ( ITAPE .LE. NSTEP ) CLOSE ( UNIT = TAPE )

        STOP
```

The MOVE subroutine is really shorthand for any of the MD or MC algorithms mentioned in the earlier chapters, complete with any force or energy evaluation routines that may be necessary. In the course of this, the current values of the potential and kinetic energies, virial function etc. will probably be calculated. Any other instantaneous values of interest, such as order parameters and distribution function histograms, should be computed in subroutines immediately following the MOVE routine (indicated schematically above by CALL CALC). The step counter, and the various accumulators, are then incremented as shown. Following these calculations, at the appropriate intervals, information on the progress of the run is sent to the default output channel, data (possibly in condensed form) is output to the tape file, and, lastly (for reasons of safety), the current coordinates, momenta, and running totals are output to the configuration file, which will be used for a restart if need be. The whole loop is repeated until the desired number of steps have been executed.

The above scheme will only allow restarts, of course, if files (specifically the tape file) that were open at the time of a program crash remain un-corrupted and accessible afterwards. If this is not the case, then more complicated steps must be taken to maximize the safety of the information generated by the program. This may involve opening, appending to, and then closing the tape file at frequent intervals (perhaps every record).

### *5.8.3 The scheme in action*

The scheme outlined above would typically be run as a series of simulations alternating with configuration-modifying programs. Typically, the configuration handler would be used to generate the initial configuration, either from scratch (see Section 5.7) or by modifying the final configuration from an old run. This would be followed by an equilibration run, with the flag EQLBRT set to 1, to select the appropriate algorithm, and probably without any output to the tape file (so ITAPE would be zero and TPFILE would be ignored by the program). The output configuration file from the equilibration run would then be used by the configuration utility routine to generate an initial configuration for the production phase. This might involve simply setting the step counter and all the accumulators to zero. The production run would then be carried out. At any stage, should a crash occur, the program could be resubmitted with an identical set of input data, and would carry on from the last update of the configuration file. All this assumes that, after each run, the user will examine the output before setting the subsequent run in motion. This is usually to be recommended: it is always desirable to check that, for example, equilibration has proceeded satisfactorily, or that any crash that may have occurred is due to a machine failure rather than a program fault or the filling of the available file store. However, it would also be quite possible to set up a sequence of jobs to run consecutively, without intervention. It is even possible to allow for automatic recovery from program failure, by submitting several identical runs in succession. A crash of program 1 would then be picked up by program 2, and so on. Should program 1 finish normally, then, as can be seen above, the subsequent programs will simply open, read, and close the tape and configuration files, and then stop. Clearly, it is possible to concoct much more sophisticated schemes to control the running of simulations. If it is possible to set flags to indicate successful (or otherwise) completion of a program, and then to use these in a control language which can take decisions governing the submission of subsequent jobs, then the whole process can be made automatic. It is still worth emphasizing, however, the importance of looking at the raw results as soon as possible after the jobs have run.

Once a simulation is completed, the line printer output should be carefully filed. Simulations tend to produce a wealth of information, and a large quantity of output. In our experience, unless this output is processed and filed immediately, it will be difficult to retrace the work at a later date. The most valuable information from the simulation may be the configurations stored on tape (the 'tape file'). These may well be analysed and re-examined for many years and form a useful database for future work. Magnetic tapes are notoriously unreliable, and a back-up copy should be made as soon as possible.

# 6

# HOW TO ANALYSE THE RESULTS

## 6.1 Introduction

It is probably true that the moment the simulation run is finished, you will remember something important that you have forgotten to calculate. Certainly, as a series of simulation runs unfolds and new phenomena become apparent, you may wish to reanalyse the configurations to calculate appropriate averages and correlation functions. To this end, configurations generated in the run are often stored on a magnetic tape or disk, which is then used for all subsequent analysis (see Section 5.8).

In a MC simulation it would be inappropriate to store every configuration since neighbouring configurations are identical or highly correlated. Typically, the configuration at the end of every 5th or 10th cycle is stored (one cycle $= N$ attempted moves). Each stored configuration will contain the vectors describing the positions of the atoms, and in the case of a molecular fluid, each orientation. It is also convenient to store the instantaneous values of the energy, virial, and any other property of interest. Although these properties can be reconstructed from the positions of the particles this is often an expensive calculation.

Equally, in a MD simulation, successive time steps are correlated and do not contain significantly new information. In this case it is sufficient to store every 5th or 10th time step on the tape for subsequent analysis. An MD simulation produces a substantial amount of useful information, and it is normal to store vectors of the positions (orientations), velocities (angular velocities), and forces (torques) for each molecule, as well as the instantaneous values of all the calculated properties. The information stored in an MD simulation is time-ordered and can be used to calculate the time correlation functions discussed in Chapter 2. The molecular positions that are stored from the MD simulation may be for particles in the central box which have been subjected to the periodic boundary conditions. It is also useful to store trajectories which have not been adjusted in this way, but which represent the actual movement of a molecule in space. These trajectories are particularly useful in calculating the self diffusion coefficient. It is possible to convert from one representation to the other, on the assumption that molecules do not naturally move distances of the order of half a box length in the interval between stored time steps. Programs for doing this are given in F.26.

In this chapter, we discuss how to analyse a tape or disk file so as to produce structural distribution functions and time correlation functions. We then proceed to the important question of assessing the statistical errors in the simulation results. Finally, we outline some techniques used to correct, extend, or smooth the raw data.

## 6.2 Liquid structure

We have assumed that the analysis of liquid structure will take place after a simulation is completed. As mentioned in Chapter 5, it is possible to do this during the simulation run itself, using methods very similar to those described here.

The pair distribution function $g(r)$ is formally defined by eqn (2.94), but is most simply thought of as the number of atoms a distance $r$ from a given atom compared with the number at the same distance in an ideal gas at the same density.

We calculate $g(r)$ as follows. Configurations are read from the tape in turn and the minimum image separations $r_{ij}$ of all the pairs of atoms are calculated. These separations are sorted into a histogram (HIST) where each bin has a width $\delta r$ (DELR) and extends from $r$ to $r+\delta r$. A typical piece of FORTRAN code for the sorting of the $N$ atoms is

```
        DO 100 I = 1, N - 1

           DO 99 J = I + 1, N

              ... calculate minimum image distances ...

              RIJSQ = RXIJ * RXIJ + RYIJ * RYIJ + RZIJ * RZIJ
              RIJ   = SQRT ( RIJSQ )
              BIN   = INT ( RIJ / DELR ) + 1

              IF ( BIN .LE. MAXBIN ) THEN

                 HIST(BIN) = HIST(BIN) + 2

              ENDIF
99         CONTINUE

100     CONTINUE
```

In this code, BIN is an INTEGER variable, and MAXBIN is the size of the HIST array. The *ij* and *ji* separations are sorted simultaneously, and the IF statement is used to limit the calculation of $g(r)$ to distances less than some maximum, say half the box-length.

When all the configurations have been processed, the histogram HIST must be normalized to calculate $g(r)$. Suppose there are $\tau_{\text{run}}$ steps on the tape, and a particular bin $b$ of the histogram, corresponding to the interval $(r, r+\delta r)$, contains $n_{\text{his}}(b)$ pairs. Then the average number of atoms whose distance from a given atom in the fluid lies in this interval, is

$$n(b) = n_{\text{his}}(b)/(N \times \tau_{\text{run}})\,. \tag{6.1}$$

The average number of atoms in the same interval in an ideal gas at the same density $\rho$ is

$$n^{\mathrm{id}}(b) = \frac{4\pi\rho}{3}[(r+\delta r)^3 - r^3]. \tag{6.2}$$

By definition, the radial distribution function is

$$g(r+\tfrac{1}{2}\delta r) = n(b)/n^{\mathrm{id}}(b) \tag{6.3}$$

and the code for normalizing HIST is

```
        CONST = 4.0 * PI * RHO / 3.0

        DO 10 BIN = 1, MAXBIN

           RLOWER   = REAL ( BIN - 1 ) * DELR
           RUPPER   = RLOWER + DELR
           NIDEAL   = CONST * ( RUPPER ** 3 - RLOWER ** 3 )
           GR(BIN) = REAL(HIST(BIN)) / REAL(NSTEP) / REAL(N) / NIDEAL

10      CONTINUE
```

Note that NIDEAL is REAL! The appropriate distance for a particular element of our $g(r)$ histogram is at the centre of the interval $(r, r+\delta r)$, i.e. at RLOWER + DELR/2 in the above example.

The double loop code for sorting separations is quite expensive, but cannot be vectorized because the array HIST is not accessed sequentially. Fincham [1983] has discussed a method of calculating $g(r)$ by sorting over the histogram bins rather than the molecules, which is suitable for use on pipeline and parallel processors. Our code involves taking a square-root for each pair in every configuration. This aspect of the calculation can be speeded up using the technique for calculating square roots discussed in Section 5.2.2. It is also possible to sort the squared distances directly into a histogram and to calculate $g(r^2)$. A disadvantage of this is that the resulting $g(r)$ is obtained at uneven intervals in $r$ with a larger spacing at small $r$, which is just the region in which the function is required with the highest precision. Extrapolation and interpolation is difficult at small $r$ because the function is rapidly varying (see Section 6.5.3).

An identical sorting technique can be applied to the site–site pair distribution functions mentioned in Section 2.6, and to the spherical harmonic coefficients defined in eqn (2.99). In the latter case, we average in a shell as follows [Streett and Tildesley 1976; Gray and Gubbins 1984]

$$g_{ll'm}(r_{ij}) = 4\pi g_{000}(r_{ij}) \langle Y^*_{lm}(\mathbf{\Omega}_i)\, Y^*_{l'\bar{m}}(\mathbf{\Omega}_j) \rangle_{\mathrm{shell}} \tag{6.4}$$

where $\bar{m} = -m$. In this equation, $\langle \ldots \rangle_{\mathrm{shell}}$ has the following interpretation. For each pair $ij$, a particular bin of the $g_{000}(r_{ij})$ histogram, corresponding to a molecular centre–centre separation $r_{ij}$, is incremented by two, just as in the atomic case. At the same time, the corresponding bin of each $g_{ll'm}$ histogram should have $Y^*_{lm}(\mathbf{\Omega}_i)\, Y^*_{l'\bar{m}}(\mathbf{\Omega}_j) + Y^*_{lm}(\mathbf{\Omega}_j)\, Y^*_{l'\bar{m}}(\mathbf{\Omega}_i)$ added to it. At the end of the calculation, each $g_{ll'm}$ histogram bin is divided by the corresponding element of the $g_{000}$ histogram. The result is the shell average in eqn (6.4). The function $g_{000}(r)$ is then calculated from its histogram in the usual way, and used in eqn (6.4) to give the other $g_{ll'm}$ functions.

## 6.3 Time correlation functions

In this section, we consider the calculation of time correlation functions from the tape file that contains positions, velocities, and accelerations stored at regular intervals during a molecular dynamics simulation. Bearing in mind that a wide variety of correlation functions may be of interest, analysis of a tape is logistically simpler than the alternative of calculating the correlation functions during the simulation run itself. However, it is possible to do some analysis of this kind during a simulation, and we shall return to this briefly below.

### *6.3.1 The direct approach*

The direct approach to the calculation of time correlation functions is based on the definition eqn (2.105). Suppose that we are interested in a mechanical property, $\mathscr{A}(t)$, which may be expressed as a function of particle positions and velocities. $\mathscr{A}(t)$ might be a component of the velocity of a particle, or of the microscopic pressure tensor, or a spatial Fourier component of the particle density, for example. From the data in the tape file, $\mathscr{A}(t)$ will be available at equal intervals of time $\delta t$; typically $\delta t$ will be a small multiple of the time step used in the simulation. We use $\tau$ to label successive steps on tape, i.e. $t = \tau \delta t$. The definition of time-average, in a discretized form, allows us to write the non-normalized autocorrelation function of $\mathscr{A}(t)$ as

$$C_{\mathscr{A}\mathscr{A}}(\tau) = \langle \mathscr{A}(\tau)\mathscr{A}(0) \rangle = \frac{1}{\tau_{\max}} \sum_{\tau_0 = 1}^{\tau_{\max}} \mathscr{A}(\tau_0)\mathscr{A}(\tau_0 + \tau). \tag{6.5}$$

In words, we average over $\tau_{\max}$ time origins the product of $\mathscr{A}$ at a time $\tau_0 \delta t$ and $\mathscr{A}$ at a time $\tau \delta t$ later. For each value of $\tau$, the value of $\tau_0 + \tau$ must never exceed the number of values of $\mathscr{A}$, $\tau_{\text{run}}$, stored on the tape. Thus the short-time correlations, with $\tau$ small, may be determined with slightly greater statistical precision because the number of terms in the average, $\tau_{\max}$, may be larger. We return to this in Section 6.4. Again, as written, eqn (6.5) assumes that each successive data point is used as a time origin. This is not necessary, and indeed may be inefficient, since successive origins will be highly correlated. A faster calculation will result from summation over every fifth or tenth point as time origin (with a corresponding change in the normalizing factor $1/\tau_{\max}$) and with little degradation of the statistics.

The calculation may be repeated for different values of $\tau$, and the result will be a correlation function evaluated at equally spaced intervals of time $\delta t$ apart, from zero to as high a value as required. In principle, $\tau \delta t$ could extend to the entire time spanned by the data, but the statistics for this longest time would be poor, there being just one term in the summation for eqn (6.5) (the product of the first and last values of $\mathscr{A}$). In practice, $C_{\mathscr{A}\mathscr{A}}(t)$ should decay to zero in a time which is short compared with the complete run time, and it may be that only a few hundred values of $\tau$ are of interest.

A simple subroutine to calculate an autocorrelation function might take the following form.

```
        SUBROUTINE CORFUN ( TRUN, TCOR, A, ACF, NORM )

        INTEGER TRUN, TCOR
        REAL    A(TRUN), ACF(0:TCOR), NORM(0:TCOR)

        INTEGER T, TO, TTO, TTOMAX
        REAL    AO

        DO 100 T = 0, TCOR

           ACF(T)  = 0.0
           NORM(T) = 0.0

100     CONTINUE

        DO 200 TO = 1, TRUN

           AO     = A(TO)
           TTOMAX = MIN ( TRUN, TO + TCOR )

           DO 199 TTO = TO, TTOMAX

              T       = TTO - TO
              ACF(T)  = ACF(T) + AO * A(TTO)
              NORM(T) = NORM(T) + 1.0

199        CONTINUE

200     CONTINUE

        DO 300 T = 0, TCOR

           ACF(T) = ACF(T) / NORM(T)

300     CONTINUE

        RETURN
        END
```

The central loop of this routine is slightly more efficient than the more obvious alternative.

```
        DO 200 T= 0, TCOR

           TMAX = TRUN - T
           ATA0 = 0.0

           DO 199 T0 = 1, TMAX

              ATA0 = ATA0 + A(T0) * A(T0 + T)

199        CONTINUE

           ACF(T) = ATA0 / REAL ( TMAX )

200     CONTINUE
```

Because inverting the order of the loops makes the correct normalizing factor less obvious, we have included the foolproof counter NORM(T) in our example. In fact, NORM(T) should be equal to REAL(TRUN-T) in this case. To select origins less frequently, the outer DO LOOP in our example should be replaced by

```
DO 200 T0 = 1, TRUN, TGAP
```

with TGAP equal to 5 or 10 for example. The modification of the subroutine to deal with cross-correlations $\langle \mathscr{A}(t)\mathscr{B}(0)\rangle$ is straightforward.

In the previous example we have assumed for simplicity that all the values of $\mathscr{A}(\tau)$ can be stored in memory at once. On modern machines, memory is quite cheap, and so this is commonly true. Some mainframes have extensive 'secondary' storage for large arrays, to improve the efficiency of handling them. On virtual machines, even if such a large amount of data cannot all be held in memory at once, the transfers to and from the disk are handled efficiently and transparently, so that our sample subroutine should still work.

If memory limitations are severe, the next best alternative is to calculate the values of $\mathscr{A}(\tau)$ and store them in a disk file, where they can be manipulated by FORTRAN direct access I/O statements. In this case, the central part of our correlation function routine might have the form

```
        DO 200 T0 = 1, TRUN

           READ ( DISK, REC = T0 ) A0
           TT0MAX = MIN ( TRUN, T0 + TCOR )

           DO 199 TT0 = T0, TT0MAX

              T = TT0 - T0
              READ ( DISK, REC = TT0 ) AT
              ACF(T) = ACF(T) + A0 * AT
              NORM(T) = NORM(T) + 1

199        CONTINUE
200     CONTINUE
```

Here DISK is the FORTRAN INTEGER variable representing the appropriate logical unit. Note that, most of the time, we are reading data sequentially from the disk, so it would only take a few modifications to replace the direct access statements above with sequential I/O statements, with REWIND and BACKSPACE statements in the appropriate places. On some systems, tape manipulation occurs by reading the entire tape onto a large disk, and conducting all subsequent manipulations by disk access, making frequent REWINDS fairly cheap and harmless.

In the worst case, it may be necessary to analyse the data by reading directly from a magnetic tape, when the priority must be to avoid repetitive physical rewinding of the tape itself. Assuming that enough memory is available to store all the desired elements of the autocorrelation function (rather than the data), it is possible to carry out the calculation with a single sweep through the data on a tape or disk. In this method, $\tau_{cor}$ time steps are read into memory, where $(\tau_{cor} - 1)\delta t$ is the maximum time for which the correlation function is required. As each step is read from tape, the correlations with all previous steps are computed. In the example of Fig. 6.1(a), step 4 is correlated with the first three steps. When $\tau_{cor}$ steps have been read (Fig. 6.1(b)), the information in step 1 is no longer needed, and step $\tau_{cor} + 1$ is read into this location. This is correlated with all the other steps (Fig. 6.1(c)). The next step is read into location 2, and the correlation proceeds (Fig. 6.1(d)). There is no need to rewind the tape, but storage requirements once again become high as soon as several correlation functions are required (for example, all the single-particle velocity autocorrelations). A sample program is given in F.27. Essentially this same method can be used to calculate correlation functions while the run is in progress, avoiding all use of tape or disk storage.

### *6.3.2 The fast Fourier transform method*

It is possible to improve the speed of calculating time correlation functions by taking advantage of the very rapid algorithms available for computing discrete Fourier transforms. This particular application of the fast Fourier transform (FFT) was proposed by Futrelle and McGinty [1971] and some details are given by Kestemont and van Craen [1976] and by Smith [1982b, c]. The method is an application of the convolution/correlation theorem given in Appendix D. Apart from the normalizing factor $\tau_{max}$ (which may be incorporated later) the discrete correlation function, eqn (6.5), may be written as

$$C'_{\mathscr{A}\mathscr{A}}(\tau) = \sum_{\tau_0 = 1}^{2\tau_{run}} \mathscr{A}(\tau_0)\mathscr{A}(\tau_0 + \tau) \qquad 0 \leqslant \tau < 2\tau_{run}. \qquad (6.6a)$$

The prime reminds us of the dropped normalization. The sum runs over twice the actual number of data points: in this equation it is assumed that we have

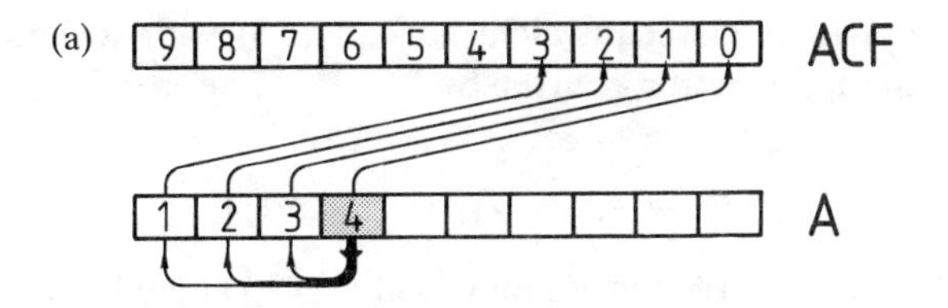


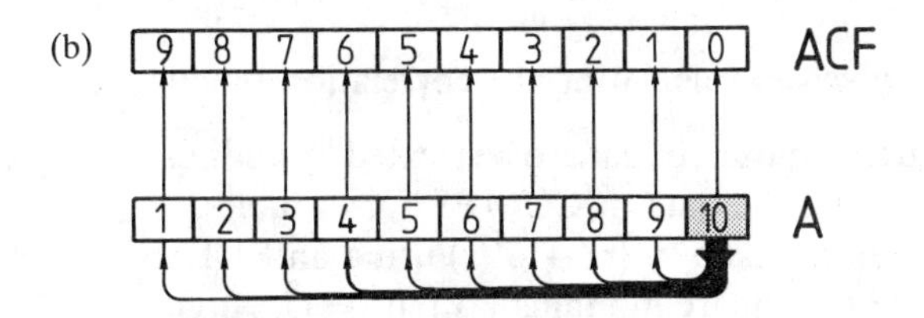


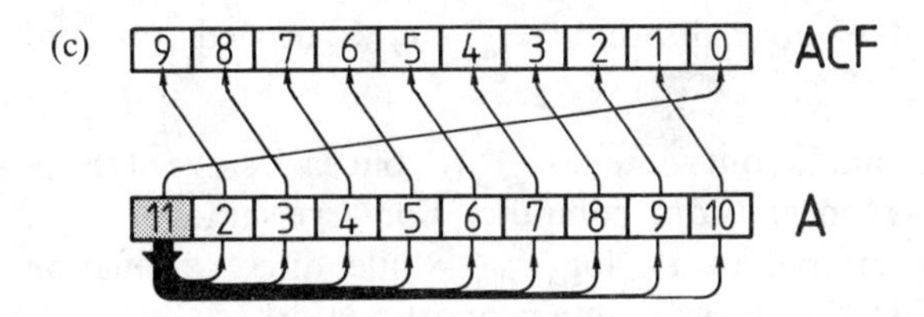


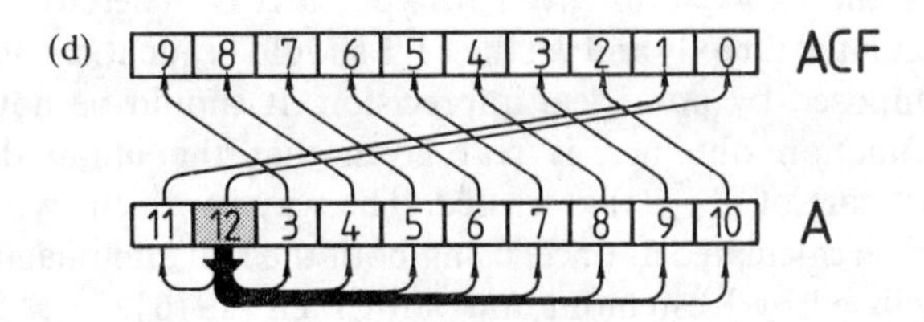


**Fig. 6.1** Calculating time correlation functions in a single sweep. In this example $\tau_{\mathrm{cor}} = 10$. The data A is correlated with itself to give the correlation function ACF. The latest data item to be read in is shaded.

appended a set of $\tau_{\mathrm{run}}$ zeroes to the end of our MD data. This allows us to treat the data as being cyclic in time, i.e. $\mathscr{A}(2\tau_{\mathrm{run}}+1) = \mathscr{A}(1)$, without introducing any spurious correlation. Physically we are only interested in $\tau = 0, \ldots \tau_{\mathrm{run}} - 1$. This is the easiest way of avoiding the spurious correlations that would otherwise arise in the FFT method [Futrelle and McGinty 1971; Kestemont and van Craen 1976]. It is convenient for this purpose to renumber the time origins starting from 0 instead of 1

$$C'_{\mathscr{A}\mathscr{A}}(\tau) = \sum_{\tau_0=0}^{2\tau_{\mathrm{run}}-1} \mathscr{A}(\tau_0)\mathscr{A}(\tau_0+\tau) \quad 0 \leqslant \tau < 2\tau_{\mathrm{run}}. \tag{6.6b}$$

Equation (6.6b) is exactly equivalent to eqn (6.5) with the normalization omitted, and the upper limit $\tau_{max}$ given by $\tau_{run} - \tau$. The equations in Appendix D give

$$\hat{C}'_{\mathscr{A}\mathscr{A}}(\nu) = \hat{\mathscr{A}}^*(\nu)\hat{\mathscr{A}}(\nu) = |\hat{\mathscr{A}}(\nu)|^2, \quad \nu = 0, 1, \ldots, 2\tau_{run} - 1 \tag{6.7}$$

where $\nu$ is the discrete frequency index, and $C'_{\mathscr{A}\mathscr{A}}(\tau)$ may be recovered from

$$C'_{\mathscr{A}\mathscr{A}}(\tau) = \frac{1}{2\tau_{run}} \sum_{\nu=0}^{2\tau_{run}-1} |\hat{\mathscr{A}}(\nu)|^2 \exp(2\pi i\nu\tau/2\tau_{run}). \tag{6.8}$$

The steps involved in calculating the correlation function are:

(a) double the amount of data to be treated by adding $\tau_{run}$ zeroes to the end of it, storing the data in COMPLEX variables;
(b) transform the data $\mathscr{A}(\tau) \to \hat{\mathscr{A}}(\nu)$ using an FFT routine;
(c) calculate the square modulus $|\hat{\mathscr{A}}(\nu)|^2 = \hat{C}'_{\mathscr{A}\mathscr{A}}(\nu)$;
(d) inverse transform the results $\hat{C}'_{\mathscr{A}\mathscr{A}}(\nu) \to C'_{\mathscr{A}\mathscr{A}}(\tau)$ using an inverse FFT routine;
(e) apply the normalization $(\tau_{run} - \tau)^{-1}$ needed to convert $C'_{\mathscr{A}\mathscr{A}}(\tau) \to C_{\mathscr{A}\mathscr{A}}(\tau)$.

This seems a roundabout route to $C_{\mathscr{A}\mathscr{A}}(\tau)$, but each stage of the process may be carried out very speedily on a computer. For large values of $\tau_{run}$, the FTT takes a time proportional to $\tau_{run}\log_2\tau_{run}$, while direct evaluation of the full correlation function takes a time proportional to $\tau_{run}^2$.

It is worth emphasizing that the above equations are exact, and may be verified using the expressions given in Appendix D. Therefore correlation functions calculated directly and via the FFT should be identical, subject to the limitations imposed by numerical imprecision. It should be noted that the correlation function obtained is real given that the initial data is real; the imaginary part of $C_{\mathscr{A}\mathscr{A}}(\tau)$ is wasted. The way in which two correlation functions can be calculated at once, using both the real and imaginary values, has been discussed by Kestemont and van Craen [1976].

When should we use direct calculation and when FFT? The FFT method requires that the entire set of data $\mathscr{A}(\tau)$ and an equal number of zeroes be stored in COMPLEX variables, all at once, which may cause a storage problem. Secondly, it produces the 'complete' correlation function over times up to the entire simulation run time. As mentioned earlier, such long-time information is usually not required, and is statistically not significant because of the poor averaging; when comparing speeds it should be remembered that the conventional method gains by not computing this unwanted information, taking a time proportional to $\tau_{run}$ (not $\tau_{run}^2$) at large $\tau_{run}$. Thirdly, as pointed out by Smith [1982c], the direct method may gain from vectorization on a pipeline machine when many correlation functions are required at once; the FFT method must simply compute them one at a time. Having said this, in situations where a large amount of data must be processed, if it can all be stored in memory at once, the raw speed of the FFT method should make it the preferred choice.

## 6.4 Estimating errors

Computer simulation is an experimental science in so far as the results may be subject to systematic and statistical errors. Sources of systematic error include size-dependence, the possible effects of random number generators, poor equilibration, etc. These should, of course, be estimated and eliminated where possible. It is also essential to obtain an estimate of the statistical significance of the results. Simulation averages are taken over runs of finite length, and this is the main cause of statistical imprecision in the mean values so obtained.

It is often possible to analyse statistical errors in quantities such as $\langle \mathscr{A} \rangle$, $\langle \delta \mathscr{A}^2 \rangle$, by assuming that $\mathscr{A}(t)$ is a Gaussian process. This means that all the moments of $\mathscr{A}$ are determined by the first two, the mean and the variance. Specifically,

$$\begin{aligned} \langle \delta \mathscr{A}(t_1) \delta \mathscr{A}(t_2) \ldots \delta \mathscr{A}(t_n) \rangle &= 0 \qquad (n \text{ odd}) \\ &= \sum_{\text{pairs}} \langle \delta \mathscr{A}(t_i) \delta \mathscr{A}(t_j) \rangle \langle \delta \mathscr{A}(t_k) \delta \mathscr{A}(t_l) \rangle \ldots \qquad (n \text{ even}) \end{aligned} \tag{6.9}$$

where the sum extends over all distinct pairings of the times $t_i$ etc. at which the function is evaluated. The same kind of formula applies to a discrete (rather than continuous-time) process, and so much the same analysis will hold in MC and MD simulations. For Gaussian processes, our estimates of errors in $\langle \mathscr{A} \rangle$, $\langle \delta \mathscr{A}^2 \rangle$ etc. will all be traced back to the variance, or in general to the function $\langle \delta \mathscr{A}(t) \delta \mathscr{A}(0) \rangle$, through eqn (6.9).

The Gaussian assumption is reasonable if the quantity of interest is essentially the sum of a large number of 'random' quantities (statistically independent or not). This is the central limit theorem of probability. Thus, a simulation run average may be thought of as being sampled from some limiting Gaussian distribution about the true mean because it is a sum over many steps. We would like to know the variance of this distribution. The same applies to an average taken over, for example, one-tenth of a run: a so-called block average. Instantaneous values in a simulation, or averages taken over very short intervals, are less likely to obey Gaussian statistics. Even here, however, the Gaussian assumption may not be far wrong. Any property, such as the energy, the virial etc., is a sum of contributions from different parts of the fluid. This at least is true when the potential is not long-ranged. We expect such a quantity to obey statistics that are approximately Gaussian. Of course, in the case of single-particle velocities and angular velocities taken at equal times, the distribution is exactly Gaussian.

Our problem, then, is to estimate the variance in a long (but finite) simulation run average. We consider this for simple averages, including structural distribution functions, for fluctuations, and for time-dependent correlation functions in the following sections.

### *6.4.1 Errors in equilibrium averages*

Suppose that we are analysing a tape of simulation results that contains a total of $\tau_{\text{run}}$ time steps, or configurations. The run average of some property $\mathscr{A}$ is

$$\langle \mathscr{A} \rangle_{\text{run}} = \frac{1}{\tau_{\text{run}}} \sum_{\tau=1}^{\tau_{\text{run}}} \mathscr{A}(\tau). \tag{6.10}$$

If we were to assume that each quantity $\mathscr{A}(\tau)$ were statistically independent of the others, then the variance in the mean would simply be given by

$$\sigma^2(\langle \mathscr{A} \rangle_{\text{run}}) = \sigma^2(\mathscr{A})/\tau_{\text{run}} \tag{6.11}$$

where

$$\sigma^2(\mathscr{A}) = \langle \delta\mathscr{A}^2 \rangle_{\text{run}} = \frac{1}{\tau_{\text{run}}} \sum_{\tau=1}^{\tau_{\text{run}}} (\mathscr{A}(\tau) - \langle \mathscr{A} \rangle_{\text{run}})^2 \tag{6.12}$$

(see eqn (2.43)). The estimated error in the mean is given by $\sigma(\langle \mathscr{A} \rangle_{\text{run}})$. Of course, the data points are usually not independent: we normally store configurations sufficiently frequently that they are highly correlated with each other. The number of steps for which these correlations persist must be built into eqn (6.11). For example, suppose that our $\tau_{\text{run}}$ configurations actually consist of blocks, each containing $2\tau_{\mathscr{A}}$ identical configurations. For large $\tau_{\mathscr{A}}$, this corresponds to a correlation 'time' $\tau_{\mathscr{A}}$. Then,

$$\sigma^2(\langle \mathscr{A} \rangle_{\text{run}}) = 2\tau_{\mathscr{A}}\sigma^2(\mathscr{A})/\tau_{\text{run}}. \tag{6.13}$$

This analysis is due to Jacucci and Rahman [1984]; note that these authors define a correlation time to be twice as long as ours. In general, the correlation 'time' $\tau_{\mathscr{A}}$ will be unknown before we start analysing the results.

To handle this problem the sequence of steps on the tape is broken up into blocks each of length $\tau_{\text{b}}$. Let there be $n_{\text{b}}$ blocks, so that $n_{\text{b}}\tau_{\text{b}} = \tau_{\text{run}}$. The mean value of $\mathscr{A}$ is calculated for each block

$$\langle \mathscr{A} \rangle_{\text{b}} = \frac{1}{\tau_{\text{b}}} \sum_{\tau=1}^{\tau_{\text{b}}} \mathscr{A}(\tau) \tag{6.14}$$

where the sum runs over configurations in block *b* only. The mean values for all the blocks of this kind may then be used to estimate the variance

$$\sigma^2(\langle \mathscr{A} \rangle_{\text{b}}) = \frac{1}{n_{\text{b}}} \sum_{\text{b}=1}^{n_{\text{b}}} (\langle \mathscr{A} \rangle_{\text{b}} - \langle \mathscr{A} \rangle_{\text{run}})^2. \tag{6.15}$$

We expect this quantity to be inversely proportional to $\tau_{\text{b}}$ at large $\tau_{\text{b}}$, as the blocks become large enough to be statistically uncorrelated. Our aim is to discover the constant of proportionality, which will allow us to estimate

$\sigma^2(\langle\mathcal{A}\rangle_b)$ for the single large block that constitutes the entire run. Following Friedberg and Cameron [1970], we define the statistical inefficiency, $s$, as

$$s = \lim_{\tau_b \to \infty} \frac{\tau_b \sigma^2(\langle\mathcal{A}\rangle_b)}{\sigma^2(\mathcal{A})}. \tag{6.16}$$

It is the limiting ratio of the observed variance of an average to the limit expected on the assumption of uncorrelated Gaussian statistics. Figure 6.2(a) shows a plot of $\tau_b\sigma^2(\langle\mathcal{A}\rangle_b)/\sigma^2(\mathcal{A})$ against $\tau_b^{1/2}$ for the pressure in a simulation of a molecular liquid [Fincham *et al.* 1986] ($\tau_b^{1/2}$ is simply a convenient variable for the plot). A plateau value of $s = 22$ is obtained. This means that only about one configuration in every 22 stored on tape contributes completely new information to the average. The RMS pressure fluctuation in the run is $\sigma(\mathcal{P}) = 10.9$ MPa. The run was of total length 30 000 time steps and so

$$\sigma(\langle\mathcal{P}\rangle_{\text{run}}) = (22/30\,000)^{1/2} \times 10.9 = 0.3 \text{ MPa}. \tag{6.17}$$

This is typical of the accuracy obtainable in computer simulations.

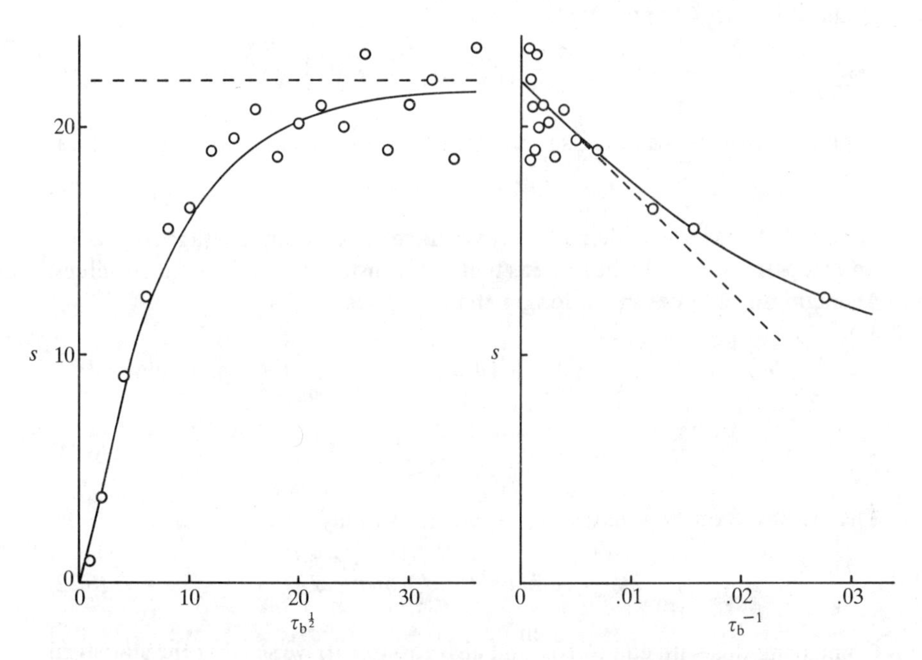


**Fig. 6.2** The calculation of the statistical inefficiency $s$. (a) The approach to the plateau. (b) The inverse-$\tau_b$ plot.

Any technique that reduces $s$ will help us to calculate more accurate simulation averages. As an example, we consider the calculation of the chemical potential in a molecular liquid, by Fincham *et al.* [1986]. These authors estimated $\mu$ with a statistical inefficiency $s \approx 20$ by inserting a test-particle lattice where the orientations of the molecules were fixed throughout the simulation. By randomizing the orientations of the test molecules on the lattice at each insertion, $s$ was reduced to $s \approx 10$. Both methods are valid, but randomizing the orientations on the test lattice allows insertion every tenth step to gain significantly new information. Inserting every tenth step in the case of fixed lattice orientations is not a significant improvement over inserting every twentieth: twice as long a run is still required to calculate $\mu$ to a given accuracy. In a similar way it is $s$ which has been used to compare the efficiency of different MC algorithms (see Section 4.3).

The above method of analysis applies to any simulation results stored on tape, but it is instructive to consider the particular case of time averages as estimated by MD. For an average

$$\langle \mathcal{A} \rangle_t = \frac{1}{t} \int_0^t \mathcal{A}(t')\,\mathrm{d}t' \tag{6.18}$$

the standard result for the variance is related to the correlation function of $\mathcal{A}$ [Papoulis 1965, Chapter 9]

$$\sigma^2(\langle \mathcal{A} \rangle_t) = \frac{2}{t} \int_0^t (1 - t'/t) \langle \delta\mathcal{A}(t') \delta\mathcal{A} \rangle \,\mathrm{d}t'. \tag{6.19}$$

Averaging over times much shorter than the correlation time $t_{\mathcal{A}}$ of $\mathcal{A}$ gives

$$\sigma^2(\langle \mathcal{A} \rangle_{t \ll t_{\mathcal{A}}}) = \sigma^2(\mathcal{A}). \tag{6.20}$$

Note that this is independent of $t$: the variance of short time averages (e.g. a few time steps) is essentially the same as that of the instantaneously sampled values. Averaging over times much longer than $t_{\mathcal{A}}$ gives

$$\begin{aligned}\sigma^2(\langle \mathcal{A} \rangle_{t \gg t_{\mathcal{A}}}) &= \frac{2}{t} \int_0^\infty \langle \delta\mathcal{A}(t') \delta\mathcal{A} \rangle \,\mathrm{d}t' - \frac{2}{t^2} \int_0^\infty t' \langle \delta\mathcal{A}(t') \delta\mathcal{A} \rangle \,\mathrm{d}t' \\ &= 2t_{\mathcal{A}} \sigma^2(\mathcal{A})/t - \frac{2}{t^2} \int_0^\infty t' \langle \delta\mathcal{A}(t') \delta\mathcal{A} \rangle \,\mathrm{d}t'. \end{aligned} \tag{6.21}$$

The leading term dominates as $t \to \infty$ and we may write

$$2t_{\mathcal{A}} = \lim_{t \to \infty} t\sigma^2(\langle \mathcal{A} \rangle_t)/\sigma^2(\mathcal{A}). \tag{6.22}$$

Comparing this with eqn (6.16) (and also eqn (6.13)) we see that the statistical inefficiency is just twice the correlation time $t_{\mathcal{A}}$ divided by the time interval between configurations on tape. Equation (6.21) also shows that the next highest term is proportional to $1/t^2$ at long time. This suggests that it is most sensible to plot $t\sigma^2(\langle \mathcal{A} \rangle_t)/\sigma^2(\mathcal{A})$ against $1/t$, or in general $\tau_b \sigma^2(\langle \mathcal{A} \rangle_b)/\sigma^2(\mathcal{A})$ against $1/\tau_b$, when a linear form at low values will be obtained [Jacucci and Rahman 1984]. Figure 6.2(b) shows the pressure data of Fincham *et al.* [1986]

plotted in this way, and exhibiting the expected dependence upon $\tau_b$. The figure also shows clearly how the estimate of $\sigma^2(\langle \mathcal{A} \rangle_b)$ becomes less precise as the length of the blocks increases and the number of blocks decreases.

Of course, it would be possible to evaluate $t_{\mathcal{A}}$ or $\tau_{\mathcal{A}}$ by integrating the time correlation function $\langle \delta\mathcal{A}(t)\delta\mathcal{A} \rangle$ in the usual fashion, and thereby estimate $\sigma^2(\langle \mathcal{A} \rangle_{run})$ through eqn (6.13) or (6.21) [Müller-Krumbhaar and Binder 1973; Swope *et al.* 1982]. Alternatively, if we can guess $t_{\mathcal{A}}$ in some other way, we can estimate the statistical inefficiency without carrying out a full analysis as described above. Smith and Wells [1984] have analysed block averages in their MC simulations, and find exponential decay (i.e. obeying a geómetric law) of the 'correlation function' of consecutive block averages. In the language of time-series analysis, the process is termed 'first-order autoregressive' i.e. Markov [Chatfield 1984]. If such behaviour is assumed, then $\tau_{\mathcal{A}}$ may be estimated from the initial correlations $\langle \delta\mathcal{A}(\tau=1)\delta\mathcal{A}(\tau=0) \rangle$. In general, it is best to carry out a full analysis to establish the form of the decay of correlations with $\tau$; once this has been done, for a given system, it may be safe to extend the results to neighbouring state points, and here the approach of Smith and Wells might save on some effort.

### 6.4.2 *Errors in fluctuations*

Errors in our estimate of fluctuation averages of the type $\langle \delta\mathcal{A}^2 \rangle$ may be estimated simply on the assumption that the process $\mathcal{A}(t)$ obeys Gaussian statistics. The resulting formula is very much like eqns (6.13) and (6.21)

$$\sigma^2(\langle \delta\mathcal{A}^2 \rangle_{run}) = 2t'_{\mathcal{A}} \langle \delta\mathcal{A}^2 \rangle^2 / t_{run} \tag{6.23}$$

where a slightly different correlation time appears

$$t'_{\mathcal{A}} = 2\int_0^\infty \mathrm{d}t \, \langle \delta\mathcal{A}(t)\delta\mathcal{A} \rangle^2 / \langle \delta\mathcal{A}^2 \rangle^2 . \tag{6.24}$$

For an exponentially decaying correlation function, $t'_{\mathcal{A}} = t_{\mathcal{A}}$, the usual correlation time; it may be reasonable to assume that this is generally true, in which case the analysis of Section 6.4.1 which yields $t_{\mathcal{A}}$ leads also to an estimate of the errors in the fluctuations through eqn (6.23).

### 6.4.3 *Errors in structural quantities*

Errors in a quantity such as $g(r)$ may be estimated by considering the histogram bins that are used in its calculation. Strictly speaking, the sum which is accumulated in a histogram bin (Section 6.2) will not obey Gaussian statistics, but provided the number of counts is large, the central limit theorem of probability applies once more, and the Gaussian approximation becomes quite good. In this case, the techniques described in Section 6.4.1 may be used to estimate the standard error in any histogram bin average. When this quantity is normalized to give a particular value of $g(r)$, the standard error is divided by exactly the same normalizing factor. Carrying out a full block-average analysis for each point in $g(r)$ would be very time-consuming, and not essential. It would be sufficient in most cases to select a few points, near the first and second peaks and in the intervening minimum for example, and estimate the statistics there. A further estimate should be made at large distances: remember that statistics should be much improved as $r$ increases, due to the increasing volume of spherical shells.

### 6.4.4 *Errors in time correlation functions*

The time correlation functions calculated in MD simulations are subject to the same kind of random errors as described for static quantities and fluctuations in the previous sections. We denote the run average by

$$C^{\text{run}}_{\mathscr{A}\mathscr{A}}(t) = \langle \mathscr{A}(t)\mathscr{A}(0) \rangle_{\text{run}} = \frac{1}{t_{\text{run}}}\int_0^{t_{\text{run}}} \mathrm{d}t'\,\mathscr{A}(t')\mathscr{A}(t'+t) \tag{6.25}$$

where we have assumed for simplicity that $\langle \mathscr{A} \rangle$ vanishes. The error we wish to estimate is that in

$$\begin{aligned}\delta C(t) &= C^{\text{run}}_{\mathscr{A}\mathscr{A}}(t) - C_{\mathscr{A}\mathscr{A}}(t)\\ &= \langle \mathscr{A}(t)\mathscr{A}(0) \rangle_{\text{run}} - \langle \mathscr{A}(t)\mathscr{A}(0) \rangle\\ &= \frac{1}{t_{\text{run}}}\int_0^{t_{\text{run}}} \mathrm{d}t'\,(\mathscr{A}(t')\mathscr{A}(t'+t) - \langle \mathscr{A}(t')\mathscr{A}(t'+t) \rangle)\end{aligned} \tag{6.26}$$

where $\langle \ldots \rangle$ denotes the true, infinite time or ensemble average. The mean value $\langle \delta C(t) \rangle$ should vanish of course, but the variance of the mean is given by [Zwanzig and Ailawadi 1969; Frenkel 1980]

$$\sigma^2(\langle \mathscr{A}(t)\mathscr{A} \rangle_{\text{run}}) = \frac{1}{t^2_{\text{run}}}\int_0^{t_{\text{run}}}\int_0^{t_{\text{run}}} \mathrm{d}t'\,\mathrm{d}t''$$
$$(\langle \mathscr{A}(t')\mathscr{A}(t'+t)\mathscr{A}(t'')\mathscr{A}(t''+t) \rangle - \langle \mathscr{A}(t)\mathscr{A}(0) \rangle^2). \tag{6.27}$$

The four-variable correlation function in this equation may be simplified if we make the assumption that $\mathscr{A}(t)$ obeys Gaussian statistics, using eqn (6.9). After some straightforward manipulations described in detail by Frenkel [1980] the variance reduces to

$$\sigma^2(\langle \mathscr{A}(t)\mathscr{A} \rangle_{\text{run}}) \approx 2t'_{\mathscr{A}}C_{\mathscr{A}\mathscr{A}}(0)^2/t_{\text{run}} \tag{6.28}$$

where $t'_{\mathscr{A}}$ is the correlation time defined by eqn (6.24). The standard error in the normalized correlation function is thus independent of time and is given by

$$\sigma(\langle \mathscr{A}(t)\mathscr{A} \rangle_{\text{run}})/\langle \mathscr{A}^2 \rangle \approx (2t'_{\mathscr{A}}/t_{\text{run}})^{1/2} \tag{6.29}$$

which has the usual appearance. As an example, for $t_{\mathscr{A}}$ of the order of ten time steps, it would be necessary to conduct a run of $10^5$ steps in order to obtain a relative precision of $\sim 1$ per cent in $C_{\mathscr{A}\mathscr{A}}(t)$. If we use the simulation average $C^{\text{run}}_{\mathscr{A}\mathscr{A}}(0) = \langle \mathscr{A}^2 \rangle_{\text{run}}$ instead of the exact ensemble average in eqn (6.29), then the error at short times is reduced due to cancellation in the random fluctuations [Zwanzig and Ailawadi 1969]

$$\sigma(\langle \mathscr{A}(t)\mathscr{A} \rangle_{\text{run}})/\langle \mathscr{A}^2 \rangle_{\text{run}} \approx (2t'_{\mathscr{A}}/t_{\text{run}})^{1/2}\,(1 - c_{\mathscr{A}\mathscr{A}}(t)) \qquad (6.30)$$

where $c_{\mathscr{A}\mathscr{A}}(t) = \langle \mathscr{A}(t)\mathscr{A}(0) \rangle / \langle \mathscr{A}^2 \rangle$. Thus the error is zero at $t = 0$, but it tends to $(2t'_{\mathscr{A}}/t_{\text{run}})^{1/2}$ at long times.

This looks rather depressing, but the gloom is lightened when we turn to the calculation of single-particle correlation functions, such as the velocity autocorrelation function. The final result is then an average over $N$ separate functions for each axis direction

$$C_{vv}(t) = \frac{1}{N} \sum_{i=1}^{N} \langle v_{i\alpha}(t) v_{i\alpha}(0) \rangle \qquad (6.31)$$

(and in this case a further average over the equivalent axes could be carried out). The analysis of this situation follows the above pattern, and the estimated error is eventually found to be $\approx (2t'_{\mathscr{A}}/Nt_{\text{run}})^{1/2}$ at long times. The extra factor of $N^{1/2}$ in the denominator suggests that 1 per cent accuracy in the velocity autocorrelation function might be achieved with $10^4$ time steps for a 100-particle system. This argument is simplistic, since the velocities of neighbouring particles at different times are not statistically independent, but single-particle correlation functions are still generally found to be less noisy than their collective counterparts. The precision with which a time correlation function may be estimated depends upon the spatial range of correlations in the fluid; the size of statistically independent regions may depend upon the range of the potential and on the state point. Some of these ideas are discussed further by Frenkel [1980].

In principle, a block analysis of time correlation functions could be carried out in much the same way as that applied to static averages. However, the block lengths would have to be substantial to make a reasonably accurate estimate of the errors, and this type of analysis may be impractical.

We have not included in the above analysis the point raised in Section 6.3, namely that the number of time origins available for the averaging of long-time correlations may be significantly less than the number of origins for short-time correlations. This limitation is imposed by the finite run length, and it means that $t_{\text{run}}$ in the previous discussion should be replaced by $t_{\text{run}} - t$ for correlations $\langle \mathscr{A}(t)\mathscr{A} \rangle$. Thus, an additional time-dependence, leading to slightly poorer statistics for longer times, enters into the formulae.

One possible source of systematic error in time correlation functions should be mentioned. The usual periodic boundary conditions mean that any disturbance, such as a sound wave, may propagate through the box, leaving through one side and re-entering through the other, so as to arrive back at its starting point. This would happen in a time of order $L/v_s$ where $L$ is the box length and $v_s$ the speed of sound. With typical values of $L = 2$ nm and $v_s = 1000\ \mathrm{m\,s^{-1}}$, this 'recurrence time' is about 2 ps, which is certainly well within the range of correlation times of interest. It is sensible, and has become the recommended practice, to inspect time correlation functions for signs of anomalous behaviour, possibly increased noise levels, at times greater than this. It is doubtful that a periodic system would correctly reproduce the correlations arising in a macroscopic liquid sample at such long times. The phenomenon was originally reported by Alder and Wainwright [1970] and more recently by Schoen, Vogelsang, and Hoheisel [1984]. The latter workers found it hard to reproduce their results for the Lennard-Jones liquid. We would expect to see much more significant effects in solids, where sound waves are well-developed, whereas phonons are much more strongly damped in liquids. Nonetheless, it is obviously a good idea to keep the possibility of correlation recurrence effects in mind, particularly if 'long-time tail' behaviour (see Chapter 11) is under study.

## 6.5 Correcting the results

When the results of a simulation have been calculated, and the errors estimated, they may still not be in the form most suitable for interpretation. The run averages may not correspond to exactly the desired state point, the structural or time-dependent properties may require extrapolation or smoothing, and it may be necessary to do some time integration or Fourier transformation to obtain the desired quantities. In this section, we discuss all these points.

### *6.5.1 Correcting thermodynamic averages*

In constant-$NVE$ molecular dynamics, the kinetic temperature fluctuates around its mean value. It is difficult to preset a desired value of $T$ in a simulation and this is inconvenient for comparison of results with other simulations, real experiments, and theory. The determination of isotherms is useful, for example, in the calculation of a coexistence curve. Powles *et al.* [1982] have suggested a useful method for the correction of thermodynamic results to the desired temperature. For a particular property $\mathscr{A}$, obtained in a simulation at a mean temperature $T_{\mathrm{run}} = \langle \mathscr{T} \rangle_{\mathrm{run}}$, the results can be corrected to the desired temperature $T$ using

$$\mathscr{A}(T) = \mathscr{A}(T_{\mathrm{run}}) + (T - T_{\mathrm{run}}) \left( \frac{\partial \mathscr{A}}{\partial T} \right)_\rho + \ldots . \tag{6.32}$$

If the temperature difference is small, the Taylor series can be truncated at the first term. For the energy $E$, the appropriate thermodynamic derivative is of course $C_V$. In the case of the chemical potential and the pressure, convenient expressions for the derivatives are

$$\left(\frac{\partial P}{\partial T}\right)_\rho = \left(P - \rho^2 \left(\frac{\partial (E/N)}{\partial \rho}\right)_T\right) \Big/ T \tag{6.33}$$

$$\left(\frac{\partial \mu}{\partial T}\right)_\rho = -\left(\rho \left(\frac{\partial (E/N)}{\partial \rho}\right)_T + (E/N) - \mu\right) \Big/ T \tag{6.34}$$

where $E/N$ is the total energy per molecule, which is known exactly in the simulation. A series of simulation runs is carried out by varying the density, while the mean temperature of each run is kept as close to the desired value $T$ as possible. This is achieved by using one of the methods described in Chapter 7 during the equilibration period. $E/N$ is almost a linear function of $\rho$, and the derivative $\partial(E/N)/\partial\rho$ is easily calculated from this series of runs. Strictly speaking we require the derivative at fixed $T$ (the desired temperature). In practice, the errors in the derivative arising from the small temperature differences between runs are small and can be ignored. Thus, by using eqn (6.32), values of $E$, $P$, and $\mu$ along an isotherm may be calculated. The technique is easily extended to other thermodynamic quantities.

### 6.5.2 *Extending $g(r)$ to large $r$*

The range of $g(r)$ that can be calculated in the simulation is limited by the length $L$ of the simulation box, which for a given number of molecules is determined by the liquid density. $g(r)$ can only be calculated for $r \leqslant L/2$ to be consistent with the minimum image convention. This truncation of $g(r)$ at such a small value of $r$ may prevent its accurate Fourier transformation to the structure factor $S(k)$ defined in eqn (2.103).

In principle, the long-range behaviour of $g(r)$ may be deduced from its behaviour at short distances. This idea is embodied in the Ornstein–Zernike equation [Hansen and McDonald 1986]

$$h(r) = c(r) + \rho \int \mathrm{d}\mathbf{r}'\, h(|\mathbf{r}-\mathbf{r}'|) c(|\mathbf{r}'|)\,. \tag{6.35}$$

Eqn (6.35) just defines the direct correlation function $c(r)$ in terms of the total correlation functions $h(r) = g(r) - 1$. While $h(r)$ is long-range in normal liquids, $c(r)$ has approximately the same range as the potential, and is expected to be small after $\approx 2\sigma$ in many cases. Verlet [1968] exploited this property in using the Percus–Yevick approximation [Hansen and McDonald 1986]

$$c(r) = (h(r) + 1)\,(1 - \exp(\beta v(r))) \tag{6.36}$$

to extend $g(r)$ beyond $r = L/2$. Subsequent workers have attempted to improve on Verlet's method. If $c(r)$ is zero beyond a certain cutoff distance $r_0$ then Baxter [1970] employs a Weiner–Hopf factorization to obtain a pair of equations:

$$r\,c(r) = -Q'(r) + 2\pi\rho \int_r^{r_0} \mathrm{d}r'\,Q'(r')Q(r'-r) \qquad 0 \leqslant r \leqslant r_0$$
$$= 0 \qquad r \geqslant r_0 \tag{6.37}$$

and

$$r\,h(r) = -Q'(r) + 2\pi\rho \int_0^{r_0} \mathrm{d}r'(r-r')h(|r-r'|)Q(r'). \tag{6.38}$$

where the function $Q(r)$ is zero for $r > r_0$ and continuous at $r_0$, and $Q'(r) = \mathrm{d}Q(r)/\mathrm{d}r$. $r_0$ is not the potential cutoff ($r_c$), but rather the distance beyond which $c(r) = 0$ (most likely $r_0 \geqslant r_c$). The continuity of $Q(r)$ means that

$$Q(r) = -\int_r^{r_0} \mathrm{d}r'\,Q'(r'). \tag{6.39}$$

The remarkable property of this factorization is that if we know $h(r)$ on the range $0 \leqslant r \leqslant r_0$, then eqn (6.37) can be solved to produce $c(r)$ over its complete range, and hence $h(r)$ over its complete range through eqn (6.35). The structure factor can be obtained directly from the relationship

$$S(k) = 1 + \rho\hat{h}(k) = (1 - \rho\hat{c}(k))^{-1} \tag{6.40}$$

(see eqn (2.103)).

Jolly, Freasier, and Bearman [1976] have proposed a method to extend $g(r)$ based on Baxter's factorization. Suppose $h(r) = g(r) - 1$ has been calculated during the simulation out to some distance $r_h \leqslant L/2$. Initially, $h(r)$ for $r_h < r < r_0$ is set to zero; $h(r)$ for $r < r_h$, as calculated in the simulation, remains fixed throughout the procedure. Jolly *et al.* [1976] actually choose $r_0 = 4\sigma$, and $r_h = 2.5\sigma$. Initially, $Q(r)$ and $Q'(r)$ are set to zero for $0 < r < r_0$.$c(r)$ is estimated over its whole range using the Percus–Yevick approximation, (eqn (6.36)). The following iterative procedure is proposed:

(a) calculate $Q'(r)$, $r < r_h$, from eqn (6.38);
(b) calculate $Q'(r)$, $r_h < r < r_0$, from eqn (6.37);
(c) calculate $Q(r)$, $0 < r < r_0$, from eqn (6.39);
(d) calculate $h(r)$, $r_h < r$, from eqn (6.38);
(e) estimate $c(r)$, $r_h < r < r_0$, using the Percus–Yevick eqn (6.36);
(f) return to step (a), repeating until convergence is achieved.

The number of iterations required is typically between 50 and 100. Convergence can be speeded up by mixing the old and new iterations in steps (a) and (b), i.e.

$$Q'_{\text{new}}(r) = m\,Q'_{\text{out}}(r) + (1-m)\,Q'_{\text{in}}(r) \tag{6.41}$$

with $m = 1.25$ being typical. The method requires simple one-dimensional numerical integration and no Fourier transforms.

Dixon and Hutchinson [1977] describe an alternative use of Baxter's factorization to extend $g(r)$. They make $r_0 < r_h$, and choose a value that minimizes any discontinuity in $h(r)$ at $r = r_0$. They assume that $c(r) = 0$ for $r > r_0$, but avoid any explicit use of a model such as the Percus–Yevick approximation. The reader is referred to the original paper for the computational details, but the general scheme is very similar to that used by Jolly *et al.*

Figure 6.3 shows the direct correlation function calculated from a simulated $g(r)$ for the Lennard-Jones fluid at state point $\rho^* = 0.88$, $T^* = 1.1$ [Verlet 1968]. The method of Dixon and Hutchinson was used, with a value of $r_0 = 2.6\sigma$. This technique reveals a minimum in $c(r)$ in the same region as the minimum in the potential, and also shows how quickly $c(r)$ decays with distance.

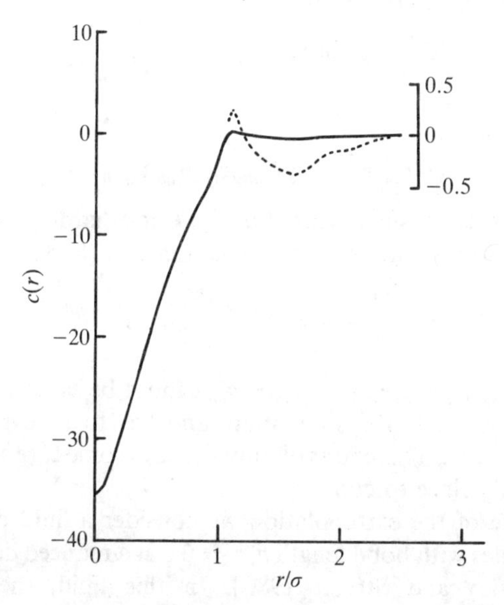


**Fig. 6.3** The direct correlation function for the Lennard-Jones fluid calculated using the factorization method of Dixon and Hutchinson [1977] (solid line). The region around the potential minimum (dashed line) is shown on an expanded scale.

### 6.5.3 *Extrapolating g(r) to contact*

For a fluid with smooth repulsive interactions (such as the Lennard-Jones fluid), $g(r)$ has a maximum which corresponds to the minimum in the potential. At lower values of $r$, $g(r)$ falls rapidly to zero. For a hard-core fluid (such as a fluid of hard spheres or hard dumb-bells), $g(r)$, or more generally $g_{ab}(r)$, is discontinuous at $r = \sigma_{ab}$, and is zero inside the core. The value of $g_{ab}$ at contact, $g_{ab}(\sigma_{ab}^+)$, is directly related to the pressure and the other thermodynamic properties of the hard-core fluid.

For a site–site hard-core fluid the potential between two molecules $i$ and $j$ can be written in terms of the unit step function, $\theta(x)$.

$$\exp(-\beta v(r_{ij}, \mathbf{\Omega}_i, \mathbf{\Omega}_j)) = \prod_{a,b} \exp(-\beta v_{ab}(r_{ab})) = \prod_{a,b} \theta(r_{ab} - \sigma_{ab}). \quad (6.42)$$

The product is over the independent site–site distances $r_{ab}$ between the pair of molecules. Differentiating eqn (6.42) gives the virial for the fluid,

$$w(r_{ij}, \mathbf{\Omega}_i, \mathbf{\Omega}_j) = -\beta^{-1} r_{ij} \sum_a \sum_b \exp(\beta v_{ab}(r_{ab}))\delta(r_{ab} - \sigma_{ab}) \left(\frac{\partial r_{ab}}{\partial r_{ij}}\right)_{\mathbf{\Omega}_i, \mathbf{\Omega}_j}. \quad (6.43)$$

This virial can be used in eqns (2.54), (2.61) to obtain the pressure [Nezbeda 1977; Aviram, Tildesley, and Streett 1977]

$$\frac{P}{\rho k_B T} = 1 + \frac{2\pi\rho}{3} \sum_a \sum_b \tau_{ab}(\sigma_{ab}^+)\sigma_{ab}^2 g_{ab}(\sigma_{ab}^+) \quad (6.44)$$

where

$$\tau_{ab}(r_{ab}) = \langle (\mathbf{r}_{ab} \cdot \mathbf{r}_{ij})/r_{ab} \rangle_{\text{shell}} \quad (6.45)$$

and the average is for a shell centred at $r_{ab}$. For a hard-sphere fluid $\tau_{ab}(r_{ab}) = r_{ab}$ and there is only one term in the sum in eqn (6.44),

$$\frac{P}{\rho k_B T} = 1 + \frac{2\pi\rho\sigma^3}{3} g(\sigma^+). \quad (6.46)$$

The product $\tau_{ab}(r_{ab})g_{ab}(r_{ab})$ for $r_{ab} = \sigma_{ab}^+$ cannot be calculated directly in a standard constant-$NVT$ MC simulation, and has to be extrapolated from values close to contact. This extrapolation requires some care since $g_{ab}(r_{ab})$ can rise or fall rapidly close to contact.

As an example of the extrapolation we consider a fluid of homonuclear diatomic molecules with bond length $d/\sigma = 0.6$ at a reduced density $\rho^* = \rho\sigma^3 = 0.446$ [Tildesley and Streett 1980]. In this fluid, the four site–site distribution functions are equivalent and eqn (6.44) reduces to

$$\begin{aligned} \frac{P}{\rho k_B T} &= 1 + \frac{8\pi\rho^*}{3} \lim_{r\to\sigma^+} (\tau^*(r)g(r)) \\ &= 1 + \lim_{r\to\sigma^+} f(r) \end{aligned} \quad (6.47)$$

where $\tau^* = \tau/\sigma$ and we have defined the function $f(r)$. This function is shown in Fig. 6.4. This calculation was performed for shells of thickness $\delta r$ centred at $\sigma + n\delta r$, where $n$ is a positive integer. There is an additional half-shell of thickness $\delta r/2$ centred at $\sigma + \delta r/4$. The contact value is shown to be independent of the shell thickness $\delta r$ for the two values used in this study ($\delta r = 0.025\sigma$ and $\delta r = 0.01\sigma$). The care needed in performing extrapolations of this kind is highlighted by the initial disagreement between two sets of simulations of the hard dumb-bell fluid [Aviram *et al.* 1977; Freasier, Jolly, and Bearman 1976] which now seems to have been resolved [Tildesley and Streett 1980; Freasier 1980].

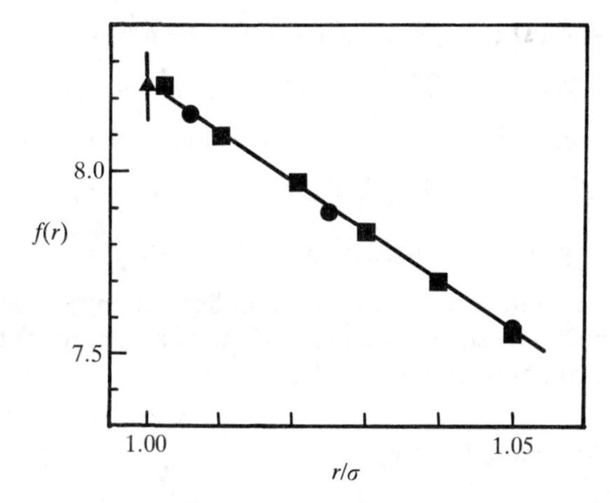


**Fig. 6.4** Extrapolating $f(r)$ to contact. Shells of thickness $0.025\sigma$ (circles) and $0.01\sigma$ (squares) were used. The triangle with the error bar indicates the extrapolated value.

A trick, which is sometimes useful in calculating the contact value, is to extrapolate $(r_{ab}/\sigma_{ab})^{\nu} f(r_{ab})$, where $\nu$ is an integer, to $r_{ab} = \sigma_{ab}$. This extrapolation produces $f(\sigma_{ab}^{+})$ regardless of the value of $\nu$. If the function is steeply varying, an appropriate choice of $\nu$ can facilitate the extrapolation.

Freasier [1980] has reported a suggestion due to D. J. Evans, that such an extrapolation procedure be employed during the simulation run itself. The problems involved in estimating the pressure, particularly for hard molecular systems, lead to a general preference for constant-$NPT$ MC simulation or to molecular dynamics where possible, when obtaining the pressure is straightforward (see Section 5.6).

### 6.5.4 *Smoothing g(r)*

The radial distribution function and any of the angular correlation functions, such as the spherical harmonic coefficients, are subject to statistical noise. For the purposes of comparing with theoretical approximations or in order to calculate accurate Fourier transforms (see Appendix D), it is sometimes useful to smooth this data. Smoothing can be achieved by fitting a least-squares polynomial in $r$. However, it is difficult to find appropriate functional forms to fit a variety of correlation functions, over a wide range of temperature and density. A useful compromise is to use a smoothing formula to replace each tabulated value by a least-squares polynomial which fits a sub-range of points [Stark 1970]. The appropriate formula for a third-degree, five-point smoothing is,

$$\tilde{g}_{n-2} = \tfrac{1}{70}(69g_{n-2} + 4g_{n-1} - 6g_n + 4g_{n+1} - g_{n+2})$$
$$\tilde{g}_{n-1} = \tfrac{1}{35}(2g_{n-2} + 27g_{n-1} + 12g_n - 8g_{n+1} + 2g_{n+2})$$
$$\tilde{g}_n = \tfrac{1}{35}(-3g_{n-2} + 12g_{n-1} + 17g_n + 12g_{n+1} - 3g_{n+2})$$
$$\tilde{g}_{n+1} = \tfrac{1}{35}(2g_{n+2} + 27g_{n+1} + 12g_n - 8g_{n-1} + 2g_{n-2})$$
$$\tilde{g}_{n+2} = \tfrac{1}{70}(69g_{n+2} + 4g_{n+1} - 6g_n + 4g_{n-1} - g_{n-2}). \tag{6.48}$$

The $\tilde{g}_n$ are the smoothed values. For most of the table entries the symmetrical formula $\tilde{g}_n$ should be used; the other four formulae are appropriate for the first two and last two points of the function.

### *6.5.5 Calculating transport coefficients*

The numerical integration of time correlation functions to obtain transport coefficients and correlation times is formally a straightforward exercise, given data at regularly spaced times. Simpson's rule, for example, would be quite satisfactory. However, there are a number of pitfalls to be avoided. Firstly, there are several correlation functions that are believed to decay to zero only slowly, having a limiting algebraic dependence $t^{-\nu}$ with exponent $\nu = 3/2$ for example (see Chapter 11). Such a tail may extend significantly beyond the range of times for which $C(t)$ has been computed, and, as has been mentioned, statistical errors will become more severe as $t$ increases. The integral under such a tail may nonetheless make a significant contribution to the total integral, and so the tail cannot be completely ignored. In estimating the tail it becomes necessary to attempt some kind of fit to the long-time behaviour of the correlation function, and then to use this to extrapolate to $t \to \infty$ and estimate a long-time tail correction. The importance of this correction is illustrated by the estimation, by MD, of the bulk and shear viscosities of the Lennard-Jones fluid near the triple point [Levesque, Verlet, and Kurkijarvi 1973; Hoover, Evans, Hickman, Ladd, Ashurst, and Moran 1980b]. In all cases, the long-time behaviour of a correlation function should be examined closely before an attempt is made to calculate the time integral.

As discussed in Chapter 2, the Einstein relation provides an alternative route to transport coefficients, which is formally equivalent to the integration of a time correlation function. This relies on the identities, for stationary processes,

$$\begin{aligned}\gamma &= \frac{1}{2t}\langle (\mathscr{A}(t_0+t) - \mathscr{A}(t_0))^2 \rangle \\ &\approx \frac{1}{2}\frac{\mathrm{d}}{\mathrm{d}t}\langle (\mathscr{A}(t_0+t) - \mathscr{A}(t_0))^2 \rangle \\ &= \int_0^t \mathrm{d}t' \langle \dot{\mathscr{A}}(t_0+t')\dot{\mathscr{A}}(t_0) \rangle .\end{aligned} \tag{6.49}$$

The approximate equality becomes exact at long times. Here the average is taken over time origins $t_0$. Thus, the diffusion coefficient may be estimated by observing the mean-squared displacement $\langle |\mathbf{r}_i(t)-\mathbf{r}_i(0)|^2 \rangle$ as a function of time. Alder, Gass, and Wainwright [1970] have pointed out that the transport coefficients may be more readily calculated from the gradient $\frac{1}{2}\mathrm{d}/\mathrm{d}t \langle (\mathscr{A}(t)-\mathscr{A}(0))^2 \rangle$ than from the expression $1/2t \langle (\mathscr{A}(t)-\mathscr{A}(0))^2 \rangle$. For variables with an exponentially decaying correlation function, $\langle \dot{\mathscr{A}}(t)\dot{\mathscr{A}} \rangle = \langle \dot{\mathscr{A}}^2 \rangle \exp(-t/t_{\mathscr{A}})$ we have

$$\frac{1}{2}\frac{\mathrm{d}}{\mathrm{d}t}\langle (\mathscr{A}(t)-\mathscr{A}(0))^2 \rangle = \int_0^t \mathrm{d}t' \langle \dot{\mathscr{A}}(t')\dot{\mathscr{A}} \rangle$$
$$= \langle \dot{\mathscr{A}}^2 \rangle t_{\mathscr{A}}(1-\exp(-t/t_{\mathscr{A}})) \qquad (6.50)$$

but

$$\frac{1}{2t}\langle (\mathscr{A}(t)-\mathscr{A}(0))^2 \rangle = \langle \dot{\mathscr{A}}^2 \rangle t_{\mathscr{A}}(1-t_{\mathscr{A}}/t+(t_{\mathscr{A}}/t)\exp(-t/t_{\mathscr{A}})) . \qquad (6.51)$$

In the first case, the correct result $\langle \dot{\mathscr{A}}^2 \rangle t_{\mathscr{A}}$ is approached exponentially quickly as $t$ increases, but the second equation has a slower inverse-$t$ dependence. From the above equations it should be obvious that the correlation function at any time $t$ may be recovered from $\langle (\mathscr{A}(t)-\mathscr{A}(0))^2 \rangle$ by numerical differentiation.

This leads us to ask when the route via the Einstein relation might be preferred to the calculation of a correlation function. The latter method is by far the most common, possibly because of the interest in the correlation functions themselves. However, there is much to be said for the Einstein relation route. In integrating the equations of motion, we use (at least) the known first and second derivatives of molecular positions and orientations: this order of numerical accuracy is 'built in' to computed mean-square displacements and the like. When we numerically integrate a correlation function using, say, Simpson's rule, especially if we have only stored data every 5 or 10 time steps, we are introducing additional sources of inaccuracy. In addition, there is the tendency to stop calculating and integrating time correlation functions when the signal seems to have disappeared in the noise. This is dangerous because of the possibility of missing contributions from the small, but systematic, long-time correlations.

A graphic illustration of this is shown in Fig. 6.5. The diffusion coefficient of $CS_2$ may be estimated [Tildesley and Madden 1983] by computing mean-squared displacements for each molecule, in a direction parallel to the molecular axis at time $t = 0$ and in two directions perpendicular to the axis. The axis system does not move or rotate with the molecule, and at long times the Einstein plots (Fig. 6.5(a)) show identical limiting behaviour. Thus, as expected, the diffusion coefficient turns out the same in each direction. The velocity autocorrelation functions may also be resolved into components along each of the three directions, and these are shown in Fig. 6.5b. Two

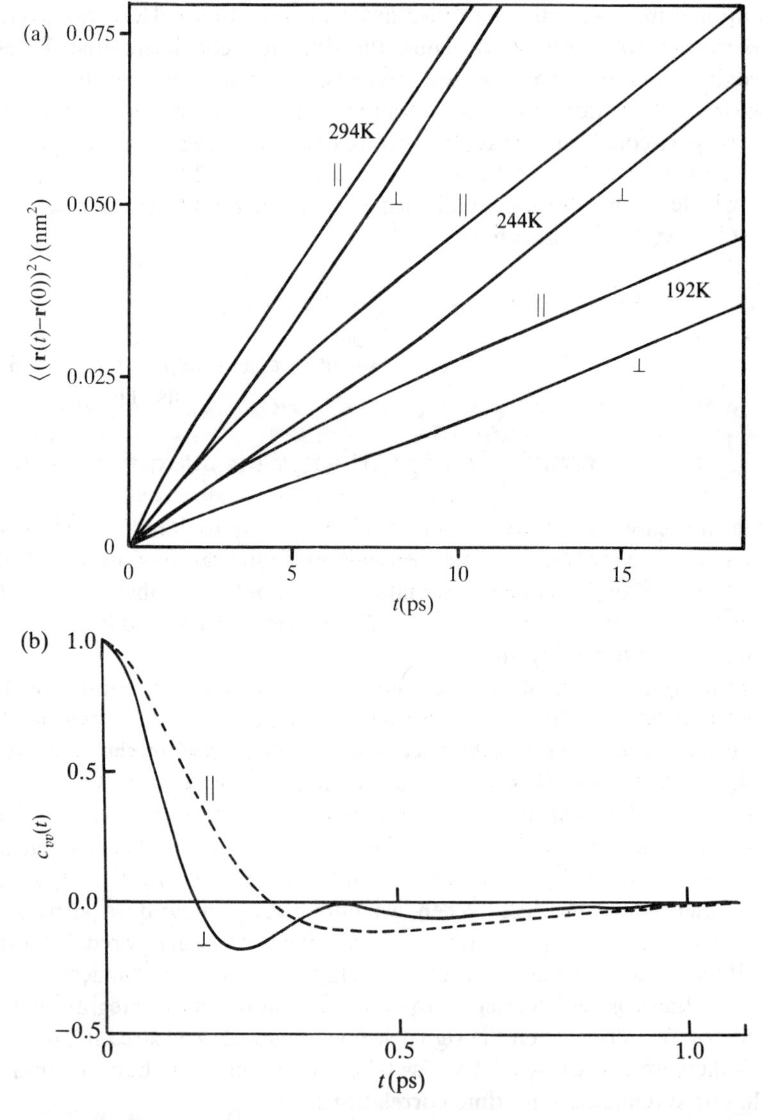


**Fig. 6.5** Calculating the diffusion coefficient in $CS_2$. (a) Mean square displacements at $T = 192\,\mathrm{K}$, 244 K, 294 K. (b) Velocity autocorrelation functions at $T = 192\,\mathrm{K}$. In each case we show components parallel and perpendicular to the molecular axis system at $t = 0$ [Tildesley and Madden 1983]

different kinds of behaviour (along and perpendicular to the axis) are seen: this is to be expected. More seriously, the correlation functions seem to have decayed to zero at $t \approx 1.8$ ps, and integration up to this point gives different results for the longitudinal and transverse cases. These estimates of $D$ are wrong. Despite appearances, there is still some residual structure in the velocity autocorrelation functions, which persists until molecular reorientations have completely relaxed. This effect can be seen on the Einstein plot, which has been extended to long enough times. In principle, if the velocity autocorrelation function were computed sufficiently accurately at long times, the same effect would be seen, and the diffusion coefficient would be correctly determined.

Alder *et al.* [1970] have described in some detail one situation in which the Einstein expression is more convenient even when the correlation function itself is of direct interest, namely the molecular dynamics of hard systems. For some dynamic quantities there will be a 'potential' contribution involving intermolecular forces, which, for hard systems act instantaneously only at collisions. Thus such contributions will be entirely absent from data stored on a tape at regular intervals for correlation function analysis. The problem is exactly analogous to that of estimating the pressure in a hard system (see Section 5.6), and occurs when we wish to calculate shear or bulk viscosities from stress (pressure) tensor correlations, and thermal conductivities from local energy fluctuations. The collisional contributions to these dynamical quantities must be taken into account during the simulation run itself. Moreover, because the forces act impulsively, the appropriate dynamical quantities $\dot{\mathscr{A}}(t)$ will contain delta functions, which would make the usual correlation function analysis rather awkward.

The Einstein relation variables $\mathscr{A}(t)$ are easier to handle: they merely change discontinuously at collisions. Following Alder *et al.* [1970] we take as our example the calculation of the shear viscosity $\eta$ via off-diagonal elements of the pressure tensor. The dynamical variable is (assuming equal masses)

$$\mathscr{Q}_{xy} = \frac{1}{V}\sum_i m r_{ix} \dot{r}_{iy} \,. \tag{6.52}$$

For systems undergoing free flight between collisions (say at times $t_1$ and $t_2$), the change in $\mathscr{Q}_{xy}$ is just

$$\mathscr{Q}_{xy}(t_2) - \mathscr{Q}_{xy}(t_1) = \frac{1}{V}\left[\sum_i m\dot{r}_{ix}\dot{r}_{iy}\right](t_2 - t_1)\,. \tag{6.53}$$

After a collision the term in square brackets changes, but this change is easy to compute, involving just the velocities of the colliding molecules. At a collision there is also a change in $\mathscr{Q}_{xy}$:

$$\mathscr{Q}_{xy}(t^+) - \mathscr{Q}_{xy}(t^-) = \frac{1}{V} m r_{ijx} \delta\dot{r}_{iy} \tag{6.54}$$

where $i$ and $j$ are the colliding pair, $r_{ijx} = r_{ix} - r_{jx}$, and $\delta\dot{r}_{iy}$ is the collisional change in the velocity of $i$ ($= -\delta\dot{r}_{jy}$). Thus the total change in $\mathscr{Q}_{xy}$ over any period of time is obtained by summing all the terms of the type shown in eqn (6.53), for all the inter-collisional intervals in that period (including the times before the first and after the last collision) and adding in all the terms of the type shown in eqn (6.54) for the collisions occurring in that interval. These values of $\mathcal{Q}_{xy}(t) - \mathcal{Q}_{xy}(0)$ may then be used in the Einstein expressions. Finally, the correlation function is recovered by numerical differentiation.

It should be noted that purely 'kinetic' correlation functions, such as the velocity autocorrelation function, and correlation functions involving molecular positions and orientations, not 'potential' terms, can be calculated in the normal way even for hard systems, and this is the preferred method where possible.

### *6.5.6 Smoothing a spectrum*

Often we wish to transform a time correlation function into the frequency domain. There are several methods of doing this, two of which are discussed in Appendix D. Just as in the case of spatial correlations, the truncation of $C(t)$ after a finite time, and the presence of random statistical errors, make the evaluation of the Fourier transform difficult. Spurious features in $\hat{C}_{\text{run}}(\omega)$, which is obtained by transforming a truncated $C_{\text{run}}(t)$, can obscure features present in the complete spectrum, $\hat{C}(\omega)$. In particular, the truncation causes spectral leakage, which often results in rapidly varying side lobes around a peak, and loss of resolution.

Windowing functions are weighting functions applied to the raw $C_{\text{run}}(t)$ to reduce the order of the discontinuity at the truncation point ($t_{\text{max}}$). Harris [1978] presents a thorough discussion of over thirty windowing functions and we illustrate their properties by considering the Blackman window. Suppose we have calculated the correlation function at regular intervals $C_{\text{run}}(\tau)$ with $\tau = 0, 1, \ldots, \tau_{\text{max}}$. Then, each value of $C_{\text{run}}(\tau)$ is multiplied by the windowing function

$$W(\tau) = 0.42 - 0.5\cos\left(\frac{\pi\tau}{\tau_{\text{max}}}\right) + 0.08\cos\left(\frac{2\pi\tau}{\tau_{\text{max}}}\right). \tag{6.55}$$

The result is then ready for Fourier transformation. Alternatively, the Fourier transform of the windowing function, $\hat{W}(\omega)$, is convoluted with $\hat{C}_{\text{run}}(\omega)$ to produce the windowed spectrum, $\hat{C}_W(\omega)$:

$$\hat{C}_W(\omega) = \int_{-\infty}^{+\infty} \frac{d\omega'}{2\pi} \hat{C}_{\text{run}}(\omega')\hat{W}(\omega - \omega'). \tag{6.56}$$

The coefficients in the windowing function are chosen so that $\hat{W}$ is sharply peaked, which leads to a good resolution in the windowed spectrum. The window function of eqn (6.55) reduces the side lobes by a factor of 58 dB from those of a rectangular window (which is equivalent to a cutoff at $t_{\text{max}}$). An example of a windowed spectrum is shown in Fig. 6.6. We consider Fourier transforming the model correlation function

$$C(\tau) = \exp(-0.01\tau)\cos(\tau) \qquad \tau = 0, 1, \ldots 34. \tag{6.57}$$

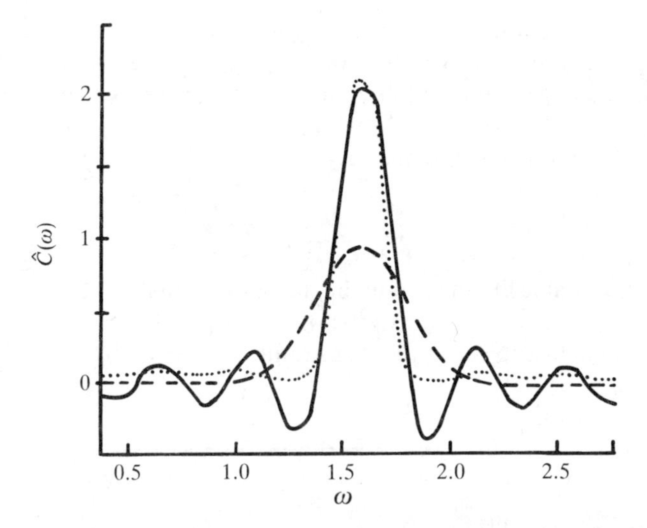


**Fig. 6.6** The spectrum of the model correlation function given in eqn (6.57): truncated data (solid line), Blackman window (dashed line), and maximum entropy method (dotted line).

This truncated function gives side lobes in the Fourier transform. These are removed on application of the Blackman window which gives a smooth curve with a maximum at the correct frequency but a lower intensity than $\hat{C}(\omega)$ from the truncated data. This is a serious consideration if band areas or peak heights are of interest. Berens and Wilson [1981] use a four-term Blackman–Harris window in computing the rotational spectrum of liquid CO in a CO/Ar mixture by simulation. They note that multiplying $\hat{C}_W(\omega)$ by the inverse sum of the squares of the windowing function makes it possible to correct the spectral band areas for the scaling effects of the windowing function.

The maximum entropy method is a technique for computing the most uniform spectrum consistent with a set of data [Guiasu and Shenitzer 1985]. To visualize this method, we imagine a team of monkeys producing an enormous number of random spectra. Without being biased, the monkeys are likely to produce more spectra of certain forms (e.g. flat, featureless ones) than others (containing specific sharp peaks). The spectra are transformed and passed to a dedicated theoretician for sorting. If a particular transform is inconsistent with the given correlation function $C_{\mathrm{run}}(t)$ then it is discarded. If the transform is consistent with $C_{\mathrm{run}}(t)$, (i.e. it agrees to within the estimated error) it is sorted into an appropriate pile: different piles for different spectra. After a very large number of compatible spectra have been sorted in this way, the piles are examined. The maximum entropy, or most likely, form of spectrum is represented by the largest pile.

In practice, the method works as follows. We make the assumption that every discrete point $C_{\mathrm{run}}(\tau)$ in the correlation function has a Gaussian error associated with it which is described by a variance $\sigma^2(C_{\text{run}}(\tau))$, or for short $\sigma^2(\tau)$. The quantity that we shall be varying is a trial fit spectrum $\hat{C}_{\text{fit}}(\nu)$ evaluated at a large number of discrete frequencies $\nu$. We can easily transform $\hat{C}_{\text{fit}}(\nu)$ to obtain the trial correlation function $C_{\text{fit}}(\tau)$ (see Appendix D), and the measure of a good fit is the quantity

$$\chi^2 = \sum_{\tau=1}^{\tau_{\max}} \frac{|C_{\text{fit}}(\tau) - C_{\text{run}}(\tau)|^2}{\sigma^2(\tau)}. \tag{6.58}$$

In fact, a reasonable fit (one within the statistical errors) would have $\chi^2 = \tau_{\max}$, and the technique will aim to fix $\chi^2$ at this value [Gull and Daniell 1978]. The most probable fit subject to this constraint is obtained by maximizing

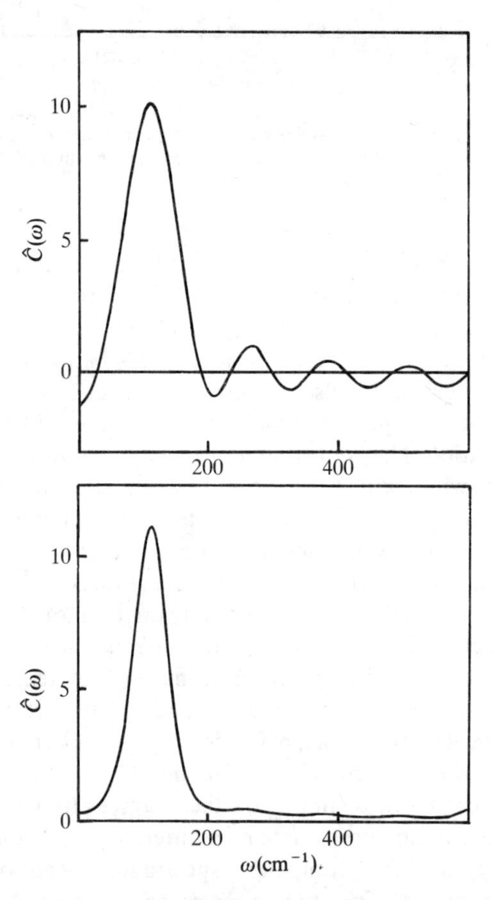


**Fig. 6.7** The spectrum of the angular velocity autocorrelation function of a nitrite ion in a simulation of solid $NaNO_2$. (a) Fourier transform of truncated data. (b) Maximum entropy spectrum. (This diagram was supplied by Dr R. M. Lynden-Bell, Cambridge, who performed the maximum entropy transform using a package due to J. Skilling.)

$$-\sum_{\nu}\hat{C}_{\mathrm{fit}}(\nu)\ln\hat{C}_{\mathrm{fit}}(\nu)+\frac{\lambda}{2}\sum_{\tau=1}^{\tau_{\max}}\frac{|C_{\mathrm{fit}}(\tau)-C_{\mathrm{run}}(\tau)|^2}{\sigma^2(\tau)}. \tag{6.59}$$

$\lambda$ is a Lagrange multiplier which constrains $\chi^2$ to be a constant. The first term in eqn (6.59) is the information–theoretical entropy of the spectrum. On differentiating we obtain

$$\hat{C}_{\mathrm{fit}}(\nu)=\exp\left\{-1+\lambda\left(\frac{1}{\tau_{\max}}\sum_{\tau=1}^{\tau_{\max}}\frac{(C_{\mathrm{run}}(\tau)-C_{\mathrm{fit}}(\tau))}{\sigma^2(\tau)}\exp(2\pi i\nu\tau/\tau_{\max})\right)\right\} \tag{6.60}$$

which can be solved iteratively. For a particular $\lambda$, we begin with a uniform $\hat{C}_{\mathrm{fit}}(\nu)$ to produce by transformation $C_{\mathrm{fit}}(\tau)$. This is used in eqn (6.60) to recalculate $\hat{C}_{\mathrm{fit}}(\nu)$, and the process repeated to convergence. This whole procedure is carried out for a number of $\lambda$ values until we obtain a consistent $\hat{C}_{\mathrm{fit}}(\nu)$ which has $\chi^2=\tau_{\max}$. In practice the solution is unchanged for a large range of $\lambda$. Note that the method completely avoids transforming the data, but it does rely on having reliable estimates of the statistical errors $\sigma^2(\tau)$ for each point in the correlation function. A package for performing the maximum entropy transform has been developed by J. Skilling.

The maximum entropy transform of eqn (6.57) is shown in Fig. 6.6: the method does an excellent job of smoothing the transform of the truncated data while maintaining the peak height. An example of the maximum entropy method on real data is shown in Fig. 6.7. Here, the aim is to Fourier transform one of the independent components of the angular velocity autocorrelation function for a model of the nitrite ion in a simulation of crystalline $NaNO_2$ (see Section 11.6) [Klein and McDonald 1982]. Figure 6.7 shows the spectrum of $c_{\omega_y\omega_y}(t)$ where the $y$-axis of the ion bisects the ONO bond. The improvement is expected to be somewhat less pronounced for a liquid.

# 7
# ADVANCED SIMULATION TECHNIQUES

## 7.1 Introduction

The MD and MC methods described in Chapters 3 and 4 may not be the most efficient ways of estimating certain statistical averages. The Metropolis prescription, eqn (4.20), for example, generates simulation trajectories that are naturally weighted to favour thermally populated states of the system, i.e. with Boltzmann-like weights. There are a number of important properties, such as free energies, that are difficult to calculate using this approach (see Sections 2.4, 5.6); direct calculation of the free energy really requires more substantial sampling over higher-energy configurations. In such non-Boltzmann sampling, $\rho_n/\rho_m$ is no longer simply $\exp(-\beta\delta\mathscr{V}_{nm})$ but the two states $n$ and $m$ are additionally weighted by a suitable function. This weighting function is designed to encourage the system to explore regions of phase space not frequently sampled by the Metropolis method. The weighted averages may be estimated more accurately than in conventional Monte Carlo, and are then corrected, giving the desired ensemble averages, at the end of the simulation. We describe this technique in Section 7.2.

A second extension involves changing the underlying stochastic matrix $\boldsymbol{\alpha}$ to make Monte Carlo 'smarter' at choosing its trial moves. In the conventional method, $\boldsymbol{\alpha}$ is symmetric and trial moves are selected randomly according to eqn (4.26). However, it is possible to choose an $\boldsymbol{\alpha}$ that is unsymmetric and that still satisfies the condition of microscopic reversibility. This may be used in the Monte Carlo method to sample preferentially in the vicinity of a solute molecule or a cavity in the fluid. Another application is to move particles preferentially in the direction of the forces acting on them. We describe these techniques in Section 7.3.

Why should these methods be more efficient? In essence, they introduce some of the character of MD into MC simulations. MD is totally deterministic, and intrinsically many-body in nature. By contrast, Metropolis MC is entirely stochastic and usually entails single-particle moves. Unfavourable energy configurations are avoided simply by rejecting them and standing still. It would be more satisfying, and perhaps more efficient, to guide the system in its search for favourable configurations, particularly if collective motions are important in avoiding 'barriers' in phase space. This is the role of the forces in MD: to lead downhill, with the kinetic terms being a counterbalancing effect. That this might be more efficient than simple MC is suggested by the analogy of Brownian motion: the diffusion coefficient is inversely proportional to the friction coefficient, which measures the strength of the random forces (see Chapter 9).

These considerations are also a motivation for adapting the constant-$NVE$

MD method so as to probe constant-temperature and constant-pressure ensembles. The analogous generalizations of MC were described in Chapter 4. Apart from possibly improving the efficiency with which we sample different ensembles, such simulations might represent more faithfully than the conventional MD method the behaviour of a fluid element, whose volume and energy would fluctuate. We describe these methods in Sections 7.4 and 7.5.

We must be careful before completely dismissing conventional MC in favour of modified MD, however. MD may be strikingly 'non-ergodic' in some circumstances. Energy transfer between nearly harmonic oscillators is very inefficient, for example. At low temperatures, the harmonic approximation holds quite well, and the system may easily become 'mode-locked' in a few quasi-harmonic oscillations [M. L. Klein, private communication]. This behaviour may persist on heating up an initially crystalline or glassy configuration, for example by scaling the velocities, and the introduction of some 'randomness' into the MD equations may assist equilibration in these circumstances [H. C. Andersen, private communication]. Some of the methods to be described in Sections 7.4 and 7.5 include stochastic elements, and some do not. The choice of the best technique, possibly intermediate between MD and MC, may vary dramatically from one system to another.

## 7.2 Free energy estimation

### *7.2.1 Introduction*

In this section we discuss the ways in which conventional simulations can be extended to facilitate the calculation of free energies. The methods of thermodynamic integration and direct particle insertion have been introduced in Section 2.4, and some technical details of the particle insertion method are given in Section 5.6. Grand canonical MC, which is another direct method for calculating the free energy, is discussed in Section 4.6. Now we take the opportunity to go further into the free-energy problem. We present a summary of the available methods and comment on their usefulness in Section 7.2.4.

### *7.2.2 Non-Boltzmann sampling*

Considerable effort has been expended on developing novel MC methods which allow determination of the 'statistical' properties (e.g. $A$ and $S$) of fluids. Such properties can be calculated from the configurational partition function $Q_{NVT}^{\text{ex}}$ defined in eqn (2.25), which may also be written as a configurational average

$$Q_{NVT}^{\text{ex}} = 1/\langle \exp(\beta\mathcal{V}) \rangle_{NVT}. \tag{7.1}$$

In principle, the denominator in eqn (7.1) can be calculated in a conventional simulation. Unfortunately, Metropolis Monte Carlo is designed to sample regions in which the potential energy is negative, or small and positive. These regions make little contribution to the average in eqn (7.1), and this route to $A$ is impractical. In this section, we discuss a more general method: umbrella sampling.

Let us begin with a less taxing problem than that of estimating eqn (7.1), specifically the calculation of a free energy difference. Consider two fluids characterized by potentials $\mathscr{V}(\mathbf{r})$ and $\mathscr{V}_0(\mathbf{r})$. If the free energy of the reference fluid, $A_0$, is known, then the free energy of the fluid of interest, $A$, can be determined from

$$A - A_0 = -k_B T \ln (Q/Q_0) = -k_B T \ln (\langle \exp - \beta \Delta \mathscr{V} \rangle_0) \tag{7.2}$$

where $\Delta\mathscr{V}(\mathbf{r}) = \mathscr{V}(\mathbf{r}) - \mathscr{V}_0(\mathbf{r})$ and the ensemble average $\langle \ldots \rangle_0$ is taken in the reference system $\mathscr{V}_0$. Unless the two fluids are very similar, and $\beta\Delta\mathscr{V}$ is small for all the important configurations in this ensemble, the average in eqn (7.2) is difficult to calculate accurately. The reason for this becomes clear if we rewrite the configurational density function $\rho_0(\mathbf{r})$ as a function, $\rho_0(\Delta\mathscr{V})$, of the energy difference. Then

$$Q/Q_0 = \int_{-\infty}^{\infty} \mathrm{d}(\Delta\mathscr{V}) \exp(-\beta\Delta\mathscr{V}) \rho_0(\Delta\mathscr{V}) . \tag{7.3}$$

$\rho_0(\Delta\mathscr{V})$ is the density (per unit $\Delta\mathscr{V}$) of configurations $\mathbf{r}$ in the reference ensemble which satisfy $\mathscr{V}(\mathbf{r}) = \mathscr{V}_0(\mathbf{r}) + \Delta\mathscr{V}$ for the specified $\Delta\mathscr{V}$. $\rho_0$ contains the Boltzmann factor $\exp(-\beta\mathscr{V}_0)$ and a factor associated with the change from $3N$ variables $(\mathbf{r})$ to one $(\Delta\mathscr{V})$. Fig. 7.1(a) shows the density $\rho_0(\Delta\mathscr{V})$ and $\exp(-\beta\Delta\mathscr{V})$ at a particular temperature. The density $\rho_0(\Delta\mathscr{V})$ decreases rapidly away from the mean value. In a simulation run of finite length, very low values of $\Delta\mathscr{V}$ are not sampled accurately. Indeed, in a histogram recording the potential energies which arise in such a simulation, there will be no entries at all for $\Delta\mathscr{V}$ less than some value $\Delta\mathscr{V}_c$ (see Fig. 7.1(b)). The true distribution (i.e. that obtained from an infinite run) would be small but non-zero below $\Delta\mathscr{V}_c$. For estimating most properties, this would not matter. However, when multiplied by the rapidly growing value of $\exp(-\beta\Delta\mathscr{V})$, these low energy points should make a substantial contribution to the integral in eqn (7.3). This contribution is the shaded area in Fig. 7.1(b), which in the finite-length simulation is incorrectly reckoned to be zero.

The solution to this problem is to sample on a non-Boltzmann distribution which favours configurations with large negative values of $\Delta\mathscr{V}$. This bias must be introduced so that it can subsequently be removed. Torrie and Valleau [1974, 1977a] sample from a general density function

$$\rho_W(\mathbf{r}) = W(\mathbf{r}) \exp(-\beta\mathscr{V}_0(\mathbf{r})) / \int \mathrm{d}\mathbf{r}\, W(\mathbf{r}) \exp(-\beta\mathscr{V}_0(\mathbf{r})) . \tag{7.4}$$

Here $W(\mathbf{r}) = W(\Delta\mathscr{V}(\mathbf{r}))$ is a positive-valued weighting function which is specified at the beginning of a simulation run. The method described in Chapter 4 is used to produce a Markov chain of states with a limiting

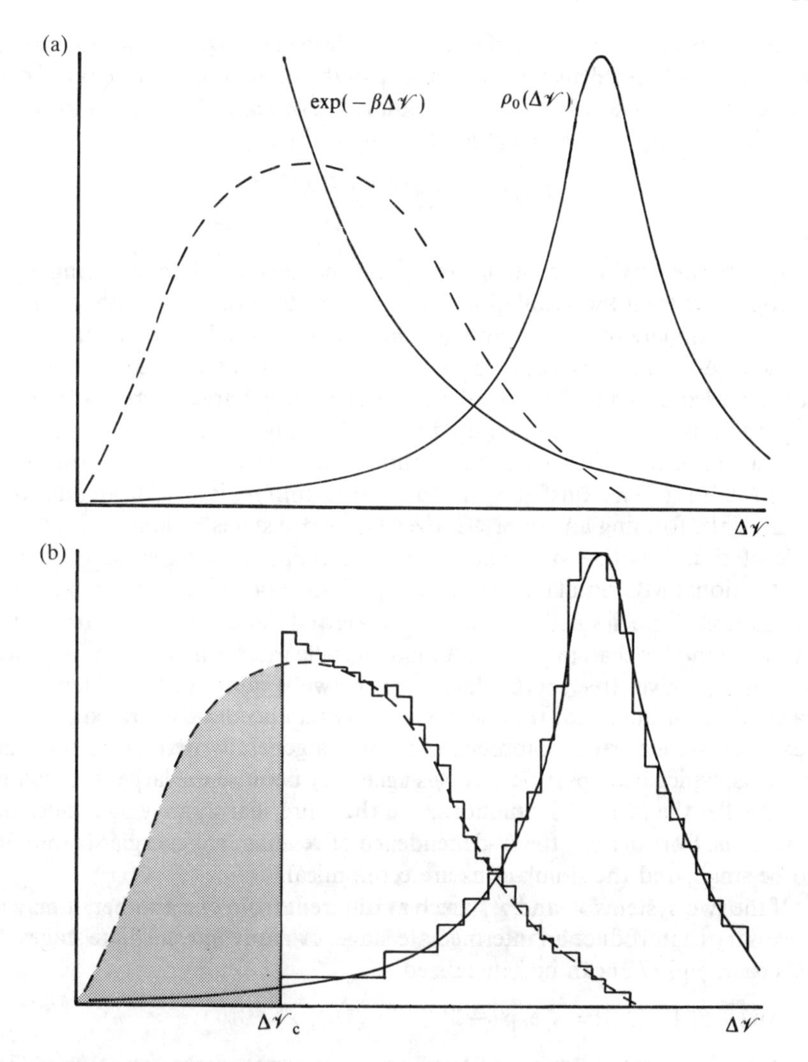


**Fig. 7.1** The problem in estimating free energy differences. (a) The functions $\rho_0(\Delta\mathscr{V})$ and $\exp(-\beta\Delta\mathscr{V})$ are shown as solid lines. The product of these two functions, the integrand in eqn (7.3), is shown as a dashed line.

(b) The way in which these functions are estimated in a finite-length simulation.

distribution given by eqn (7.4). Specifically, a trial move, from state $m$ to state $n$, is accepted with a probability given by $\min(1, (W_n/W_m)\exp(-\beta(\delta\mathscr{V}_0)_{nm}))$. The average of any property in the reference ensemble, $\langle \mathscr{A} \rangle_0$, can be related to averages taken over MC trials, i.e. in the weighted ensemble, using

$$\langle \mathscr{A} \rangle_0 = \frac{\langle \mathscr{A}/W \rangle_{\text{trials}}}{\langle 1/W \rangle_{\text{trials}}} = \frac{\langle \mathscr{A}/W \rangle_{\text{W}}}{\langle 1/W \rangle_{\text{W}}} \tag{7.5}$$

where the notation $\langle \ldots \rangle_{\mathrm{W}}$ reminds us of the weighting. This means that the ratio $\mathscr{A}(\mathbf{r})/W(\mathbf{r})$ is calculated for each step of the simulation, and averaged over the run; the average of $1/W(\mathbf{r})$ is also required in order to obtain the final result. The densities $\rho_0(\Delta\mathscr{V})$ and $\rho_{\mathrm{W}}(\Delta\mathscr{V})$ are related by

$$\rho_0(\Delta\mathscr{V}) = \frac{\rho_{\mathrm{W}}(\Delta\mathscr{V})/W(\Delta\mathscr{V})}{\langle 1/W(\Delta\mathscr{V}) \rangle_{\mathrm{W}}}. \tag{7.6}$$

Thus, the density function $\rho_0$ itself may be calculated by building up a histogram during the simulation. An appropriate choice of $W(\Delta\mathscr{V})$ with an accurate estimate of the denominator in eqn (7.6) gives $\rho_0$ over a much wider range of $\Delta\mathscr{V}$ than is possible in a conventional simulation. The improved $\rho_0$ can be used in eqn (7.3) to calculate the required free energy difference. Equivalently, eqn (7.5) can be used with $\mathscr{A} = \exp(-\beta\Delta\mathscr{V})$.

One of the difficulties with this method is that there is no *a priori* recipe for $W(\Delta\mathscr{V})$. It is often adjusted by trial and error until $\rho_{\mathrm{W}}$ is as wide and uniform as possible, forming an 'umbrella' over the two systems $\mathscr{V}$ and $\mathscr{V}_0$. A useful rule of thumb is that it should extend the range of energies sampled in a conventional MC simulation by a factor of three or more, allowing accurate calculation of much smaller $\rho_0$ values [Torrie and Valleau 1977a]. A limitation of the method is that, in practice, unlike particle insertion or grand canonical MC, it only gives free energy differences between quite similar systems. The calculation of absolute free energies requires an accurate knowledge of the reference system value. Umbrella sampling is generally performed on small systems, typically 32 particles. This is necessary because the larger the system, the smaller the relative fluctuations, and the more sharply varying the density functions. Fortunately, the $N$-dependence of relative free energies is thought to be small, and the simulations are economical.

If the two systems $\mathscr{V}$ and $\mathscr{V}_0$ are very different from one another, it may be necessary to introduce an intermediate stage, or many intermediate stages. In this case, eqn (7.2) can be generalized to

$$\exp(-\beta(A - A_0)) = \langle \exp(-\beta(\mathscr{V} - \mathscr{V}_n)) \rangle_n \times \\ \langle \exp(-\beta(\mathscr{V}_n - \mathscr{V}_{n-1})) \rangle_{n-1} \ldots \langle \exp(-\beta(\mathscr{V}_1 - \mathscr{V}_0)) \rangle_0 \tag{7.7}$$

where systems $\mathscr{V}_1 \ldots \mathscr{V}_n$ have been introduced with properties intermediate between those of $\mathscr{V}$ and $\mathscr{V}_0$. This multistage sampling [Valleau and Card 1972] has been employed directly to calculate the free-energy difference between hard spheres and Coulombic hard spheres. Each of the separate averages in eqn (7.7) can be evaluated with the help of umbrella sampling, which reduces the number of intermediate stages [Torrie and Valleau 1977a].

As an illustration of the umbrella sampling technique, Torrie and Valleau [1977a] have related the free energy of the Lennard-Jones fluid to that of the inverse 12th power fluid, and these same authors also found it useful in the study of liquid mixtures [Torrie and Valleau 1977b]. Umbrella sampling has also been used to calculate the surface tension (i.e. excess surface free energy) of a model of water [Lee and Scott 1980]. These authors suggest

$$W(\Delta\mathscr{V}) = \exp(-\beta\Delta\mathscr{V}/2) \tag{7.8}$$

as an appropriate choice for the weighting function.

Shing and Gubbins [1981, 1982] used umbrella sampling in conjunction with test particle insertion to calculate the chemical potential. We describe the second of their two methods, which is the more generally applicable. A single test particle is inserted in the fluid, at intervals during the normal simulation. It moves through the fluid using a non-Boltzmann sampling algorithm which favours configurations of high $\exp(-\beta\mathscr{V}_{\text{test}})$ (see Section 2.4, eqn (2.68)). Each configuration is weighted by a factor $W(\mathscr{V}_{\text{test}})$. One particularly simple form for $W$ is

$$\begin{aligned} W(\mathscr{V}_{\text{test}}) &= 1 \qquad \mathscr{V}_{\text{test}} \leqslant \mathscr{V}_{\text{max}} \\ &= 0 \qquad \mathscr{V}_{\text{test}} > \mathscr{V}_{\text{max}}\,. \end{aligned} \tag{7.9}$$

In the simulation of a Lennard-Jones fluid, $\mathscr{V}_{\text{max}}$ was taken to be $200\varepsilon$, and the weighting function rejected all moves which led to a significant overlap. The test particle has no real interaction with the atoms in the fluid, and, in general, a test particle move from position $\mathbf{r}^m_{\text{test}}$ to $\mathbf{r}^n_{\text{test}}$ is accepted if

$$W(\mathbf{r}^n_{\text{test}})/W(\mathbf{r}^m_{\text{test}}) \geqslant \xi \tag{7.10}$$

where $\xi$ is a random number in the range (0, 1). During the run the distribution of test particle energies, $\rho_{\text{W}}(\mathscr{V}_{\text{test}})$ is calculated. The distribution is proportional to the unweighted distribution, $\rho_0(\mathscr{V}_{\text{test}})$ for $\mathscr{V}_{\text{test}} \leqslant \mathscr{V}_{\text{max}}$ (see eqn (7.6)). The constant of proportionality is most easily obtained in this case by performing a parallel set of unweighted test-particle insertions, and comparing the two distributions in the region where they are both well-known. Once $\rho_0(\mathscr{V}_{\text{test}})$ is known accurately over its complete range, then the chemical potential can be calculated from

$$\mu^{\text{ex}} = -k_{\text{B}}T\ln\left(\int_{-\infty}^{+\infty} \rho_0(\mathscr{V}_{\text{test}})\exp(-\beta\mathscr{V}_{\text{test}})\mathrm{d}\mathscr{V}_{\text{test}}\right). \tag{7.11}$$

The usual problem with the insertion method, namely the high probability of finding overlaps at high densities, is controlled by the weighted sampling. Shing and Gubbins [1982] have also proposed a method which concentrates the sampling on the configurations that exhibit suitable 'holes' for the insertion. A recent modification of the particle insertion method has been to 'turn on' the test particle interaction gradually [Mon and Griffiths 1985]. The idea of using a variable coupling parameter has been used to estimate solubilities [Swope and Andersen 1984].

### 7.2.3 *Acceptance ratio method*

An interesting extension of the ideas introduced in the previous section is the work of Bennett [1976]. In the canonical ensemble, the ratio of the partition functions of two fluids is given in terms of an arbitrary weighting function $W(\mathbf{r})$

$$\frac{Q_1}{Q_0} = \frac{Q_1 \int \mathrm{d}\mathbf{r}\, W(\mathbf{r}) \exp(-\beta(\mathscr{V}_1 + \mathscr{V}_0))}{Q_0 \int \mathrm{d}\mathbf{r}\, W(\mathbf{r}) \exp(-\beta(\mathscr{V}_1 + \mathscr{V}_0))} = \frac{\langle W \exp(-\beta\mathscr{V}_1) \rangle_0}{\langle W \exp(-\beta\mathscr{V}_0) \rangle_1}. \tag{7.12}$$

The choice $W = \exp(\beta\mathscr{V}_0)$ or $W = \exp(\beta\mathscr{V}_1)$ leads to eqn (7.2), but, as we have seen, this is likely to be impractical. Bennett shows that a particular choice of $W$ will minimize the variance in the estimation of $Q_1/Q_0$. The best choice is

$$W = \text{constant} \times \left( \frac{Q_0}{(\tau_0/s_0)} \exp(-\beta\mathscr{V}_1) + \frac{Q_1}{(\tau_1/s_1)} \exp(-\beta\mathscr{V}_0) \right)^{-1} \tag{7.13}$$

where $(\tau_0/s_0)$ and $(\tau_1/s_1)$ are the number of statistically independent configurations generated in each of the two Markov chains (see eqn (6.16)). Substitution of eqn (7.13) into eqn (7.12) gives

$$\frac{Q_1}{Q_0} = \frac{\langle \mathscr{F}(+\beta\Delta\mathscr{V} + C) \rangle_0}{\langle \mathscr{F}(-\beta\Delta\mathscr{V} - C) \rangle_1} \exp(C) \tag{7.14}$$

where $C = \ln(Q_1 s_1 \tau_0 / Q_0 s_0 \tau_1)$, $\Delta\mathscr{V} = (\mathscr{V}_1 - \mathscr{V}_0)$ and $\mathscr{F}$ is the Fermi function

$$\mathscr{F}(x) = (1 + \exp(x))^{-1}. \tag{7.15}$$

Writing eqn (7.14) in terms of energy distributions, we obtain

$$\frac{Q_1}{Q_0} = \frac{\int \mathrm{d}(\Delta\mathscr{V}) \mathscr{F}(+\beta\Delta\mathscr{V} + C) \rho_0(\Delta\mathscr{V})}{\int \mathrm{d}(\Delta\mathscr{V}) \mathscr{F}(-\beta\Delta\mathscr{V} - C) \rho_1(\Delta\mathscr{V})} \exp(C). \tag{7.16}$$

The constant $C$ acts as a shift in potential, so as to bring the two systems into as close a correspondence as possible. The method works as follows. A simulation of each fluid is performed, and the density functions $\rho_1$ and $\rho_0$ calculated by constructing histograms as functions of $\Delta\mathscr{V}$. A value of $C$ is guessed, and the ratio $Q_1/Q_0$ calculated from eqn (7.16). $C$ is recalculated from

$$C \approx \ln(Q_1/Q_0). \tag{7.17}$$

In eqn (7.17) we have assumed that $\tau_1/s_1 \approx \tau_0/s_0$. This can be checked by direct calculation using the methods described in Section 6.4. An iterative solution of eqns (7.16) and (7.17) gives a value for $C$ and $Q_1/Q_0$. Bennett [1976] presents an alternative graphical solution. The method works well if there is any overlap between the functions $\rho_1$ and $\rho_0$. The overlap can be improved using umbrella or multi-stage sampling. Equation (7.16) has been used to calculate the free energy of a model of liquid nitrogen from that of the hard dumb-bell fluid [Jacucci and Quirke 1980b].

A slightly modified form of eqn (7.14) is applicable in the constant-$NVE$ ensemble [Frenkel 1986]

$$\frac{Q_1}{Q_0} = \frac{\langle \mathscr{F}[+(\Delta\mathscr{V} - \Delta E)/k_B\mathscr{T}_0 + C] \rangle_0}{\langle \mathscr{F}[-(\Delta\mathscr{V} - \Delta E)/k_B\mathscr{T}_1 - C] \rangle_1} \exp(C) \tag{7.18}$$

where $C = \ln(Q_1 s_1 \tau_0 / Q_0 s_0 \tau_1)$, $Q$ means $Q_{NVE}$, and $\Delta E = E_1 - E_0$ is the difference between the total energies in the two simulations. $\mathscr{T}_0$ and $\mathscr{T}_1$ are the instantaneous values of the temperature and eqn (7.18) assumes that $\langle \mathscr{T}_1 \rangle = \langle \mathscr{T}_0 \rangle$. The microcanonical partition function can be related to the entropy through eqn (2.18). In the MD simulations the two density functions $\rho_1((\Delta\mathscr{V} - \Delta E)/k_B\mathscr{T}_1)$ and $\rho_0((\Delta\mathscr{V} - \Delta E)/k_B\mathscr{T}_0)$ are calculated and eqn (7.18) is solved iteratively for $C$.

### 7.2.4 *Summary*

Statistical properties can be calculated directly by simulating in the grand canonical ensemble. Grand canonical simulations have not yet been performed using the MD method and a purpose-built MC program is required. Even with this effort, GCMC simulations are not useful at high density without some biased sampling trick.

The umbrella sampling method does give a useful route to free energy differences. However, it cannot give absolute free energies, and there is always a subjective element in choosing the appropriate weighting function. Two systems that are quite different can only be linked by performing several intermediate simulations, even with the use of umbrella sampling at each stage. If there is any overlap between the distributions of configurational energy in the two systems, then Bennett's method is a useful route to the free energy differences. It can be easily enhanced by the use of umbrella sampling.

Perhaps the most direct attack is to calculate the chemical potential by the particle insertion method in the canonical or microcanonical ensemble (using the appropriate formula). This method is easy to program and fits neatly into an existing code. The additional time required for the calculation is approximately 20 per cent of the normal run time. This method may also fail at densities close to the triple point, although there is some disagreement about its precise range of validity. A useful check is to calculate the distribution of test particle energies and real molecule energies during a run. When the logarithm of the ratio of these distributions is plotted against $\beta\mathscr{V}_{\text{test}}$ it should be a straight line of slope one, and the intercept should be $-\beta\mu^{\text{ex}}$ [Powles *et al.* 1982]. If this method gives a different result from the straightforward average of the Boltzmann factor of the test particle energy, then there is a problem with convergence. In this case the particle insertion should be enhanced by umbrella sampling [Shing and Gubbins 1982].

The internal energy can be accurately calculated by simulation in the canonical ensemble, and the temperature can be accurately calculated in the microcanonical ensemble. This makes the thermodynamic integration of eqn (2.65) an accurate route to free energy differences. One possible disadvantage is that a large number of simulations may be required to span the integration range. This is not a problem if the aim of the simulation is an extensive exploration of the phase diagram, and one short cut is to plan simulations at appropriate temperatures along the integration range to enable you to perform a Gauss–Legendre quadrature of eqn (2.65) without the need for interpolation [Frenkel 1986]. One other possible difficulty is the requirement of finding a reversible path between the state of interest and some reference state. Ingenious attempts have been made to integrate along a thermodynamic path linking the liquid with the ideal gas [Hansen and Verlet 1969] or with the harmonic lattice [Hoover and Ree 1968] without encountering the irreversibility associated with the intervening phase transitions. In the solid state, it may be necessary to apply an external field to reach the Einstein crystal [Frenkel and Ladd 1984] and a similar technique may be used to calculate the free energy of a liquid crystal [Frenkel, Mulder, and McTague 1985] (see Chapter 11).

### 7.3 Smarter Monte Carlo

In the conventional MC method, all the molecules are moved with equal probability, in directions chosen at random. This may not be the most efficient way to proceed: we might wish to attempt moves for some molecules more often than others, or to bias the moves in preferred directions.

This preferential sampling can be accomplished using an extension of the Metropolis solution (eqn (4.20)) of the following form:

$$\begin{aligned}
\pi_{mn} &= \alpha_{mn} & \alpha_{nm}\rho_n &\geqslant \alpha_{mn}\rho_m \quad m \neq n \\
\pi_{mn} &= \alpha_{mn}\left(\frac{\alpha_{nm}\rho_n}{\alpha_{mn}\rho_m}\right) & \alpha_{nm}\rho_n &< \alpha_{mn}\rho_m \quad m \neq n \\
\pi_{mm} &= 1 - \sum_{n \neq m} \pi_{mn} . & &
\end{aligned} \tag{7.19}$$

We recall that $\pi_{mn}$ is the one-step transition probability of going from state $m$ to state $n$. In this case it is easy to show that microscopic reversibility is satisfied, even if $\alpha_{mn} \neq \alpha_{nm}$. The Markov chain can be easily generated by making random trial moves from state $m$ to state $n$ according to $\alpha_{mn}$. The trial move is accepted with a probability given by $\min(1, \alpha_{nm}\rho_n/\alpha_{mn}\rho_m)$. The details of this type of procedure are given in Section 4.4. We make use of the above prescription in the following.

#### *7.3.1 Preferential sampling*

In a dilute solution of an ion in water, for example, the most important interactions are often those between solute and solvent, and between solvent molecules in the primary solvation shell. The solvent molecules further from the ion do not play such an important role. It is sensible to move the molecules in the first solvation shell more frequently than the more remote molecules. Let us define a region $\mathscr{R}_{\text{sol}}$ around the solute molecule, solvent molecules within the region being designated 'in', and the remainder being 'out'. A parameter $p$ defines how often we wish to move the 'out' molecules relative to the 'in' ones: $p$ lies between 0 and 1, values close to 0 corresponding to much more frequent moves of the 'in' molecules. A move consists of the following steps [Owicki and Scheraga 1977a]:

(a) Choose a molecule at random.
(b) If it is 'in', make a trial move.
(c) If it is 'out', generate a random number uniformly on (0, 1). If $p$ is greater than the random number then make a trial move. If not, then return to step (a).

In step (c), if it is decided not to make a trial move, we return to step (a) immediately, and select a new molecule, without accumulating any averages etc. This procedure will attempt 'out' molecule moves with probability $p$ relative to 'in' molecule moves. Trial moves are accepted with a probability $\min(1, \alpha_{nm}\rho_n/\alpha_{mn}\rho_m)$ and the problem is to calculate the ratio $\alpha_{nm}/\alpha_{mn}$ for this scheme. Consider a configuration $m$ with $N_{\text{in}}$ 'in' molecules and $N_{\text{out}}$ 'out' molecules. The chance of selecting an 'in' molecule is

$$\begin{aligned} p_{\text{in}} &= \frac{N_{\text{in}}}{N} + (1-p)\frac{N_{\text{out}}}{N}\frac{N_{\text{in}}}{N} + \left((1-p)\frac{N_{\text{out}}}{N}\right)^2 \frac{N_{\text{in}}}{N} + \ldots \\ &= \frac{N_{\text{in}}}{N'} \end{aligned} \tag{7.20}$$

where $N' = pN + (1-p)N_{\text{in}}$. Note how, in eqn (7.20) we count all the times that we look at 'out' molecules, decide not to try moving them, and return to step (a), eventually selecting an 'in' molecule.

Once we have decided to attempt a move, there are four distinct cases, corresponding to the moving molecule in states $m$ and $n$ being 'in' or 'out' respectively. Let us consider the case in which we attempt to move a molecule which was initially 'in' the region $\mathscr{R}_{\text{sol}}$ to a position outside that region (see Fig. 7.2). Suppose that trial moves may occur to any of $N_{\mathscr{R}}$ positions within a cube $\mathscr{R}$ centred on the initial position of the molecule. Then $\alpha_{mn}$ is the probability of choosing a specific 'in' molecule, and attempting to move it to one of these sites as shown in Fig. 7.2:

$$\alpha_{mn} = \frac{1}{N_{\text{in}}}\frac{N_{\text{in}}}{N'}\frac{1}{N_{\mathscr{R}}} = \frac{1}{N'N_{\mathscr{R}}}. \tag{7.21}$$

The chance of attempting the reverse move, from a state containing $N_{\text{out}}+1 = N - N_{\text{in}} + 1$ 'out' molecules and $N_{\text{in}} - 1$ 'in' molecules, is

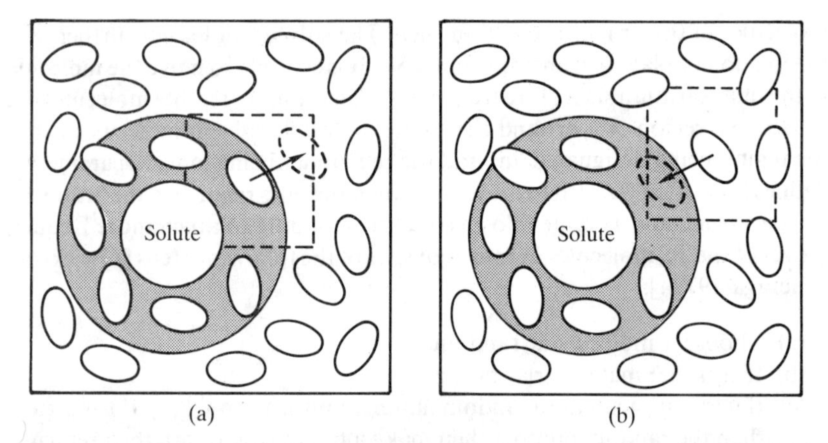


**Fig. 7.2** (a) Attempting to move a solvent molecule out of the region around the solute (shaded). (b) Attempting the reverse move.

$$\alpha_{nm} = \frac{1}{N - N_{\text{in}} + 1}\left(1 - \frac{N_{\text{in}} - 1}{pN + (1-p)(N_{\text{in}} - 1)}\right)\frac{1}{N_{\mathscr{R}}}$$
$$= \frac{p}{N' N_{\mathscr{R}}}\left(1 - \frac{(1-p)}{N'}\right)^{-1} \tag{7.22}$$

and the desired ratio can be obtained. Summarizing for all four cases we have

$$(m \to n)$$
$$\alpha_{nm}/\alpha_{mn} = p(1-(1-p)/N')^{-1} \qquad (\text{in} \to \text{out}) \tag{7.23a}$$
$$\alpha_{nm}/\alpha_{mn} = 1 \qquad (\text{out} \to \text{out}) \tag{7.23b}$$
$$\alpha_{nm}/\alpha_{mn} = 1 \qquad (\text{in} \to \text{in}) \tag{7.23c}$$
$$\alpha_{nm}/\alpha_{mn} = [p(1+(1-p)/N')]^{-1} \qquad (\text{out} \to \text{in}). \tag{7.23d}$$

where $N'$ is calculated for $N_{\text{in}}$ molecules in state $m$. In the simulation, $p$ is chosen so that the initial probability of attempting an 'in' molecule move is typically 1/2. In an unweighted simulation, the probability of moving 'in' molecules obviously depends on the system size, but would be much lower, say 10–20 per cent.

Owicki [1978] has suggested an alternative method for preferential sampling which has been used in the simulation of aqueous solutions [Mehrotra, Mezei, and Beveridge 1983]. In this method, the probability of choosing a solvent molecule decays monotonically with its distance from the solute. We define a weighting function, which typically takes the form

$$W'(r_{i0}) = r_{i0}^{-\nu} \tag{7.24}$$

where $\nu$ is an integer. Here $r_{i0}$ is the distance of molecule $i$ from the solute, which we label 0. At any instant, a properly normalized weight may be formed

$$W(r_{i0}) = W'(r_{i0}) / \sum_j W'(r_{j0}) \tag{7.25}$$

and used to define a probability distribution for the current configuration. A molecule $i$ is chosen from this distribution using a rejection technique as described in Appendix G. An attempted move is then made to any of the $N_{\mathscr{R}}$ neighbouring positions. Denoting $W(r_{i0})$ in the initial and final states simply by $W_m$ and $W_n$ respectively, the required ratio of underlying transition matrix elements is

$$\alpha_{nm}/\alpha_{mn} = W_n/W_m\,. \tag{7.26}$$

In the above, we have tacitly assumed that the solute molecule is fixed. This is permissible, but relaxation of the first neighbour shell will be enhanced if the solute is allowed to move as well. This will, of course, change all the interactions with solvent molecules. In the scheme described above, the solute may be moved as often as desired, with $\alpha_{nm}/\alpha_{mn} = 1$, without any additional modifications.

A useful example of preferential sampling is the cavity-biased GCMC method [Mezei 1980]. GCMC becomes less useful at high densities because of the difficulty of making successful creation and destruction attempts. In the cavity-biased method insertion is only allowed at points where a cavity of a suitable radius, $r_c$, exists. The probabilities of accepting a creation or destruction attempt (eqns (4.40) and (4.41)) are modified by an additional factor $p_N$, the probability of finding a cavity of radius $r_c$, or larger, in a fluid of $N$ molecules. A creation attempt is accepted with a probability given by

$$\min(1, \exp[-\beta\delta\mathscr{V}_{nm} + \ln(zVp_N/N+1)]) \tag{7.27a}$$

and a destruction attempt is accepted with a probability given by

$$\min(1, \exp[-\beta\delta\mathscr{V}_{nm} + \ln(N/zVp_{N-1})])\,. \tag{7.27b}$$

The simulation is realized by distributing a number of test sites uniformly throughout the fluid. During the run each site is tested to see whether it is at the centre of a suitable cavity or not. In this way $p_N$ is calculated with a steadily improving reliability and at the same time it is possible to locate points in the fluid suitable for an attempted creation. In the event that no cavity is available, we can continue with the next move or use a scheme which mixes the cavity sampling with the more conventional GCMC method. Details of the mixed scheme, which requires complicated book-keeping to ensure microscopic reversibility, are given in the original paper. Mezei reports an eightfold increase in the efficiency of creation/destruction attempts in a simulation of a supercritical Lennard-Jones fluid.

### *7.3.2 Force-bias Monte Carlo*

In real liquids, the movement of a molecule is biased in the direction of the forces acting on it. It is possible to build this bias into the underlying stochastic matrix $\boldsymbol{\alpha}$ of the Markov chain. The reason for adopting a force-bias scheme is to improve convergence to the limiting distribution, and steer the system more efficiently around the bottlenecks of phase space (see Section 4.3 and Fig. 2.1).

Pangali, Rao, and Berne [1978] adopt the following prescription for the underlying Markov chain:

$$\begin{aligned} \alpha_{mn} &= \exp\left(+\lambda\beta\left(\mathbf{f}_i^m \cdot \delta\mathbf{r}_i^{nm}\right)\right)/C\left(\mathbf{f}_i^m, \lambda, \delta r_{\max}\right) \quad & n \in \mathscr{R} \\ \alpha_{mn} &= 0 & n \notin \mathscr{R}. \end{aligned} \tag{7.28}$$

Here we have assumed that just one atom $i$ is to be moved, $\mathbf{f}_i^m$ is the force on this atom in state $m$, $\delta\mathbf{r}_i^{nm} = \mathbf{r}_i^n - \mathbf{r}_i^m$ is the displacement vector in a trial move to state $n$, $\lambda$ is a constant and $C$ is a normalizing factor. Typically, $\lambda$ lies between 0 and 1. When $\lambda = 0$, eqn (7.28) reduces to eqn (4.26) for the conventional transition probability. As usual, $\mathscr{R}$ is a cube of side $2\delta r_{\max}$ centred on the initial position $\mathbf{r}_i^m$ (see Fig. 4.3). A little manipulation shows that

$$C(\mathbf{f}_i^m, \lambda, \delta r_{\max}) = \frac{8\sinh(\lambda\beta\delta r_{\max} f_{ix}^m)\sinh(\lambda\beta\delta r_{\max} f_{iy}^m)\sinh(\lambda\beta\delta r_{\max} f_{iz}^m)}{\lambda^3\beta^3 f_{ix}^m f_{iy}^m f_{iz}^m}. \tag{7.29}$$

It is clear from eqn (7.28) that this prescription biases $\delta\mathbf{r}_i^{nm}$ in the direction of the force on the atom.

The FB method is implemented as follows. An atom $i$ is chosen at random and given a trial random displacement $\delta\mathbf{r}_i^{nm}$ selected using a rejection technique (see Appendix G) from the probability distribution determined by eqn (7.28). The trial move is accepted with a probability given by $\min(1, \alpha_{nm}\rho_n/\alpha_{mn}\rho_m)$. The ratio appearing here is given by

$$\frac{\alpha_{nm}\rho_n}{\alpha_{mn}\rho_m} = \exp\left(-\beta\left(\delta\mathscr{V}_{nm} + \lambda\delta\mathbf{r}_i^{nm}\cdot(\mathbf{f}_i^m + \mathbf{f}_i^n) + \delta W^{\mathrm{FB}}\right)\right) \tag{7.30}$$

where

$$\delta W^{\mathrm{FB}} = -k_{\mathrm{B}}T\ln\left(\frac{C(\mathbf{f}_i^m, \lambda, \delta r_{\max})}{C(\mathbf{f}_i^n, \lambda, \delta r_{\max})}\right) \tag{7.31}$$

can be calculated using eqn (7.29). For small values of $\delta r_{\max}$

$$\delta W^{\mathrm{FB}} \approx \tfrac{1}{6}\lambda^2\beta\delta r_{\max}^2\left((\delta\mathbf{f}_i^{nm})^2 + 2\delta\mathbf{f}_i^{nm}\cdot\mathbf{f}_i^m\right) \tag{7.32}$$

where $\delta\mathbf{f}_i^{nm} = \mathbf{f}_i^n - \mathbf{f}_i^m$. The two parameters in the method, $\lambda$ and $\delta r_{\max}$, can be adjusted to maximize the root-mean-square displacement of the system through phase space (a simple, though not unique, measure of efficiency) [Rao and Berne 1979; D'Evelyn and Rice 1981].

The FB method is particularly powerful when dealing with hydrogen-bonded liquids such as water, which are susceptible to bottlenecks in phase space. Molecular translation is handled as described above, and the extension to include torque-biased rotational moves is straightforward. The analogous equation to eqn (7.28) is

$$\alpha_{mn} = C^{-1} \exp\left(+\lambda\beta \mathbf{f}_i^m \cdot \delta\mathbf{r}_i^{nm} + \lambda\beta \boldsymbol{\tau}_i^m \cdot \delta\boldsymbol{\phi}_i^{nm}\right) \quad n \in \mathscr{R} \tag{7.33}$$

where $\boldsymbol{\tau}_i^m$ is the torque on molecule $i$ in state $m$, and $\delta\boldsymbol{\phi}_i^{nm}$ is the trial angular displacement, i.e. $\delta\boldsymbol{\phi}_i^{nm} = \delta\phi_i^{nm}\mathbf{e}$ where $\mathbf{e}$ is the axis of rotation. A study of a model of water using the force-bias method [Rao, Pangali, and Berne 1979] demonstrated clear advantages over conventional MC methods, and better agreement with MD results for this system. A further study [Mehrotra *et al.* 1983] showed an improvement in convergence by a factor 2–3 over conventional MC.

### *7.3.3 Smart Monte Carlo*

Force-bias Monte Carlo involves a combination of stochastic and systematic effects on the choice of trial moves. A similar situation applies to the motion of a Brownian molecule in a fluid: it moves around under the influence of random forces (from surrounding solvent molecules) and systematic forces (from other nearby Brownian molecules). We will turn to the simulation of Brownian motion in detail in Chapter 9, and simply give here the application of this method to the 'Smart Monte Carlo' (SMC) scheme devised by Rossky, Doll, and Friedman [1978]. The trial displacement of a molecule $i$ from state $m$ to state $n$ may be written

$$\delta\mathbf{r}_i^{nm} = \beta A\, \mathbf{f}_i^m + \delta\mathbf{r}_i^{\mathrm{G}}. \tag{7.34}$$

$\delta\mathbf{r}_i^{\mathrm{G}}$ is a random displacement whose components are chosen from a Gaussian distribution with zero mean and variance $\langle (\delta r_{ix}^{\mathrm{G}})^2 \rangle = \langle (\delta r_{iy}^{\mathrm{G}})^2 \rangle = \langle (\delta r_{iz}^{\mathrm{G}})^2 \rangle = 2A$. The quantity $A$ is an adjustable parameter (equal to the diffusion coefficient multiplied by the time step in Brownian dynamics simulation). The underlying stochastic matrix for this procedure is

$$\alpha_{mn} = (4A\pi)^{-3/2} \exp\left(-(\delta\mathbf{r}_i^{nm} - \beta A\, \mathbf{f}_i^m)^2/4A\right). \tag{7.35}$$

In practice a trial move consists of selecting a random vector $\delta\mathbf{r}_i^{\mathrm{G}}$ from a Gaussian distribution as described in Appendix G, and using it to displace a molecule chosen at random according to eqn (7.34). The move is accepted with probability $\min(1, \alpha_{nm}\rho_n/\alpha_{mn}\rho_m)$ (see eqn (7.19)) and the required ratio is

$$\frac{\alpha_{nm}\rho_n}{\alpha_{mn}\rho_m} = \exp\left(-\beta\left(\delta\mathscr{V}_{nm} + \tfrac{1}{2}(\mathbf{f}_i^n + \mathbf{f}_i^m)\cdot \delta\mathbf{r}_i^{nm} + \delta W^{\mathrm{SMC}}\right)\right) \tag{7.36}$$

where

$$\delta W^{\mathrm{SMC}} = \frac{\beta A}{4}\left((\delta\mathbf{f}_i^{nm})^2 + 2\mathbf{f}_i^m \cdot \delta\mathbf{f}_i^{nm}\right) \tag{7.37}$$

and the notation is the same as for eqn (7.32). Rossky *et al.* [1978] tested the method by simulating ion clusters, and Northrup and McCammon [1980] have used Smart Monte Carlo to study protein structure fluctuations.

There are clear similarities, and slight differences, between the FB and SMC methods. One difference is that eqn (7.35) puts no upper limit on the displacement that a given molecule may suffer at any step, using a Gaussian probability distribution instead of a cubic trial displacement region. However, if we write eqn (7.35) in the form

$$\alpha_{mn} = (4A\pi)^{-3/2} \exp\left(-\beta^2 A \mathbf{f}_i^{m^2}/4 - \delta\mathbf{r}_i^{nm^2}/4A\right) \exp\left(+\tfrac{1}{2}\beta \mathbf{f}_i^m \cdot \delta\mathbf{r}_i^{nm}\right) \quad (7.38)$$

and compare with eqn (7.28), we note that the distributions are particularly similar for $\lambda = \frac{1}{2}$. For this choice, the two ratios governing acceptance of a move are identical if [Rao and Berne 1979]

$$\delta W^{\mathrm{SMC}} = \delta W^{\mathrm{FB}} \quad (7.39)$$

and for small step sizes, this holds for

$$A = \delta r_{\max}^2 / 6\,. \quad (7.40)$$

Comparisons between the two techniqu/s are probably quite system-dependent. Both offer a substantial improvement over conventional MC on a step-by-step basis in many cases, but they are comparable with molecular dynamics in complexity and expense since they involve calculation of forces and torques. Both methods improve the acceptance rate of moves. The most efficient method, in this sense, would make $\alpha_{nm}/\alpha_{mn} = \rho_m/\rho_n$, when every move would be accepted, but of course $\rho_n$ is not known before a move is tried. SMC, and FB with $\lambda = 1/2$, both approach 100 per cent acceptance rates quadratically as the step size is reduced. This makes multi-molecule moves more feasible. In fact, Smart Monte Carlo simulations with $N$-molecule moves and small step sizes are almost identical with the Brownian dynamics (Schmoluchowski equation) simulations of Chapter 9. The extra flexibility of SMC and FB methods lies in the possibility of using larger steps (and rejecting some moves) and also in being able to move any number of molecules from 1 to $N$.

### *7.3.4 Virial-bias Monte Carlo*

FB and SMC simulations are not limited to the canonical ensemble. Mezei [1983] has used a similar technique in the constant-$NPT$ ensemble. For a volume perturbation the analogue of the force is (see Section 4.5)

$$f_V = \frac{\partial}{\partial V}(\mathcal{V} + PV - Nk_{\mathrm{B}}T \ln V) = \frac{\partial \mathcal{V}}{\partial V} + P - \rho k_{\mathrm{B}}T = P - \mathcal{P}' \quad (7.41)$$

(see eqn (2.56)). A virial-biased volume move is attempted using an underlying stochastic matrix given by

$$\begin{aligned}\alpha_{mn} &= \exp(+\lambda\beta(f_V^m \delta V_{nm})/C(f_V, \lambda, \delta V_{\max}). \quad && n \in \mathscr{R}_V \\ &= 0 && n \notin \mathscr{R}_V \end{aligned} \tag{7.42}$$

where the new volume after the attempted move, $V_n$, is in a region $\mathscr{R}_V$ defined by $V_m \pm \delta V_{\max}$, and $C$ is a normalizing constant. A trial move is accepted with a probability given by min $(1, \alpha_{nm}\, \rho_n / \alpha_{mn}\, \rho_m)$. This algorithm is analogous to the FB method. There is a similar technique based on the SMC method [Mezei 1983]. Both techniques appear to give a significant improvement in the sampling of the constant-$NPT$ ensemble.

## 7.4 Constant-temperature molecular dynamics

The FB and SMC techniques may be viewed as attempts to introduce some dynamic features into constant-$NVT$ Monte Carlo. Now we approach the problem from the other direction, and seek to adapt MD so as to sample a constant-temperature ensemble. Several different methods of prescribing the temperature in a molecular dynamics simulation exist. A recent review [Andersen, Allen, Bellemans, Board, Clarke, Ferrario, Haile, Nosé, Opheusden, and Ryckaert 1984] has attempted to summarize these methods and highlight their advantages and disadvantages. Here we outline the ones in common use, and discuss practical points in their implementation.

### *7.4.1 Stochastic methods*

A physical picture of a system corresponding to the canonical ensemble involves weak 'stray interactions' between the molecules of the system and the particles of a heat bath at a specified temperature [Tolman 1938]. This leads to a straightforward adaptation of the MD method [Andersen 1980]. At intervals, the velocity of a randomly selected molecule is chosen afresh from the Maxwell–Boltzmann distribution (see Appendix G). This corresponds to a collision with an imaginary heat-bath particle. The system moves through phase space on a constant-energy surface, until the velocity of one molecule is changed; the system then jumps onto a different energy surface and the hamiltonian motion proceeds. In this way, the system samples all of the important regions of phase space, generating an irreducible Markov chain. The limiting probability density of the chain can be shown to be the canonical one of Section 2.2 [Andersen 1980].

In the original description of the method, times between collisions with the bath are chosen from a Poisson distribution, with a specified mean collision time, but this choice does not affect the final phase-space distribution. If the collisions take place infrequently, energy fluctuations will occur slowly, but kinetic energy (temperature) fluctuations will occur much as in conventional MD. If the collisions occur very frequently, then kinetic temperature fluctuations are dominated by them, rather than by the systematic dynamics. Too high a collision rate will slow down the speed at which the molecules in the system explore configuration space, whereas too low a rate means that the canonical distribution of energies will only be sampled slowly. If it is intended that the system mimic a volume element in a real liquid, in thermal contact with its surroundings, then Andersen [1980] suggests a collision rate given by

$$\text{rate per particle} \propto \frac{\lambda_T}{\rho^{1/3} N^{2/3}} \tag{7.43}$$

where $\lambda_T$ is the thermal conductivity. Note that this decreases as the system size goes up. In suitable circumstances, the collisions have only a small effect on single-particle time correlation functions [Haile and Gupta 1983], but too high a collision rate will lead to exponentially decaying correlation functions [Evans and Morriss 1984a].

An alternative to altering the velocity of one particle at a time is to reselect the velocities of all particles at once, rather less frequently, at equally spaced intervals of time [Andrea *et al.* 1983]. In between these 'massive stochastic collisions', time correlation functions may be calculated in the usual way. This allows us to obtain correlation functions with Newtonian dynamics, but averaged over an initial canonical distribution.

Heyes [1983b] has suggested a method in which an attempt is made to scale all the velocities systematically up or down and the attempt accepted or rejected using a MC technique. Every (say) tenth step in an MD simulation, a random number $\zeta$ is chosen uniformly in a small range, e.g. $[-0.05, 0.05]$. The new trial velocities are given by

$$\dot{\mathbf{r}}^n = (1+\zeta)\,\dot{\mathbf{r}}^m \tag{7.44}$$

and the ratio of the probabilities of the new and old states in momentum space is

$$\rho_n/\rho_m = \exp(-A) \tag{7.45}$$

where

$$A = \tfrac{1}{2} m\beta \sum_i |\dot{\mathbf{r}}^m|^2 \left((1+\zeta)^2 - 1\right) - 3N \ln(1+\zeta). \tag{7.46}$$

The second term is associated with the change in the velocity volume element. The trial move is accepted with a probability given by $\min(1, \exp(-A))$. This method also samples the canonical ensemble.

### 7.4.2 *Extended system methods*

A second way to treat the dynamics of a system in contact with a thermal reservoir is to include a degree of freedom which represents that reservoir, and carry out a simulation of this 'extended system'. Energy is allowed to flow dynamically from the reservoir to the system and back; the reservoir has a certain 'thermal inertia' associated with it, and the whole technique is rather like controlling the volume of a sample by using a piston (see Section 7.5). Nosé [1984] has described the implementation of this method. The extra degree of freedom is denoted $s$, and it has a conjugate momentum $p_s$. The real particle velocities are related to the time-derivatives of position by

$$\mathbf{v} = s\dot{\mathbf{r}} = \mathbf{p}/ms. \tag{7.47}$$

An extra potential energy term is associated with $s$:

$$\mathscr{V}_s = (f+1)k_B T \ln s \tag{7.48}$$

where $f$ is the number of degrees of freedom ($3N-3$ if the total momentum is fixed) and $T$ is the specified temperature. There is also a kinetic energy term

$$\mathscr{K}_s = \tfrac{1}{2}Q\dot{s}^2 = p_s^2/2Q \tag{7.49}$$

where $Q$ is the thermal inertia parameter with dimension (energy) (time)$^2$ which controls the rate of temperature fluctuations.

The Lagrangian of the system is

$$\mathscr{L}_s = \mathscr{K} + \mathscr{K}_s - \mathscr{V} - \mathscr{V}_s \tag{7.50}$$

where $\mathscr{K} = \sum_i \frac{1}{2}m_i v_i^2$ and $\mathscr{V}$ is evaluated as a function of $\mathbf{r}$ in the usual way. The equations of motion can be readily derived

$$\ddot{\mathbf{r}} = \mathbf{f}/ms^2 - 2\dot{s}\dot{\mathbf{r}}/s \tag{7.51a}$$

$$Q\ddot{s} = \sum_i m\dot{r}_i^2 s - (f+1)k_B T/s\,. \tag{7.51b}$$

The extended system Hamiltonian $\mathscr{H}_s = \mathscr{K} + \mathscr{K}_s + \mathscr{V} + \mathscr{V}_s$ is conserved, and the extended system density function is microcanonical

$$\rho_{NVE_s}(\mathbf{r}, \mathbf{p}, s, p_s) = \frac{\delta(\mathscr{H}_s - E_s)}{\int \mathrm{d}\mathbf{r}\mathrm{d}\mathbf{p}\mathrm{d}s\mathrm{d}p_s\,\delta(\mathscr{H}_s - E_s)}\,. \tag{7.52}$$

Manipulation of the delta functions and an integration over the variables $s$ and $p_s$ gives a canonical distribution of the variables $\mathbf{r}$ and $\mathbf{p}/s$. Nosé [1984] shows how this result depends upon the logarithmic dependence of $\mathscr{V}_s$ on $s$. The equations of motion are solved using standard predictor–corrector methods (see Appendix E) and the conservation of $\mathscr{H}_s$ acts as a useful check on the programming. A sample program is given in F.28.

Nosé [1984] discusses the choice of the adjustable parameter $Q$. Too high a value of $Q$ results in slow energy flow between the system and reservoir, and in the limit $Q \to \infty$ we regain conventional MD. On the other hand, if $Q$ is too low, long-lived, weakly damped oscillations of the energy occur, resulting in poor equilibration. It may be necessary to choose $Q$ by trial and error, so as to achieve satisfactory damping of these correlations.

### 7.4.3 *Constraint methods*

A simple method of fixing the kinetic temperature of a system in MD is to rescale the velocities at each time step by a factor of $(T/\mathscr{T})^{1/2}$ where $\mathscr{T}$ is the current kinetic temperature and $T$ is the desired thermodynamic temperature. This method has been used extensively in the equilibration phase of MD simulation (see Chapter 5) and has also been suggested as a means of performing 'isothermal molecular dynamics' [Woodcock 1971]. Velocity rescaling turns out [Andersen *et al.* 1984] to be a crude method of solving a set of equations of motion that differ from the Newtonian ones. Newtonian mechanics implies that the energy and momentum are the conserved variables of the motion. Constant kinetic temperature dynamics is generated by the equations of motion [Hoover, Ladd, and Moran 1982; Evans 1983a]

$$\dot{\mathbf{r}} = \mathbf{p}/m \tag{7.53a}$$

$$\dot{\mathbf{p}} = \mathbf{f} - \xi(\mathbf{r}, \mathbf{p})\mathbf{p} . \tag{7.53b}$$

The quantity $\xi(\mathbf{r}, \mathbf{p})$ is a kind of 'friction coefficient' which varies so as to constrain $\mathscr{T}$ to a constant value, i.e. to guarantee that (in an atomic system)

$$\dot{\mathscr{T}} \propto \frac{\mathrm{d}}{\mathrm{d}t}\left(\sum_i p_i^2\right) \propto \sum_i \dot{\mathbf{p}}_i \cdot \mathbf{p}_i = 0 . \tag{7.54}$$

The constraint is chosen so as to perturb as little as possible the classical equations of motion, by making $\xi$ a Lagrange multiplier which minimizes the difference (in a least-squares sense) between the constrained and Newtonian trajectories. This principle of least constraint is due to Gauss [Hoover 1983a; Evans and Morriss 1984a]. The resulting expression is

$$\xi = \frac{\sum_i \mathbf{p}_i \cdot \mathbf{f}_i}{\sum_i |\mathbf{p}_i|^2} . \tag{7.55}$$

These equations generate a path which samples the constant-$NV\mathscr{T}$ ensemble [Evans and Morriss 1983a] with a density function $\rho_{NV\mathscr{T}}$ proportional to

$$\delta(\mathscr{T} - T)\delta(\mathbf{P})\exp(-\mathscr{V}/k_{\mathrm{B}}\mathscr{T}) .$$

Here $\mathbf{P}$ is the total linear momentum, which is usually set to zero. $\mathscr{T}$ is obtained from the kinetic energy using the correct number of degrees of freedom, which is $3N-4$ in a fluid of $N$ atoms, given that the kinetic energy and three momentum components are fixed. The delta functions do not depend on configurational coordinates, and so this method generates configurational properties in the canonical ensemble. The momentum distribution is not canonical, but the equivalence of ensembles guarantees that the differences in most simple averages will be of order $\mathcal{O}(1/N)$. Usually, configurational properties are calculated during a simulation, and the exactly known kinetic properties are added in separately, as in constant-$NVT$ Monte Carlo.

Equations (7.53a, b) can be solved using the Gear predictor–corrector method discussed in Chapter 3 and Appendix E. Alternatively, a variant of the leap-frog scheme (Section 3.2.1) has been proposed [Brown and Clarke 1984]. This takes the form of a modified velocity equation

$$\dot{\mathbf{r}}(t+\tfrac{1}{2}\delta t) = \dot{\mathbf{r}}(t-\tfrac{1}{2}\delta t) + (\mathbf{f}(t)/m - \xi\dot{\mathbf{r}}(t))\delta t \tag{7.56}$$

and it is implemented as follows.

(a) Make an unconstrained half step

$$\dot{\mathbf{r}}'(t) = \dot{\mathbf{r}}(t-\tfrac{1}{2}\delta t) + \tfrac{1}{2}\mathbf{f}(t)\delta t/m \tag{7.57}$$

(b) Calculate $\chi = (T/\mathcal{T})^{1/2}$ where $T$ is the desired temperature and $\mathcal{T}$ is calculated from these unconstrained velocities $\dot{\mathbf{r}}'(t)$. $\chi^{-1}$ is equal to $1+\tfrac{1}{2}\xi\delta t$.

(c) Complete the full step using

$$\dot{\mathbf{r}}(t+\tfrac{1}{2}\delta t) = (2\chi-1)\dot{\mathbf{r}}(t-\tfrac{1}{2}\delta t) + \chi\mathbf{f}(t)\delta t/m\,. \tag{7.58}$$

A copy of a program to implement this method is given in F.29. In molecular fluids, the rotational and translational kinetic energies can be constrained separately using this technique [Fincham *et al.* 1986].

### 7.4.4 *Other methods*

There are several additional techniques for performing constant temperature molecular dynamics. A number of these use Brownian dynamics algorithms to solve the many-particle Langevin equation [Ermak and Yeh 1974; Schneider and Stoll 1978] and some of these will be discussed in Chapter 9. Instead of reselecting Maxwellian velocities at random, it is possible to do so when particles cross a 'thermal wall' [Ciccotti and Tenenbaum 1980]. A further refinement of the velocity rescaling approach has been proposed [Berendsen, Postma, van Gunsteren, DiNola, and Haak 1984]. At each time step, velocities are scaled by a factor

$$\chi = \left(1 + \frac{\delta t}{t_T}\left(\frac{T}{\mathcal{T}} - 1\right)\right)^{1/2} \tag{7.59}$$

where $\mathcal{T}$ is the current kinetic temperature and $t_T$ is a preset time constant. This method forces the system towards the desired temperature at a rate determined by $t_T$, while only slightly perturbing the forces on each molecule. In a simulation of water, Berendsen *et al.* [1984] found a relaxation time of

$t_T = 0.4$ ps to be appropriate. This method does not generate states in the canonical ensemble, but seems to be very useful for purposes of changing state and equilibrating a system at the new temperature.

Hoover [1985] has extended the analysis of Nosé (see Section 7.4.2). He derives a slightly different set of equations which dispense with the time-scaling parameter $s$.

$$\dot{\mathbf{r}} = \mathbf{p}/m$$

$$\dot{\mathbf{p}} = \mathbf{f} - \xi\mathbf{p}\,. \tag{7.60}$$

In this case the friction coefficient $\xi$ is given by the first-order differential equation

$$\dot{\xi} = \frac{f}{Q}(k_B\mathcal{T} - k_B T) \tag{7.61}$$

where $Q$ is the thermal inertia parameter and $f$ the number of degrees of freedom. These equations do generate states in the canonical ensemble and eqn (7.61) is unique in this respect [Hoover 1985]. The method of Berendsen *et al.* [1984] can be cast in a similar form but in this case

$$\xi = \frac{1}{2t_T k_B\mathcal{T}}(k_B\mathcal{T} - k_B T)\,. \tag{7.62}$$

Note that it is $\xi$ and not $\dot{\xi}$ that is constrained here. Equation (7.61) steers the temperature towards the required value in a much gentler way than eqn (7.62). They are both in sharp contrast to eqn (7.55) which constrains the kinetic energy to a constant value.

## 7.5 Constant-pressure molecular dynamics

The various schemes for prescribing the pressure of a molecular dynamics simulation have also been reviewed in the recent CECAM report [Andersen *et al.* 1984]. Once more, we summarize the popular methods and discuss practical points. In all of these approaches it is inevitable that the system box must change its volume (as it does in constant-pressure MC simulations). No stochastic methods for constant-pressure MD seem to have been developed, although it would probably be feasible to incorporate MC-like box-size 'moves' at intervals in a conventional MD simulation. Analogues of the extended system and constraint methods of temperature control do, however, exist.

### *7.5.1 Extended system methods*

Andersen [1980] originally proposed a method for constant pressure MD, which involves coupling the system to an external variable $V$, the volume of the simulation box. This coupling mimics the action of a piston on a real system.

The piston has a 'mass' $Q$ (which actually has the units of (mass) (length)$^{-4}$) and is associated with a kinetic energy

$$\mathscr{K}_V = \tfrac{1}{2} Q \dot{V}^2 . \tag{7.63}$$

The potential energy associated with the additional variable is

$$\mathscr{V}_V = PV \tag{7.64}$$

where $P$ is the specified pressure. The potential and kinetic energies associated with the molecules are written with $\mathbf{r}$ and $\mathbf{v}$ given in terms of scaled variables

$$\mathbf{r} = V^{1/3} \mathbf{s} \tag{7.65a}$$

$$\mathbf{v} = V^{1/3} \dot{\mathbf{s}} \tag{7.65b}$$

so that $\mathscr{V} = \mathscr{V}(V^{1/3}\mathbf{s})$ and $\mathscr{K} = \frac{1}{2} m \sum_i v_i^2 = \frac{1}{2} m V^{2/3} \sum_i \dot{s}_i^2$. The equations of motion can be readily obtained from the Lagrangian

$$\mathscr{L}_V = \mathscr{K} + \mathscr{K}_V - \mathscr{V} - \mathscr{V}_V \tag{7.66}$$

and are

$$\ddot{\mathbf{s}} = \mathbf{f}/(mV^{1/3}) - (2/3)\dot{\mathbf{s}}\dot{V}/V \tag{7.67a}$$

$$\ddot{V} = (\mathscr{P} - P)/Q \tag{7.67b}$$

where the forces $\mathbf{f}$ and the pressure function $\mathscr{P}$ (eqn (2.55)) are calculated using normal, unscaled, coordinates and momenta. The Hamiltonian of this system $\mathscr{H}_V = \mathscr{K} + \mathscr{K}_V + \mathscr{V} + \mathscr{V}_V$ is conserved, being equal to the enthalpy of the fluid plus an additional factor of $\frac{1}{2} k_B T$ associated with the kinetic energy of the volume fluctuation. The conservation law is a useful check of a properly functioning program. These equations of motion generate trajectories which sample the isobaric–isoenthalpic ensemble. This is not one of the common ensembles discussed in Chapter 2, but its properties have been described [Ray, Graben, and Haile 1981].

Haile and Graben [1980] describe a method of implementing this type of MD simulation. This essentially solves the equations of motion in terms of the scaled positions and momenta in a box of unit length. The equations are typically solved using a Gear predictor–corrector with coefficients given in Appendix E (as appropriate for a second-order differential equation with the first derivatives appearing on the right). The differential equation for the volume of the box is coupled to the equations of motion for the molecules, since the forces and pressure are evaluated using unscaled coordinates. When comparing the instantaneous pressure $\mathscr{P}$ to the desired pressure $P$, it is essential to add the long-range correction (eqn (2.134)) at each time step as the simulation proceeds, since the density is continually changing. Unscaled trajectories can be obtained from eqn (7.65a) and the unscaled velocities obtained by differentiation

$$\dot{\mathbf{r}} = V^{1/3}\dot{\mathbf{s}} + \tfrac{1}{3} V^{-2/3} \dot{V} \mathbf{s} . \tag{7.68}$$

The method does not fit easily into the leap-frog scheme since $\dot{\mathbf{s}}$ is not available at each time step $t$. A modified form of leap-frog has been described [Brown and Clarke 1984] which avoids this problem, and which is constructed in terms of unscaled positions and velocities. A modification of the velocity Verlet scheme has also been described [Fox and Andersen 1984]. An example of a constant-$NPH$ MD program is given in F.30.

The parameter $Q$, the 'piston mass', is an adjustable parameter in Andersen's method. A low 'mass' will result in rapid box size oscillations, which are not damped very efficiently by the random motions of the molecules. A large 'mass' will give rise to slow exploration of volume-space, and an infinite mass restores normal MD. To mimic events in a small volume element of a liquid, Andersen [1980] recommends that the time scale for box-volume fluctuations should be roughly the same as those for a sound wave to cross the simulation box. Brown and Clarke [1984] suggest a value of $Q\sigma^4/m = 0.0027$ in their simulations of 256 Lennard-Jones atoms, and Fox and Andersen [1984] use a similar value.

Interesting though the isobaric–isoenthalpic ensemble may be, it is unusual. Quite commonly, the constant pressure method of this section is combined with one of the constant temperature methods described in Section 7.4 (see [Andersen 1980]) so as to simulate the constant-$NPT$ ensemble. If the stochastic approach is used, the distribution from which the scaled velocities are chosen is proportional to $\exp(-mV^{2/3}\dot{s}_i^2/2k_BT)$. Fox and Andersen [1984] recommend also making stochastic impacts on the piston, and describe a method of doing this.

### 7.5.2 *Constraint methods*

In the same way as seen in Section 7.4.3, equations of motion may be devised which make the instantaneous pressure $\mathscr{P}$ (as defined by eqn (2.55)) a constant of the motion. By using a Lagrange multiplier, and applying Gauss's principle, this can be done in such a way as to minimize the change in dynamics [Evans and Morriss 1983b, 1984a]

$$\dot{\mathbf{r}} = \mathbf{p}/m + \chi(\mathbf{r}, \mathbf{p})\mathbf{r} \tag{7.69a}$$

$$\dot{\mathbf{p}} = \mathbf{f} - \chi(\mathbf{r}, \mathbf{p})\mathbf{p} \tag{7.69b}$$

$$\dot{V} = 3V\chi(\mathbf{r}, \mathbf{p})\,. \tag{7.69c}$$

Here $\chi(\mathbf{r}, \mathbf{p})$ is the Lagrange multiplier, which can also be thought of as the rate of dilation of the system. An expression for $\chi$ can be obtained by differentiating eqn (2.55) which defines $\mathscr{P}$:

$$3\dot{\mathscr{P}}V + 3\mathscr{P}\dot{V} = \sum_i (2/m)\mathbf{p}_i\cdot\dot{\mathbf{p}}_i + \dot{\mathbf{r}}_i\cdot\mathbf{f}_i + \dot{\mathbf{f}}_i\cdot\mathbf{r}_i\,. \tag{7.70}$$

If the instantaneous value of the pressure is constant then we can set $\mathscr{P} = P$ and $\dot{\mathscr{P}} = 0$. Then eqns (7.69) and 7.70) give

$$\chi = \frac{(2/m)\sum_i \mathbf{p}_i \cdot \mathbf{f}_i - (1/m)\sum_i \sum_{j>i} (\mathbf{r}_{ij}\cdot \mathbf{p}_{ij})x(r_{ij})/r_{ij}^2}{(2/m)\sum_i p_i^2 + \sum_i \sum_{j>i} x(r_{ij}) + 9PV} \tag{7.71}$$

where $\mathbf{p}_{ij} = \mathbf{p}_i - \mathbf{p}_j$ and the function $x(r_{ij})$ is defined in eqn (2.78). In calculating the denominator of eqn (7.71), we must include long-range corrections for the $x(r_{ij})$ term

$$9\mathscr{X}_{\mathrm{LRC}} = \left(\sum_i \sum_{j>i} x(r_{ij})\right)_{\mathrm{LRC}} = 2\pi\rho N \int_{r_c}^{\infty} x(r_{ij}) r_{ij}^2 \, \mathrm{d}r_{ij} \tag{7.72}$$

and a similar long-range correction (eqn (2.134)) for the $PV$ term. As in the last section, this is because the density of the system varies during the simulation.

The equations of motion are solved, as written above, by a standard predictor–corrector technique. Some care must be taken whenever a particle crosses a periodic boundary, since all the derivatives of position change discontinuously when the position is shifted (see eqn (7.69a)). For example,

$$\ddot{\mathbf{r}} = \dot{\mathbf{p}} + \chi \mathbf{r} + \dot{\chi}\mathbf{r}. \tag{7.73}$$

To calculate these, we need the derivatives of $\chi$ which are obtained by repeated differentiation of eqn (7.69c)

$$\dot{\chi} = (\ddot{V}V - \dot{V}^2)/3V^2 \tag{7.74}$$

and so on.

The ensemble generated by these equations has constant mechanical pressure and enthalpy functions, and so is not one of the usual ones. It is perhaps more useful to simulate the 'isothermal–isobaric' ensemble, by which we mean constant-$N\mathscr{P}\mathscr{T}$. To do this, eqn (7.69b) is modified to include an additional Lagrange multiplier, as in Section 7.4.3 [Evans and Morriss 1983b]

$$\dot{\mathbf{p}} = \mathbf{f} - \chi(\mathbf{r}, \mathbf{p})\mathbf{p} - \xi(\mathbf{r}, \mathbf{p})\mathbf{p}. \tag{7.75}$$

In this case, the expression for $\chi$ is

$$\chi = -\frac{(1/m)\sum_i \sum_{j>i} (\mathbf{r}_{ij}\cdot \mathbf{p}_{ij})x(r_{ij})/r_{ij}^2}{\sum_i \sum_{j>i} x(r_{ij}) + 9PV} \tag{7.76}$$

and $(\chi + \xi)$ is given by the expression on the right of eqn (7.55). The equations of motion are solved as before. The phase space density function for the ensemble is now proportional to [Evans and Morriss 1983a]

$$\delta(\mathscr{T} - T)\,\delta(\mathscr{P} - P)\exp(-(\mathscr{H} + \mathscr{P}V)/k_{\mathrm{B}}\mathscr{T})$$

(compare the constant-$NPT$ case discussed in Section 2.2). A sample program appears in F.31.

### 7.5.3 *Other methods*

Berendsen *et al.* [1984] have described a particularly simple technique for coupling to a 'pressure bath'. An extra term is added to the equations of motion to produce a pressure change. The system is made to obey the equation

$$\mathrm{d}\mathscr{P}/\mathrm{d}t = (P - \mathscr{P})/t_P \tag{7.77}$$

where $P$ is the desired pressure and $t_P$ is a time constant. At each step, the volume of the box is scaled by a factor $\chi$, and the molecular centre-of-mass coordinates by a factor $\chi^{1/3}$

$$\mathbf{r}' = \chi^{1/3}\mathbf{r} \tag{7.78}$$

where

$$\chi = 1 - \beta_T \frac{\delta t}{t_P}(P - \mathscr{P}). \tag{7.79}$$

Here, $\beta_T$ is the isothermal compressibility and $\delta t$ the simulation time step. An exact knowledge of $\beta_T$ is not necessary, since this factor can be included in the time constant $t_P$. In simulations of water, values of $t_P = 0.01$ ps and $t_P = 0.1$ ps were found suitable [Berendsen *et al.* 1984]. The method does not drastically alter the dynamic trajectories and is easy to program, but the appropriate ensemble has not been identified. Hoover [1985] has given a set of equations that probe the constant-$NPT$ ensemble in the spirit of Andersen and Nose. His equations are

$$\begin{aligned}
\dot{\mathbf{s}} &= \mathbf{p}/mV^{1/3} \\
\dot{\mathbf{p}} &= \mathbf{f} - (\chi + \xi)\mathbf{p} \\
\dot{\xi} &= \left( \sum_i |\mathbf{p}_i|^2/m - fk_\mathrm{B}T \right)/Q \\
\chi &= \dot{V}/3V \\
\dot{\chi} &= (\mathscr{P} - P)\,V/t_P^2 k_\mathrm{B}T
\end{aligned} \tag{7.80}$$

where $t_P$ is a relaxation time for the pressure fluctuations.

### 7.5.4 *Changing box-shape*

The constant-pressure method of Andersen [1980] allows for isotropic changes in the volume of the simulation box. Parrinello and Rahman [1980, 1981, 1982] have extended this method to allow the simulation box to change shape as well as size. This is not of great use in liquid-state simulations, since the box may become quite elongated in the absence of elastic restoring forces, but we describe it briefly here for completeness. The technique is particularly helpful in the study of solids, since it allows for phase changes in the simulation which may involve changes in the unit cell dimensions and angles. The scaled coordinates are now introduced through the equation

$$\mathbf{r} = \mathbf{H}\mathbf{s} \tag{7.81}$$

where $\mathbf{H} = (\mathbf{h}_1, \mathbf{h}_2, \mathbf{h}_3)$ is a transformation matrix whose columns are the three vectors $\mathbf{h}_\alpha$ representing the sides of the box. $V$, the volume of the box, is given by

$$V = |\mathbf{H}| = \mathbf{h}_1 \cdot \mathbf{h}_2 \times \mathbf{h}_3 . \tag{7.82}$$

The changing box-shape is represented in Fig. 7.3.

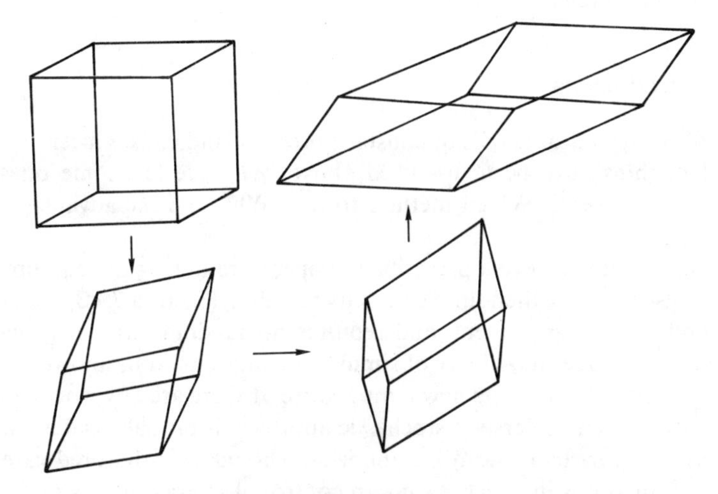

**Fig. 7.3** Changing box-shape.

The 'potential energy' associated with the box is once more

$$\mathscr{V}_V = PV \tag{7.83}$$

and the extra 'kinetic energy' term is

$$\mathscr{K}_V = \tfrac{1}{2} Q \sum_\alpha \sum_\beta \dot{H}_{\alpha\beta}^2 \tag{7.84}$$

where $Q$ is the box 'mass'. The equations of motion are obtained in the usual way from the Lagrangian $\mathscr{L} = \mathscr{K} + \mathscr{K}_V - \mathscr{V} - \mathscr{V}_V$ and they are

$$m\ddot{\mathbf{s}} = \mathbf{H}^{-1}\mathbf{f} - m\mathbf{G}^{-1}\dot{\mathbf{G}}\dot{\mathbf{s}} \tag{7.85a}$$

$$Q\ddot{\mathbf{H}} = (\mathscr{P} - \mathbf{1}P)V(\mathbf{H}^{-1})^{\mathrm{T}} \tag{7.85b}$$

where $\mathbf{G} = \mathbf{H}^{\mathrm{T}}\mathbf{H}$ is a metric tensor and T stands for transpose. Here the pressure tensor $\mathscr{P}$ (see eqn (2.114)) plays the same role as did $\mathscr{P}$ in the Andersen method. In scaled variables it is

$$\mathscr{P}_{\alpha\beta} = \frac{1}{V}\left( \sum_i m(\mathbf{H}\dot{\mathbf{s}}_i)^{\mathrm{T}}_\alpha (\mathbf{H}\dot{\mathbf{s}}_i)_\beta + \sum_i \sum_{j>i} (\mathbf{H}\mathbf{s}_{ij})_\alpha (\mathbf{f}_{ij})_\beta \right) \tag{7.86}$$

where $\mathbf{f}_{ij}$ is the force on $i$ due to $j$ in unscaled form. A simple expression for $\mathbf{H}^{-1}$ can be obtained from the reciprocal lattice vectors of the MD box. Once more the extended system Hamiltonian is conserved and the equations can be solved using a predictor–corrector technique. The method has been extended to molecular systems by Nosé and Klein [1983], who discuss many interesting technical aspects including the question of the overall rotation of the box. The method has also been applied in connection with constraint dynamics [Ferrario and Ryckaert 1985].

## 7.6 Practical points

Many of the technical details of constant-pressure and constant-temperature MD algorithms have been discussed above. We have left some questions unanswered however. Which method to use? What are the advantages and drawbacks?

If the aim is to achieve a prescribed temperature during the equilibration phase of a simulation, then simple velocity rescaling (Section 7.4.3) is certainly quick and easy to implement, and requires no random number generator. However, it is crude, and plenty of simulation time should be allowed for the system to 'settle down' to its new temperature. If there are any worries about equilibration, then Andersen's stochastic approach is probably safer, and the progress of the system towards equilibrium should be monitored carefully. There are situations in which a smooth control of temperature is preferable: for example, in Chapter 8 we shall meet non-equilibrium methods which would otherwise cause the system to heat up. Here, the more sophisticated constraint method [Hoover *et al.* 1982; Evans 1983a] is the easiest to employ, and has no adjustable parameters. For the generation of true canonical ensemble averages, the method of Nosé is an interesting possibility, but Hoover's version [1985] and the stochastic approach [Andersen 1980] are easier to use: both require a sensible selection of parameters.

To equilibrate a system at a new pressure, the method of Berendsen *et al.* [1984] has much to recommend it on the grounds of simplicity. The method of Hoover [1985] looks like the best candidate for conducting equilibrium constant-pressure simulations. It combines naturally with the constant temperature algorithm given in the same paper for simulating the constant-$NPT$ ensemble. The extended system methods of Section 7.5.1 are slightly more complicated to program.

It should be noted that the constraint methods, eqns (7.53) and (7.69), are only designed to conserve $\mathscr{T}$ and $\mathscr{P}$ respectively, not to fix these quantities at a prescribed value. The desired values $T$ and $P$ do not, in fact, appear in the algorithms. This means that, due to accumulation of algorithm error, $\mathscr{T}$ and $\mathscr{P}$

will tend to drift from their initial values, and it is necessary to make small corrections, say every time step. It also means that special measures must be taken whenever it is required to change temperature or pressure. This is an easy matter in the former case, since the kinetic temperature can be changed by simple velocity scaling, but the adjustment of pressure to a prescribed value is a little awkward [Evans and Morriss 1983b]. For this reason, other methods may be preferable.

## 7.7 The Gibbs Monte Carlo method

Recently, Panagiotopoulos [1987 *a*] has devised a new simulation technique, the Gibbs Monte Carlo method, for the direct simulation of fluid phase equilibria. The method uses two basic simulation boxes which are located within the two coexisting phases. The boxes are surrounded by the normal periodic images and there is no attempt to simulate the interface between the phases. The Monte Carlo technique uses three types of move. There are independent particle displacements in each box which are made using the normal Metropolis algorithm. There is a combined attempted volume-move in which the volume of one box changes by $\Delta V$ while the volume of the other box changes by $-\Delta V$. The pressure in the two boxes is equal, but its precise value is not required in the algorithm. Finally there is a combined attempted creation/destruction-move in which a randomly chosen particle is extracted from one box and placed at random in the other box. The chemical potential in the two boxes is equal, but its precise value is not required. The simulation is started from a temperature and an overall density, for both boxes, which is in the mechanically unstable two-phase region. The volume and number of particles in each box changes rapidly from the arbitrary starting configuration to values characteristic of the two coexisting phases. Details of the technique can be found in the original paper. The method has been extended to mixture and membrane equilibria [Panagiotopoulos, Quirke, Stapleton, and Tildesley, 1988] and to inhomogeneous systems [Panagiotopoulos 1987*b*]. The method has significant advantages in speed over conventional free energy calculations of phase boundaries, but in its present form it will not simulate equilibria involving solid phases.

# 8
# NON-EQUILIBRIUM MOLECULAR DYNAMICS

## 8.1 Introduction

So far in this book, we have considered the computer simulation of systems at equilibrium. Even the introduction, into the molecular dynamics equations, of terms representing the coupling to external systems (constant-temperature reservoirs, pistons etc.) have preserved equilibrium: the changes do not induce any thermodynamic fluxes. In this chapter, we examine adaptations of MD that sample non-equilibrium ensembles. One motivation for this is to improve the efficiency with which transport coefficients are calculated, the route via linear response theory and time correlation functions (eqns (2.109)–(2.122)) being subject to significant statistical error. This has been discussed in part in Chapter 6. Another is to examine directly the response of a system to a large perturbation lying outside the regime of linear response theory.

One problem with time correlation functions is that they represent the average response to the naturally occurring (and hence fairly small) fluctuations in the system properties. The signal-to-noise ratio is particularly unfavourable at long times, where there may be a significant contribution to the integral defining a transport coefficient. Moreover, the finite system size imposes a limit on the maximum time for which reliable correlations can be calculated. The idea behind non-equilibrium methods is that a much larger fluctuation may be induced artificially, and the signal-to-noise level of the measured response improved dramatically. By measuring the steady state response to such a perturbation, problems with long-time behaviour of correlation functions are avoided. NEMD measurements are made in much the same way as that used to estimate simple equilibrium averages such as the pressure and temperature. These methods correspond much more closely to the procedure adopted in experiments: shear and bulk viscosities, and thermal conductivities, are measured by creating a flow (of momentum, energy etc.) in the material under study.

Early attempts to induce momentum or energy flow in a molecular dynamics simulation have been reviewed by Hoover and Ashurst [1975]. One possibility is to introduce boundaries, or boundary regions, where particles are made to interact with external momentum or energy reservoirs [Ashurst and Hoover 1972, 1973, 1975; Hoover and Ashurst 1975; Tenenbaum, Ciccotti, and Gallico 1982]. These methods have the disadvantage of being incompatible with periodic boundary conditions, and so they introduce surface effects into the simulation. The approaches we shall describe avoid this, by being designed for consistency with the usual periodic boundaries, or by modifying these boundaries in a homogeneous way, preserving translational invariance and periodicity. Non-equilibrium molecular dynamics simulations have grown in popularity over the last few years, and several excellent reviews [Ciccotti, Jacucci, and McDonald 1979; Hoover 1983a, 1983b; Evans and Morriss 1984a] may be consulted for further details.

The methods we are about to describe all involve perturbing the usual equations of motion in some way. Such a perturbation may be switched on at time $t = 0$, remaining constant thereafter, in which case the measured responses will be proportional to time-integrated correlation functions. The long-time steady-state responses (the infinite time integrals) may then yield transport coefficients. Alternatively, the perturbation may be applied as a delta function pulse at time $t = 0$ with subsequent time evolution occurring normally. In this case, the responses are typically proportional to the correlation functions themselves: they must be measured at each time step following the perturbation, and integrated numerically to give transport coefficients. Finally, a sinusoidally oscillating perturbation may be applied. After an initial transient period, the measured responses will be proportional to the real and imaginary parts of the Fourier–Laplace transformed correlation functions, at the applied frequency. To obtain transport coefficients, several experiments at different frequencies must be carried out, and the results extrapolated to zero frequency. The advantages and disadvantages of these different techniques will be discussed following a description of some of the perturbations applied.

The perturbations appear in the equations of motion as follows [Evans and Morriss 1984a]

$$\dot{\mathbf{q}} = \mathbf{p}/m + \mathscr{A}_{\mathbf{p}} \cdot \mathscr{F}(t) \tag{8.1a}$$

$$\dot{\mathbf{p}} = \mathbf{f} - \mathscr{A}_{\mathbf{q}} \cdot \mathscr{F}(\mathbf{t}) . \tag{8.1b}$$

The condensed notation disguises the complexity of these equations in general. Here, $\mathscr{F}(t)$ is a $3N$-component vector representing a time-dependent applied field. It can be thought of as applying to each molecule, in each coordinate direction, separately. The quantities $\mathscr{A}_{\mathbf{p}}(\mathbf{q}, \mathbf{p})$ and $\mathscr{A}_{\mathbf{q}}(\mathbf{q}, \mathbf{p})$ are functions of particle positions and momenta. They describe the way in which the field couples to the molecules, perhaps through a term in the system Hamiltonian. Each can be a $3N \times 3N$ matrix in the general case, but usually many of the components vanish. The perturbation can often be thought of as coupling separately to some property of each molecule (for example, its momentum), in which case $\mathscr{A}_{\mathbf{p}}$ and $\mathscr{A}_{\mathbf{q}}$ become very simple indeed. However, some properties (e.g. the energy density, the pressure tensor), while being formally broken down into molecule-by-molecule contributions, actually depend on intermolecular interactions, and so $\mathscr{A}_{\mathbf{p}}$ and $\mathscr{A}_{\mathbf{q}}$ must be functions of all particle positions and momenta in the general case.

In standard linear response theory, the perturbation is represented as an additonal term in the system Hamiltonian

$$\mathscr{H}^{\text{ne}} = \mathscr{H} + \mathscr{A}(\mathbf{q}, \mathbf{p}) \cdot \mathscr{F}(t)$$
$$= \mathscr{H} + \sum_i \mathscr{A}_i(\mathbf{q}, \mathbf{p}) \cdot \mathscr{F}_i(t) \tag{8.2}$$

in which case we simply have

$$\mathscr{A}_{\mathbf{p}} = \nabla_{\mathbf{p}} \mathscr{A} \tag{8.3a}$$
$$\mathscr{A}_{\mathbf{q}} = \nabla_{\mathbf{q}} \mathscr{A} . \tag{8.3b}$$

The average value $\langle \mathscr{B} \rangle_{\text{ne}}$ of any phase function $\mathscr{B}(\mathbf{q}, \mathbf{p})$ in the non-equilibrium ensemble generated by the perturbation is given by

$$\langle \mathscr{B}(t) \rangle_{\text{ne}} = -\frac{1}{k_B T} \int_0^t \mathrm{d}t' \, \langle \mathscr{B}(t-t') \, \dot{\mathscr{A}}(0) \rangle \cdot \mathscr{F}(t') \tag{8.4}$$

assuming that the equilibrium ensemble average $\langle \mathscr{B} \rangle$ vanishes and that the perturbation is switched on at time $t = 0$. However, it has long been recognized that the perturbation need not be derived from a Hamiltonian [Jackson and Mazur 1964]. Provided that

$$\nabla_{\mathbf{q}} \cdot \dot{\mathbf{q}} + \nabla_{\mathbf{p}} \cdot \dot{\mathbf{p}} = (\nabla_{\mathbf{q}} \cdot \mathscr{A}_{\mathbf{p}} - \nabla_{\mathbf{p}} \cdot \mathscr{A}_{\mathbf{q}}) \cdot \mathscr{F}(t) = 0 \tag{8.5}$$

the incompressibility of phase space still holds, and eqn (8.4) may still be derived. In this case, however, $\dot{\mathscr{A}}$ cannot be regarded as the time derivative of a variable $\mathscr{A}$. Rather, it is simply a function of $\mathbf{q}$ and $\mathbf{p}$, defined by the rate of change of internal energy

$$\dot{\mathscr{H}} = -((\mathbf{p}/m) \cdot \mathscr{A}_{\mathbf{q}} + \mathbf{f} \cdot \mathscr{A}_{\mathbf{p}}) \cdot \mathscr{F}(t) = -\dot{\mathscr{A}} \cdot \mathscr{F}(t) . \tag{8.6}$$

Thus $\mathscr{A}_{\mathbf{q}}$ and $\mathscr{A}_{\mathbf{p}}$ are sufficient to define $\dot{\mathscr{A}}$ in eqn (8.4). These equations have been developed and extended by Evans and co-workers [Evans and Morriss 1984a].

When a perturbation is applied in molecular dynamics, typically the system heats up. This heating may be controlled by techniques analogous to those employed in constant-temperature MD, as discussed in Chapter 7. In the following sections, we shall omit the extra terms in the equations of motion which serve this purpose. This choice corresponds to a perturbation which is applied adiabatically, i.e. the work done on the system exactly matches the increase in the internal energy. We shall return to this in Section 8.7. Moreover, in Sections 8.2–8.5, the perturbations are assumed to apply to an atomic system, or to the centres of mass of a system of molecules. Accordingly, we shall revert to the notation $\mathbf{r}, \mathbf{p}$ rather than $\mathbf{q}, \mathbf{p}$.

## 8.2 Shear flow

Some of the earliest non-equilibrium simulations attempted to measure the shear viscosity of an atomic Lennard-Jones fluid. One technique, which maintains conventional cubic periodic boundary conditions, is to use a spatially periodic perturbation to generate an oscillatory velocity profile [Gosling, McDonald, and Singer 1973; Ciccotti, Jacucci, and McDonald 1975, 1979]. At each time step in an otherwise conventional MD simulation, an external force in the $x$-direction is applied to each molecule. The magnitude of the force depends upon the molecule's $y$-coordinate as follows:

$$f_{ix}^{\text{ext}} = \mathscr{F} \cos(2\pi n r_{iy}/L) = \mathscr{F} \cos k r_{iy} \tag{8.7}$$

where $\mathscr{F}$ is a constant and the wavevector $\mathbf{k} = (0, k, 0) = (0, 2\pi n/L, 0)$, with $n$ an integer, is commensurate with the side $L$ of the simulation box. This force field is illustrated in Fig. 8.1 for the lowest wavelength case, $n = 1$. On applying this perturbation and waiting, a spatially periodic velocity profile develops. Specifically, at a given $y$-coordinate $r_y$, the mean $x$-velocity of a molecule should be

$$\langle v_x(r_y) \rangle_{\text{ne}} \approx (\rho/k^2\eta)\, \mathscr{F} \cos k r_y \,. \tag{8.8}$$

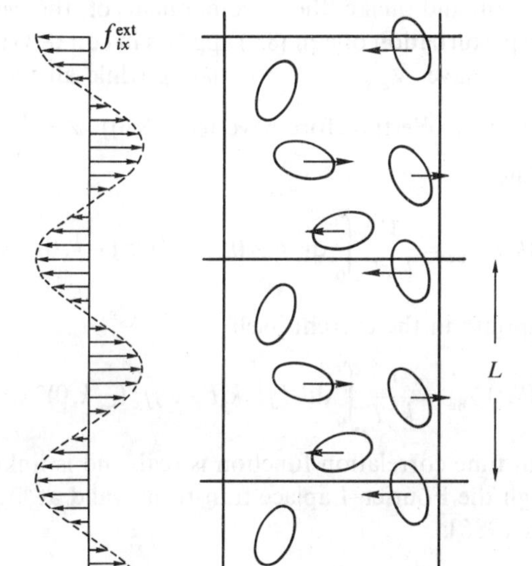


**Fig. 8.1** Spatially oscillating shear flow perturbation.

By fitting their results to this equation, Gosling *et al.* [1973] were able to estimate the shear viscosity $\eta$ with significantly less computational effort than that required using equilibrium methods.

It is worth examining carefully the origins of eqn (8.8). The perturbation of eqn (8.7) is a non-Hamiltonian one, but falls into the general scheme of the last section. Writing in a slightly more general form

$$f_{ix}^{\text{ext}}(t) = \mathscr{F}(t)\exp(-ikr_{iy}) \tag{8.9}$$

we can show that the response in any $k$-dependent quantity $\mathscr{B}(k)$ is related to a correlation function involving the transverse current $j_x^{\perp}(k,t)$ or the transverse momentum $p_x^{\perp}(k,t)$:

$$j_x^{\perp}(k,t) = \frac{1}{V}\sum_{i=1}^{N} v_{ix}(t)\exp(ikr_{iy}(t)) \tag{8.10a}$$

$$p_x^{\perp}(k,t) = \frac{1}{V}\sum_{i=1}^{N} p_{ix}(t)\exp(ikr_{iy}(t)). \tag{8.10b}$$

These equations are analogous to eqn (2.130), except that we take the $k$-vector in the $y$ direction and make the $x$-component of the velocity explicit. Specifically, the perturbation of eqn (8.9) appears in eqn (8.1) in the following way. For all $i$, we have $\mathscr{A}_{qix} = -\exp(-ikr_{iy})$, while all the remaining $\mathscr{A}_q$ and $\mathscr{A}_p$ terms vanish. We therefore have (eqn (8.6)) $\dot{\mathscr{A}} = \sum_i (p_{ix}/m)\mathscr{A}_{qix} = -Vj_x^{\perp}(-k)$. Thus

$$\langle \mathscr{B}(k,t)\rangle_{\text{ne}} = \frac{V}{k_{\text{B}}T}\int_0^t \text{d}t' \, \langle \mathscr{B}(k,t-t')j_x^{\perp}(-k,0)\rangle \mathscr{F}(t') \tag{8.11}$$

and for the response in the current itself

$$\langle j_x^{\perp}(k,t)\rangle_{\text{ne}} = \frac{V}{k_{\text{B}}T}\int_0^t \text{d}t' \, \langle j_x^{\perp}(k,t-t')j_x^{\perp}(-k,0)\rangle \mathscr{F}(t'). \tag{8.12}$$

The equilibrium time correlation function is real, and is linked to the shear viscosity through the Fourier–Laplace transform, valid at low $k, \omega$ [Hansen and McDonald 1986]:

$$\frac{V}{k_{\text{B}}T}\langle j_x^{\perp}(k,\omega)j_x^{\perp}(-k)\rangle \approx \frac{\rho/m}{i\omega + k^2\eta(k,\omega)/\rho m}. \tag{8.13}$$

This equation may be used to define a $k,\omega$-dependent shear viscosity $\eta(k,\omega)$, which goes over to the transport coefficient $\eta$ on taking the limit $\omega \to 0$ followed by the limit $k \to 0$. Taking the zero-frequency limit here means time-integrating from $t = 0$ to $t = \infty$, so

$$\frac{V}{k_{\text{B}}T}\int_0^{\infty} \text{d}t \, \langle j_x^{\perp}(k,t)j_x^{\perp}(-k)\rangle = \frac{\rho^2}{k^2\eta(k)}. \tag{8.14}$$

If the perturbation remains constant, $\mathscr{F}(t) = \mathscr{F}$ from $t = 0$ onwards, this is essentially the quantity appearing on the right of eqn (8.12) as the integration limit goes to infinity and the steady state is obtained. Thus

$$\langle j_x^{\perp}(k, t \to \infty) \rangle_{\text{ne}} = \frac{\mathscr{F}\rho^2}{k^2 \eta(k)} . \tag{8.15}$$

Apart from a factor of $\rho$ linking the current and velocity profile, and the explicit appearance of the $\cos kr_y$ term reflecting the fact that a real perturbation (the real part of eqn (8.9)) yields a real, in-phase response, this is eqn (8.8). Note how the method relies on a $k$-dependent viscosity going smoothly to $\eta$ as $k \to 0$. This means that in a real application, several different $k$-vectors should be chosen, all orthogonal to the $x$-direction, and an extrapolation to zero $k$ should be undertaken [Ciccotti *et al.* 1975].

The pitfalls inherent in using finite-$k$ perturbations are well illustrated by another attempt to measure $\eta$ via coupling to the transverse momentum density [Ciccotti *et al.* 1979]. In this case, the perturbation is of Hamiltonian form, eqn (8.2), with

$$\mathscr{A} \cdot \mathscr{F}(t) = V p_x^{\perp}(-k, t) \mathscr{F}(t) \tag{8.16}$$

and $\mathbf{k} = (0, k, 0)$ as before. The responses in this case are given by

$$\langle \mathscr{B}(k,t) \rangle_{\text{ne}} = \frac{V}{k_{\text{B}}T} \int_0^t \mathrm{d}t' \langle \mathscr{B}(k, t-t') \mathscr{P}_{yx}(-k) \rangle ik \mathscr{F}(t') \tag{8.17}$$

where

$$\mathscr{P}_{yx}(k,t) = \frac{1}{V} \sum_i m v_{ix} v_{iy} \exp(ikr_{iy}) + \frac{1}{V} \sum_i \sum_{j>i} r_{ijy} f_{ijx} \left( \frac{\exp(ikr_{iy}) - \exp(ikr_{jy})}{ikr_{ijy}} \right) \tag{8.18}$$

is defined so that $\dot{p}_x^{\perp}(k,t) = ik \mathscr{P}_{yx}(k,t)$ (compare eqn (2.114)). Specifically

$$\langle \mathscr{P}_{yx}(k,t) \rangle_{\text{ne}} = \frac{V}{k_{\text{B}}T} \int_0^t \mathrm{d}t' \langle \mathscr{P}_{yx}(k, t-t') \mathscr{P}_{yx}(-k) \rangle ik \mathscr{F}(t') . \tag{8.19}$$

Now the infinite time integral of $\langle \mathscr{P}_{yx}(k,t) \mathscr{P}_{yx}(-k) \rangle$ can be calculated by applying a steady perturbation and measuring the long-time response of the out-of-phase component of $\mathscr{P}_{yx}(k)$ (i.e. the $\sin kr_y$ component if a real $\cos kr_y$ field is applied, because of the $ik$ factor in eqn (8.19)). Unfortunately, this quantity vanishes identically for finite $k$. This is because in the equation

$$\eta = \lim_{\omega \to 0} \lim_{k \to 0} \frac{V}{k_{\text{B}}T} \int_0^{\infty} \mathrm{d}t \, \mathrm{e}^{-i\omega t} \langle \mathscr{P}_{yx}(k,t) \mathscr{P}_{yx}(-k) \rangle \tag{8.20}$$

the $k \to 0$ limit must be taken first (yielding eqn (2.111)). It is not possible to define a quantity $\eta(k, \omega)$ which has sensible behaviour at low $\omega$ simply by omitting the limiting operations in eqn (8.20). Of course the same difficulty applies in any attempt to measure $\eta$ via an equilibrium MD calculation of eqn (8.20). A way around this problem has been given by Evans [1981a], but in either case (equilibrium or non-equilibrium calculations) the wavevector and frequency-dependent correlation function in eqn (8.20) is required before the extrapolation to give $\eta$ can be undertaken.

If a zero-wavevector transport coefficient is required, then a zero-wavevector technique is preferred, and this involves a modification of the periodic boundary conditions. Such a modification was proposed by Lees and Edwards [1972] and is illustrated in Fig. 8.2. In essence, the infinite periodic system is subjected to a uniform shear in the $xy$ plane. The simulation box and its images centred at $(x, y) = (\pm L, 0)$, $(\pm 2L, 0)$, etc. (for example, A and E in Fig. 8.2.) are taken to be stationary. Boxes in the layer above, $(x, y) = (0, L)$, $(\pm L, L)$, $(\pm 2L, L)$, etc. (e.g. B,C,D) are moving at a speed $(\mathrm{d}v_x/\mathrm{d}r_y)\,L$ in the positive $x$ direction ($\mathrm{d}v_x/\mathrm{d}r_y$ is the shear rate). Boxes in the layer below, $(x, y) = (0, -L)$, $(\pm L, -L)$, $(\pm 2L, -L)$, etc. (e.g. F,G,H) move at a speed $(\mathrm{d}v_x/\mathrm{d}r_y)\,L$ in the negative $x$ direction. In the more remote layers, boxes are moving proportionally faster relative to the central one. Box B, for example, starts off adjacent to $A$ but, if $(\mathrm{d}v_x/\mathrm{d}r_y)$ is a constant, it will move away in the positive $x$-direction throughout the simulation, possibly ending up hundreds of box lengths away. This is a transformation of the same kind as seen in the Parrinello–Rahman method described in Section 7.5.4, but corresponding to a simple shear in one direction only (see Fig. 8.2). It is most convenient to

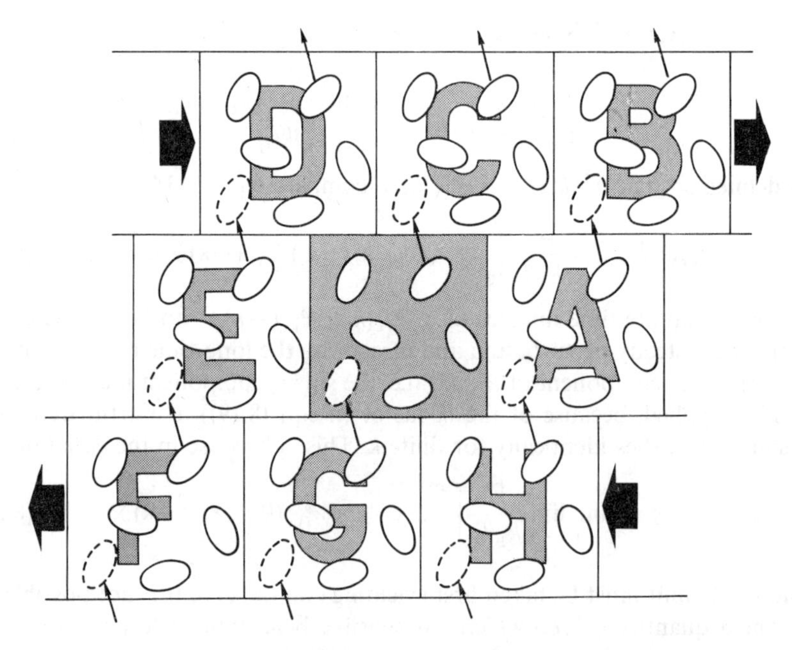


**Fig. 8.2** Homogeneous shear boundary conditions.

represent this using shifted cubic boxes, rather than by deforming the box, since a steady shear can then be maintained without the box angles becoming extremely acute (compare Fig. 7.3). Of course it is also possible to make $(dv_x/dr_y)$ vary sinusoidally in time, in which case the box B will oscillate about its initial position.

The periodic minimum image convention must be modified in this case. Suppose the upper layer (BCD in Fig. 8.2) is displaced relative to the central box by an amount $\delta r_x$.

```
CORY = ANINT ( RYIJ / YBOX )
RXIJ = RXIJ - CORY * DELRX
RXIJ = RXIJ - ANINT ( RXIJ / XBOX ) * XBOX
RYIJ = RYIJ - CORY * YBOX
RZIJ = RZIJ - ANINT ( RZIJ / ZBOX ) * ZBOX
```

where DELRX stores the displacement $\delta r_x$. Thus, an extra correction is applied to the $x$ component, which depends upon the number of boxes separating the two molecules in the $y$ direction. Note that adding or subtracting whole box lengths to or from $\delta r_x$ makes no difference to this correction and it is convenient to take $-L/2 \leqslant \delta r_x \leqslant L/2$. It is advisable to keep replacing molecules in the central box as they cross the boundaries, especially if a steady-state shear is imposed, to prevent the build-up of substantial differences in the $x$ coordinates. When this is done, the $x$ velocity of a molecule must be changed as it crosses the box boundary in the $y$-direction, for consistency with the applied velocity gradient. Let the variable DELVX store the value of the difference in box velocities between adjacent layers, i.e. $(dv_x/dr_y)\,L$. Then periodic boundary crossing is handled as follows.

```
CORY  = ANINT ( RY(I) / YBOX )
RX(I) = RX(I) - CORY * DELRX
RX(I) = RX(I) - ANINT ( RX(I) / XBOX ) * XBOX
RY(I) = RY(I) - CORY * YBOX
RZ(I) = RZ(I) - ANINT ( RZ(I) / ZBOX ) * ZBOX
VX(I) = VX(I) - CORY * DELVX
```

For large simulations, it may be desirable to speed up the calculation by using a neighbour list. If so, it makes more sense to use a link-list cell method rather than the simple Verlet approach, because of the steadily changing box geometry [Evans and Morriss 1984a]. There is a subtlety here, since the shifting layers of boxes may necessitate searching more cells than is the case in a conventional simulation. The way this is done is shown in program F.32.

The Lees–Edwards boundary conditions alone can be used to set up and maintain a steady linear velocity profile, with gradient $dv_x/dr_y$. The shear viscosity is then estimated from the steady-state non-equilibrium average of $\mathscr{P}_{yx}(k=0)$

$$\langle \mathscr{P}_{yx}(t \to \infty) \rangle_{\text{ne}} = -\eta (dv_x/dr_y)\,. \tag{8.21}$$

This technique was used by Naitoh and Ono [1976, 1979], Evans [1979a,b], and by many others subsequently. It is a satisfactory way to mimic steady Couette flow occurring in real systems [Schlichting 1979]. However, the modified boundaries alone are not sufficient to drive the most general time-dependent perturbations.

Now we wish to apply a shear perturbation to each molecule, instead of just relying on the modified boundaries. Suitable equations of motion which generate a time-dependent shear were proposed by Hoover, Evans, and co-workers [Hoover *et al.* 1980b]. The extra perturbation term is of hamiltonian form, eqn (8.2), with

$$\mathcal{A}\cdot\mathcal{F}(t) = \left(\sum_i r_{iy}p_{ix}\right)\mathcal{F}(t)\,. \tag{8.22}$$

$\mathcal{F}$ is the instantaneous rate of strain, i.e. $\mathcal{F} = (\mathrm{d}v_x/\mathrm{d}r_y)$. This gives equations of motion of the form (eqns (8.1), (8.3))

$$\begin{aligned}\dot{r}_{ix} &= p_{ix}/m + r_{iy}\mathcal{F}(t)\\ \dot{r}_{iy} &= p_{iy}/m\\ \dot{r}_{iz} &= p_{iz}/m\end{aligned} \tag{8.23a}$$

$$\begin{aligned}\dot{p}_{ix} &= f_{ix}\\ \dot{p}_{iy} &= f_{iy} - p_{ix}\mathcal{F}(t)\\ \dot{p}_{iz} &= f_{iz}\,.\end{aligned} \tag{8.23b}$$

These equations are implemented in conjunction with the periodic boundary conditions of Lees and Edwards (consider replacing $r_{iy}$ with $r_{iy}\pm L$ in eqn (8.23a)). We will show that they are a consistent low-$k$ limit of eqns (8.16), (8.17). If the perturbation in eqn (8.16) is divided by $-ik$ to give instead

$$\mathcal{A}\cdot\mathcal{F}(t) = Vp_x^{\perp}(-k,t)\mathcal{F}(t)/(-ik) = \sum_i p_{ix}\exp(-ikr_{iy})\mathcal{F}(t)/(-ik) \tag{8.24}$$

the exponential may be expanded and the first term dropped because of momentum conservation. Taking the limit $k\to 0$ then gives eqn (8.22). $\mathcal{F}(t)$ is the instantaneous rate of strain. The analogue of eqn (8.17) is then

$$\langle\mathcal{B}(t)\rangle_{\mathrm{ne}} = \frac{-V}{k_{\mathrm{B}}T}\int_0^t \mathrm{d}t'\,\langle\mathcal{B}(t-t')\mathcal{P}_{yx}\rangle\mathcal{F}(t') \tag{8.25}$$

and eqn (8.19) becomes

$$\langle\mathcal{P}_{yx}(t)\rangle_{\mathrm{ne}} = \frac{-V}{k_{\mathrm{B}}T}\int_0^t \mathrm{d}t'\,\langle\mathcal{P}_{yx}(t-t')\mathcal{P}_{yx}\rangle\mathcal{F}(t') \tag{8.26}$$

where zero-$k$ values are implied throughout.

In fact, a slight modification of eqns (8.23) is now preferred [Evans and Morriss 1984a, 1984b; Ladd 1984]:

$$\dot{r}_{ix} = p_{ix}/m + r_{iy}\mathscr{F}(t)$$
$$\dot{r}_{iy} = p_{iy}/m$$
$$\dot{r}_{iz} = p_{iz}/m \tag{8.27a}$$
$$\dot{p}_{ix} = f_{ix} - p_{iy}\mathscr{F}(t)$$
$$\dot{p}_{iy} = f_{iy}$$
$$\dot{p}_{iz} = f_{iz}\,. \tag{8.27b}$$

These equations are non-Hamiltonian, but generate the same linear responses as eqns (8.23)–(8.26). They are preferred because they are believed to give correct non-linear properties [Evans and Morriss 1984b] and the correct distribution in the dilute gas [Ladd 1984] where eqn (8.23) fails. They also generate trajectories identical to the straightforward Lees–Edwards boundary conditions when $\mathscr{F}(t)$ is a constant. This follows from an elimination of momenta

$$\ddot{r}_{ix} = f_{ix}/m + r_{iy}\dot{\mathscr{F}}(t)$$
$$\ddot{r}_{iy} = f_{iy}/m$$
$$\ddot{r}_{iz} = f_{iz}/m\,. \tag{8.28}$$

If a step function perturbation is applied, i.e. $\mathscr{F}(t) = \text{constant}$ for $t > 0$, eqns (8.28) are integrated over an infinitesimal time interval at $t = 0$ (this sets up the correct initial velocity gradient) and evolution thereafter occurs with normal Newtonian mechanics ($\dot{\mathscr{F}} = 0$) plus the modified boundaries. For the more general case (e.g. $\mathscr{F}(t)$ oscillating in time) step-by-step integration of eqns (8.27) or (8.28) is needed.

Couette flow, as simulated through the preceding equations, contains some rotational character: the equations are not symmetrical in the $x$ and $y$ coordinates. A symmetrized set of equations may be devised by combining the perturbations for $\mathrm{d}v_x/\mathrm{d}r_y$ and $\mathrm{d}v_y/\mathrm{d}r_x$ and applying the shear in both directions simultaneously. This generates irrotational flow. Alternatively, the shear viscosity may be investigated by combining a compression in the $x$-direction with an expansion in the $y$-direction (thus maintaining rectangular symmetry), and measuring $\mathscr{P}_{xx} - \mathscr{P}_{yy}$ [K. Singer, unpublished, acting on a suggestion of W. Hoover; Heyes 1983c]. Unfortunately, in neither case is it possible to apply these perturbations continuously and indefinitely in time, because the simulation box becomes extremely elongated.

## 8.3 Expansion and contraction

Ciccotti *et al.* [1979] have outlined schemes to compute, by non-equilibrium methods analogous to those of the previous section, the longitudinal current correlation function and the autocorrelation of diagonal elements of the pressure tensor, at finite $k$. Because of the hydrodynamic coupling between the particle density, the longitudinal current, and the energy density, the expressions linking these with the bulk viscosity also involve the thermal conductivity and the compressibility [Hansen and McDonald 1986]. Consequently, to estimate the bulk viscosity, we must turn to the zero-$k$ expression, eqn (2.116), and use a homogeneous NEMD technique for simplicity. Such a technique, very closely related to those used for shear viscosity, has been developed by Hoover and co-workers [Hoover, Ladd, Hickman, and Holian 1980a; Hoover *et al.* 1980b]. The modified equations of motion are

$$\dot{\mathbf{r}}_i = \mathbf{p}_i/m + \mathbf{r}_i \mathscr{F}(t) \tag{8.29a}$$

$$\dot{\mathbf{p}}_i = \mathbf{f}_i - \mathbf{p}_i \mathscr{F}(t) \tag{8.29b}$$

and they correspond to a homogeneous dilation or contraction of the system. They are combined with uniformly expanding or contracting periodic boundary conditions, of the kind described for Andersen's method [Andersen 1980] of constant-pressure dynamics (Section 7.5.1). In fact, there is a close connection between the two methods, the difference lying in the quantities held fixed and those allowed to vary.

## 8.4 Heat flow

The development of a non-equilibrium method of determining the thermal conductivity has been a non-trivial exercise. At finite $k$, it is straightforward to introduce a Hamiltonian perturbation along the lines previously described, which couples to the energy density and which can yield the energy current autocorrelation function [Ciccotti, Jacucci, and McDonald 1978, 1979]. The link with the transport coefficient, however, suffers from the same drawback as in most of the previous cases: the limit $k \to 0$ must be taken before $\omega \to 0$ (i.e. before a steady-state time integration) or the function vanishes. Essentially identical homogeneous algorithms which avoid this problem, and which are still compatible with periodic boundaries, were developed independently by Evans [Evans 1982a; Evans and Morriss 1984a] and by Gillan and Dixon [1983]. The modified equations are

$$\dot{\mathbf{r}}_i = \mathbf{p}_i/m \tag{8.30a}$$

$$\dot{\mathbf{p}}_i = \mathbf{f}_i + \delta\varepsilon_i \boldsymbol{\mathscr{F}}(t) + \tfrac{1}{2}\sum_j \mathbf{f}_{ij}(\mathbf{r}_{ij}\cdot\boldsymbol{\mathscr{F}}(t)) - \frac{1}{2N}\sum_j\sum_k \mathbf{f}_{jk}(\mathbf{r}_{jk}\cdot\boldsymbol{\mathscr{F}}(t)) . \tag{8.30b}$$

Here, $\boldsymbol{\mathscr{F}}(t)$ is a three-component vector chosen to lie (say) in the $x$-direction: $\boldsymbol{\mathscr{F}} = (\mathscr{F},0,0)$. The term $\delta\varepsilon_i = \varepsilon_i - \langle\varepsilon_i\rangle$ is the deviation of the 'single-particle energy' from its average value (see eqn (2.121)). The last term in eqn (8.30b) ensures that momentum is conserved (it redistributes non-conservative terms, with a negative sign, equally amongst all the particles). These equations are non-Hamiltonian, but satisfy the condition laid down in eqn (8.5), and so allow linear response theory to be applied in the usual way. The responses are related to correlations with the zero-$k$ energy flux $j_x^\varepsilon$:

$$\langle \mathscr{B}(t) \rangle_{\mathrm{ne}} = \frac{V}{k_{\mathrm{B}}T} \int_0^t \mathrm{d}t' \, \langle \mathscr{B}(t-t') j_x^\varepsilon \rangle \mathscr{F}(t') \tag{8.31}$$

where

$$j_x^\varepsilon = \frac{1}{V} \left( \sum_i \delta\varepsilon_i v_{ix} + \sum_i \sum_{j>i} (\mathbf{v}_i \cdot \mathbf{f}_{ij}) r_{ijx} \right). \tag{8.32}$$

In particular,

$$\langle j_x^\varepsilon(t) \rangle_{\mathrm{ne}} = \frac{V}{k_{\mathrm{B}}T} \int_0^t \mathrm{d}t' \, \langle j_x^\varepsilon(t-t') j_x^\varepsilon \rangle \mathscr{F}(t') \tag{8.33}$$

so (compare with eqn (2.119)) the thermal conductivity is given by a steady-state experiment, with $\mathscr{F}(t') = \mathscr{F}$ after $t = 0$,

$$\lambda_T T = \langle j_x^\varepsilon(t \to \infty) \rangle_{\mathrm{ne}} / \mathscr{F}. \tag{8.34}$$

The method induces an energy flux, without requiring a temperature gradient which would not be compatible with periodic boundaries. Note that in mixtures, the formulae are more complicated, and involve the heat flux rather than the energy flux (these two are identical if all the molecules have the same mass [Hansen and McDonald 1986]).

## 8.5 Diffusion

Non-equilibrium methods to measure the diffusion coefficient or mobility are most closely connected with the original linear-response theory derivations [Kubo 1957, 1966; Luttinger 1964; Zwanzig 1965]. The mobility of a single molecule in a simulation may be measured by applying an additional force to that molecule and measuring its drift velocity at steady state [Ciccotti and Jacucci 1975]. This is a useful approach when a single solute molecule is present in a solvent. The generalization of this approach to measure mutual diffusion in a binary mixture was considered by Ciccotti *et al.* [1979], and it is simplest to consider in the context of measuring the electrical conductivity in a binary electrolyte. A Hamiltonian perturbation (eqn (8.2)) is applied with

$$\mathscr{A} \cdot \mathscr{F}(t) = -\sum_i z_i r_{ix} \mathscr{F}(t) \tag{8.35}$$

so the equations of motion are conventional except for

$$\dot{p}_{ix} = f_{ix} + z_i \mathscr{F}(t) . \tag{8.36}$$

Here, $z_i = \pm 1$ (say) is the charge on each ion. Responses are then related to correlations with the charge current

$$j_x^z(t) = \frac{1}{V} \sum_i z_i v_{ix}(t) \tag{8.37}$$

and in particular

$$\langle j_x^z(t) \rangle_{\mathrm{ne}} = \frac{V}{k_B T} \int_0^t \mathrm{d}t' \, \langle j_x^z(t-t') j_x^z \rangle \, \mathscr{F}(t') . \tag{8.38}$$

Applying a steady-state field for $t > 0$, and measuring the steady-state induced current, gives the electrical conductivity. Ciccotti *et al.* [1979] made it clear, however, that it is not necessary for the particles to be charged; the quantities $z_i$ simply label different species. In the case of a neutral 1 : 1 binary mixture, the steady-state response in $j_x^z$ is simply related to the mutual diffusion coefficient $D_m$ [Jacucci and McDonald 1975]:

$$D_m = \frac{V}{\rho} \int_0^\infty \mathrm{d}t \, \langle j_x^z(t) j_x^z(0) \rangle \tag{8.39}$$

where $\rho$ is the total number density. Hence

$$D_m = \frac{k_B T}{\rho \mathscr{F}} \langle j_x^z(t \to \infty) \rangle_{\mathrm{ne}} \tag{8.40}$$

if we apply $\mathscr{F}$ at $t = 0$.

Recently, this approach has been taken to its natural conclusion, when the two components become identical [Evans, Hoover, Failor, Moran, and Ladd 1983; Evans and Morriss 1984a]. Now the $z_i$ are simply labels without physical meaning, in a one-component system: half the particles are labelled $+1$ and half labelled $-1$ at random. When the perturbation of eqn (8.35) is applied, eqn (8.40) yields the self-diffusion coefficient. Evans *et al.* [1983] compare the Hamiltonian algorithm described here with one derived from Gauss's principle of least constraint, and find the latter to be more efficient in establishing the desired steady state. For a one-component fluid, of course, there is less motivation to develop non-equilibrium methods of measuring the diffusion coefficient, since equilibrium simulations give this quantity with reasonable accuracy compared with the other transport coefficients.

Before leaving this section, we should mention that $k$-dependent perturbations which induce charge currents may also be applied, much as for the other cases considered previously [Ciccotti *et al.* 1979]. We should also mention an elegant technique due to Holian [Hoover and Ashurst 1975; Erpenbeck and Wood 1977] which measures the diffusion coefficient by 'colouring' particles in a conventional MD simulation.

## 8.6 Other perturbations

There is plenty of scope to extend NEMD methods to study quantities other than the transport coefficients of hydrodynamics and their associated correlation functions. In an atomic fluid, an example is the direct measurement of the dynamic structure factor $S(k, \omega)$ [W.A.B. Evans, unpublished]. Here, a spatially periodic perturbation is applied which couples to the number density. The method yields a response function that may be converted directly to $S(k, \omega)$ in a manner preserving the quantum mechanical detailed balance condition, eqn (2.146), whereas conventional methods of obtaining $S(k, \omega)$ (through eqn (2.128)) yield the symmetrical, classical function.

When it comes to molecular fluids, many more quantities are of interest. All the methods described previously can be applied, but there is the choice of applying the perturbations to the centres of mass of the molecules, or to other positions such as the atomic sites (if any) [Allen 1984b; Ladd 1984; Allen and Maréchal 1986]. This choice does not affect the values of hydrodynamic transport coefficients, but differences can be seen at finite wavevector and frequency. Shear-orientational coupling can be measured like this [Evans 1981b; Allen and Kivelson 1981] as can the way in which internal motions of chain molecules respond to flows [Brown and Clarke 1983].

In addition, totally new NEMD techniques may be applied to molecular fluids. Evans and Powles [1982] have investigated the dielectric properties of polar liquids by applying a suitable electric field and observing the response in the dipole moment. Normal periodic boundaries are employed in this case. Evans [1979c] has described a method of coupling to the antisymmetric modes of a molecular liquid via an imposed 'sprain rate'. No modification of periodic boundaries is necessary, merely the uniform adjustment of molecular angular velocities. The transport coefficient measured in this way is the vortex viscosity: its measurement by equilibrium simulations is fraught with danger [Evans and Streett 1978; Evans and Hanley 1982]. Evans and Gaylor [1983] have proposed a method for coupling to second-rank molecular orientation variables. This is useful for determining transport coefficients which appear in the theory of Rayleigh light scattering and flow birefringence experiments. Indeed, predicted Rayleigh and Raman spectra may be generated directly by this method.

## 8.7 Practical points

Now we turn to the practical implementation of NEMD simulations. Much discussion of the methods has appeared in a recent conference report [Hanley 1983] and in the review articles mentioned earlier.

In the previous sections, we have given the modified equations of motion which generate the desired response. These equations may be solved by standard methods as discussed in Chapter 3. It is possible to adapt one of the

Verlet methods for this purpose, but, since most of the perturbations involve the equivalent of velocity-dependent forces, it is probably more convenient to use one of the general purpose Gear algorithms [Gear 1966, 1971] (see Appendix E).

There are three common ways in which the applied perturbation may vary in time: as a delta function at time $t = 0$, as a step function, switched on at time $t = 0$, or as a sinusoidally varying function, beginning at time $t = 0$. The most straightforward way of measuring a transport coefficient is to apply a step-function perturbation, since the desired quantity is usually the $t \to \infty$ limiting steady-state response. In a few cases, notably bulk viscosity and the symmetrized shear viscosity calculations, the modified periodic boundary conditions do not allow a steady perturbation to be applied for a long time, because the cell dimensions would become extreme. In these cases, a delta-function or a frequency-dependent perturbation may be of use; the frequency-dependence of correlation functions may, in any case, be of interest.

The $\omega$-dependent calculations can be extremely lengthy, since an extrapolation to zero frequency is entailed [Hoover *et al.* 1980a, b]. Following a few cycles which are discarded to remove any transient effects of switching on the perturbation at $t = 0$, several hundred cycles of oscillation may be required at each frequency. Most of the work should be concentrated at the lowest frequencies, where the response is weakest. The total may amount to a million integration steps or more, which makes small system sizes obligatory. The extrapolation process is made more difficult by the interesting low-frequency cusp-like behaviour which seems to be exhibited by many correlation functions. Typically, responses proportional to $\omega^{1/2}$, apparently resulting from long-time $t^{-3/2}$ tails, are observed. We return to this in Chapter 11.

In straightforward applications, large perturbations, much larger than those employed in real experiments on molecular systems, are needed to achieve a reasonable signal-to-noise ratio. This leads to measurably non-linear responses, which may themselves be of interest, but which necessitate an extrapolation to zero applied field if a transport coefficient is to be measured. Once again, a square-root dependence of response upon field seems to be suggested by NEMD [Naitoh and Ono 1976; Evans 1979b] although this has been questioned [Erpenbeck 1983]. Needless to say, the functional form assumed at low applied field significantly affects the extrapolated value. If linear response theory applied throughout, it would be possible to apply several perturbations simultaneously (e.g. at different frequencies or wavelengths) without the responses interfering with each other. In practice, non-linear effects act to mix up the responses, making it inadvisable to attempt such multiple experiments in general.

When a large perturbation is applied to a system, there are three common ways of allowing the system to respond: adiabatically, isoenergetically, or isothermally.

In an adiabatic experiment, all the work done by the external forces appears as a change in the internal energy of the system, which will therefore 'heat up'. This heating was allowed to proceed in early work [Gosling *et al.* 1973], but carries with it the annoyance that the state point under study is steadily changing. More recently, adiabatic evolution has been allowed to occur during the course of cycles of an oscillatory perturbation [Hoover *et al.* 1980a, b], with the state point being reset (by velocity scaling) before the start of each new cycle. In this work, the strain amplitudes were kept sufficiently small (5–20 per cent) to avoid excessive heating during the course of a cycle.

More usually, the state point is controlled at each step of the simulation. This used to be done by velocity rescaling to maintain constant kinetic temperature [Evans 1979a, b] but now, more sophisticated modifications to the equations of motion, as described in Chapter 7, may be used to guarantee isothermal (i.e. constant-$\mathscr{T}$) or isoenergetic (constant-$\mathscr{H}$) evolution. To be specific, a term $-\xi(\mathbf{r}, \mathbf{p})\,\mathbf{p}$ is added to the momentum equations given in the previous sections, with $\xi$ chosen by a Lagrange multiplier technique so as to constrain $\mathscr{T}$ or $\mathscr{H}$ to the desired value.

To avoid all the problems associated with system heating and non-linearity, Ciccotti *et al.* [1979] have suggested the use of a 'subtraction technique'. Perturbed and unperturbed trajectories are run simultaneously, from the same initial configuration. The desired response is computed, not merely as a value in the perturbed ensemble, but as the difference between perturbed and unperturbed values. In the thermodynamic limit, given that the equilibrium ensemble average of the property of interest vanishes, this makes no difference, but in practical terms this procedure cuts out most of the non-systematic 'statistical' noise, at least in the short-time response. In turn, this means that small perturbations, as used in real experiments, may be employed. The drawback, of course, is that trajectories evolving according to slightly different equations of motion will diverge exponentially from one another, and become statistically uncorrelated, in a way similar to that shown in Fig. 3.1. Only in the initial portion of the trajectory is a linear, systematic response difference seen; at long times, noise dominates. Thus, there is no point in attempting to establish a long-term, 'steady-state', response, or a long-term response to an oscillatory perturbation. The only way of measuring a transport coefficient is to apply a delta function (or step function) perturbation at time $t = 0$, and hope that the correlation function of interest has decayed to zero (or its time integral has reached a plateau value) before the noise sets in. To achieve acceptable statistical averaging over the initial conditions, the procedure is repeated several (50–200) times and the results accumulated. Generally, this is done by running the unperturbed simulation continuously, and starting perturbed segments from time to time in the manner shown in Fig. 8.3.

The subtraction technique may be combined with single-particle [Ciccotti and Jacucci 1975], $k$-dependent [Ciccotti *et al.* 1979], and spatially homogeneous [Singer, Singer, and Fincham 1980; Evans and Powles 1982] perturbations. It is unfortunate that, in general, just as the response seems to be

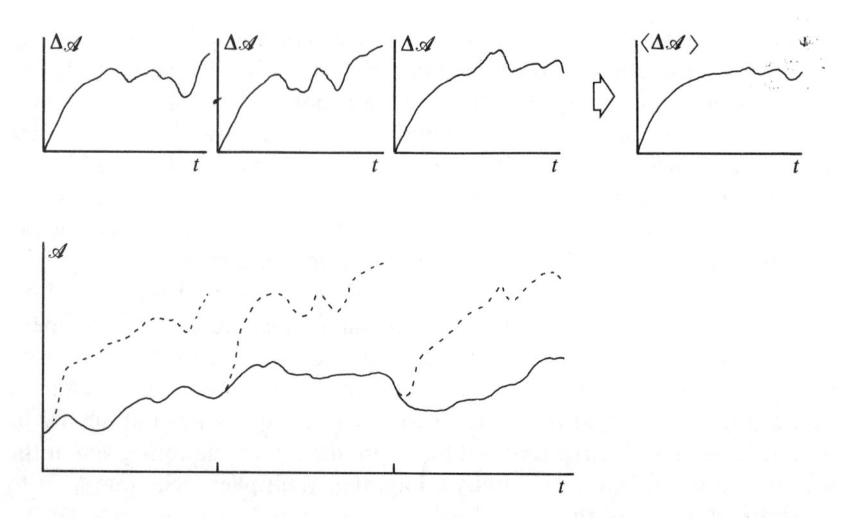


**Fig. 8.3** Schematic view of the subtraction technique. The solid line is the unperturbed run, and the dashed lines the perturbed segments. The accumulation of several differential responses is illustrated schematically.

approaching its limiting value, it becomes overwhelmed in noise. An approach similar to that used in Chapter 6 to estimate statistical errors in correlation functions may be applied to this question, and it may be shown that indeed the noise level will begin to dominate after one or two correlation times [G. Ciccotti, unpublished].

Why should we use non-equilibrium methods? It is still an open question as to whether or not they are more efficient at estimating transport coefficients than equilibrium simulations. It should be remembered that the techniques described in Sections 6.3 and 6.5.5 can produce an entire range of correlation functions and transport coefficients from the output of a single equilibrium run. Non-equilibrium simulations are normally able to provide only one fundamental transport coefficient, plus the coefficients describing cross-coupling with the applied perturbation, at once. If the basic simulation is expensive (e.g. for a complicated molecular model) compared with the correlation function analysis, then equilibrium methods should be used. On the other hand, NEMD is claimed to be much more efficient when the comparison is made for individual transport coefficients. By applying a steady-state perturbation, the problem of integrating a correlation function which has a long-time tail (or similarly examining an Einstein plot) is avoided, but it is replaced with the need to extrapolate, in a possibly ill-defined manner, to zero applied perturbation. It is believed that there is much less $N$-dependence in some transport coefficients determined by NEMD methods than in those obtained by equilibrium simulations [Ashurst and Hoover 1977; Evans 1979b; Hoover *et al.* 1980b; Holian and Evans 1983]. Arguments for and against non-equilibrium simulations continue in the literature [Hanley 1983].

# 9
# BROWNIAN DYNAMICS

## 9.1 Introduction

The simulation techniques we shall describe in this chapter are not motivated in quite the same way as those treated in previous chapters. The problem we wish to address is timescale separation, which occurs when one form of motion in the system is much faster than another. This can be a serious problem in MD, and similar difficulties arise in MC simulations. The short time-steps needed to handle the fast motion and the very long runs needed to allow evolution of the slower modes make the simulations very expensive. This is especially irritating if the fast motions are not of great interest in themselves. In some cases (Sections 3.4, 4.7) it is possible to replace such fast degrees of freedom by rigid constraints, but this is not always feasible. Consider the simulation of a large molecule (e.g. a protein) or a collection of heavy ions in a solvent. Even though the motion of the solvent molecules is of little interest, they will be present in large numbers, and they will make a full MD simulation very expensive. In such a case, an approximate approach may be adopted. The solvent particles are omitted from the simulation, and their effects upon the solute represented by a combination of random forces and frictional terms. Newton's equations of motion are thus replaced by some kind of Langevin equation.

## 9.2 Projection operators

The theoretical basis for simplifying the equations of motion, and removing rapidly varying degrees of freedom, is given by Zwanzig [1960, 1961a, b] and by Mori [1965a, b]. These authors introduced 'projection operators' into the equations of motion to obtain a 'reduced description'. Mori's derivation begins with the dynamical variables of interest, $\mathscr{A}_i$, each of which evolves according to eqn (2.8). Consider a set of dynamical variables $\mathscr{A} = (\mathscr{A}_1, \mathscr{A}_2, \ldots \mathscr{A}_n)$ which is a subset of the complete set of all such functions. For example, $\mathscr{A}$ may consist of the phase space coordinates of the molecules of interest, the solvent particles being omitted. The aim is an equation of motion involving just the set $\mathscr{A}$, not the other variables. To this end, Mori introduces operators $\mathbb{P}$ and $\mathbb{Q}$ with $\mathbb{P} + \mathbb{Q} = 1$, defined to project out from any dynamical variable those parts which, respectively, lie within, and are orthogonal to, the state space defined by $\mathscr{A}$. In this context, two variables $\mathscr{A}_i$ and $\mathscr{B}_i$ are orthogonal if the equilibrium average $\langle \mathscr{A}_i \mathscr{B}_i \rangle$ vanishes. The exact evolution equation for $\mathscr{A}$ is then divided up as follows:

$$\dot{\mathscr{A}}(t) = iL\mathscr{A}(t) = \mathbb{P}iL\mathscr{A}(t) + \mathbb{Q}iL\mathscr{A}(t)\,. \tag{9.1}$$

The succeeding steps may be found in the original references [Mori 1965a, b] and elsewhere [Berne and Pecora 1976; McQuarrie 1976; Hansen and McDonald 1986]. The result is still an exact equation of motion

$$\dot{\mathscr{A}}(t) = i\mathbf{\Omega}\mathscr{A}(t) - \int_0^t \mathbf{M}(t')\,\mathscr{A}(t-t')\,\mathrm{d}t' + \mathring{\mathscr{A}}(t) \tag{9.2}$$

which is often termed a 'generalized Langevin equation' (see below). The so-called 'frequency matrix' $i\mathbf{\Omega}$ is simply a dyadic

$$i\mathbf{\Omega} = \langle \dot{\mathscr{A}}\mathscr{A} \rangle \langle \mathscr{A}\mathscr{A} \rangle^{-1} . \tag{9.3}$$

The quantity $\mathring{\mathscr{A}}(t)$ is termed the 'random force'. It is that part of $\dot{\mathscr{A}}$ which is initially orthogonal to $\mathscr{A}$

$$\mathring{\mathscr{A}}(0) = \mathbb{Q}iL\mathscr{A}(0) = \mathbb{Q}\dot{\mathscr{A}}(0) \tag{9.4}$$

and it evolves in a non-standard fashion (compare eqn (2.9))

$$\mathring{\mathscr{A}}(t) = \mathrm{e}^{\mathbb{Q}iLt}\mathring{\mathscr{A}}(0) = \mathrm{e}^{\mathbb{Q}iLt}\mathbb{Q}iL\mathscr{A}(0) \tag{9.5}$$

so as to remain orthogonal to $\mathscr{A}(0)$ at all subsequent times

$$\langle \mathring{\mathscr{A}}(t)\mathscr{A}(0) \rangle = 0 . \tag{9.6}$$

$\mathbf{M}(t)$ is the normalized autocorrelation matrix of this projected random force, and is usually called the 'memory function' matrix

$$\mathbf{M}(t) = \langle \mathring{\mathscr{A}}(t)\mathring{\mathscr{A}}(0) \rangle \langle \mathscr{A}\mathscr{A} \rangle^{-1} . \tag{9.7}$$

Multiplication of eqn (9.2) by $\mathscr{A}(0)$ and ensemble averaging yields a relation between $\mathbf{M}(t)$ and the autocorrelation matrix $\mathbf{C}(t) = \langle \mathscr{A}(t)\mathscr{A}(0) \rangle$

$$\dot{\mathbf{C}}(t) = i\mathbf{\Omega}\mathbf{C}(t) - \int_0^t \mathbf{M}(t')\mathbf{C}(t-t')\,\mathrm{d}t' . \tag{9.8}$$

Like eqn (9.2) this equation is exact, but formal, in that it merely defines the properties of $\mathring{\mathscr{A}}(t)$ and $\mathbf{M}(t)$ in terms of those of $\mathscr{A}(t)$ and $\mathbf{C}(t)$.

The utility of these equations lies in the hope that it will be possible to model $\mathring{\mathscr{A}}(t)$ in some simple way, as a stochastic process with specified statistical properties, and that this will yield an approximate equation of motion for $\mathscr{A}(t)$. It may be that $\mathbf{M}(t)$ will decay more rapidly than $\mathbf{C}(t)$, i.e. that the chosen variables $\mathscr{A}(t)$ are slow compared with $\mathring{\mathscr{A}}(t)$. In the simplest case, the elements of $\mathbf{M}(t)$ may be taken to be proportional to delta functions in the time, and the convolutions in eqns (9.2), (9.8) may be performed immediately. The variables $\mathscr{A}(t)$ then have exponentially decaying, possibly oscillatory, correlation functions, and the time evolution is a Markov process, i.e. one without memory. In the more general case, specification of $\mathbf{M}(t)$ is sufficient (through the Laplace transform of eqn (9.8)) to define $\mathbf{C}(t)$. If $\mathbf{M}(t)$ does not decay rapidly compared to $\mathbf{C}(t)$, it may be possible to improve matters by including additional time derivatives of the dynamical variables in the set $\mathscr{A}(t)$.

Zwanzig [1960, 1961a, b] originally presented an analogous derivation in which projection operators are introduced into the equation of motion, eqn (2.5), for the phase space distribution function $\rho(\Gamma, t)$, rather than into the dynamic variable equations. This yields a generalized Fokker–Planck equation for the reduced distribution function. The approach is in all respects equivalent to that of Mori, the relationship between the two being akin to that between the Heisenberg and Schrödinger pictures in quantum mechanics. An elegant unified treatment has been presented by Nordholm and Zwanzig [1975].

### 9.3 Brownian dynamics

The classical application of projection operators, and the one of interest to us, is when $\mathscr{A}$ consists of a single component $p_{i\alpha}(t)$ of the momentum of a single molecule. In Cartesian coordinates, for simplicity, the projected equations of motion take the form

$$\dot{r}_{i\alpha}(t) = p_{i\alpha}(t)/m \tag{9.9a}$$

$$\dot{p}_{i\alpha}(t) = -\int_0^t M(t')p_{i\alpha}(t-t')\,\mathrm{d}t' + \mathring{p}_{i\alpha}(t)\,. \tag{9.9b}$$

In the delta function approximation we have

$$\langle \mathring{p}_{i\alpha}(t)\mathring{p}_{i\alpha}(0)\rangle = M(t)\langle p_{i\alpha}^2\rangle = 2\xi\langle p_{i\alpha}^2\rangle\,\delta(t) = 2mk_{\mathrm{B}}T\xi\delta(t) \tag{9.10}$$

where $\xi$ is the friction constant. Integration over the delta function from time $t = 0$ in eqn (9.9b) gives a factor $\frac{1}{2}$ and leads to the classical Langevin equation

$$\dot{p}_{i\alpha}(t) = -\xi p_{i\alpha}(t) + \mathring{p}_{i\alpha}(t) \tag{9.11}$$

and exponential decay of the momentum autocorrelation function

$$\langle p_{i\alpha}(t)p_{i\alpha}(0)\rangle = \langle p_{i\alpha}^2\rangle \mathrm{e}^{-\xi t}. \tag{9.12}$$

The first term on the right of eqn (9.11) is a frictional force. It is the role of $\mathring{p}_{i\alpha}(t)$ in eqn (9.11) which has led to the designation 'random force' for $\mathring{\mathscr{A}}$, in the more general case eqn (9.2). Equation (9.11) generates classical Brownian motion as expected for a particle under the influence of rapid, random, buffeting from its neighbours in a liquid [Chandrasekhar 1943]. The short-time dynamics is unphysical: the momentum autocorrelation function does not decay exponentially in a real fluid (see Fig. 2.3), and a realistic memory function is not even particularly short-lived in time by comparison [Hansen and McDonald 1986]. Nonetheless, at long times, molecular displacements generated by eqns (9.9)–(9.12) conform to Einstein's relation, eqn (2.110), with $\xi$ related to the diffusion coefficient $D$ by

$$\xi = k_{\mathrm{B}}T/mD \tag{9.13}$$

as can be seen by integrating eqn (9.12) and using eqn (2.109).

In order to realize eqn (9.11) in a simulation, the statistical properties of $\mathring{p}_{i\alpha}(t)$ must be completely specified. Almost invariably, $p_{i\alpha}(t)$ is assumed to be a Gaussian random process (see Section 6.4) whose moments are defined by $\langle \mathring{p}_{i\alpha}(t)\mathring{p}_{i\alpha}(0)\rangle$ [Wang and Uhlenbeck 1945]. The linear form of eqn (9.9) then leads directly to a Maxwellian velocity distribution. The presence of the delta function in eqn (9.10) makes it awkward to consider the explicit form of these moments. There are some hidden subtleties in the Langevin equation [Doob 1942] which may be circumvented by always considering time integrals of $\mathring{p}_{i\alpha}$ rather than the random force itself. For a Gaussian random process of the type described above, and for any function $f(t)$, the variable

$$\delta p_{i\alpha}^{\mathrm{G}} = \int_t^{t+\delta t} f(t')\,\mathring{p}_{i\alpha}(t')\,\mathrm{d}t' \tag{9.14}$$

is a random variable with a Gaussian distribution

$$\rho(\delta p_{i\alpha}^{\mathrm{G}}) = \frac{1}{\sigma_p(2\pi)^{1/2}}\exp\{-\tfrac{1}{2}(\delta p_{i\alpha}^{\mathrm{G}}/\sigma_p)^2\} \tag{9.15}$$

with zero mean, and variance given by

$$\sigma_p^2 = 2\xi m k_{\mathrm{B}}T\int_t^{t+\delta t} f^2(t')\,\mathrm{d}t' \tag{9.16}$$

[Chandrasekhar 1943]. As we shall see shortly, these results allow the construction of a simulation algorithm, based on the integration of eqn (9.11) over a succession of time steps $\delta t$, in the usual MD fashion.

For a system of independent particles, the above equations would apply to each momentum component separately. To be useful, they must be generalized to the case of interacting particles. This introduces some subtleties, to which we shall return below. The simplest approach is to add the forces of interaction directly into eqn (9.11), which we now write as a $3N$-dimensional vector equation

$$\dot{\mathbf{p}}(t) = -\xi\mathbf{p}(t) + \mathbf{f}(t) + \mathring{\mathbf{p}}(t) \tag{9.17}$$

where $\mathbf{f}$ is derived from a potential in the usual way. The friction coefficient may be different for different types of molecule (we shall confine ourselves to a single component for simplicity) but it is assumed to be independent of particle positions and momenta. Frictional effects are taken to be isotropic, so $\xi$ is a scalar. The random forces $\mathring{\mathbf{p}}(t)$ on different molecules are taken to be independent of each other, with each vectorial component satisfying eqns (9.10), (9.14)–(9.16). It should be emphasized that the projection operator approach does not lead directly to eqn (9.17): the full effects of solvent-mediated interparticle interactions are neglected in the above *ad hoc* treatment.

Equation (9.17), together with $\dot{\mathbf{r}} = \mathbf{p}/m$, is equivalent to a Fokker–Planck equation for the phase space density function [see e.g. Stratonovich 1963, 1967; Haken 1975, Sections 11C and 12B]

$$\frac{\partial}{\partial t}\rho(\mathbf{r}, \mathbf{p}, t) + \frac{\mathbf{p}}{m}\cdot\nabla_{\mathbf{r}}\rho(\mathbf{r}, \mathbf{p}, t) + \mathbf{f}\cdot\nabla_{\mathbf{p}}\rho(\mathbf{r}, \mathbf{p}, t) =$$

$$\xi\nabla_{\mathbf{p}}\cdot(\mathbf{p}\rho(\mathbf{r}, \mathbf{p}, t) + mk_BT\nabla_{\mathbf{p}}\rho(\mathbf{r}, \mathbf{p}, t)) \quad (9.18a)$$

or, written out in full,

$$\frac{\partial}{\partial t}\rho(\mathbf{r}, \mathbf{p}, t) + \sum_i \left(\frac{\mathbf{p}_i}{m}\cdot\nabla_{\mathbf{r}_i}\rho(\mathbf{r}, \mathbf{p}, t) + \mathbf{f}_i\cdot\nabla_{\mathbf{p}_i}\rho(\mathbf{r}, \mathbf{p}, t)\right) =$$

$$\xi\sum_i \nabla_{\mathbf{p}_i}\cdot(\mathbf{p}_i\rho(\mathbf{r}, \mathbf{p}, t) + mk_BT\nabla_{\mathbf{p}_i}\rho(\mathbf{r}, \mathbf{p}, t)) \quad (9.18b)$$

(compare eqns (2.4) and (2.5)). This equation has the canonical ensemble distribution as a stationary solution.

A straightforward method of conducting 'Brownian dynamics' simulations based on eqn (9.17) has been developed by Ermak [Ermak 1976; Ermak and Buckholtz 1980]. Somewhat different schemes have been employed elsewhere [Morf and Stoll 1977; Schneider and Stoll 1978; Turq, Lantelme, and Friedman 1977; Adelman 1979]. In Ermak's approach, the equations of motion are integrated over a time interval $\delta t$ under the assumption that the systematic forces $\mathbf{f}(t)$ remain approximately constant. The result is an algorithm resembling those of Chapter 3, based on stored positions, velocities, and accelerations. For a one-component atomic system, the algorithm may be written

$$\mathbf{r}(t+\delta t) = \mathbf{r}(t) + c_1\delta t\mathbf{v}(t) + c_2\delta t^2\mathbf{a}(t) + \delta\mathbf{r}^{\mathrm{G}} \quad (9.19a)$$

$$\mathbf{v}(t+\delta t) = c_0\mathbf{v}(t) + c_1\delta t\mathbf{a}(t) + \delta\mathbf{v}^{\mathrm{G}}. \quad (9.19b)$$

The numerical coefficients in these equations are

$$c_0 = \mathrm{e}^{-\xi\delta t} \approx 1 - (\xi\delta t) + \tfrac{1}{2}(\xi\delta t)^2 - \tfrac{1}{6}(\xi\delta t)^3 + \tfrac{1}{24}(\xi\delta t)^4 - \ldots \quad (9.20a)$$

$$c_1 = (\xi\delta t)^{-1}(1 - c_0) \approx 1 - \tfrac{1}{2}(\xi\delta t) + \tfrac{1}{6}(\xi\delta t)^2 - \tfrac{1}{24}(\xi\delta t)^3 + \ldots \quad (9.20b)$$

$$c_2 = (\xi\delta t)^{-1}(1 - c_1) \approx \tfrac{1}{2} - \tfrac{1}{6}(\xi\delta t) + \tfrac{1}{24}(\xi\delta t)^2 - \ldots \quad (9.20c)$$

where a low-$\xi\delta t$ expansion is also given. The random variables $\delta\mathbf{r}^{\mathrm{G}}$ and $\delta\mathbf{v}^{\mathrm{G}}$ are defined as stochastic integrals

$$\delta\mathbf{r}^{\mathrm{G}} = \int_t^{t+\delta t} \xi^{-1}(1 - \mathrm{e}^{-\xi(t+\delta t-t')})m^{-1}\mathring{\mathbf{p}}(t')\,\mathrm{d}t'$$

$$= \int_t^{t+\delta t} f_r(t')m^{-1}\mathring{\mathbf{p}}(t')\,\mathrm{d}t' \quad (9.21a)$$

$$\delta \mathbf{v}^{\mathrm{G}} = \int_{t}^{t+\delta t} \mathrm{e}^{-\xi(t+\delta t-t')} m^{-1} \mathring{\mathbf{p}}(t')\,\mathrm{d}t'$$

$$= \int_{t}^{t+\delta t} f_v(t') m^{-1} \mathring{\mathbf{p}}(t')\,\mathrm{d}t' \qquad (9.21\mathrm{b})$$

and we have defined the functions $f_r(t')$ and $f_v(t')$. Equations (9.19a) and (9.19b) are stochastic, in the sense that they equate the statistical properties of $\delta \mathbf{r}^{\mathrm{G}}$ and $\delta \mathbf{v}^{\mathrm{G}}$ with those of the other terms in the equations. Thus, unlike conventional dynamics, there is no 'unique' trajectory: we can only generate a representative trajectory, i.e. produce a realization of the stochastic process. In a simulation, each pair of vectorial components of $\delta \mathbf{r}^{\mathrm{G}}$, $\delta \mathbf{v}^{\mathrm{G}}$, i.e. $\delta r_{i\alpha}^{\mathrm{G}}$, $\delta v_{i\alpha}^{\mathrm{G}}$, is sampled from a bivariate Gaussian distribution defined by equations similar to eqns (9.14)–(9.16) [Chandrasekhar 1943]

$$\rho(\delta r_{i\alpha}^{\mathrm{G}}, \delta v_{i\alpha}^{\mathrm{G}}) = \frac{1}{2\pi\sigma_r\sigma_v(1-c_{rv}^2)^{1/2}}$$

$$\times \exp\left\{ -\frac{1}{2(1-c_{rv}^2)} \left( \left(\frac{\delta r_{i\alpha}^{\mathrm{G}}}{\sigma_r}\right)^2 + \left(\frac{\delta v_{i\alpha}^{\mathrm{G}}}{\sigma_v}\right)^2 - 2c_{rv}\left(\frac{\delta r_{i\alpha}^{\mathrm{G}}}{\sigma_r}\right)\left(\frac{\delta v_{i\alpha}^{\mathrm{G}}}{\sigma_v}\right) \right) \right\} \qquad (9.22)$$

with zero mean values, variances given by

$$\sigma_r^2 = \langle (\delta r_{i\alpha}^{\mathrm{G}})^2 \rangle = 2\xi \frac{k_{\mathrm{B}}T}{m} \int_{t}^{t+\delta t} f_r^2(t')\,\mathrm{d}t' = 2\xi \frac{k_{\mathrm{B}}T}{m} \int_{0}^{\delta t} f_r^2(t'')\,\mathrm{d}t''$$

$$= \delta t^2 \frac{k_{\mathrm{B}}T}{m} (\xi\delta t)^{-1} (2 - (\xi\delta t)^{-1}(3 - 4\mathrm{e}^{-\xi\delta t} + \mathrm{e}^{-2\xi\delta t})) \qquad (9.23\mathrm{a})$$

$$\sigma_v^2 = \langle (\delta v_{i\alpha}^{\mathrm{G}})^2 \rangle = 2\xi \frac{k_{\mathrm{B}}T}{m} \int_{t}^{t+\delta t} f_v^2(t')\,\mathrm{d}t'$$

$$= 2\xi \frac{k_{\mathrm{B}}T}{m} \int_{0}^{\delta t} f_v^2(t'')\,\mathrm{d}t'' = \frac{k_{\mathrm{B}}T}{m} (1 - \mathrm{e}^{-2\xi\delta t}) \qquad (9.23\mathrm{b})$$

and a correlation coefficient $c_{rv}$ determined by

$$c_{rv}\sigma_r\sigma_v = \langle \delta r_{i\alpha}^{\mathrm{G}} \delta v_{i\alpha}^{\mathrm{G}} \rangle = 2\xi \frac{k_{\mathrm{B}}T}{m} \int_{t}^{t+\delta t} f_r(t') f_v(t')\mathrm{d}t'$$

$$= 2\xi \frac{k_{\mathrm{B}}T}{m} \int_{0}^{\delta t} f_r(t'') f_v(t'')\,\mathrm{d}t''$$

$$= \delta t \frac{k_{\mathrm{B}}T}{m} (\xi\delta t)^{-1} (1 - \mathrm{e}^{-\xi\delta t})^2 . \qquad (9.23\mathrm{c})$$

The way in which this sampling is carried out is outlined in Appendix G. Following the selection of values of $\delta \mathbf{r}^{\mathrm{G}}$ and $\delta \mathbf{v}^{\mathrm{G}}$, $\mathbf{r}(t+\delta t)$ and $\mathbf{v}(t+\delta t)$ are made to satisfy eqns (9.19a, b) and the algorithm proceeds step-by-step in the usual way. At each stage it is essential that correlated values $\delta r_{i\alpha}^{G}$, $\delta v_{i\alpha}^{G}$ are sampled as described above, since they are integrals involving the same random process $\mathring{p}_{i\alpha}(t)$ over the same time interval (eqns (9.21a), (9.21b)); different particles, and different vectorial components, are sampled independently.

Ermak's algorithm is an attempt to treat properly both the systematic dynamic and stochastic elements of the Langevin equation. At low values of the friction coefficient $\xi$, the dynamical aspects dominate, and Newtonian mechanics is recovered as $\xi \to 0$. Equations (9.19) and (9.20) then become a simple Taylor series predictor algorithm. As discussed in Chapter 3, this is not a particularly accurate method of conducting MD simulations, and the same is true of Brownian dynamics at low friction: what is needed here is a stochastic generalization, with friction, of a predictor–corrector or Verlet-like algorithm. A simple algorithm of this type, which reduces to the velocity Verlet algorithm of Section (3.2.1), is obtained if, on integrating the velocity equation, the systematic force is assumed to vary linearly with time:

$$\mathbf{r}(t+\delta t) = \mathbf{r}(t) + c_1 \delta t \mathbf{v}(t) + c_2 \delta t^2 \mathbf{a}(\mathbf{t}) + \delta \mathbf{r}^{G} \tag{9.24a}$$

$$\mathbf{v}(t+\delta t) = c_0 \mathbf{v}(t) + (c_1 - c_2)\delta t \mathbf{a}(t) + c_2 \delta t \mathbf{a}(t+\delta t) + \delta \mathbf{v}^{G}. \tag{9.24b}$$

After the selection of the random components $\delta \mathbf{r}^{G}$ and $\delta \mathbf{v}^{G}$ for a given step, the algorithm is implemented in the usual way. Other Verlet-like algorithms have been proposed [Allen 1980, 1982; van Gunsteren and Berendsen 1982] and, although there is no unique way of generalizing the method, these are all closely related to each other and provide a similar measure of improvement over the simple predictor of Ermak, at low friction. At high values of $\xi$, the dynamical aspects become less important, and there is little to choose between the different methods.

If long-time configurational dynamics are of interest, then the momentum variables may be dropped from the equations of motion, in the spirit of time-scale separation implicit in the projection-operator method. The 'position Langevin equation' is a simplified version of equations given by Lax [1966] and by Zwanzig [1969]:

$$\dot{\mathbf{r}}(t) = \frac{D}{k_{\mathrm{B}}T}\mathbf{f}(t) + \mathring{\mathbf{r}}(t) = \mathbf{f}(t)/m\xi + \mathring{\mathbf{r}}(t). \tag{9.25}$$

Here, $D$ is the diffusion coefficient and $\mathbf{f}$ the instantaneous systematic force, as usual. The quantity $\mathring{\mathbf{r}}(t)$ is a 'random velocity process' which may be taken to have a delta function correlation for each molecule

$$\langle \mathring{r}_{i\alpha}(t)\mathring{r}_{j\beta}(0)\rangle = 2D\delta(t)\delta_{ij}\delta_{\alpha\beta}. \tag{9.26}$$

The associated equation of motion for the configurational distribution function is the well-known Schmoluchowski equation

$$\frac{\partial}{\partial t}\rho(\mathbf{r}, t) + \frac{D}{k_{\mathrm{B}}T}\nabla_{\mathbf{r}} \cdot (\mathbf{f}\rho(\mathbf{r}, t)) = D\nabla_{\mathbf{r}}^{2}\rho(\mathbf{r}, t) \tag{9.27a}$$

or

$$\frac{\partial}{\partial t}\rho(\mathbf{r},t)+\frac{D}{k_{\mathrm{B}}T}\sum_i \nabla_{\mathbf{r}_i}\cdot(\mathbf{f}_i\rho(\mathbf{r},t)) = D\sum_i \nabla^2_{\mathbf{r}_i}\rho(\mathbf{r},t). \tag{9.27b}$$

At long times, these equations lead to Einstein's diffusion law and a steady-state canonical distribution for the positions. An algorithm based on eqn (9.25) [Ermak and Yeh 1974; Ermak 1975] is

$$\begin{aligned}\mathbf{r}(t+\delta t) &= \mathbf{r}(t)+\frac{D}{k_{\mathrm{B}}T}\mathbf{f}(t)\,\delta t+\delta\mathbf{r}^{\mathrm{G}} \\ &= \mathbf{r}(t)+\delta t\mathbf{f}(t)/m\xi+\delta\mathbf{r}^{\mathrm{G}}\end{aligned} \tag{9.28}$$

where each component $\delta r^{\mathrm{G}}_{i\alpha}$ is chosen independently from a Gaussian distribution with zero mean and variance $\langle(\delta r^{\mathrm{G}}_{i\alpha})^2\rangle = 2D\delta t$ (compare eqn (9.23a) for large $\xi\delta t$). As usual, these equations apply to each component of $\mathbf{r}$. The short-time dynamics generated by these equations are even more unrealistic than those resulting from the Langevin equation. In fact, the method is very much more closely related to the force-bias and smart MC methods of Section (7.3) than to MD. However, some realism in the long-time dynamics may be restored by the inclusion of hydrodynamic effects.

In the above, we have concentrated on atomic systems. The formalism may be generalized to include rigid and non-rigid molecules, and the incorporation of constraints into Brownian dynamics is straightforward [van Gunsteren and Berendsen 1982] although the usual care should be taken in their application [van Gunsteren 1980; van Gunsteren and Karplus 1982].

## 9.4 Hydrodynamic and memory effects

The projection operator approach has been the basis of the dynamic techniques discussed in the previous section, but the extension to interacting many-body systems has involved additional assumptions. In particular, the only coupling between molecules in eqns (9.17), (9.25), is that due to direct intermolecular forces derived from a potential. The effect of one molecule on another through the flow of solvent molecules around them is completely neglected, as is any modification of the interaction between them due to solvent structure. The inclusion of such effects makes Brownian dynamics more realistic and self-consistent, as indeed does the incorporation of memory effects into the time evolution. We discuss these extensions in this section.

In principle, the inclusion of a specified memory function of finite duration in the Brownian dynamics algorithm is straightforward. The method is based on the integration of eqns (9.9a, b) over a time step $\delta t$, with the convolution term being evaluated by quadrature. Of course, it is necessary to store the values of the momenta at previous time steps in order to evaluate the convolution integral. The longer the 'memory' of the system, the more of each molecule's prior history must be stored, and this makes the method difficult and expensive to implement in practice [Ciccotti, Orban, and Ryckaert 1976; Doll and Dion 1976]. Alternative approaches are possible when the memory function can be approximated by a simple exponential decay in time [Ermak and Buckholtz 1980] or indeed as a sum of a finite number of exponentials [Ciccotti and Ryckaert 1980]. The simplest way of treating this is true to the spirit of the projection operator formalism, adding more time derivatives to the set of dynamic variables under consideration, in order to improve the description, particularly at short times. The equations of motion may be written (compare eqns (9.9a, b) and (9.11))

$$
\begin{aligned}
\dot{\mathbf{r}} &= \mathbf{p}/m \\
\dot{\mathbf{p}} &= \mathbf{p}^{(1)} \\
\dot{\mathbf{p}}^{(1)} &= \mathbf{p}^{(2)} \\
\dot{\mathbf{p}}^{(2)} &= \mathbf{p}^{(3)} \\
&\vdots \\
\dot{\mathbf{p}}^{(n-1)} &= \mathbf{p}^{(n)} \\
\dot{\mathbf{p}}^{(n)} &= -\xi^{(n)}\mathbf{p}^{(n)} + \mathring{\mathbf{p}}^{(n)}
\end{aligned}
\qquad (9.29)
$$

where the set of equations is truncated at the $n$th level. This corresponds to Mori's 'continued fraction' representation [Mori 1965b], so-called because of the form of the equation linking the Laplace-transformed momentum autocorrelation function and the friction coefficient $\xi^{(n)}$. This last quantity is the only dynamical parameter of the method, and it may be linked to the value of the diffusion coefficient through eqn (9.29). By a procedure of this kind [Ciccotti and Ryckaert 1980; Adelman 1979] Brownian dynamics of a single molecule may be carried out in a manner exactly analogous to the simulations discussed in the previous section, so as to give a momentum autocorrelation function which is a sum of $n$ exponentials in the time; the only difference is that a vector Markov process, rather than a scalar one, is being realized. The method generalizes in the usual way to the many-body case.

Now we turn to the effects of molecular interactions on the stochastic equations of motion. It should be realized at the outset that a formal projection operator derivation of the generalized Langevin equation for interacting molecules [Ciccotti and Ryckaert 1981] introduces some unavoidable difficulties. The random force term is no longer orthogonal to the initial momentum, and the time-scale separation arguments leading to the adoption of a delta-function memory are no longer plausible, when the systematic forces depend non-linearly on the molecular positions. The equation linking the magnitude of the random forces with the temperature (e.g. eqn (9.10)) may also be altered [Bossis, Quentrec, and Boon 1982].

If an equation such as eqn (9.17) is to apply to a system of interacting molecules, then the friction coefficient $\xi$ must depend upon all the molecular positions and momenta. The physical reason for this is that solvent flow, induced by one molecule, must have an effect through the frictional forces on other molecules. In fact, the problem is only tractable if we drop all dependence of the friction coefficient on momenta, and concentrate on configuration-dependent effects. This means that the incorporation of hydrodynamic flow into Brownian dynamics is done at the position-Langevin/Schmoluchowski level of description, rather than at the momentum-Langevin/Fokker–Planck level. The Schmoluchowski equation for interacting molecules is [Murphy and Aguirre 1972; Wilemski 1976]

$$\frac{\partial}{\partial t}\rho(\mathbf{r}, t) + \nabla_{\mathbf{r}}\cdot\frac{\mathbf{D}(t)}{k_{\mathrm{B}}T}\cdot\mathbf{f}\rho(\mathbf{r}, t) = \nabla_{\mathbf{r}}\cdot\mathbf{D}(t)\cdot\nabla_{\mathbf{r}}\rho(\mathbf{r}, t) \tag{9.30a}$$

or, in full,

$$\frac{\partial}{\partial t}\rho(\mathbf{r}, t) + \sum_i\sum_j \nabla_{\mathbf{r}_i}\cdot\frac{\mathbf{D}_{ij}(t)}{k_{\mathrm{B}}T}\cdot\mathbf{f}_j\rho(\mathbf{r}, t) = \sum_i\sum_j \nabla_{\mathbf{r}_i}\cdot\mathbf{D}_{ij}(t)\cdot\nabla_{\mathbf{r}_j}\rho(\mathbf{r}, t)\,. \tag{9.30b}$$

The associated Langevin-type of equation depends upon the convention adopted for stochastic differentials. In the Stratonovich interpretation it is [Lax 1966; Hess and Klein 1978]

$$\dot{\mathbf{r}}(t) = \frac{\mathbf{D}(t)}{k_{\mathrm{B}}T}\cdot\mathbf{f}(t) + (\nabla_{\mathbf{r}}\cdot\boldsymbol{\sigma}(t))\cdot\boldsymbol{\sigma}(t) + \mathring{\mathbf{r}}(t) \tag{9.31a}$$

or

$$\dot{\mathbf{r}}_i(t) = \sum_j \frac{\mathbf{D}_{ij}(t)}{k_{\mathrm{B}}T}\cdot\mathbf{f}_j(t) + \sum_j\sum_k (\nabla_{\mathbf{r}_j}\cdot\boldsymbol{\sigma}_{jk}(t))\cdot\boldsymbol{\sigma}_{ik}(t) + \mathring{\mathbf{r}}_i(t) \tag{9.31b}$$

while in the Itô calculus it is [Tough, Pusey, Lekkerkerker, and van den Broek, 1986]

$$\dot{\mathbf{r}}(t) = \frac{\mathbf{D}(t)}{k_{\mathrm{B}}T}\cdot\mathbf{f}(t) + \nabla_{\mathbf{r}}\cdot\mathbf{D}(t) + \mathring{\mathbf{r}}(t) \tag{9.32a}$$

or

$$\dot{\mathbf{r}}_i(t) = \sum_j \frac{\mathbf{D}_{ij}(t)}{k_{\mathrm{B}}T}\cdot\mathbf{f}_j(t) + \nabla_{\mathbf{r}_j}\cdot\mathbf{D}_{ij}(t) + \mathring{\mathbf{r}}_i(t)\,. \tag{9.32b}$$

These equations have been seen in skeletal form as eqn (9.25), but the full forms are much more complicated. Here the diffusion 'constant' $\mathbf{D}$ is in fact a $3N \times 3N$ tensor or matrix, whose components depend upon the molecular positions. It can be regarded as a set of $3 \times 3$ matrices $\mathbf{D}_{ij}$ for each pair of molecules. The $\boldsymbol{\sigma}$ matrix is the 'square root' of $\mathbf{D}$, i.e. $\mathbf{D} = \boldsymbol{\sigma}^2$. Both $\mathbf{D}$ and $\boldsymbol{\sigma}$ are symmetric. The formal differences between eqns (9.31) and (9.32) and the link with Zwanzig's equation [Zwanzig 1969] have been discussed by Tough *et al.* [1986]. Any ambiguity disappears when a finite time step is considered. An integration algorithm based on these equations may be written in much the same way as eqn (9.28) [Ermak and McCammon 1978]

$$\mathbf{r}_i(t+\delta t) = \mathbf{r}_i(t) + \sum_j \frac{\mathbf{D}_{ij}(t)}{k_B T}\cdot \mathbf{f}_j(t)\delta t + \nabla_{\mathbf{r}_j}\cdot \mathbf{D}_{ij}(t)\delta t + \delta \mathbf{r}_i^G \tag{9.33}$$

where the components of $\delta\mathbf{r}_i^G$ are random variables selected from the $3N$-variate Gaussian distribution with zero means, and covariance matrix

$$\langle \delta\mathbf{r}_i^G \delta\mathbf{r}_j^G \rangle = 2\mathbf{D}_{ij}\,\delta t. \tag{9.34}$$

Note that the components of $\delta\mathbf{r}_i^G$ for different molecules are no longer statistically independent. In the techniques discussed so far, generating the random numbers is almost always less time-consuming than evaluating the systematic forces, but the reverse may well be true in this case. Sampling the $3N$ correlated values of $\delta\mathbf{r}_i^G$ is discussed in Appendix G, but we note here that this will involve some expensive manipulations of the $3N \times 3N$ matrix $\mathbf{D}$ at each time step. The way in which the diffusion tensor depends upon molecular positions is not known exactly, but is clearly important. On commonly adopted form, suggested by macroscopic hydrodynamics, is the Oseen tensor, defined by the following $3 \times 3$ matrix for each pair of molecules $i$ and $j$:

$$\begin{aligned} \mathbf{D}_{ij} &= \frac{k_B T}{6\pi\eta r_0}\mathbf{1} && (i=j) \\ &= \frac{k_B T}{8\pi\eta r_{ij}}\left(\mathbf{1} + \frac{\mathbf{r}_{ij}\mathbf{r}_{ij}}{r_{ij}^2}\right) && (i \neq j) \end{aligned} \tag{9.35}$$

where $\eta$ is the viscosity and $r_0$ an estimate of the molecular radius. This tensor has the property that $\nabla_{\mathbf{r}}\cdot \mathbf{D} = 0$ so this term may be dropped from eqn (9.33). Ermak and McCammon [1978] give a detailed description of the application of these techniques. A sample program is given in F.33.

We should make a few comments regarding $\mathbf{D}_{ij}$. Firstly, we should emphasize that this is part of the input to a simulation, not the output. It represents diffusive or frictional effects in the dilute system, not the measured diffusion coefficients in the more concentrated systems, which may be studied by Brownian dynamics. The term 'mobility tensor' is sometimes used for $\mathbf{D}_{ij}/k_B T$, and this may be less confusing. Secondly, eqn (9.35) is only the first term in an expansion of the pair diffusion coefficient for an incompressible fluid in inverse powers of $r_{ij}$; higher-order terms have been given [Felderhof 1977; Schmitz and Felderhof 1982] and these could be incorporated in a Brownian dynamics simulation. The leading term, eqn (9.35), is very long-ranged, and it may be possible to handle this, in a periodic system, by Ewald summation [Beenakker 1986]. However, it is not clear that periodic boundary conditions (included, for example, in program F.33) are consistent with the use of equations such as eqn (9.35) and its extensions, with or without an Ewald-like summation [Smith 1987]. Thirdly, the use of a pairwise expression for diffusive effects is itself an approximation. Expressions which include three- and four-body effects have been published [Mazur 1982; Mazur and van

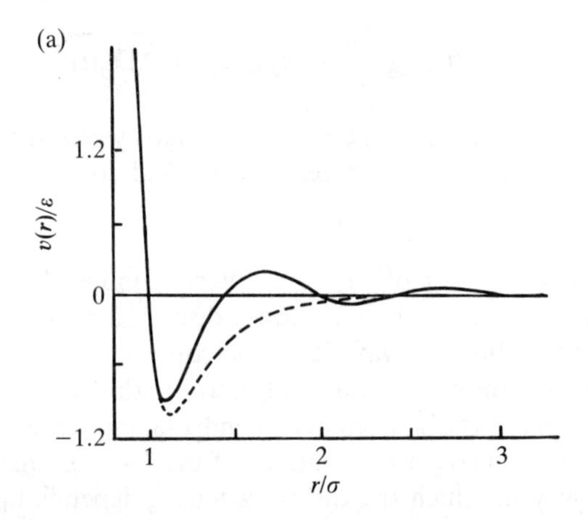


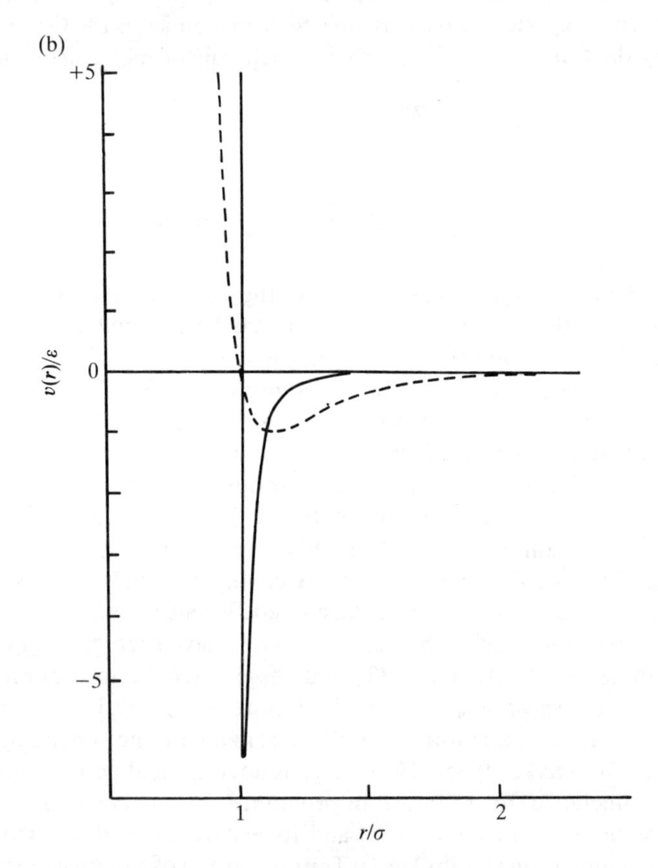


**Fig. 9.1** (a) The Lennard-Jones potential (dashed line) and the potential of mean force (solid line) $T^* = 0.76$, $\rho^* = 0.85$). (b) The Lennard-Jones potential (dashed line) and the DLVO potential between two colloidal particles (solid line).

Saarloos 1982; van Saarloos and Mazur 1983] but to implement these in a simulation would be considerably more expensive than the already costly pairwise form for **D**. In an attempt to incorporate many-body effects, Snook, van Megen, and Tough [1983] have used effective screened pair diffusion tensors.

One final modification to the basic Brownian dynamics method may be made, in order to represent more faithfully the effect of the solvent. The bare interaction between molecules should be replaced by a solvent-averaged effective potential (sometimes loosely called a 'potential of mean force'). This makes sure that the effect of solvent structure is included in the simulation. The simplest possible form, which caters for pairwise effects, would be

$$v^{\text{eff}}(r) \approx -k_{\text{B}}T \ln g(r) \tag{9.36}$$

where $g(r)$ is an estimate of the pair distribution function for the two molecules in the fluid of interest (eqns (2.93) and (2.94)). The theoretical foundation for this modification lies in the work of Chandler and Pratt [1976]; the required distribution functions may be calculated theoretically, or could be obtained from a preliminary MD simulation incorporating the solvent molecules explicitly. In most cases, a simple functional form for the effective pair potential will not be available, so a tabulated potential (Section 5.2.3) will have to be used in the Brownian dynamics simulation. We illustrate in Fig. 9.1 the potential of mean force expected for a pair of atoms in a Lennard-Jones fluid close to the triple point. Also shown is the so-called DLVO potential [Verwey and Overbeek 1948], which plays the same role as an effective interaction between two charged species in a colloidal suspension. Of course, if the aim is not to mimic the motion of solute molecules in a solvent, but just their thermodynamic and structural properties, then solvent-averaged effective potentials can be used in conventional MD and MC simulations.

# 10
# QUANTUM SIMULATIONS

## 10.1 Introduction

In Section 2.9 we described the way in which first-order quantum corrections may be applied to classical simulations, once the results are available. The corrections to the thermodynamic properties arise from the Wigner–Kirkwood expansion of the phase-space distribution function [Green 1951; Oppenheim and Ross 1957] which may in turn be treated as an extra term in the Hamiltonian [Singer and Singer 1984]

$$\mathscr{H}^{\mathrm{qu}} = \mathscr{H}^{\mathrm{cl}} + \frac{\hbar^2\beta}{24m}\left[-\frac{\beta}{m}\left(\sum_i \mathbf{p}_i \cdot \nabla_{\mathbf{r}_i}\right)^2 \mathscr{V}^{\mathrm{cl}} + 3\sum_i \nabla^2_{\mathbf{r}_i}\mathscr{V}^{\mathrm{cl}} - \beta\sum_i(\nabla_{\mathbf{r}_i}\mathscr{V}^{\mathrm{cl}})^2\right] \tag{10.1}$$

Of course, as an alternative, the quantum potential can be obtained by integrating over the momenta

$$\mathscr{V}^{\mathrm{qu}} = \mathscr{V}^{\mathrm{cl}} + \frac{\hbar^2\beta}{24m}\left(2\sum_i \nabla^2_{\mathbf{r}_i}\mathscr{V}^{\mathrm{cl}} - \beta\sum_i(\nabla_{\mathbf{r}_i}\mathscr{V}^{\mathrm{cl}})^2\right). \tag{10.2}$$

This potential may be used in a conventional Monte Carlo simulation to generate quantum-corrected configurational properties. It is the treatment of this additional term by thermodynamic perturbation theory that gives rise to the quantum corrections mentioned in Section 2.9. A molecular dynamics simulation of a model of neon, based on the Hamiltonian of eqn (10.1), and using normal Hamiltonian mechanics, has been reported [Singer and Singer 1984]. Apart from measures to cope with numerical instabilities resulting from derivatives of the repulsive part of the potential, the technique is quite standard.

In this chapter, we consider techniques not based on an expansion in $\hbar^2$ such as that of eqn (10.1). We start with the non-normalized quantum-mechanical density operator

$$\rho = \exp(-\beta\mathscr{H}) = 1 - \beta\mathscr{H} + \frac{\beta^2}{2}\mathscr{H}\mathscr{H} + \dots \tag{10.3}$$

which satisfies the Bloch equation

$$\partial\rho/\partial\beta = -\mathscr{H}\rho\,. \tag{10.4}$$

In the coordinate representation of quantum mechanics, we may define the density matrix as

$$\begin{aligned}\rho(\mathbf{r},\mathbf{r}';\beta) &= \langle\mathbf{r}|\rho|\mathbf{r}'\rangle = \langle\mathbf{r}|\exp(-\beta\mathscr{H})|\mathbf{r}'\rangle \\ &= \sum_n \Psi_n^*(\mathbf{r})\rho\Psi_n(\mathbf{r}')\end{aligned} \tag{10.5}$$

where $n$ is a quantum state of the system. The time-dependent Schrödinger equation is

$$i\hbar\partial\Psi/\partial t = \mathscr{H}\Psi\,. \tag{10.6}$$

which means that the quantum-mechanical propagator from time 0 to time $t$ has the formal solution

$$U(t) = \exp(\mathscr{H}t/i\hbar)\,. \tag{10.7}$$

This propagator converts $\Psi(0)$ to $\Psi(t)$. Thus we see an analogy between the propagator of eqn (10.7) and the definition of $\rho$, eqn (10.3), or similarly between eqn (10.6) and the Bloch equation (10.4). This isomorphism is achieved by $\beta \leftrightarrow it/\hbar$. Some of the techniques discussed in this chapter use high-temperature or short-time approximations to the quantum-mechanical density matrix or propagator. Nonetheless, these techniques are often useful in low temperature simulations where the $\hbar^2$-expansions might fail.

By taking the trace of $\rho(\mathbf{r}, \mathbf{r}'; \beta)$, i.e. by setting $\mathbf{r} = \mathbf{r}'$ and then integrating over $\mathbf{r}$, we obtain the quantum partition function.

$$Q_{NVT}(\beta) = \int \mathrm{d}\mathbf{r}\,\rho(\mathbf{r}, \mathbf{r}; \beta) = \int \mathrm{d}\mathbf{r} \sum_n \Psi_n^*(\mathbf{r})\rho\Psi_n(\mathbf{r})\,. \tag{10.8}$$

As usual $\mathbf{r}$ represents the position of all the molecules. Using this relationship, thermodynamic, static structural properties, and dynamic properties may, in some circumstances, be estimated by the temperature–time analogy already discussed.

As an alternative to the simple expansion in powers of $\hbar^2$ the path integral formulation of quantum mechanics [Feynman and Hibbs 1965] has been a fruitful source of theoretical approaches and simulation techniques. In Section 10.2 we describe a simulation algorithm which has arisen directly out of this formalism, and in Section 10.3 we turn to an attempt to solve the time-dependent Schrödinger equation approximately, using Gaussian wavepackets. All these approaches are semiclassical finite-temperature techniques, which may provide a measure of improvement over quantum-corrected classical results (e.g. liquid neon), but which are not applicable in the strongly quantum-mechanical low-temperature limit (e.g. liquid helium). For the latter case, special techniques have been developed, and as an example we consider a random walk estimation of the ground state in Section 10.4.

This chapter is not an extensive review of all the quantum-mechanical simulation techniques that have been employed. For example, Green's function Monte Carlo is beyond the scope of this book. Excellent reviews of the simulation of liquids of bosons and fermions are available [Ceperley and Kalos 1979; Schmidt and Kalos 1984; Kalos 1984].

## 10.2 Semiclassical path-integral simulations

One of the most straightforward of the semiclassical simulation techniques is that based on a discretization of the path-integral form of the density matrix [Feynman and Hibbs 1965], because the method essentially reduces to performing a classical simulation. Since the early simulation work [Fosdick and Jordan 1966; Jordan and Fosdick 1968] and the work of Barker [1979], the technique has become popular, in part because the full implications of the quantum-classical isomorphism have become clear [Chandler and Wolynes 1981; Schweizer, Stratt, Chandler, and Wolynes 1981].

The development proceeds as follows, considering initially a single molecule. Starting with eqn (10.5) and eqn (10.8) we divide the exponential into $P$ equal parts

$$Q_{1VT}(\beta) = \int \mathrm{d}\mathbf{r}_1 \, \langle \mathbf{r}_1 | \mathrm{e}^{-\beta\mathscr{H}/P} \ldots \mathrm{e}^{-\beta\mathscr{H}/P} \ldots \mathrm{e}^{-\beta\mathscr{H}/P} | \mathbf{r}_1 \rangle \tag{10.9}$$

and inserting unity in the form

$$1 = \int \mathrm{d}\mathbf{r} \, |\mathbf{r}\rangle \, \langle \mathbf{r}| \tag{10.10}$$

between each exponential gives

$$\begin{aligned} Q_{1VT}(\beta) &= \int \mathrm{d}\mathbf{r}_1 \mathrm{d}\mathbf{r}_2 \ldots \mathrm{d}\mathbf{r}_P \langle \mathbf{r}_1 | \mathrm{e}^{-\beta\mathscr{H}/P} | \mathbf{r}_2 \rangle \langle \mathbf{r}_2 | \mathrm{e}^{-\beta\mathscr{H}/P} | \mathbf{r}_3 \rangle \\ &\quad \ldots \langle \mathbf{r}_{P-1} | \mathrm{e}^{-\beta\mathscr{H}/P} | \mathbf{r}_P \rangle \langle \mathbf{r}_P | \mathrm{e}^{-\beta\mathscr{H}/P} | \mathbf{r}_1 \rangle . \qquad (10.11a) \\ &= \int \mathrm{d}\mathbf{r}_1 \mathrm{d}\mathbf{r}_2 \mathrm{d}\mathbf{r}_P \rho(\mathbf{r}_1, \mathbf{r}_2; \beta/P) \rho(\mathbf{r}_2, \mathbf{r}_3; \beta/P) \\ &\quad \ldots \rho(\mathbf{r}_P, \mathbf{r}_1; \beta/P) \qquad (10.11b) \end{aligned}$$

We seem to have complicated the problem; instead of one integral over diagonal elements of $\rho$, we now have many integrals involving offdiagonal elements. However, each term $\rho(\mathbf{r}_a, \mathbf{r}_b;\ \beta/P)$ involves, effectively, a higher temperature (or a weaker Hamiltonian) than the original. At sufficiently large values of $P$, the following approximation becomes applicable:

$$\rho(\mathbf{r}_a, \mathbf{r}_b; \beta/P) \approx \rho_{\text{free}}(\mathbf{r}_a, \mathbf{r}_b; \beta/P) \exp\left(-(\beta/2P)\left[\mathscr{V}^{\text{cl}}(\mathbf{r}_a) + \mathscr{V}^{\text{cl}}(\mathbf{r}_b)\right]\right) \tag{10.12}$$

where $\mathscr{V}^{\text{cl}}(\mathbf{r}_a)$ is the classical potential energy as a function of the configurational coordinates, and where the free-particle density matrix is known exactly. For a single molecule it is

$$\rho_{\text{free}}(\mathbf{r}_a, \mathbf{r}_b; \beta/P) = \left(\frac{Pm}{2\pi\beta\hbar^2}\right)^{3/2} \exp\left\{-\frac{Pm}{2\beta\hbar^2} r_{ab}^2\right\}. \tag{10.13}$$

Now the expression for $Q$ is

$$Q_{1VT}(\beta) \approx \left(\frac{Pm}{2\pi\beta\hbar^2}\right)^{3P/2} \int d\mathbf{r}_1 \ldots d\mathbf{r}_P$$

$$\exp\left\{-\frac{Pm}{2\beta\hbar^2}(r_{12}^2 + r_{13}^2 + \ldots + r_{P1}^2)\right\} \exp\left\{-\frac{\beta}{P}(\mathcal{V}^{\text{cl}}(\mathbf{r}_1)\right.$$

$$\left. + \mathcal{V}^{\text{cl}}(\mathbf{r}_2) + \ldots + \mathcal{V}^{\text{cl}}(\mathbf{r}_P))\right\}. \tag{10.14}$$

The above formulae are almost unchanged when we generalize to a many molecule system. For $N$ atoms,

$$Q_{NVT}(\beta) \approx \frac{1}{N!}\left(\frac{Pm}{2\pi\beta\hbar^2}\right)^{3NP/2} \int d\mathbf{r}_1 \ldots d\mathbf{r}_P$$

$$\exp\left\{-\frac{Pm}{2\beta\hbar^2}(r_{12}^2 + r_{23}^2 + \ldots + r_{P1}^2)\right. \exp\left\{-\frac{\beta}{P}(\mathcal{V}^{\text{cl}}(\mathbf{r}_1)\right.$$

$$\left. + \mathcal{V}^{\text{cl}}(\mathbf{r}_2) + \ldots + \mathcal{V}^{\text{cl}}(\mathbf{r}_P))\right\}. \tag{10.15}$$

We must consider carefully what eqn (10.15) represents. Each vector $\mathbf{r}_a$ represents a complete set of $3N$ coordinates, defining a system like our $N$-atom quantum system of interest. The function $\mathcal{V}(\mathbf{r}_a)$ is the potential energy function for each one of these systems, calculated in the usual way. Imagine a total of $P$ such systems, which are more or less superimposed on each other. Each atom in system $a$ is quite close to (but not exactly on top of) the corresponding atom in systems $b$, $c$, ... etc. Each contributes a term $\mathcal{V}^{\text{cl}}(\mathbf{r}_a)$ to the Boltzmann factors in eqn (10.15), but the total is divided by $P$ to obtain, in a sense, an averaged potential. The systems interact with each other through the first exponential term in the integrand of eqn (10.15). Each vector $\mathbf{r}_{ab}$ ($\mathbf{r}_{12}$, $\mathbf{r}_{23}$ etc.) represents the complete set of $N$ separations between corresponding atoms of the two separate systems $a$ and $b$. Specifically the squared terms appearing in eqn (10.15) are

$$r_{ab}^2 = \sum_{i=1}^{N} |\mathbf{r}_{ia} - \mathbf{r}_{ib}|^2 \tag{10.16}$$

where $\mathbf{r}_{ia}$ is the position of atom $i$ in system $a$. These interactions are of a harmonic form, i.e. the systems are coupled by springs.

There is an alternative and very fruitful way of picturing our system of $NP$ atoms [Chandler and Wolynes 1981]. It can be regarded as set of $N$ molecules, each consisting of $P$ atoms which are joined together to form a classical ring polymer. This is illustrated in Fig. 10.1. We write the integral in eqn (10.15) in the form of a classical configurational integral

$$Z_{NVT} = \int \exp(-\beta\mathcal{V}(\mathbf{r}))\, d\mathbf{r}_{11} \ldots d\mathbf{r}_{ia} \ldots d\mathbf{r}_{NP} \tag{10.17}$$

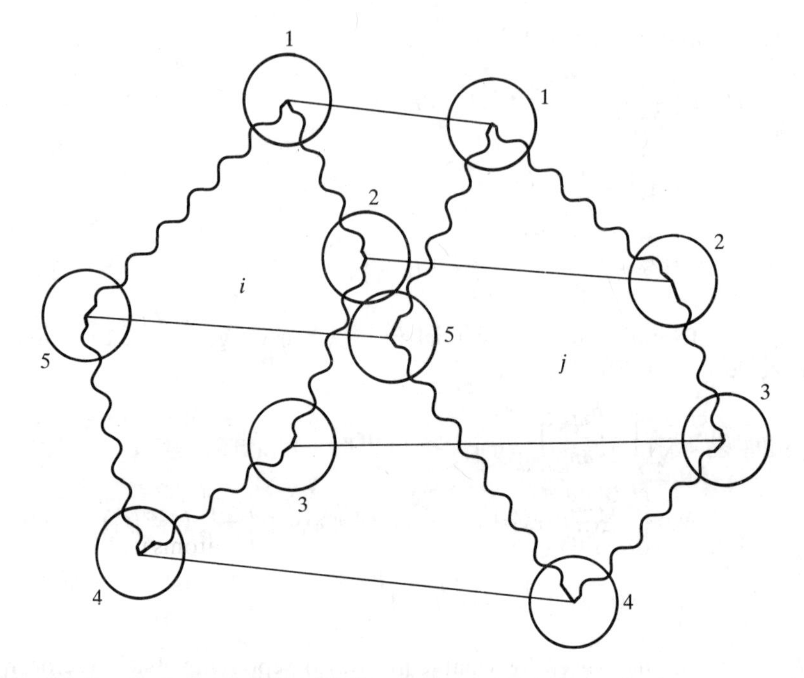


**Fig. 10.1** Two ring-polymer 'molecules' ($P = 5$) representing the interaction between two atoms in a path-integral simulation. The straight lines are the intermolecular potential interactions, the wavy lines represent the intramolecular spring potentials.

with the configurational energy consisting of two parts

$$\mathscr{V}(\mathbf{r}) = \mathscr{V}^{\mathrm{cl}}(\mathbf{r}) + \mathscr{V}^{\mathrm{qu}}(\mathbf{r}). \tag{10.18}$$

The classical part is

$$\begin{aligned}\mathscr{V}^{\mathrm{cl}} &= \frac{1}{P}(\mathscr{V}^{\mathrm{cl}}(\mathbf{r}_1) + \mathscr{V}^{\mathrm{cl}}(\mathbf{r}_2) + \ldots \mathscr{V}^{\mathrm{cl}}(\mathbf{r}_P)) \\ &= \frac{1}{P}\sum_{a=1}^{P}\sum_{i<j}^{N} v^{\mathrm{cl}}(\mathbf{r}_{ia} - \mathbf{r}_{ja}) \\ &= \frac{1}{P}\sum_{a=1}^{P}\sum_{i<j}^{N} v^{\mathrm{cl}}(r_{iaja}).\end{aligned} \tag{10.19}$$

We have assumed pairwise additivity here and in Fig. 10.1 for simplicity, although this is not essential. The quantum part of the potential is

$$\begin{aligned}\mathscr{V}^{\mathrm{qu}} &= \left(\frac{Pm}{2\beta^2\hbar^2}\right)(|\mathbf{r}_{12}|^2+|\mathbf{r}_{23}|^2+\ \ldots\ +|\mathbf{r}_{P1}|^2)\\ &= \left(\frac{Pm}{2\beta^2\hbar^2}\right)\sum_{a=1}^{P}\sum_{i=1}^{N}|\mathbf{r}_{ia}-\mathbf{r}_{ia+1}|^2\\ &= \sum_i\sum_a v^{qu}(r_{iaia+1})\end{aligned} \tag{10.20}$$

where we take $a+1$ to equal 1 when $a = P$. Note how the interactions between molecules only involve correspondingly numbered atoms $a$ (i.e. atom 1 on molecule $i$ only sees atom 1 on molecule $j$, $k$, etc.), while the interactions within molecules just involve atoms with adjacent labels. The system is formally a set of polymer molecules, but an unusual one: the molecules cannot become entangled, because of the form of eqn (10.19), and the equilibrium atom–atom bond lengths in each molecule, according to eqn (10.20), are zero.

The term outside the integral of eqn (10.15) may be regarded as the kinetic contribution to the partition function, if the mass of the atoms in our system is chosen appropriately. Actually, this choice is not critical, since it is the configurational averaging which is the key problem to solve. Nonetheless it proves convenient (see below) to use an MD-based simulation method, and De Raedt, Sprik, and Klein [1984] recommend making each atom of mass $Pm = M$. Then the kinetic energy of the system becomes

$$\mathscr{K} = \frac{1}{2}\sum_{ia}(Pm)|\mathbf{v}_{ia}|^2 = \frac{1}{2}\sum_{ia}|\mathbf{p}_{ia}|^2/(Pm) = \frac{1}{2}\sum_{ia}|\mathbf{p}_{ia}|^2/M\,. \tag{10.21}$$

and the integration over momenta, in the usual quasi-classical way, yields

$$Q_{NVT}(\beta) \approx \frac{1}{(NP)!}\left(\frac{M}{2\pi\beta\hbar^2}\right)^{3NP/2}\int d\mathbf{r}\exp\{-\beta(\mathscr{V}^{\mathrm{cl}}+\mathscr{V}^{\mathrm{qu}})\}\,. \tag{10.22}$$

Apart from the indistinguishability factors, which may be ignored as far as the calculation of averages is concerned, this is the approximate quantum partition function eqn (10.15) for our $N$-particle system. (Note that, although we are free to choose $M$, the $m$ appearing in $\mathscr{V}^{\mathrm{qu}}$ eqn (10.20) is not adjustable.)

Thus a Monte Carlo simulation of the classical ring polymer system with potential energy $\mathscr{V}$ given by eqn (10.18), or a molecular dynamics simulation with Hamiltonian $\mathscr{K}+\mathscr{V}$ specified by eqns (10.18) and (10.21), may be used to generate averages in an ensemble whose configurational distribution function approximates that of a quantum system. The simulation of isolated quantum atoms in a classical solvent bath is straightforward: the classical atoms behave like polymers contracted to a point.

As the number of particles $P$ in our ring polymer grows we obtain a better approximation to the quantum partition function; these equations become formally exact as $P \to \infty$, going over to the Feynman path-integral representation [Feynman and Hibbs 1965]. How well do we expect to do at finite $P$? Some idea of what we may expect can be obtained by studying the quantum harmonic oscillator for which exact solutions are available in the classical $P = 1$ case, in the quantum-mechanical $P \to \infty$ limit, and for all intermediate $P$ [Schweizer *et al.* 1981]. The computed average energy is plotted in Fig. 10.2. It can be seen that the finite-$P$ energy curves deviate from the true result as the temperature decreases, leaving the zero-point level ($\frac{1}{2}\hbar\omega$) and dropping to the classical value at $T = 0$, $\langle E \rangle = 0$. The region of agreement may be extended to lower temperatures by increasing $P$.

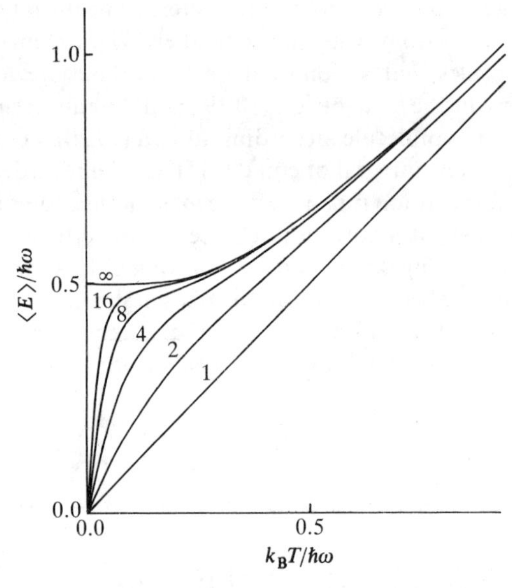


**Fig. 10.2** The average energy of the path-integral approximation to the quantum harmonic oscillator of frequency $\omega$ as a function of temperature. We give the results for various values of $P$, the number of 'atoms' in the ring polymer. $P = 1$ is the classical result, and $P \to \infty$ is the quantum mechanical limit.

Now to the practical matters. The classical polymer model is easily simulated by standard techniques, such as constant-$NVT$ MC or MD. There are, however, some technical problems to be expected as $P$ increases. According to eqn (10.20) the internal spring constant increases with $P$ as well as with temperature, while the external forces felt by each individual atom decrease as $P$ increases. In an MC simulation this might mean that separate attention would have to be given to moves which altered intramolecular distances and those involving the molecule as a whole. In some cases, a normal mode analysis of the polymer may help in choosing MC step sizes [Herman, Bruskin, and Berne 1982]. More directly, one can abandon intramolecular Metropolis moves and build the polymer from scratch, by sampling from the internal, free-molecule, distribution [Jacucci and Omerti 1983]. An important development is to construct the molecule in stages [Sprik, Klein, and Chandler 1985]. In MD simulations there is the corresponding danger that the time-scale separation of the internal and external motions may become well separated at high $P$ and this will necessitate a shorter time step. A sensible choice of the dynamic mass $M$ helps to overcome this problem: with the choice of eqn (10.21) the stiffest internal modes of the polymer will be characterized by a frequency $k_B T/\hbar$; alternative choices of $M$ may be made so as to match the internal frequencies with the external time scales [De Raedt *et al.* 1984]. Nonetheless, the danger of slow energy exchange between internal and external modes, leading to non-ergodic behaviour, is a serious concern. The 'randomness' of the simulation may be increased by adopting one of the stochastic constant-$NVT$ MD algorithms described in Section 7.4, and this is the recommended method.

There are some subtleties in the way a semiclassical path integral simulation is used to estimate ensemble averages. The most obvious concerns the energy. This is obtained in the usual way by forming $-Q_{NVT}^{-1}(\partial Q_{NVT}/\partial\beta)$; however, the 'quantum spring' potential is temperature-dependent, and the result is

$$\begin{aligned}\langle E\rangle &= \langle \mathcal{V}^{\mathrm{cl}}\rangle + \tfrac{3}{2}NPk_B T - \langle \mathcal{V}^{\mathrm{qu}}\rangle \\ &= \langle \mathcal{V}^{\mathrm{cl}}\rangle + \langle \mathcal{K}\rangle - \langle \mathcal{V}^{\mathrm{qu}}\rangle .\end{aligned} \tag{10.23}$$

Note the sign of the $\langle \mathcal{V}^{\mathrm{qu}}\rangle$ term. This might cause some confusion in an MD simulation, where the total energy $\mathcal{V}^{\mathrm{cl}} + \mathcal{V}^{\mathrm{qu}} + \mathcal{K}$ is the conserved variable (between stochastic collisions if these are applied). This quantity is not the desired estimator for the quantum energy. There is yet a further wrinkle. The 'quantum kinetic energy' part of eqn (10.23) is the difference between two quantities, $\tfrac{3}{2}NPk_B T$ and $\langle \mathcal{V}^{\mathrm{qu}}\rangle$, both of which increase with $P$. This gives rise to loss of statistical precision: in fact the relative variance in this quantity increases linearly with $P$, making the estimate worse as the simulation becomes more accurate [Herman *et al.* 1982]. The solution to this is to use the virial theorem for the harmonic potentials to replace eqn (10.23) by the following

$$\begin{aligned}\langle E\rangle &= \langle \mathcal{V}^{\mathrm{cl}}\rangle + \frac{3}{2}Nk_B T + \frac{1}{2P}\langle \mathbf{r}\cdot\nabla_{\mathbf{r}}\mathcal{V}^{\mathrm{cl}}\rangle \\ &= \langle \frac{1}{P}\sum_a\sum_{i<j} v^{\mathrm{cl}}(r_{iaja})\rangle + \frac{3}{2}Nk_B T + \frac{1}{2}\langle \frac{1}{P}\sum_a\sum_{i<j} \mathbf{r}_{iaja}\cdot\nabla_{\mathbf{r}_{iaja}} v^{\mathrm{cl}}(r_{iaja})\rangle .\end{aligned} \tag{10.24}$$

Actually, it is not clear that this will give a significant improvement. Jacucci [1984] has pointed out that the statistical inefficiency (see Section 6.4.1) may be worse for the quantity in eqn (10.24) than for that in eqn (10.23), due to the persistence of correlations, thus wiping out the advantage. Other configurational properties are estimated in a straightforward way. For example, the pressure may be obtained from the volume derivative of $Q$. The pair distribution function, which is a partially integrated form of $\rho(\mathbf{r}, \mathbf{r}; \beta)$, becomes essentially a site–site $g(r)$ for atoms with the same atom label [Chandler and Wolynes 1981]. It is accumulated in the normal way. The 'size' of each quantum molecule is also of interest [De Raedt *et al.* 1984; Parrinello and Rahman 1984]. This may be measured by the radius of gyration $R_i$

$$R_i^2 = \frac{1}{P} \sum_{a=1}^{P} |\mathbf{r}_{ia} - \mathbf{r}_i|^2 \tag{10.25}$$

where $\mathbf{r}_i = (1/P) \sum_{a=1}^{P} \mathbf{r}_{ia}$ is the centre of mass.

Time-dependent properties are a different matter. Even if the Hamiltonian $\mathscr{V}^{\text{cl}} + \mathscr{V}^{\text{qu}} + \mathscr{K}$ is used in an MD simulation, the resulting dynamics do not have any physical meaning. An indirect approach, based on the time–temperature analogy, has been advanced [Thirumalai and Berne, 1983, 1984]. This involves measuring the internal 'spatial' correlation functions of the polymer chain.

The extension of the path integral method to molecular systems is possible and desirable when, in a case such as water, translational motion may be regarded as classical while rotation is quantum-mechanical [Kuharski and Rossky 1984]. There are additional complications in the case of asymmetric tops.

The path integral method described in this section is often termed the 'primitive algorithm'; it uses the most crude approximation to the density matrix. Other improved approximations have been advocated [Schweizer *et al.* 1981] with the aim of reducing the number of polymer units required and possibly improving the convergence. At least one straightforward improvement, namely

$$\rho(\mathbf{r}_a, \mathbf{r}_b; \beta/P) \approx$$
$$\rho_{\text{free}}(\mathbf{r}_a, \mathbf{r}_b; \beta/P) \exp\left\{ -\frac{\beta}{P}\left(\frac{1}{2}(\mathscr{V}(\mathbf{r}_a) + \mathscr{V}(\mathbf{r}_b)) + \frac{\hbar^2}{24m} \nabla_{\mathbf{r}}^2 \mathscr{V}(\mathbf{r}_a)\right)\right\} \tag{10.26}$$

has been tried using the MC method [Thirumalai and Berne 1983].

Stratt and Miller [1977] have suggested an interesting extension of the path integral techniques discussed in this section. They write the quantum density matrix, semiclassically, in terms of an action integral $\mathscr{S}$

$$\mathscr{S} = \int_0^t \mathrm{d}t' \, \mathscr{L}(t') \tag{10.27}$$

where $\mathscr{L}(t) = \mathscr{L}(\mathbf{r}(t), \mathbf{p}(t))$ is the Lagrangian, but, in this case, in imaginary time. Adopting a short-time (high-temperature) approximation for the action would lead to the type of expressions seen earlier in this section. Rather than do this, Stratt and Miller suggest estimating the density matrix by following the classical path along a trajectory for an imaginary time $\hbar\beta/2i$. The imaginary-time equations of motion are of Hamiltonian form, but involve an imaginary momentum, and a potential that is the negative of the real one. These equations are solved using standard MD techniques. The average of the Hamiltonian along the trajectory can be used, for example, in the Boltzmann weighting factor of a Monte Carlo simulation.

## 10.3 Semiclassical Gaussian wavepackets

Interest in dynamic properties has lead to attempts to solve the time-dependent Schrödinger equation, eqn (10.6), for many-particle systems, by computer simulation. A direct, full, treatment of this problem is well beyond current capabilities, but useful approximate treatments exist. Using the work of Heller [1975, 1976] as a basis, Singer and co-workers [Corbin and Singer 1982; Singer and Smith 1986] have suggested MD methods using Gaussian wavepackets. The total wavefunction is taken to be a simple product, (i.e. one without the correct quantum symmetry)

$$\Psi(\mathbf{r}, t) = \prod_{i=1}^{N} \psi_i(\mathbf{r}_i, t) \tag{10.28}$$

where each single-particle wavefunction takes the form of a generalized complex Gaussian

$$\psi_i(\mathbf{r}_i, t) = \exp\left\{\frac{i}{\hbar}\left(a_i(t)|\mathbf{r}_i - \mathbf{r}_i(t)|^2 + \mathbf{p}_i(t)\cdot(\mathbf{r}_i - \mathbf{r}_i(t)) + b_i(t)\right)\right\} \tag{10.29}$$

specified by the parameters $\mathbf{r}_i(t)$, $\mathbf{p}_i(t)$, $a_i(t)$, and $b_i(t)$. We distinguish $\mathbf{r}_i(t)$ from the argument $\mathbf{r}_i$ of the wavefunction, by retaining the explicit time dependence throughout. $\mathbf{r}_i(t)$ is a real vector representing the centre of the wavepacket. It satisfies the equation

$$\mathbf{r}_i(t) = \langle \psi_i(\mathbf{r}_i, t)|\mathbf{r}_i|\psi_i(\mathbf{r}_i, t)\rangle \tag{10.30}$$

where the brackets represent the quantum mechanical expectation value, (i.e. integrating over all $\mathbf{r}_i$). In the same way $\mathbf{p}_i$ is a real vector which measures the expectation value of the momentum

$$\begin{aligned}\mathbf{p}_i(t) &= \langle \psi_i(\mathbf{r}_i, t)|\mathbf{p}_i|\psi_i(\mathbf{r}_i, t)\rangle \\ &= \langle \psi_i(\mathbf{r}_i, t)|\frac{\hbar}{i}\nabla_{\mathbf{r}_i}|\psi_i(\mathbf{r}_i, t)\rangle .\end{aligned} \tag{10.31}$$

The quantities $a_i(t)$ and $b_i(t)$ are complex. $a_i(t)$ essentially determines the 'width' of the wavepacket. As defined by eqn (10.29), $\psi_i(\mathbf{r}_i, t)$ is spherically symmetrical; non-spherical wavepackets may be treated by making $a_i(t)$ a matrix, but this is not thought to be advantageous [Singer and Smith, 1986]. The phase and normalization of the wavepacket are determined by $b_i(t)$. If real and imaginary parts are defined by

$$a_i(t) = a_i^r(t) + i a_i^i(t) \tag{10.32a}$$

$$b_i(t) = b_i^r(t) + i b_i^i(t) \tag{10.32b}$$

then the imaginary components are related by

$$\langle \psi_i(\mathbf{r}_i, t) | \psi_i(\mathbf{r}_i, t) \rangle = 1 = \left( \frac{\hbar \pi}{2 a_i^i(t)} \right)^{3/2} \exp(-2 b_i^i(t)/\hbar). \tag{10.33}$$

The time evolution of the system, i.e. that of the parameters $\mathbf{r}_i(t)$, $\mathbf{p}_i(t)$, $a_i(t)$, $b_i(t)$, is determined by the time-dependent Schrödinger equation, and by the ansatz that the wavepackets retain the Gaussian form at all times. The Gaussian form is indeed preserved for a wavepacket in an isotropic harmonic potential, and Heller's first suggestion [Heller 1975] is to make a locally quadratic approximation in the region of each wavepacket. This leads to classical motion of the wavepacket centre, i.e. $\mathbf{r}_i(t)$ and $\mathbf{p}_i(t)$, with the parameters $a_i(t)$, $b_i(t)$ obeying a subsidiary set of differential equations. The method seems to fail for liquids where anharmonic terms in the potential are important [Corbin and Singer 1982]. The second suggestion of Heller [1976] is that the parameters defining the wavepacket evolve so as to minimize the difference between the left and right sides of the time-dependent Schrödinger equation, eqn (10.6), in a least-squares sense. This is an analogous procedure to that used to modify the classical equations of motion, via Gauss's principle, as described in Section 7.4.3.

The development proceeds as follows [Corbin and Singer 1982; Singer and Smith 1986]. Assuming that the potential takes a pairwise-additive form, we may define an effective one-body potential for each molecule at each instant

$$v_i(\mathbf{r}_i, t) = \sum_j \langle \psi_j(\mathbf{r}_j, t) | v(\mathbf{r}_i - \mathbf{r}_j) | \psi_j(\mathbf{r}_j, t) \rangle . \tag{10.34}$$

This quantity is pre-averaged over the neighbouring wavepackets. Because of the simple form of eqn (10.28), each wavepacket should obey the equation of motion

$$\begin{aligned} i\hbar \frac{\partial}{\partial t} \psi_i(\mathbf{r}_i, t) &= -\frac{\hbar^2}{2m} \nabla^2_{\mathbf{r}_i} \psi_i(\mathbf{r}_i, t) + v_i(\mathbf{r}_i, t) \psi_i(\mathbf{r}_i, t) \\ &= \mathscr{H}_i \psi_i(\mathbf{r}_i, t) . \end{aligned} \tag{10.35}$$

The variational principle which has to be satisfied is that [McLachlan 1964]

$$\mathscr{I}(\Psi, \dot{\Psi}) = \int \mathrm{d}\mathbf{r} \, | i\hbar \dot{\Psi} - \mathscr{H} \Psi |^2 \tag{10.36}$$

shall be minimized with respect to the variation in $\dot{\Psi}$, i.e. with respect to the time derivatives of the parameters in each $\psi_i$. With eqn (10.28) we obtain for each molecule $i$, the following expression involving $a_i(t)$

$$\int \mathrm{d}\mathbf{r}_i \left( \frac{\partial \psi_i}{\partial a_i(t)} \right)^* \left( i\hbar \frac{\partial \psi_i}{\partial t} - \mathscr{H}_i \psi_i \right) = 0 \tag{10.37}$$

and similar equations for the other parameters $b_i(t)$, $\mathbf{r}_i(t)$, $\mathbf{p}_i(t)$. From eqn (10.29) it can be seen that the parameter derivatives are proportional to $\psi_i^*(\mathbf{r}_i, t)$, or to $\psi_i^*(\mathbf{r}_i, t)(\mathbf{r}_i - \mathbf{r}_i(t))$, or to $\psi_i^*(\mathbf{r}_i, t)|\mathbf{r}_i - \mathbf{r}_i(t)|^2$ so eqn (10.37) and its analogues lead directly to

$$\int d\mathbf{r}_i \left( i\hbar \frac{\partial \psi_i}{\partial t} - \mathcal{H}_i \psi_i \right) \psi_i^* = 0 \tag{10.38a}$$

$$\int d\mathbf{r}_i \left( i\hbar \frac{\partial \psi_i}{\partial t} - \mathcal{H}_i \psi_i \right) (\mathbf{r}_i - \mathbf{r}_i(t)) \psi_i^* = 0 \tag{10.38b}$$

$$\int d\mathbf{r}_i \left( i\hbar \frac{\partial \psi_i}{\partial t} - \mathcal{H}_i \psi_i \right) |\mathbf{r}_i - \mathbf{r}_i(t)|^2 \psi_i^* = 0 . \tag{10.38c}$$

Knowing the forms of all the quantities appearing in these equations, it is easy to obtain equations of motion for the parameters. After some manipulation these become

$$\dot{\mathbf{r}}_i(t) = \mathbf{p}_i(t)/m \tag{10.39a}$$

$$\dot{\mathbf{p}}_i(t) = -\langle \psi_i(\mathbf{r}_i, t) | \nabla_{\mathbf{r}_i} v_i(\mathbf{r}_i, t) | \psi_i(\mathbf{r}_i, t) \rangle \tag{10.39b}$$

$$\dot{a}_i(t) = -2a_i(t)^2/m - \tfrac{1}{6} \langle \psi_i(\mathbf{r}_i, t) | \nabla^2_{\mathbf{r}_i} v_i(\mathbf{r}_i, t) | \psi_i(\mathbf{r}_i, t) \rangle \tag{10.39c}$$

$$\dot{b}_i(t) = 3i\hbar a_i(t)/m + |\mathbf{p}_i(t)|^2/2m - \langle \psi(\mathbf{r}_i, t) | v_i(\mathbf{r}_i, t) | \psi(\mathbf{r}_i, t) \rangle$$
$$+ \left( \frac{\hbar}{8a_i^{\text{i}}(t)} \right) \langle \psi_i(\mathbf{r}_i, t) | \nabla^2_{\mathbf{r}_i} v_i(\mathbf{r}_i, t) | \psi_i(\mathbf{r}_i, t) \rangle . \tag{10.39d}$$

These equations have a simple form. They involve the interaction potential, its gradient and its Laplacian, averaged over the wavepacket centred on $\mathbf{r}_i$ (having already been averaged over all the neighbouring wavepackets). Heller's locally quadratic approximation evaluates these terms as if the wavepacket around $i$ had no extent; thus $\langle \nabla v_i \rangle$ simply becomes $(\nabla v_i)_{\mathbf{r}_i = \mathbf{r}_i(t)}$ and the first two equations generate classical motion. The last term in eqn (10.39d), which is actually proportional to $\langle |\mathbf{r}_i - \mathbf{r}_i(t)|^2 \rangle$, also vanishes in this limit. In practice, it seems that the full eqns (10.39) are preferable. The solution of these equations in an MD simulation is almost a standard exercise. To avoid instabilities in eqns (10.39c,d), $a_i$ and $b_i$ may be replaced by two new parameters [Corbin and Singer 1982; Reimers, Wilson, and Heller 1983], $z_i$ and $p_{z_i}$. These are defined by

$$a_i(t) = p_{z_i}(t)/2z_i(t) \tag{10.40a}$$

and

$$\dot{z}_i(t) = p_{z_i}(t)/m \tag{10.40b}$$

and the first-order differential equations solved by, for example, a Gear method (Appendix E). An equation for $b_i(t)$ could also be obtained in terms of these parameters. In practice, the phase information that appears in the real part of $b_i(t)$ is not required, while the imaginary part may be determined by eqn (10.33), so $b_i(t)$ may be dropped from the algorithm altogether. Because of the form of the integrals $\langle \psi_i | v_i | \psi_i \rangle$ etc., and the use of Gaussian wavepackets, it is necessary to use an approximate form of the pair potential that avoids a divergence at $\mathbf{r}_i = \mathbf{r}_i(t)$. The Lennard-Jones potential can be approximated quite well by a sum of two Gaussian functions [Corbin and Singer 1982], with the additional advantage that the Gaussian integrals have simple functional forms.

## 10.4 Quantum random walk simulations

The semiclassical methods discussed in the previous sections are suitable for the simulation of liquid neon, where the quantum effects are significant but not dominant. For liquids whose behaviour is essentially quantum mechanical, such as liquid helium, a number of MC methods have been developed, and in this section we discuss solving the many-body Schrödinger equation by generating a random walk in imaginary time.

The adoption of an imaginary time evolution converts the Schrödinger equation into one of a diffusional kind.

$$-\frac{\partial \Psi(\mathbf{r}, s)}{\partial s} = (-D\nabla_{\mathbf{r}}^2 + \mathscr{V}(\mathbf{r}) - E_T)\Psi(\mathbf{r}, s) \tag{10.41}$$

where $s = it/\hbar$, $\mathscr{V}$ is the potential, and $E_T$ is an arbitrary zero of the energy which is useful in this problem. The 'diffusion coefficient' is defined to be

$$D = \hbar^2/2m. \tag{10.42}$$

The simulation of this equation to solve the quantum many-body problem is a very old idea, possibly dating back to Fermi [Metropolis and Ulam 1949], but it is the modern implementation of Anderson [1975, 1976] that interests us here.

The diffusive part of eqn (10.41) can clearly be simulated by a 'position-Langevin' equation approach, as discussed in Section 9.3; if we interpret $\Psi(\mathbf{r}, s)$ (note: not $|\Psi|^2$!) as a probability density, it is essentially the Schmoluchowski equation, eqn (9.27) without the systematic forces. The additional complication is that the $(\mathscr{V}(\mathbf{r}) - E_T)\Psi$ term acts as a birth and death process (or a chemical reaction) which changes the weighting (probability) of configurations with time. To incorporate this in a Langevin simulation means allowing the creation and destruction of whole systems of molecules. Simulations of many individual systems are run in parallel with one another. Although this sounds cumbersome, in practice it is a feasible route to the properties of the quantum liquid. That such a simulation may yield a ground-state stationary solution of the Schrödinger equation may be seen by the following argument. Any time-dependent wavefunction can be expanded as a set of stationary states, $\Psi_n$, when the time evolution becomes

$$\Psi(\mathbf{r}, t) = \sum_n c_n \exp\left( -\frac{i}{\hbar} t(E_n - E_T) \right) \Psi_n(\mathbf{r})$$

$$\Psi(\mathbf{r}, s) = \sum_n c_n \exp(-s(E_n - E_T)) \Psi_n(\mathbf{r}) \quad (10.43)$$

where the $c_n$ are the initial condition coefficients. In the imaginary time formalism, the state with the lowest energy $E_0$ becomes the dominant term at long times. If we have chosen $E_T < E_0$, then the ground-state exponential decays the least rapidly with time, while if $E_T > E_0$ the ground-state function grows faster than any other. If we are lucky enough to choose $E_T = E_0$, then the function $\Psi(\mathbf{r}, s)$ tends to $\Psi_0(\mathbf{r})$ at long times while the other state contributions decay away. For $\Psi(\mathbf{r}, s)$ to be properly treated as a probability density, it must be everywhere positive (or negative) and this will be true for the ground state of a liquid of bosons.

The reaction part of the 'reaction–diffusion' equation is treated as usual, by integrating over a short time step $\delta s$. Formally

$$\Psi(s + \delta s) = \Psi(s) \exp(-(\mathscr{V}(\mathbf{r}) - E_T)\delta s). \quad (10.44)$$

This enters into the simplest quantum Monte Carlo algorithm as follows.

Begin with a large number (100–1000) of replicas of the $N$-body system of interest.

(a) Perform a Brownian dynamics step for the Schmoluchowski equation, with $D$ given by eqn (10.42), on each system. (Note that the temperature does not enter into the random walk algorithm since there are no systematic forces.)

(b) For each system, evaluate $\mathscr{V}(\mathbf{r})$, compute $\exp(-(\mathscr{V}(\mathbf{r}) - E_T)) = K$, and replace the system by $K$ identical copies of itself (see below).

(c) Return to step (a).

Step (b) requires a little more explanation, since in general $K$ will be a real number between 0 and $\infty$. If $K > 1$, say equal to $\mathrm{int}(K) + K'$ then the system is replaced by $\mathrm{int}(K)$ replicas of itself, and a further copy is added with a probability $K'$ (using a random number generated uniformly on the range $(0, 1)$). If $K < 1$, then the current system is deleted from the simulation with probability $(1 - K)$ (again using a random number).

The above scheme is fairly crude. Clearly, depending on $E_T$, the number of systems still under consideration may grow or fall dramatically, and the value of this parameter is continually adjusted during the simulation to keep the current number approximately constant [Anderson 1975]. Hopefully in the course of a run conducted in this way $E_T \rightarrow E_0$. The fluctuation in the number of systems is substantial and this makes the estimate of $E_T$ subject to a large statistical error. A number of ways around this difficulty have been proposed

[Anderson 1980; Mentch and Anderson, 1981] and we shall concentrate on one approach [Kalos, Levesque, and Verlet 1974; Reynolds, Ceperley, Alder, and Lester 1982] which uses importance sampling. Suppose we multiply $\Psi(\mathbf{r}, s)$ by a specified trial wavefunction $\Psi_T(\mathbf{r}, s)$ and use the result

$$\Phi(\mathbf{r}, s) = \Psi(\mathbf{r}, s)\Psi_T(\mathbf{r}, s) \tag{10.45}$$

in the Schrödinger equation. Then we obtain

$$\begin{aligned} -\frac{\partial \Phi}{\partial s} &= -D\nabla_{\mathbf{r}}^2 \Phi + (E_T(\mathbf{r}) - E_T)\Phi + D\nabla_{\mathbf{r}} \cdot (\Phi \nabla_{\mathbf{r}} \ln |\Psi_T(\mathbf{r})|^2) \\ &= -D\nabla_{\mathbf{r}}^2 \Phi + (E_T(\mathbf{r}) - E_T)\Phi + D\nabla_{\mathbf{r}} \cdot (\Phi \mathscr{F}) \end{aligned} \tag{10.46}$$

where the local energy is defined by

$$E_T(\mathbf{r}) = \Psi_T^{-1} \mathscr{H} \Psi_T, \tag{10.47}$$

and should not be confused with $E_T$. We have also defined the quantum force $\mathscr{F}$, which is derived from the pseudo-potential $u(r_{ij})$ if $\Psi_T$ is given (as is common) by

$$\Psi_T(\mathbf{r}) = \exp\{-\tfrac{1}{2}\sum_i \sum_{j>i} u(r_{ij})\} = \prod_{i<j} \exp\{-\tfrac{1}{2} u(r_{ij})\}. \tag{10.48}$$

Compare these equations with eqn (9.27). Note that we formally set $k_B T = 1$ throughout. All the techniques described in this section are now applied to the function $\Phi$ rather than to $\Psi$. The procedure for duplicating or deleting systems now depends on $(E_T(\mathbf{r}) - E_T)$, where $E_T(\mathbf{r})$ is evaluated for each system. This process is controlled more easily by a sensible choice of $\Psi_T(\mathbf{r})$ as discussed by Reynolds *et al.* [1982]. The quantum force appears in these simulations just as the classical force appears in the smart MC method described in Chapter 7, or the Brownian dynamics of Chapter 9. This force guides the system in its search for low 'energy' i.e. high $\Psi_T^2$. If $\Psi_T$ is a good approximation to the ground state $\Psi_0$, then the energy $E_T(\mathbf{r})$ tends to $E_0$ independently of $\mathbf{r}$, and so $\langle E_T(\mathbf{r}) \rangle$ is subject to little uncertainty. If $E_T$ is adjusted so as to maintain the steady-state population of systems, then this will also tend to $E_0$. As Reynolds *et al.* [1982] point out, the average $\langle E_T(\mathbf{r}) \rangle$ obtained without any system creation/destruction attempts would correspond to a variational estimate based on $\Psi_T$. Identical variational estimates can also be calculated using the Monte Carlo method [McMillan 1965; Schmidt and Kalos 1984]. In the random walk techniques the MC simulation is replaced by a Brownian dynamics technique. The inclusion of destruction and creation allows $\Psi(\mathbf{r})$ to differ from $\Psi_T$ and the simulation probes the improved $\Psi(\mathbf{r})$. Of course making $\Psi_T$ more complicated and hence more complete, adds to the computational expense.

The particular problems of fermion systems are discussed by Reynolds *et al.* [1982]. The essential point is that the ground-state fermion wavefunction must contain multidimensional nodal surfaces. Each region of configuration space bounded by these 'hypersurfaces', in which $\Psi_T$ may be taken to have one sign throughout, may be treated separately by the random walk technique. The nodal positions themselves are essentially fixed by the predetermined form of $\Psi_T(\mathbf{r})$. This introduces a further variational element into the calculation. The fixed-node approximation, and alternative approaches for fermion systems are described in detail by Reynolds *et al.* [1982].

# 11
# SOME APPLICATIONS

## 11.1 Introduction

In this chapter, we shall present several examples illustrating the application of the simulation techniques described earlier. It is not our intention to give a complete survey of each field. Rather, we have chosen a few case studies which emphasize the role of computer simulation as depicted in Fig. 1.2, namely to test theories and explain experimental results by providing a microscopic view of liquids not obtainable by any other means. This is a personal selection of topics which interest us, and which illustrate specific points; more detail, which we cannot include for reasons of space, will be found in the references.

## 11.2 The liquid drop

The statistical mechanics of inhomogeneous systems, such as the gas–liquid interface, is an active area of current research [Rowlinson and Widom 1982]. There are many properties of small liquid drops, such as the radial dependence of the pressure and the size-dependence of the surface tension, which are of fundamental interest and which are not readily available from experiment. A small drop, in equilibrium with its vapour, is an obvious candidate for computer simulation. This section discusses new technical problems associated with the preparation and equilibration of stable systems in the two-phase region, and highlights some of the important results that have emerged from the recent studies.

The main thrust of this work has been to explore fundamental properties of drops rather than to make a connection with the scant experimental results. For this reason, this simulations employ simple models such as the truncated Lennard-Jones potential, or the Stockmayer potential. The three major studies of the Lennard-Jones drop [Rusanov and Brodskaya, 1977; Powles, Fowler, and Evans 1983a; Thompson, Gubbins, Walton, Chantry, and Rowlinson 1984] use the molecular dynamics method. Unbiased MC methods can lead to bottlenecks, which have caused artificial structure in the density profile of a planar interface [Lee *et al.* 1974].

Large system sizes and long runs are required to obtain useful results for drops. The largest systems studied contained 2048 atoms [Thompson *et al.* 1984]. In the same study a drop and vapour containing 138 atoms was simulated for $350 \times 10^3$ time steps. Powles *et al.* [1983a] studied 1300 atoms for $15 \times 10^3$ time steps. Their simulations employed an unusually large potential cutoff of $10\sigma$ which made long-range corrections unnecessary, but which increased the computing time significantly. The earlier simulations of

Rusanov and Brodskaya [1977] were only 5000 time steps, which is probably too short to produce accurate results for the pressure tensor **P** or for the surface tension $\gamma_s$, although their results are in qualitative agreement with the recent studies.

The simulation of the drop begins by performing a normal bulk simulation of the Lennard-Jones fluid using periodic boundary conditions. The drop is excised from the bulk and placed either at the centre of a new periodic system with a larger central box [Powles *et al.* 1983] or in a spherical container [Rusanov and Brodskaya 1977; Thompson *et al.* 1984]. The size of the central box or the spherical container must be large enough so that two periodic images of the drop, or the drop and the wall of the container, do not interfere with one another. On the other hand, if the system size is chosen to be too large, the liquid drop will evaporate to produce a uniform gas. The difficulty of choosing a suitable starting density can only be resolved by trial and error. In practice the distance between the outside of the two periodic images of the drop should be at least a drop diameter. In the case of a container, its radius should be two to three times larger than the radius of the drop.

The spherical container is best thought of as a static external field which confines the molecules to a constant volume. Thompson *et al.* [1984] use the repulsive Lennard-Jones potential $v^{\mathrm{RLJ}}(d)$ (eqn. (1.10a)) to model this wall; $d$ is the distance along a radius from the molecule to the wall. Solving Newton's equations for this system will conserve energy and angular momentum. The drop moves around inside the spherical container as atoms evaporate from the surface of the liquid and subsequently rejoin the drop. In another variant of this technique the external field moves so that it is always centred on the centre of mass of the system. Solution of Newton's equations in a time-dependent external field does not conserve energy; in this particular instance [Thompson *et al.* 1984] the simulation was also performed at constant temperature using momentum scaling (see Section 7.4.3) and the equilibrium results were shown to be equivalent to those obtained in the more conventional microcanonical ensemble.

Figure 11.1 shows a snapshot of part of a drop after equilibration. At any instant the drop is non-spherical, but on average the structure is spherical and the drop is surrounded by a uniform vapour. The radius of the drop, defined shortly in eqn (11.4), changes by less than 1 per cent during the production phase of the run. The temperature profile through the drop is constant.

The principal structural property of the drop is the density profile, $\rho(r)$. It is defined as the average number of atoms per unit volume a distance $r$ from the centre of the drop. Since the drop moves during the run, it is necessary to recalculate its centre of mass as an origin for $\rho(r)$ at each step. This is defined, assuming equal mass atoms, by

$$\mathbf{r}'_{\mathrm{cm}} = \frac{1}{N'} \sum_{i=1}^{N'} \mathbf{r}_i(t) \tag{11.1}$$

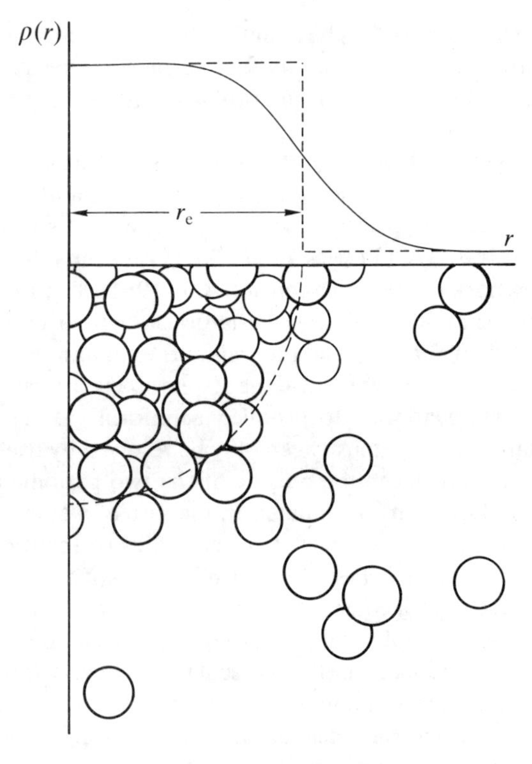


**Fig. 11.1** A snapshot of one octant of a drop with its centre at the origin. We also show the density profile and the Gibbs dividing surface, which defines the radius $r_e$ [after Thompson *et al.* 1984].

where $N'$ is the number of atoms in the drop at time $t$. This has to be defined in some way. Powles, Fowler, and Evans [1983b] have implemented the nearest-neighbour distance criterion of Stoddard [1978] for identifying atoms in the drop. The method makes use of a clustering algorithm. This begins by picking an atom $i$. All atoms $j$ that satisfy

$$r_{ij} < r_{cl} \tag{11.2}$$

where $r_{cl}$ is a critical atom separation, are defined to be in the same cluster as $i$. Each such atom $j$ is added to the cluster, and is subsequently used in the same way as $i$, to identify further members. When this first cluster is complete, an atom outside the cluster is picked, and the process repeated to generate a second cluster, and so on. The whole procedure partitions the complete set of atoms into mutually exclusive clusters. In the case of the liquid drop system, the largest cluster is the drop itself and the algorithm works most efficiently if the first atom $i$ is near the centre of the drop. The atoms not in the drop cluster are defined to be in the vapour. Fowler [1984] has described in detail the implementation of the method. An efficient clustering algorithm is given in

F.34. $r_{cl}$ has to be chosen sensibly. Studying the dependence of $N'$ upon $r_{cl}$ provides guidance in the choice of this parameter. Powles *et al.* [1983b] recommend a value of $r_{cl} = 1.9\sigma$; Thompson *et al.* [1984] suggest a range of $1.3$–$1.8\sigma$.

The simulated $\rho(r)$ can be fitted using a hyperbolic tangent functional form

$$\rho(r) = \tfrac{1}{2}(\rho_l + \rho_g) - \tfrac{1}{2}(\rho_l - \rho_g)\tanh(2(r - r_0)/d_s) \tag{11.3}$$

where $\rho_l$ and $\rho_g$ are the density of the liquid well inside the drop and the density of the vapour, respectively. $d_s$ is a measure of the thickness of the interface and $r_0$ is an estimate of the drop radius. It is likely to be quite close to the radius of the equimolar dividing surface (Gibbs surface)

$$r_e^3 = \frac{1}{\rho_g - \rho_l}\int_0^\infty \frac{d\rho(r)}{dr} r^3 \, dr . \tag{11.4}$$

$r_e$ is defined so that if the limiting densities of the two phases were constant up to $r = r_e$ and changed discontinuously at $r = r_e (d_s \to 0$ in eqn (11.3)), the system would contain the same number of molecules. This is illustrated in Fig. 11.1.

Both simulation studies show that the width of the surface increases rapidly with temperature, and that drops disintegrate at temperatures below the normal critical temperature. At $T^* = 0.71$ the surface thickness is almost independent of the drop size, but at $T^* = 0.80$, $d_s$ increases for smaller drop size. The liquid density of the drop increases above the planar limit as the drop size decreases. This is predicted by the Laplace equation, where the pressure difference, $\Delta P = P_l - P_g$, is inversely proportional to the radius of the surface of tension, $r_s$

$$\Delta P = 2\gamma_s / r_s . \tag{11.5}$$

$\gamma_s$ is the surface tension, which is defined to act at the surface $r = r_s$. Hence, for a moderately small drop, we see a large pressure difference and a high $\rho_l$. For very small drops, eqn (11.5) does not apply; the opposite trend is observed and $\rho_l$ decreases with drop size as the attractive cohesive forces decrease.

Powles *et al.* [1983b] show that the capillary wave contribution to the surface thickness can be estimated using linearized hydrodynamics. If this broadening due to capillary waves is added to the intrinsic profiles calculated from integral equation theories, the results agree well with the simulated widths at all but the highest temperatures.

The pressure tensor in a spherical drop can be written in terms of two independent components; the normal, $P^N(r)$, and the transverse, $P^T(r)$. The condition for mechanical equilibrium, $\nabla \cdot \mathbf{P} = 0$, relates these components through a differential equation.

$$P^T(r) = P^N(r) + \frac{r}{2}\frac{dP^N(r)}{dr} . \tag{11.6}$$

This is in contrast to the planar interface where the components must be calculated separately. For the drop, the calculation of $P^N(r)$ is sufficient to describe **P**, and to calculate $\gamma_s$ through a thermodynamic or a mechanical route.

The pressure tensor has the following form (compare eqns (2.113), (2.114))

$$P_{\alpha\beta}(\mathbf{r}) = k_B T\rho(\mathbf{r})\delta_{\alpha\beta} - \tfrac{1}{2}\sum_i \sum_{j\neq i} \left\langle \frac{r_{ij\alpha}}{r_{ij}} \frac{dv(r_{ij})}{dr_{ij}} \int_{C_{ij}} dr'_\beta\, \delta(\mathbf{r}-\mathbf{r}') \right\rangle$$

$$= k_B T\rho(\mathbf{r})\delta_{\alpha\beta} + \tfrac{1}{2}\sum_i \sum_{j\neq i} \left\langle f_{ij\alpha} \int_{C_{ij}} dr'_\beta\, \delta(\mathbf{r}-\mathbf{r}') \right\rangle \tag{11.7}$$

where $C_{ij}$ is any contour joining atom $i$ to $j$ [Schofield and Henderson 1982]. The simplest choice of contour, due to Irving and Kirkwood (IK), is to use the straight line between the two atoms. There are an infinite number of other choices, and this results from the lack of uniqueness in the definition of the microscopic stress tensor. It should be stressed that $P_l$ and $P_g$, the pressure well inside the drop and in the vapour respectively, are independent of the choice of contour. The IK normal component of the pressure tensor calculated in the simulation can be described by a tanh formula similar to eqn (11.3). The corresponding $P^T$, calculated from eqn (11.6), generally displays a large negative region. For all but the smallest drops, $P^T(r) < P^N(r)$, indicating that the surface is under tension. Thompson *et al.* [1984] have compared the IK pressure tensor results with an alternative choice due to Harasima, but find that the latter is subject to greater statistical fluctuations.

There are a number of possible routes to the surface tension. The Laplace equation, eqn (11.5), can be combined with the Tolman equation for the variation of $\gamma_s$ with drop size

$$\frac{\gamma_s}{\gamma_\infty} = 1 - \frac{2(r_e - r_s)}{r_s}. \tag{11.8}$$

$\gamma_\infty$ is the surface tension of the fluid in the planar limit. A rearrangement of eqns (11.5) and (11.8) gives

$$r_s = \frac{3\gamma_\infty - [9\gamma_\infty^2 - 4\gamma_\infty r_e (P_l - P_g)]^{1/2}}{P_l - P_g} \tag{11.9}$$

and the thermodynamic expression for $\gamma_s$

$$\gamma_s = \tfrac{1}{2} r_s (P_l - P_g). \tag{11.10}$$

$P_l$, $P_g$, and $r_e$ are unambiguously determined in the simulation and $\gamma_s$ can be calculated in terms of $\gamma_\infty$. A mechanical route to $\gamma_s$ gives

$$\gamma_s^3 = -\tfrac{1}{8}(P_l - P_g)^2 \int_0^\infty r^3 \frac{dP^N(r)}{dr}\, dr. \tag{11.11}$$

This requires $P^N(r)$ over its whole range, which is not uniquely determined.

The thermodynamic formula (eqn (11.10)) for $\gamma_s$ gives slightly larger values than eqn (11.11) using the IK value of $P^N(r)$. Despite the uncertainties in $\gamma_\infty$ for the model, the agreement is good. The simulations show that $\gamma_s$ decreases with increasing curvature and that $r_e$ exceeds $r_s$ by $\sim 0.4$–$0.6\sigma$. The fall in $\gamma_s$ with temperature is greater than in the case of the planar interface.

Powles *et al.* [1983a] suggest the Kelvin equation as a route to the surface tension:

$$\rho_l k_B T \ln\left[\frac{P_g(r_s)}{P_g(\infty)}\right] = \frac{2\gamma_s}{r_s} \tag{11.12}$$

where $P_g(r_s)$ is the vapour pressure around a particular drop and $P_g(\infty)$ is the vapour pressure above the planar interface at the same temperature. Powles *et al.* [1983a] have used this equation to estimate the surface tension $\gamma_\infty$. For a large drop, the Kelvin equation can be written in its limiting form

$$\rho_l k_B T \ln\left(\frac{P_g(r_e)}{P_g(\infty)}\right) = \frac{2\gamma_\infty}{r_e}. \tag{11.13}$$

If the density of the drop, the vapour pressures, and the equimolar radius can be measured for a large enough drop, then eqn (11.13) gives a route to $\gamma_\infty$, which is independent of the liquid pressure. It should be stressed that eqn (11.12) and eqn (11.13) assume a drop of incompressible fluid surrounded by an ideal vapour, which is a useful approximation close to the triple point. Equation (11.12) does not constitute an independent route to $\gamma_s$ since there is no straightforward way to obtain $r_s$ from $r_e$ without knowing the liquid pressure. Powles *et al.* [1983a] estimate $P_g$ for their drops by calculating $\rho_g$ and assuming that the vapour is ideal. The largest drop studied gave a value of $\gamma_\infty^* = 1.55 \pm 0.20$ compared with an estimate from the planar simulation of $\gamma_\infty^* = 1.31 \pm 0.02$. Interestingly, the pressure of the vapour can also be obtained by the normal virial calculation, eqn (2.55), in the case of a drop simulated using periodic boundary conditions. The only requirement is that the surface of the central box is always in the vapour part of the system. The virial equation, using the mean density of the sample, simply gives the average of the normal pressure over the surface of the box. This result is not true for simulations using a spherical external field. In the case of a hard-wall spherical container, the conventional virial calculation, using the mean density, yields only the local kinetic contribution to the pressure at the wall of the container [Powles, Rickayzen, and Williams 1985]. Thompson *et al.* [1984] demonstrate the large uncertainties in the calculation of the vapour pressure in their simulations (which are for large system sizes and very long runs). They maintain that this makes the Kelvin route to the surface tension a much less viable proposition than eqns (11.9) and (11.10).

Recently, simulations have been extended to drops of Stockmayer molecules (Lennard-Jones atoms plus point dipoles) [Powles, Fowler, and Evans 1984].

This may be a route to the macroscopic dielectric constant that avoids the technical problems of periodic boundary conditions for systems with long-range forces (see Section 5.5).

## 11.3 Melting

The computer simulation method produced some of its most interesting results within a few years of its inception. The first hard disk MC simulations were reported in 1953 [Metropolis *et al.* 1953], hard spheres were under investigation the following year [Rosenbluth and Rosenbluth 1954], and by 1957 both MD and MC simulations indicated that hard spheres could form two distinct phases [Wood and Jacobson 1957; Alder and Wainwright 1957]. Considerable care was exercised in comparing and cross-checking these results [Wood 1986] and there was a good deal of caution in interpreting them. Later, a first-order melting transition for hard spheres was demonstrated convincingly, and the terms 'solid' and 'fluid' were used.

What did these early workers see? The hard-disk system was seen to exhibit a doubly branched equation of state, i.e. for a given $\rho$, it seemed to be possible for the system to be in equilibrium at either of two pressures $P$. On the high-pressure branch, the system appeared disordered and molecular diffusion could occur, while on the low-pressure branch, an ordered system with little molecular motion was seen. In a sufficiently long run, with $\rho$ fixed at $\approx 2/3$ of the close-packed density, the system would occasionally switch back and forth between the two branches.

Flipping between two branches of an equation of state is behaviour typical of a small system ($N = 32 - 108$) in which the free energy penalty for forming an interface prevents two-phase coexistence. For somewhat larger systems, and in two rather than three dimensions, coexistence is more easily achieved. A memorable 'time-lapse photograph' of the hard-disk system [Alder and Wainwright 1962] shows clearly distinguishable fluid-like and solid-like phases. For coexisting phases, the interface free energy produces a smooth 'small-system' loop (sometimes called, rather misleadingly, a 'van der Waals' loop) in the phase diagram. This curve connects the fluid and solid branches [Alder and Wainwright 1962; Mayer and Wood 1965] and a schematic illustration appears in Fig. 11.2(a). For larger systems, the relative contribution of the interface to the free energy diminishes, and the loop becomes flatter. In the thermodynamic limit, it reduces to a horizontal straight line. In a constant-pressure simulation [see for example Wood 1968a] coexisting phases should not be seen. Close to the phase transition, the overall density will fluctuate, taking the system from the low-density branch to the high-density branch and back. For a small system, the distribution of densities is rather broad and featureless, but as $N$ increases the density distribution develops a double-peaked structure. Eventually, for large $N$, the system density remains very close to one value or the other, flipping occasionally between the two

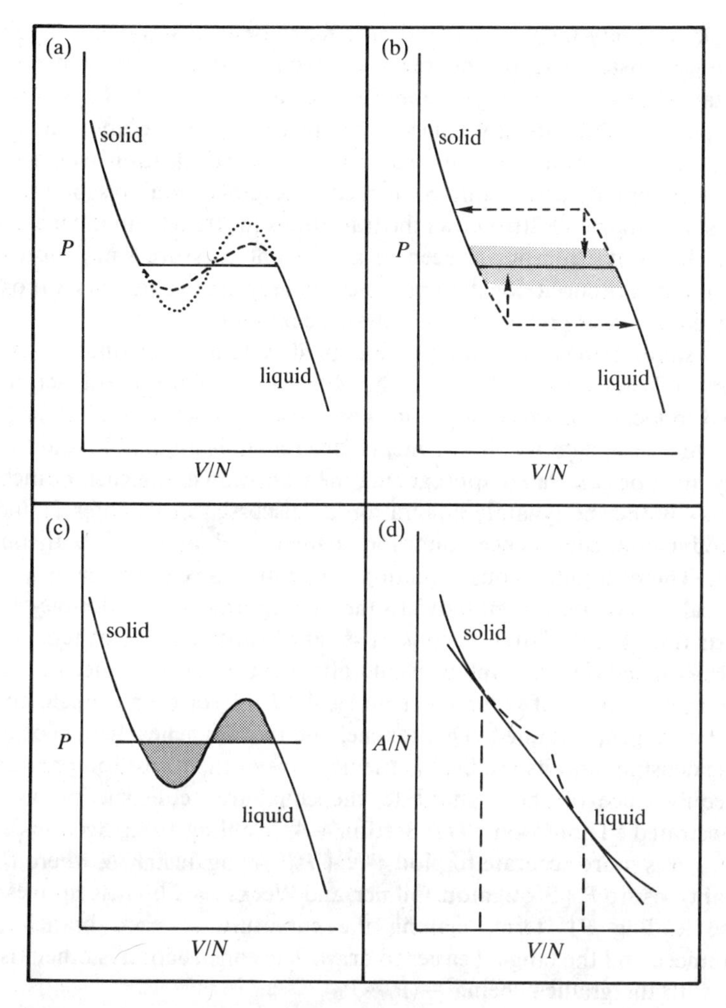


**Fig. 11.2** Thermodynamic properties near melting.
(a) The equation of state. We show the loop which appears for a small system (dots), for a larger system (dashes), and the thermodynamic limit (solid line).
(b) Metastability and hysteresis in simulations. The dotted lines represent transitions from metastable states as seen in a constant-pressure or constant-volume simulation.
(c) The Maxwell construction. The indicated areas are equal.
(d) The double tangent construction. Dotted lines indicate the metastable regions.

branches, and spending a proportion of time in each state, depending on the value of the pressure.

For fairly large systems, the dynamics close to a phase transition become quite sluggish, and the above pattern is not followed: poor equilibration results in hysteresis effects. A metastable solid will be produced by heating or expanding past the thermodynamic coexistence point (see Fig. 11.2(b)). Usually this will suddenly melt to a liquid at a more or less well-defined point of mechanical instability. In the reverse direction, it is easy to produce a metastable liquid by cooling or compressing past coexistence. Crystallization to the perfect solid will almost never take place; depending upon the rate of change of state point from one run to the next, nucleation will occur to produce either a defective solid composed of several crystallites, or a random glassy state. Figure 11.2(b) shows the transitions (dotted) from the metastable states. Horizontal jumps are seen in a constant pressure simulation while vertical jumps occur at constant volume: the grey area represents a (possibly poorly equilibrated) region of two-phase coexistence.

If the simulation runs yield a genuine small-system loop without hysteresis [Alder and Wainwright 1962], then the Maxwell equal-area construction may be used to locate the thermodynamic coexistence points (Fig. 11.2(c)). In the more usual case, data within the metastable region is unreliable, and the free energy must be calculated approaching the transition along each branch (for example by thermodynamic integration as discussed in Chapter 2) and the thermodynamic coexistence conditions (equal $T$, $P$, and $\mu$ in both phases) solved. These simultaneous equations may be solved numerically, or a graphical method may be employed. One such approach is the double-tangent construction (Fig. 11.2(d)). Suppose a series of runs at constant temperature has been carried out. A graph of Helmholtz free energy per molecule, $A/N$, is drawn as a function of volume per molecule, $V/N$, for each branch, and the common tangent obtained. This touches the two branches at the coexisting inverse densities and its gradient is the negative of the transition pressure $P_t$. The equivalence of the method to the equal-area construction is easily demonstrated [Thompson 1972, Section 4–4; Mohling 1982, Section 39]. In practice, it is more accurate to plot $A/N + P_0 V/N$ against $V/N$ where $P_0$ is a constant close to $P_t$ [Broughton, Gilmer, and Weeks 1982]. This removes most of the leading $PV$ term, making the curvature of each branch more pronounced and the tangent easier to draw. The construction is otherwise the same, with the gradient being $-(P_t - P_0)$.

It is not our intention to survey the study of melting by computer simulation. Some studies have been particularly instructive, however. Hoover and Ree [1968] determined the hard sphere melting parameters, using an interesting single-occupancy cell method to evaluate the entropy in the solid phase. As well as a series of conventional simulations, Hoover and Ree conducted simulations of a system in which each particle was restricted to lie within its own Wigner–Seitz lattice cell in configuration space. In the dense solid, this restriction has no effect on the system properties. As the density is reduced, the single-occupancy system does not melt in the conventional sense, but goes over to a type of lattice gas, whose properties can be calculated exactly. The method was devised to enable thermodynamic integration to be carried out along a reversible path linking the solid with a gaseous phase. In fact, a weak phase transition does seem to occur for the single-occupancy system, and this necessitates very careful thermodynamic integration in the transition region. The study of Hoover and Ree was noteworthy in that it clarified many ideas concerning 'communal entropy', and in particular showed that this quantity does not suddenly change on melting, but varies smoothly as the density is changed. Hoover, Gray, and Johnson [1971] conducted a survey of several soft-sphere systems, determining equations of state and melting properties. Their work demonstrates the more straightforward thermodynamic integration route to $A$, along a path to the ideal, low-temperature, harmonic lattice limit, whose properties can be calculated exactly. Note that this option does not exist for hard spheres, since the harmonic approximation is never valid for the hard-sphere crystal. Melting in the Lennard-Jones system has been studied conventionally by Hansen and Verlet [1969], and also by Ladd and Woodcock [1977, 1978]. In this latter study, a large system was set up with all three phases in coexistence, and the properties of interest measured directly, by looking at regions located well within each phase. The sluggish equilibration time-scales in such a large inhomogeneous system were a serious problem in this work.

Computer simulation has also contributed to our understanding of the mechanism of melting, with interest in the two-dimensional case being particularly stimulated by a theory due to Halperin and Nelson [1978] and Young [1979]. In this theory, the unbinding of pairs of dislocation defects was supposed to cause the solid to melt by a continuous (not first-order) transition. The resulting two-dimensional fluid was termed the 'hexatic phase', and was predicted to be translationally disordered but exhibit long-range orientational ordering (that is, ordering of the directions of vectors between neighbouring atoms). In this context, 'long-range' correlations decay algebraically as $r^{-\nu}$ with separation, while 'short-range' correlations vanish exponentially as $\exp(-r/r_c)$. A second continuous transition was required to convert this hexatic phase into a normal, isotropic fluid, with short-range orientational ordering. Some universal predictions regarding the elastic behaviour of the solid and the form of spatial correlation functions in the fluid were made. Most of these predictions were amenable to testing by computer simulation. Many simulations of different systems by MD and MC were carried out, with some workers claiming support for the theory and others favouring a conventional first-order mechanism with no intermediate phase [see e.g. Frenkel and McTague 1979; Tobochnik and Chester 1980, 1982; Barker, Henderson, and Abraham 1981; Broughton, Gilmer, and Weeks 1981, 1982; Allen, Frenkel, Gignac, and McTague 1983]. The general results seem to be that the thermodynamic properties (determined by methods such as those described above) are consistent with first-order melting, and that there is no convincing evidence of a hexatic phase. This last point is not surprising, since the hexatic phase is predicted to be 'critical', i.e. one with long-range, slow, correlations, and dynamics [Zippelius, Halperin, and Nelson 1980] which would be difficult to simulate properly. However, the elastic constant behaviour seems generally to conform to the Halperin–Nelson–Young picture, and 'orientational' correlations seem to grow as the transition is approached from the fluid phase, which would not be expected if a simple first-order mechanism operated [Allen *et al.* 1983].

Two aspects of these studies are of particular interest. Firstly, a method was developed to allow the identification of point defects in the two-dimensional system [McTague, Frenkel, and Allen 1980; Stillinger and Weber 1981], and thereby test the underlying assumptions of the melting theory. The method involves specifying the nearest neighbours of each atom by a geometrical construction called a Voronoi polygon, so unambiguously assigning each atom a coordination number. In two-dimensional periodic boundaries the average coordination number is exactly six. In a perfect lattice, each atom has six neighbours. In an imperfect system, the atoms with coordination numbers other than six are the defects. Each seven- or five-coordinate atom is a disclination: a defect which disrupts the 'orientational' ordering in the lattice (see Fig. 11.3(a)). The second transition is associated with formation of isolated disclinations by dissociation of bound pairs. A 7–5 pair of atoms is an isolated dislocation, which disrupts positional ordering (see Fig. 11.3(b)). The first transition is associated with unbinding of dislocation pairs; bound dislocation pairs appear as 7–5–7–5 quartets. Typical defect structures above and below melting in a soft-disk system [Allen *et al.* 1983] are shown in Fig. 11.3(c, d). Identification of the defects in this way gives a nice example of how computer simulation can provide geometrical and topological information, as well as simple thermodynamics and structures. In this case, very complex defect structures (not just isolated dislocations and disclinations) were seen close to melting. Incidentally, in three dimensions, the Voronoi construction is also useful in discussing freezing and nucleation phenomena [Tanemura, Hiwatari, Matsuda, Ogawa, Ogita, and Ueda 1977, 1978; Hsu and Rahman 1979]. Routines for carrying out the Voronoi construction in two and three dimensions are given in F.35. The second interesting development is a study by Saito [1982, 1983], which dispenses with the molecules altogether. Saito conducts a grand canonical ensemble simulation of a system of dislocation defects on a lattice, interacting through a hamiltonian derived from continuum elasticity theory, thus testing part of the melting theory directly. He finds that first-order melting or continuous melting can be generated, depending upon the value of the dislocation core energy.

---

**Fig. 11.3** (a, b) Two-dimensional melting defect structure. Seven-coordinate atoms are denoted by +, five-coordinate atoms by −. Lattice circuits start at the black atoms and move anticlockwise as indicated by the solid black lines. (a) Definition of a disclination. The 'compass needle' remains fixed in a local coordinate system. After a lattice circuit, it is rotated by $\pi/3$. (b) Definition of a dislocation. A lattice circuit fails to close: it would close in the absence of the dislocation.

(c, d) Defect structure in the soft-disc fluid [after Allen *et al.* 1983]. Five-coordinate atoms are labelled with −, seven-coordinate atoms with +, and eight-coordinate atoms with #. Six-coordinate atoms are omitted for clarity. (c) Solid near melting. (d) Fluid near melting.

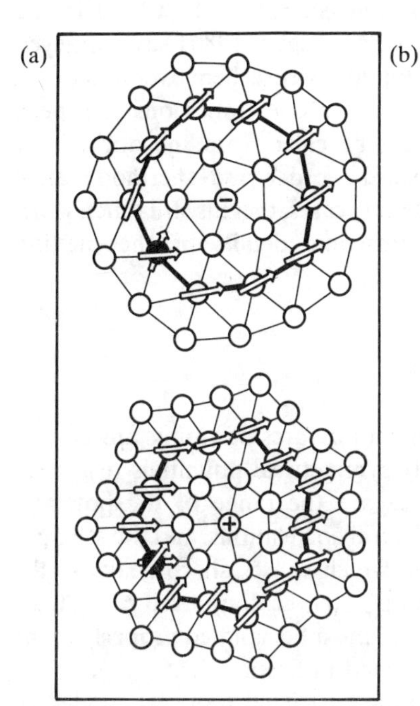
(a)

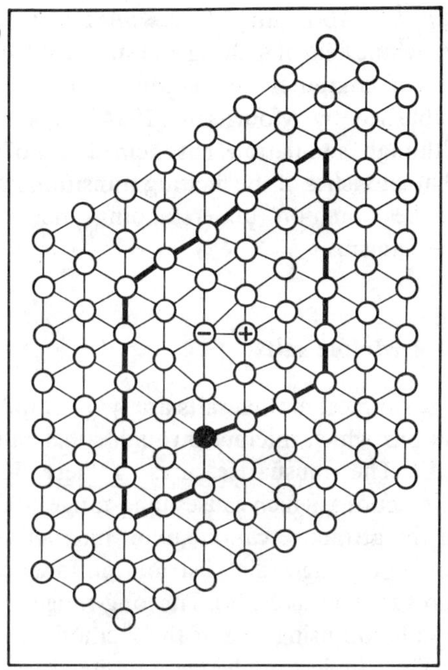
(b)

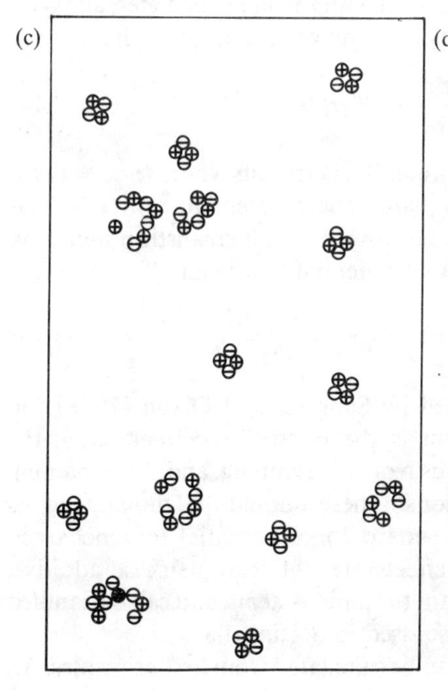
(c)

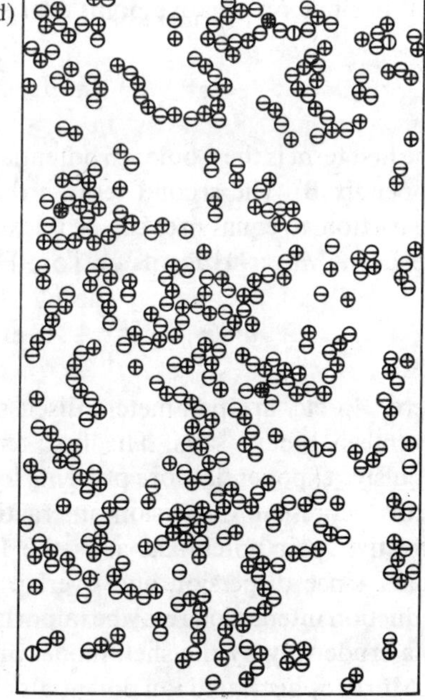
(d)

Further discussions of melting in two dimensions may be found in the review of Abraham [1982], while Frenkel and McTague [1980] have surveyed the whole field of melting and supercooled liquids. An excellent overview of the use of computer simulations in the study of phase transitions has been published by Mouritsen [1984]. It should be clear from the above that although simulation has helped establish the order and thermodynamic characteristics of the melting transition in many cases, there is still much work to be done on understanding the microscopic details of the melting mechanism.

## 11.4 Molten salts

The simplest molten salts are two-component mixtures of atomic anions and cations which interact through spherically symmetrical potentials (e.g. KCl, LiF). The unusual features of these fluids are the range of the potential interactions, which causes long-range structural correlations, and the strength of the attractive cation–anion interaction, which causes sharp peaks in the corresponding radial distribution function $g_{+-}(r)$ (i.e. long-lived cage structures around each ion). The long-range forces must be handled properly in the simulation using one of the methods described in Section 5.5.

The simplest model for a molten salt is the 'restricted primitive model', in which ions are modelled by hard spheres, all with the same diameter, and with unit positive or negative point charges $z_i$ at the centre of each sphere:

$$v^{\mathrm{RPM}}(r_{ij}) = \frac{z_i z_j}{r_{ij}} + v^{\mathrm{HS}}(r_{ij}) \,. \tag{11.14}$$

The first term is the Coulomb potential, eqn (1.11), in units where $4\pi\varepsilon_0 = 1$ (see Appendix B). The second term is the hard-sphere potential, eqn (1.7). The 'restriction' to equal sizes can be relaxed of course. A more realistic potential is the Born–Mayer–Huggins or Tosi–Fumi potential [Fumi and Tosi 1964]

$$v^{\mathrm{TF}}(r_{ij}) = \frac{z_i z_j}{r_{ij}} + A_{ij} \exp\left(-B_{ij} r_{ij}\right) - \frac{C_{ij}}{r_{ij}^6} - \frac{D_{ij}}{r_{ij}^8} \,. \tag{11.15}$$

Here, $A_{ij}$ etc. are parameters discussed by Sangster and Dixon [1976] for seventeen binary salts. The first term is the electrostatic interaction, the repulsive exponential core prevents ions from overlapping, and the remaining terms represent dispersion interactions. These should be thought of as 'effective' pair potentials, as is the Lennard-Jones potential for uncharged fluids, since dispersion and overlap effects are not truly pairwise additive. Induction interactions may be important for ionic systems, and can be handled in a crude way by the shell model discussed in Section 1.3.2.

More sophisticated ion potentials can be calculated from first principles. An example is the self-consistent-field calculation of the potential between $Na^+$ and $NO_3^-$ ions used in a simulation of molten $NaNO_3$ [Goddard, Klein, and

Ozaki 1983]. In this study, good agreement between the simulated structure and that obtained from X-ray diffraction was observed. *Ab initio* calculations are more successful for potentials with large well-depths because the inherent errors (particularly basis-set superposition) are relatively less important in these cases. These methods are expected to find greater application in the study of molten salts and polar liquids than in the case of non-polar systems.

An example of an extensive study of the non-polarizable rigid ion model is the simulation of molten NaCl using the potential $v^{TF}(r)$ [Lewis and Singer 1975]. This simulation accurately reproduces the internal energy and specific heat of the molten salt. As usual in simulation, this simple potential gives less satisfactory agreement with experimental pressures. The partial radial distributions $g_{++}(r)$, $g_{--}(r)$, and $g_{+-}(r)$, which describe structure in the melt, can be calculated by simulation and compared with the results of neutron diffraction from isotopically enriched NaCl samples [Edwards, Enderby, Howe, and Page 1975; Enderby and Neilson 1980]. Figure 11.4 compares such neutron scattering results for $g_{+-}(r)$ with the simulations of Lewis and Singer, and also with a shell model calculation [Dixon and Sangster 1976]. The two

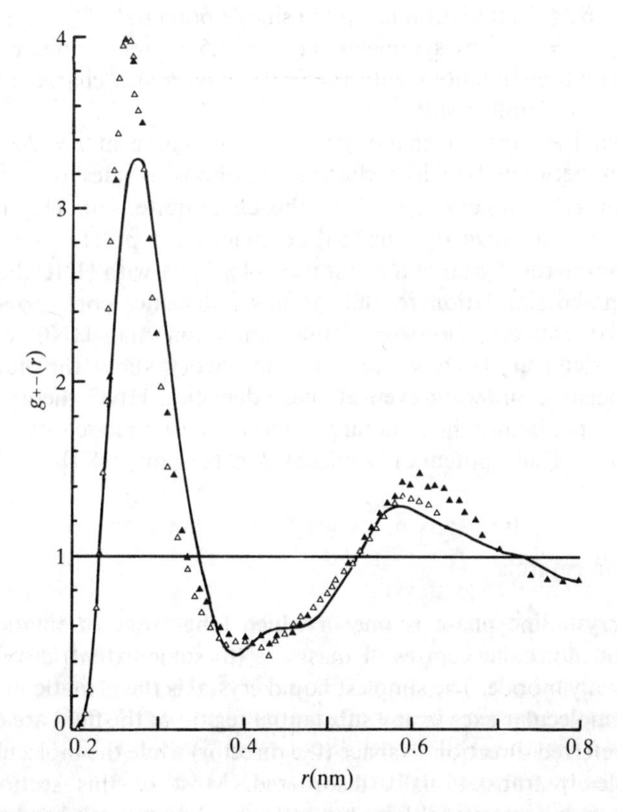


**Fig. 11.4** Comparison of experimental $g_{+-}(r)$ (line) with simulations using rigid ions (solid symbols) and shell model ions (open symbols) [after Sangster and Dixon 1976].

simulations are in excellent agreement with each other, but the small feature in the experimental $g_{+-}(r)$ at 0.45 nm is not reproduced and the simulated first peak is too sharp. The agreement between simulation and experiment for $g_{--}(r)$ and $g_{++}(r)$ is slightly better, particularly in the case of the shell model (the rigid ion model fails to distinguish sufficiently between the two functions). Although we can see that there is still room for improvement, in general potentials of the kind $v^{\mathrm{TF}}(r)$ reproduce experimental peak positions and the general shapes of distribution functions for ionic melts quite well [Enderby and Neilson 1980]. More realistic potentials, rather than improved ways of handling long-range forces, will probably lead to the next significant advances.

Apart from a direct comparison with experiment, the simulation can be used to test theories of ionic liquids. A strong candidate here is the hypernetted chain (HNC) equation which closes the Ornstein–Zernike formula (eqn (6.35)). The HNC approximation is

$$c(r) = h(r) - \beta v(r) - \ln g(r) \tag{11.16}$$

and it is expected to be more accurate for ionic fluids than the Percus–Yevick approximation, eqn (6.36) [Hansen and McDonald 1986]. Larsen [1978] has performed an MC simulation using the simple potential $v^{\mathrm{RPM}}(r)$, eqn (11.14), for which $g_{++} = g_{--}$ by symmetry. Figure 11.5 shows a comparison of the radial distribution functions with the HNC theory at a charge and density appropriate to a molten salt.

Note that the peculiar shape of the $g_{++}(r)$ curve at $r = 2\sigma$ is due to correlations between two like charges on opposite sides of a differently charged ion. HNC theory reproduces this effect quite well. The theory also works well at close range, right up to the contact value ($g_{++}(\sigma) \approx 13$). There is a discrepancy in the region of the first peak of $g_{++}(r)$, with HNC theory being flatter than the simulation results. At lower densities corresponding to a concentrated aqueous electrolyte, the simulation and HNC results are essentially identical. Both simulation and theory show the build up of significant charge ordering even at these densities. HNC theory has been successful in predicting the structure of more realistic models of ionic melts, using the Tosi–Fumi potential [Enderby and Neilson, 1980].

## 11.5 Liquid crystals

A liquid crystalline phase is one in which long-range orientational order persists but molecular centres of mass are (to some extent) disordered and translationally mobile. The simplest liquid crystal is the nematic. In a nematic phase, the molecular axes in any substantial region of the fluid are distributed about a preferred direction in space (the director) while the molecular centres are completely translationally disordered. Most of this section will be concerned with nematics, although a little simulation work has been carried

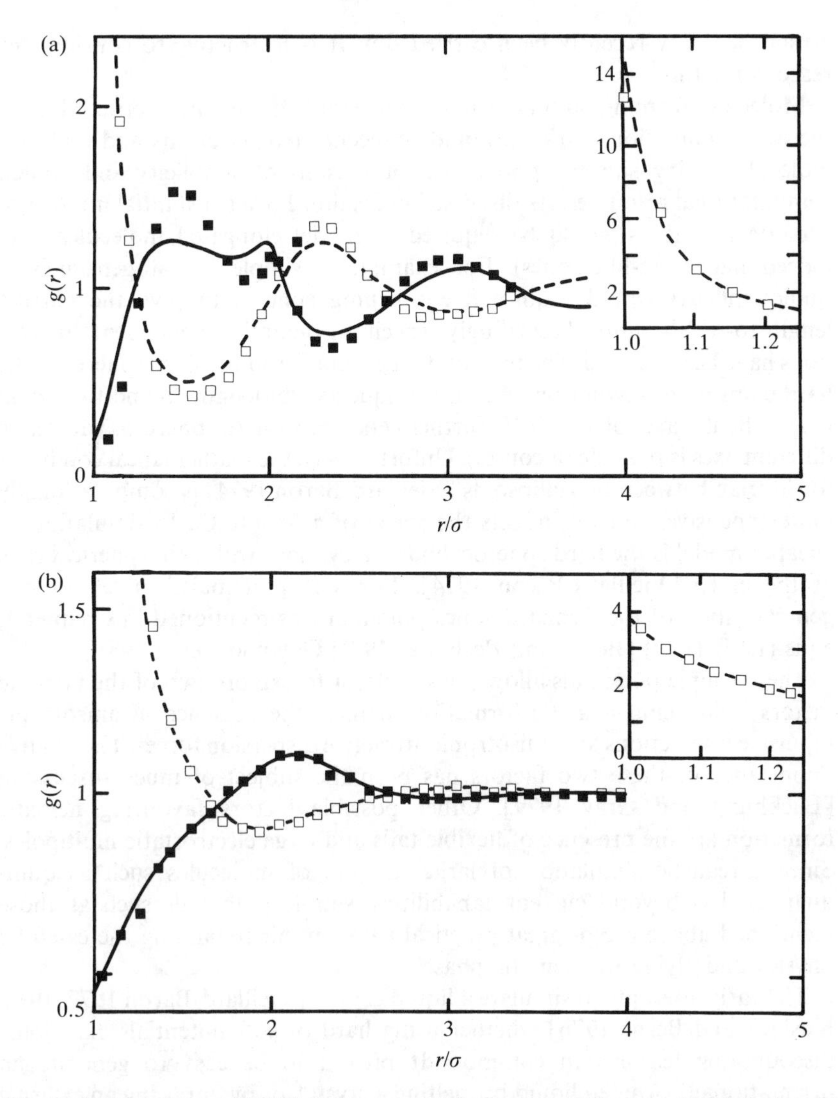


**Fig. 11.5** Simulation and HNC theory compared. Open symbols, $g_{+-}$ (simulation); solid symbols, $g_{++}$ (simulation); the lines are the corresponding HNC results. (a) Molten salt. (b) Concentrated electrolyte solution. The insets in (a) and (b) show the behaviour of $g_{+-}$ close to $r/\sigma = 1$ [after Larsen 1978].

out on other liquid crystals. The liquid crystal phases intervene between solid and liquid in the phase diagram; it is thought that all non-spherical molecules would form a liquid crystal on cooling or compressing the liquid, if freezing did not occur first. It may seem surprising, in view of the early successes of computer simulation of melting, that a convincing simulation of a liquid crystal has only recently been carried out. It is instructive to consider the reasons for this.

Molecules forming liquid crystals are generally quite anisometric. Most of the early simulation work on liquids concentrated on atoms and diatomic molecules, using site–site potentials, for reasons of simplicity and limited computational resources. As discussed in Section 1.3, a substantial number of site–site potentials would be required to model elongated molecules (and indeed flat, plate-like, ones). For example, a simple nematogen such as quinquaphenyl would require five touching spheres to give the correct length-to-width ratio. Accordingly, specific orientation-dependent interactions have been devised. The prototype, a generalization of hard spheres, is the hard ellipsoid of revolution with one unique axis of length $2a$ and two equal perpendicular axes of length $2b$ (further generalization to spheroids with three different axes is possible of course). Unfortunately, the mathematical condition for overlap between two ellipsoids [Vieillard-Baron 1974] is computationally quite expensive, and this affects the speed of a Monte Carlo simulation. A cheaper model is the hard spherocylinder (a cylinder with hemispherical caps at the ends) [Vieillard-Baron 1974]. For soft potentials, an 'ellipsoidal' generalization of the Lennard-Jones potential was mentioned in Chapter 1, eqns (1.15)–(1.17) [Berne and Pechukas 1972; Gay and Berne 1981].

These simple potentials allow the simulator to explore two of the possible factors influencing nematic formation, namely the presence of anisotropic repulsive interactions and anisotropic attractive dispersion forces. The relative importance of these two factors has been the subject of much discussion [Luckhurst and Gray 1979]. Other possible factors favouring nematic formation are the presence of flexible tails and large electrostatic multipoles. Since a realistic simulation of large samples of molecules such as quinquaphenyl is beyond current capabilities, simple potentials such as those mentioned above are of great practical value in understanding the essential physics underlying the nematic phase.

The early attempts to simulate a liquid crystal [Vieillard-Baron 1972, 1974; Kushick and Berne 1976] whether using hard or soft potentials, had some discouraging features in common. It proved to be easy to generate an orientationally ordered liquid by melting a crystal, or by imposing an external field. The orientation would persist on equilibrating in the absence of the fields for long times, but eventually, in almost every case, orientational order would disappear. Sometimes this would occur suddenly, as a jump from one 'plateau' value to another, raising the spectre of 'bottlenecks' in phase space (see Fig. 2.1). In no case, in this early work, was spontaneous ordering observed on cooling or compressing the isotropic liquid. In Monte Carlo studies of hard systems, an additional impediment to the gathering of thermodynamic evidence for any phase transition was the difficulty of measuring the pressure (see Section 5.6). Almost the only encouraging fact to come out of this early work was that (as expected) it is easier to generate a nematic phase in a two-dimensional system than in a three-dimensional one. Suitably elongated hard ellipses will form a liquid crystal in between the normal liquid and solid phases in two dimensions [Vieillard-Baron 1972]. A particularly easy case to program is the extreme one of infinitesimally thin, finite-length line segments or needles (which cannot form a solid at all). An MC program for this system is given in F.36.

Accordingly, simulations of orientationally ordered phases have been mainly restricted to lattice systems, with the molecular centres of mass being perfectly ordered. Useful information regarding liquid crystals can be obtained from these simulations despite the lattice restriction. This is because the details of the most interesting aspect of these liquid crystals, namely the nematic–isotropic transition itself, may be independent of the translational degrees of freedom. Typically, these simulations involve nearest-neighbour interactions using the Lebwohl–Lasher potential [Luckhurst and Simpson 1982; Zannoni 1979]

$$v(r_{ij}, \mathbf{\Omega}_i, \mathbf{\Omega}_j) = AP_2(\cos \gamma_{ij}) \tag{11.17}$$

where $\gamma_{ij}$ is the angle between molecules $i$ and $j$ as defined in Appendix C.3, $A$ is a negative constant, and $P_2$ a Legendre polynomial. This system shows an orientational phase transition. Early studies of system-size dependence indicated a first-order transition, with a much larger entropy change than observed in real liquid crystals. The orientational order parameter (see below) which should be infinite in the nematic phase, did not diverge sufficiently quickly in the simulation on approaching the transition from the disordered phase. Very recently, Fabbri and Zannoni (1986) have shown that these details are crucially dependent on the system size. Studying a system of 27 000 molecules, they obtained results in much closer agreement with experiment.

Recently, the lattice restriction has been removed. Using a constant-$NPT$ Monte Carlo method, Luckhurst and Romano [1980] simulated a system of Lennard-Jones atoms with an orientational potential of the form of eqn (11.17) multiplied by a distance-dependent term. In this simulation, the angle-dependent potential was introduced into a ready-equilibrated Lennard-Jones liquid system, and an orientational order–disorder transition was observed.

How can we be sure that we are really simulating a nematic-isotropic transition? To be as convincing as the simulation of melting, it would be necessary to determine the free energies of isotropic and nematic phases close to the transition, and to locate the liquid crystal transition in relation to the normal melting point, so as to ensure against simulating a metastable state of some kind. In other words we would need to map out the complete phase diagram of the system. Ideally, spontaneous ordering from isotropic to nematic phases should be seen. In addition, it might be possible to observe a specific heat anomaly consistent with the predictions of finite size scaling for first order transitions [Mouritsen 1984].

Recently, a complete phase diagram has been determined for a hard core system. This was partly motivated by Onsager's prediction [Onsager 1949] that a liquid of very long, thin, molecules would form a nematic phase. Monte Carlo studies of extremely oblate ellipsoids (hard platelets) [Frenkel and Eppenga 1982; Eppenga and Frenkel 1984] were followed by a general survey of the phase diagram for ellipsoids of various length-to-width ratios [Frenkel, Mulder, and McTague 1984, 1985; Frenkel and Mulder, 1985; Perram, Wertheim, Lebowitz, and Williams 1984]. These simulations differed from the earlier work of Vieillard–Baron in only a few details, but the improvements turned out to be crucial. Firstly, the difficulty with pressure evaluation in a constant-volume system was avoided by conducting constant-pressure MC runs (Section 4.5) and by developing a new method for estimating the pressure in constant-volume simulations (Section 5.6) [Eppenga and Frenkel 1984]. Secondly, the coexistence points were established by computing free energies in a variety of ways, including biased-orientation test-particle insertion [Eppenga and Frenkel 1984] and thermodynamic integration along paths involving the application of external fields [Frenkel and Mulder 1985]. Thirdly, an alternative way of rejecting configurations of overlapping ellipsoids was introduced [Perram *et al.* 1984]. Using the improved methods, and a relatively small system ($N \approx 100$) it proved possible to generate a nematic liquid crystal. Indeed, spontaneous nucleation from the isotropic liquid was observed. Relaxation 'times' in and around the nematic phase were quite long, but hysteresis effects near the transition were not so severe as in the solid–liquid case. The transition was identified as being first order, with a very small change in density ($\sim 1$ per cent) and hence a small transition entropy, as observed in real systems. The phase diagram for hard ellipsoids is illustrated in Fig. 11.6. The system shows normal and plastic solid phases as well as nematic and isotropic liquids. The shapes of the liquid–solid coexistence lines reflect the different entropies, i.e. different degrees of ordering, in the various phases. The diagram is approximately symmetrical about the $a/b = 1$ hard sphere line. This $a/b \leftrightarrow b/a$ symmetry also appears in the thermodynamic properties. This is partly fortuitous, in that the second virial coefficient $B$ is the same for $a/b$ as it is for $b/a$, but higher coefficients are not.

How can we identify a liquid crystal phase in a simulation? In the real nematic, the director changes slowly with time and from place to place in the liquid. The length scale of director fluctuations will be large compared to a typical simulation box size, so a single director will apply to the whole sample at any instant. Slow fluctuations in 'time', however, may well occur in MD and, particularly, MC simulations, so identifying the director is a first step in calculating the nematic order parameter. This is the ensemble (or simulation) average of

$$P_2(t) = \frac{1}{N}\sum_i P_2(\cos\theta_i'(t)) = \frac{1}{N}\sum_i P_2(\mathbf{e}_i^s \cdot \mathbf{n}) \tag{11.18}$$

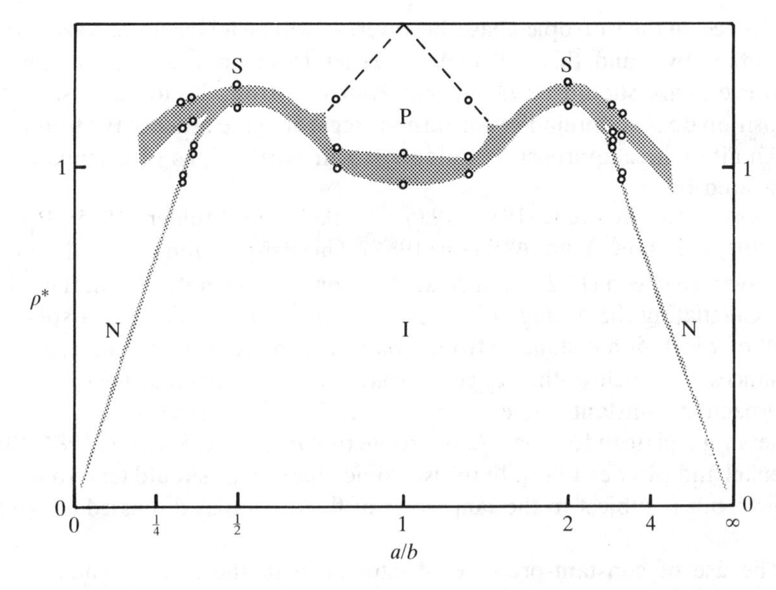


**Fig. 11.6** The phase diagram for hard ellipsoids of revolution. Note that the horizontal axis measures the length-to-width ratio $a/b$ on a non-linear scale, and that the density is in reduced units $\rho^* = \rho ab^2$. Circles are the MC results [Frenkel and Mulder 1985]. Shading denotes coexisting phases. I is the isotropic liquid, N the nematic liquid crystal, S the normal solid, and P the plastic solid. The limiting behaviour of the nematic-solid coexistence lines is not known.

where $\theta'_i$ is the angle between the molecular axis $\mathbf{e}_i^s$ and the director $\mathbf{n}$. The unknown director may be determined by maximizing $P_2$ with respect to rotations of $\mathbf{n}$. Writing $\mathbf{P}_2 = \mathbf{n}\cdot\mathbf{Q}\cdot\mathbf{n}$ where the ordering matrix $\mathbf{Q}$ is

$$Q_{\alpha\beta} = \frac{1}{2N}\sum_i 3e_{i\alpha}^s e_{i\beta}^s - \delta_{\alpha\beta} \tag{11.19}$$

reduces the problem to diagonalizing $\mathbf{Q}$. Suppose the eigenvalues of this matrix are $\lambda_+$, $\lambda_0$, and $\lambda_-$ in order of decreasing size. The largest positive eigenvalue $\lambda_+$ may be taken to be $P_2(t)$ and the corresponding eigenvector is the director [Zannoni 1979]. The other eigenvalues are $\lambda_0 \approx \lambda_- \approx -\frac{1}{2}P_2(t)$. In principle, we should diagonalize $\mathbf{Q}$ whenever we wish to calculate the order parameter. In practice, since director fluctuations are slow, the matrix elements may be block-averaged over a number of MC cycles or MD steps, before diagonalizing to obtain the director; this director may be used for several subsequent steps or cycles to calculate $P_2(t)$ through eqn (11.18). A difficulty arises when the three eigenvalues are comparable in magnitude, for example just on the isotropic side of the isotropic–nematic transition. $\lambda_+$ will then be positive, $\mathcal{O}(N^{-1/2})$. Eppenga and Frenkel [1984] consider order parameter fluctuations in some detail, and recommend taking $-2\lambda_0$ as the order parameter rather than $\lambda_+$. In the nematic phase this choice makes little difference; in the isotropic phase, however, $\lambda_0$ will be much closer to zero than the other two, and this will make it easier to distinguish the two phases. However, care should be taken that switching from $\lambda_+$ to $\lambda_0$ close to the transition does not prejudice conclusions regarding the order of the transition.

An alternative approach is to look at pair correlations of orientation, as measured by

$$g_2(r) = \langle P_2(\cos\gamma_{ij})\rangle_{\text{shell}}. \tag{11.20}$$

The average in eqn (11.20) is calculated by considering molecule $i$ at the origin and calculating the average of $P_2(\cos\gamma_{ij})$ over all molecules $j$ in a spherical shell of width $\delta r$ a distance $r$ from molecule $i$. In the nematic phase, at large distances (but below the range of spatial director fluctuations) $g_2(r)$ will approach a constant value equal to $\langle P_2\rangle^2$. Thus, measuring $g_2(r)$ and observing a plateau for $r < L/2$ is a route to $\langle P_2\rangle$ [Frenkel *et al.* 1984, 1985; Frenkel and Mulder 1985]. In the isotropic phase, $g_2(r)$ should tend to zero at large $r$, but is subject to the same sorts of fluctuation as discussed above for $\langle P_2\rangle$.

The use of constant-pressure Monte Carlo in these cases leads one to speculate that constant-volume simulations might have led to more problems with bottlenecks in the earlier work. In any case, the successful simulation of a nematic phase will no doubt encourage further work, to establish the possible roles of attractive interactions and flexibility for instance, and to calculate the elastic constants and transport properties. We may also see simulations of more complex liquid crystals (biaxials, smectics, cholesterics, for instance), and the examination of surface-induced ordering, all of which are of great practical interest.

## 11.6 Rotational dynamics

The reorientational correlation function of a single rigid molecule, $c_{lmm'}(t)$ (eqn (2.132)) is a measure of the degree of correlation between the orientation of a molecule at time $t$ and the same molecule at time 0. It answers the question about how well a molecule in a condensed phase remembers where it was pointing a time $t$ earlier. In principle, these correlation functions can be obtained experimentally from the Fourier transform of the infra-red spectrum ($l = 1$) and the Raman spectrum ($l = 2$), although there are considerable experimental and theoretical difficulties in making this connection [Yarwood 1984]. However, the comparison of the simulated correlation times with those measured experimentally, by optical spectroscopy or magnetic resonance, is a useful check on how well a particular intermolecular potential models the liquid dynamics. Apart from the comparison with experimental measurements, simulated correlation functions can be used to evaluate the numerous phenomenological theories of molecular reorientation in liquids. In addition the simulation provides a route to more fundamental correlation functions such as those of angular velocity $c_{\omega\omega}(t)$, and torque $c_{\tau\tau}(t)$, which are not available experimentally. This section begins with a description of $c_{lmm'}(t)$ and $c_{\omega\omega}(t)$ and the connection between them for various model liquids. A number of simple models are examined in the light of the MD results. The techniques for calculating a correlation function are described in detail in Chapter 6.

Consider the angular velocity of a single molecule $\boldsymbol{\omega}_i^s(t)$ measured in the space-fixed frame. We are interested in the autocorrelation function

$$c_{\omega\omega}(t) = \langle \boldsymbol{\omega}_i^s(t) \cdot \boldsymbol{\omega}_i^s(0) \rangle / \langle \omega_i^2 \rangle . \tag{11.21}$$

In addition, for symmetric and asymmetric top molecules (such as $CH_3CN$ and $H_2O$ respectively) it is also useful to decompose $\boldsymbol{\omega}_i^s(t)$ into components directed along the principal axes $\mathbf{e}_{i\alpha}^s(t)$ of the molecule at time $t$:

$$c_{\omega_\alpha\omega_\alpha}(t) = \langle \omega_{i\alpha}(0)\omega_{i\alpha}(t)\mathbf{e}_{i\alpha}^s(0) \cdot \mathbf{e}_{i\alpha}^s(t) \rangle / \langle \omega_{i\alpha}^2 \rangle . \tag{11.22}$$

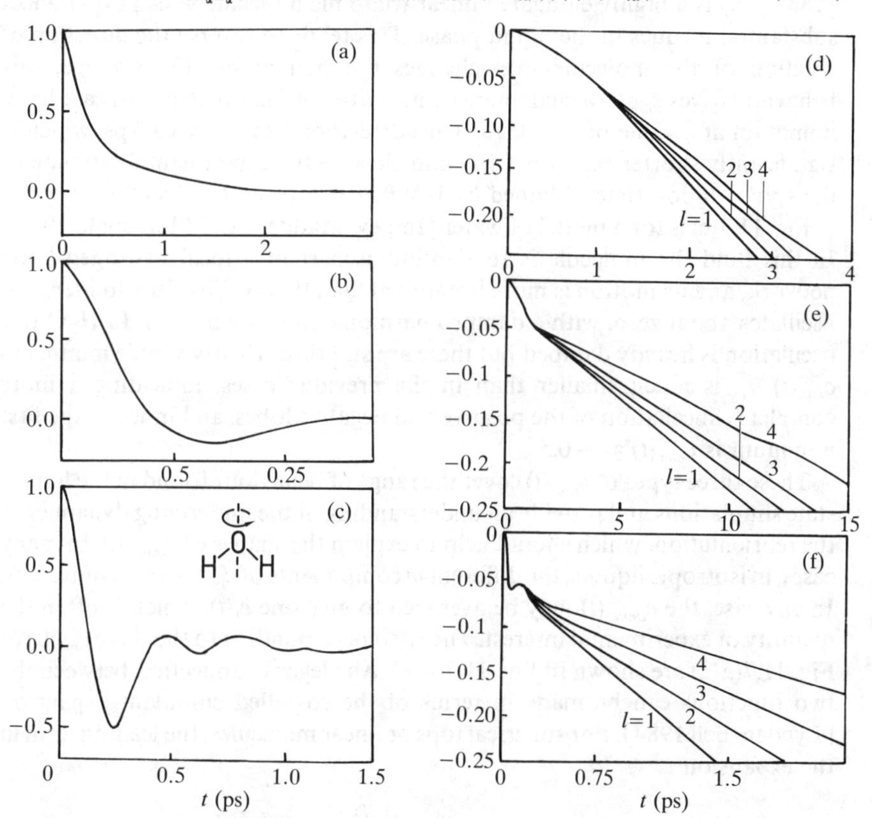


**Fig. 11.7 Rotational correlation functions.**
**(a) $c_{\omega\omega}(t)$ for model liquid $CX_4$, $T = 312$ K.**
**(b) $c_{\omega\omega}(t)$ for liquid $CS_2$, $T = 244$ K.**
**(c) $c_{\omega_z\omega_z}(t)$ for $H_2O$, $T = 286$ K.**
**(d) (e), and (f) are the orientational correlation functions, corresponding to (a), (b) and (c) respectively. We plot $(l(l+1))^{-1} \ln c_l(t)$ vs $t$ for various $l$ values.**

For spherical tops (e.g. $CH_4$) and linear molecules (e.g. $N_2$) $c_{\omega\omega}(t)$ and $c_{\omega_\alpha\omega_\alpha}(t)$ are identical. Typical shapes for these correlation functions in the liquid phase are shown in Fig. 11.7.

For a typical tetrahedral molecule $CX_4$, resembling $CBr_4$ [Lynden-Bell and McDonald 1981], at a high temperature and a low density, Fig. 11.7(a), the influence of the neighbouring molecules decorrelates the angular velocity slowly, on a time scale of $t_\omega \approx 0.418$ ps. The behaviour in Fig. 11.7(a) is quite different from the ideal gas where $c_{\omega\omega}(t)$ would remain constant at a value of unity. The observed behaviour is typical of molecules whose forward rotational motion is, on average, weakly damped (i.e. a low-torque fluid). In this case the initial decay of the correlation function must be quadratic, but at longer times it is well represented by an exponential (and so $t_\omega^{-1} \approx$ slope).

Fig. 11.7(b) is for liquid $CS_2$ on the orthobaric curve [Tildesley and Madden 1983]. $CS_2$ is a highly elongated linear triatomic molecule which experiences substantial torques in the liquid phase. This tends to reverse the direction of rotation of the molecules, i.e. changes the sign of $\boldsymbol{\omega}_i$. On average, this behaviour gives $c_{\omega\omega}(t)$ a characteristic negative region, which in this case has a minimum at a value of $\sim -0.15$. The correlation time $t_\omega \approx 0.06$ ps, which is significantly shorter than for $CBr_4$ and close to the experimental estimate of the spin rotation time obtained by NMR (0.076 ps at $T = 249$ K).

Fig. 11.7(c) is for a model of water [Impey, Madden, and McDonald 1982]. In this fluid the molecule is reorienting as part of a local hydrogen-bond network, and its motion is more librational than that of $CS_2$. In a solid, $c_{\omega\omega}(t)$ oscillates about zero, with a damped harmonic functional form. In $H_2O$ this oscillation is heavily damped but there are still three clearly visible minima in $c_{\omega\omega}(t)$. $t_\omega$ is much smaller than in the previous cases, indicating a more complete cancellation of the positive and negative lobes, and in water the first minimum is $c_{\omega\omega}(t) \approx -0.5$.

These three types of $c_{\omega\omega}(t)$ cover the range of behaviour found in the liquid-state simulations and provide an understanding of the underlying dynamics of the reorientation, which should help to explain the shapes of $c_{lmm'}(t)$. In many cases, in isotropic liquids, the different $m$ components of $c_{lmm'}(t)$ are equivalent. In any case, the $c_{lmm'}(t)$ may be averaged to give one $c_l(t)$, which is often the quantity of experimental interest. The $c_l(t)$ corresponding to the three $c_{\omega\omega}(t)$ of Fig. 11.7(a–c) are shown in Fig. 11.7(d–e). An elegant connection between the two functions can be made in terms of the so-called cumulant expansion [Lynden-Bell 1984]. For spherical tops or linear molecules, the leading term in the expansion is

$$\ln c_l(t) \approx -l(l+1)\,\langle \omega_x^2 \rangle \int_0^t (t-t')\,c_{\omega\omega}(t')\,\mathrm{d}t' \tag{11.23}$$

where $\langle \omega_x^2 \rangle = k_B T/I$, $I$ being the moment of inertia. Higher-order terms (sometimes called non-Gaussian terms) in the cumulant expansion contain functions of four and six angular velocities and the reader is referred to Lynden-Bell [1984] for more details.

Equation (11.23) can be used to obtain the limiting behaviour of $\ln c_l(t)$. At long times, when the decay of $c_{\omega\omega}(t)$ is complete, then

$$\ln c_l(t) \approx -l(l+1)\langle \omega_x^2 \rangle (t_\omega t + \text{constant}) \tag{11.24}$$

and a plot of $(l(l+1))^{-1} \ln c_l(t)$ against $t$ should be linear in time. This linear portion is present in Fig. 11.7(d–f). It is the part of $c_l(t)$ most readily available from experiment, and it gives no information about the short-time or collisional dynamics of the molecules. At short times, $c_{\omega\omega}(t) \approx 1$ and

$$\ln c_l(t) \approx -l(l+1)\langle \omega_x^2 \rangle t^2/2 \tag{11.25}$$

i.e. the function is quadratic in $t$. The nature of the change from quadratic to linear behaviour depends on the precise form of $c_{\omega\omega}(t)$. We can understand this by differentiating $\ln c_l(t)$:

$$\frac{\mathrm{d}}{\mathrm{d}t} \ln c_l(t) \approx -l(l+1)\langle \omega_x^2 \rangle \int_0^t c_{\omega\omega}(t')\,\mathrm{d}t'$$

$$\frac{\mathrm{d}^2}{\mathrm{d}t^2} \ln c_l(t) \approx -l(l+1)\langle \omega_x^2 \rangle c_{\omega\omega}(t). \tag{11.26}$$

If $c_{\omega\omega}(t)$ is always positive, then the first and second derivatives are always negative. Whenever $c_{\omega\omega}(t)$ changes sign, then there is a corresponding point of inflexion in $\ln c_l(t)$. Finally if the area of $c_{\omega\omega}(t)$ below the axis at any time $t$ exceeds the area above the axis, the slope of $\ln c_l(t)$ will be positive. Examples of these three kinds of behaviour can be seen in Fig. 11.7(d–f). Interestingly, if the cumulant expansion had converged at its leading term, eqn (11.23), the curves for the various $l$ values shown in Fig. 11.7 would be coincident. The effect of the higher-order cumulants is clear. Curves with higher $l$ values lie above those with $l = 1$, and this fanning out is a general phenomenon.

Correlation times measured from the half widths of the infra-red and Raman spectra should be associated with the limiting slope of $\ln c_l(t)$. Correlation times measured by NMR and other relaxation techniques give an estimate of the integral of $c_l(t)$. For an exponential reorientational correlation function, $c_l(t) = \exp(-t/t_l)$, these two methods give an identical $t_l$. In a low-torque liquid, $\ln c_l(t)$ approaches its limiting slope from below and $t_l(\text{slope}) > t_l(\text{integral})$. For a high-torque liquid, $\ln c_l(t)$ approaches its limiting slope from above and $t_l(\text{slope}) < t_l(\text{integral})$. This difference in correlation times, due to the different definitions, has been seen for $l = 1$ and $l = 2$ in the simulations of $CH_3CN$ and $CS_2$, and in the corresponding experiments.

## 11.7 Long-time tails

The hard-sphere molecular dynamics simulations of Alder and Wainwright [1969, 1970] provided one of the most interesting insights into liquid state dynamics. This was the revelation that the velocity autocorrelation function does not decay exponentially at long times, but instead exhibits the much slower dependence $c_{vv}(t) \sim t^{-d/2}$, where $d$ is the spatial dimensionality of the system. This algebraic long-time tail makes a substantial contribution to the diffusion coefficient in three dimensions, through eqn (2.107). In two dimensions, the integral of $c_{vv}(t)$ diverges, and so the diffusion law is not obeyed. The slow decay of the velocity correlations may be explained by kinetic [Dorfman and Cohen 1970, 1972, 1975] and mode coupling [Ernst, Hauge, and van Leeuwen 1970, 1971, 1976a, b] theories, but there is a simple underlying hydrodynamic picture [Alder and Wainwright 1970; Alder, 1986]. A moving atom compresses the liquid in front of it, rarefies the liquid behind it, and causes a vortex flow to circulate around it. The vortex creates a long-time 'push' from behind. The vortex velocity field can be thought of as occupying a volume of the fluid whose linear dimensions grow diffusively ($\sim t^{1/2}$) and whose volume increases $\sim t^{d/2}$. Momentum conservation in this region leads to the magnitude of the push felt by the central atom decreasing as $\sim t^{-d/2}$. Alder and Wainwright painstakingly calculated this velocity field in the two-dimensional case. Their results [Alder and Wainwright 1970], illustrated schematically in Fig. 11.8, agree very well with the predictions of continuum hydrodynamics for times longer than 10 collisions.

Hydrodynamic theories predict correctly the magnitude of the tail in $c_{vv}(t)$ which, as can be seen from the above argument, is positive. In dense fluids of course, $c_{vv}(t)$ shows a negative 'back-scattering' effect at moderately short times (see Fig. 2.3), but the asymptotic long-time behaviour is still as described above. The long-time tail is an example of a cooperative effect influencing single-particle motion, and breaking down the assumption of molecular chaos that underlies the simplest kinetic theory calculations. Its discovery by computer simulation is also an example of a result that would have been very difficult to obtain in any other way.

The long-time behaviour of other correlation functions, such as the off-diagonal pressure or stress autocorrelation function which determines the viscosity, has been the subject of much discussion in recent years. The stress can be divided into a kinetic or ideal part and a potential or excess contribution

$$\mathscr{P}_{\alpha\beta} = \mathscr{P}^{\text{id}}_{\alpha\beta} + \mathscr{P}^{\text{ex}}_{\alpha\beta} \tag{11.27}$$

and so its autocorrelation function can be broken down into three components: $\langle \mathscr{P}^{\text{id}}_{\alpha\beta}(t)\,\mathscr{P}^{\text{id}}_{\alpha\beta}(0)\rangle$, $\langle \mathscr{P}^{\text{ex}}_{\alpha\beta}(t)\,\mathscr{P}^{\text{ex}}_{\alpha\beta}(0)\rangle$, and $\langle \mathscr{P}^{\text{id}}_{\alpha\beta}(t)\,\mathscr{P}^{\text{ex}}_{\alpha\beta}(0)\rangle = \langle \mathscr{P}^{\text{ex}}_{\alpha\beta}(t)\;\mathscr{P}^{\text{id}}_{\alpha\beta}(0)\rangle$. An early study of the two-dimensional hard-disk fluid [Wainwright *et al.* 1971] suggested the existence of a long-time tail in the kinetic part as predicted by theory [Ernst *et al.* 1976a], but seemed also to

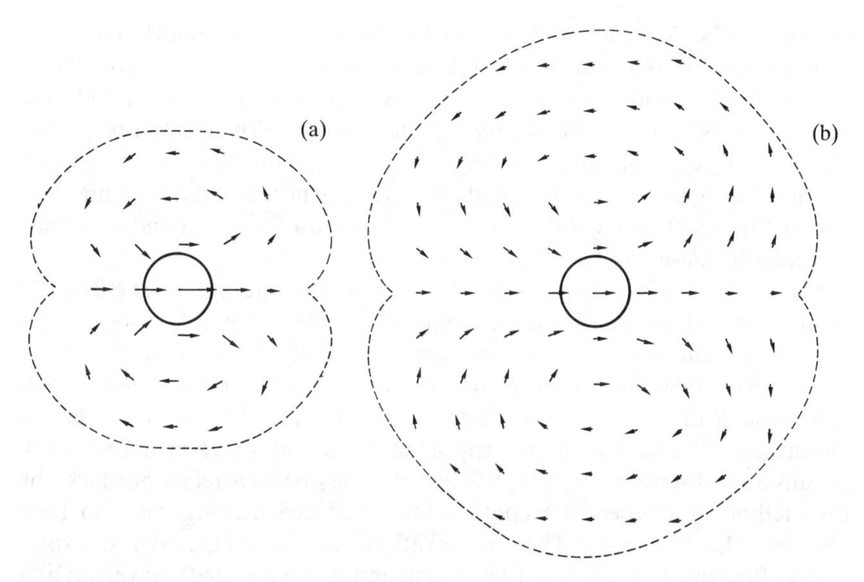


**Fig. 11.8** The velocity field surrounding a central atom in a two-dimensional fluid. The central atom has a velocity $v$ at time $t = 0$. (a) Velocity field a short time later. (b) Velocity field a long time later.

show long-time tails in the other components, which are not in agreement with theoretical predictions. This study was not conclusive, but more extensive equilibrium simulations of the hard-sphere fluid [Erpenbeck and Wood 1981], and especially the non-equilibrium oscillatory shear simulations of the Lennard-Jones fluid [Evans, 1980, 1981c, 1983b; Hoover *et al.* 1980b] seem to show a long-time tail with contributions from all three components, the potential term outweighing the others, and the overall magnitude being many times the theoretical prediction. In the NEMD simulations (see Section 8.7) a $t^{-3/2}$ long-time tail appears as a limiting square-root dependence, $\omega^{1/2}$, in the frequency-dependent shear viscosity.

Reservations have been expressed regarding these results [see for example Alder 1983, 1986]. It had been known for a long time that stress relaxation in dense liquids is sluggish [Alder *et al.* 1970; Levesque *et al.* 1973]. This has led to the term 'molasses tail' for the slow decay of the stress correlation function. There is no way to guarantee that the correlations have been followed for long enough in an MD simulation for the asymptotic behaviour to set in. In equilibrium simulations the 'sound wave' problem is a limiting factor (see Section 6.4.4), while in the fitting of NEMD data, much weight is inevitably given to the high-frequency results which are cheaper to obtain and hence usually of greater statistical precision.

The long-time tail question for $\langle \mathscr{P}_{\alpha\beta}(t)\mathscr{P}_{\alpha\beta}(0) \rangle$ also seems to be linked to the shear-rate dependence of $\eta$ as observed in NEMD. Once again, theory predicts a square-root dependence of $\eta$ on the rate of strain [Kawasaki and

Gunton 1973; Yamada and Kawasaki 1975; Ernst, Cichocki, Dorfman, Sharma, and van Beijeren 1978] while some NEMD simulations show such a dependence but with a much larger numerical coefficient [Naitoh and Ono 1976, 1978, 1979; Evans 1979b, 1981c]. Other work [Erpenbeck, 1983, 1984] shows no square-root dependence at low strain rate. This point deserves further investigation, particularly since there may be a theoretical connection between the coefficients of the square-root terms for frequency and shear-rate dependence [Zwanzig 1981].

Some of the most interesting information to come out of the NEMD simulations, which may have a bearing on the long-time tails, concerns the structural changes which occur on shearing a liquid [Heyes, Kim, Montrose and Litovitz 1980; Heyes, Montrose and Litovitz 1983; Evans 1983b]. This may even lead to phase transitions [Erpenbeck 1984; Heyes, 1986] in which ordered chains of atoms appear in the liquid; however, recent work [Evans and Morriss 1986] suggests that this may be an artefact produced by the method of temperature control. Shear-induced melting has also been observed [Evans 1982b]. These observations may be compared with experiment [Beysens, Gbadamassi 1979; Clark and Ackerson 1980; Ackerson and Clark 1983].

Long-time tails may also exist in other correlation functions. Hoover and co-workers [Hoover *et al.* 1980a, b] have carried out NEMD studies of the autocorrelation function of diagonal elements of the pressure tensor, finding the familiar square-root frequency-dependence at low frequency. They point out that for soft spheres interacting through potentials of the form of eqn (1.9), a simple proportionality exists between the kinetic, potential, and cross-contributions to the frequency-dependent bulk viscosity. Thus, if one component has a long-time tail, all components must do so; moreover the potential contributions will dominate in all cases of interest. The resolution of the apparent differences between simulation and theory in these cases is a problem of great current interest.

## 11.8 Interfaces

The computer simulation of interfaces has been thoroughly surveyed by Nicholson and Parsonage [1982]. In this section, we consider a number of studies that illustrate the different techniques used in this area.

We begin by considering the solid–liquid interface. In the simplest model, the solid surface may be represented as a structureless, smooth, rigid, hard wall; in fact, a fixed external potential acting on the fluid molecules. As we shall see, more sophistication is needed to represent a real solid surface, but rigid walls are a useful starting point for comparisons with experiment and tests of theory.

Out of the large number of simulations of this kind, we shall pick a series of studies by Henderson and van Swol [Henderson and van Swol 1984, 1985; van

Swol and Henderson 1986], using hard sphere and square-well potentials. In this work, periodic boundaries in one direction ($z$, say) are replaced by a pair of rigid confining walls, while periodicity in the other two 'transverse' directions is preserved. As well as the infinitely repulsive contributions to the potentials, either or both of the molecule–molecule and molecule–wall interactions may have an attractive square-well contribution (eqn (1.8)).

The most basic quantity of interest is the fluid density profile near a wall, and a technical question arises immediately: where is the wall? Henderson and van Swol [1984] point out that it is correctly defined as the plane beyond which molecular centres cannot penetrate, i.e. it lies half a diameter in from the intuitive position of contact with the hard-sphere surfaces. In their simulations of hard spheres near freezing between two hard walls, Henderson and van Swol show that the density profile rises as the distance (from the true wall) is decreased from $\approx \frac{1}{2}\sigma$ to zero. Further from the wall, the profile typically oscillates with distance, indicating a layered structure. The transverse pressure components, defined in one of the ways mentioned in Section 11.1, oscillate in a similar way, while the normal pressure component is independent of position. Henderson and van Swol also measured pair correlations near the wall, i.e. $g(s_{ij}, r_{iz}, r_{jz})$ where $s_{ij}^2 = r_{ijx}^2 + r_{ijy}^2$. They found that the function at contact with the wall, $g(s_{ij}, 0^+, 0^+)$, is very similar to the bulk $g(r)$, having somewhat more pronounced oscillations with longer wavelength.

The situation becomes much more interesting when attractive interactions are present: 'wetting' phenomena can occur. Consider a solid in contact with vapour as one approaches the bulk liquid–vapour coexistence curve. Liquid may be adsorbed at the solid surface: either partial wetting to give surface droplets or complete wetting to give a film covering the whole surface may occur. In the latter case, the degree of adsorption diverges as one approaches bulk coexistence. As one moves along the coexistence curve away from the bulk critical point, one expects a changeover from complete to partial wetting at some temperature $T_w$. This wetting transition may be first order (the divergence in the adsorption disappears abruptly) or continuous. We can reverse the roles of liquid and vapour here and consider instead wetting by the vapour, or 'drying'.

Henderson and van Swol have simulated the square-well fluid against hard walls [Henderson and van Swol 1985] and against square-well walls [van Swol and Henderson 1986]. For hard walls, complete wetting by vapour occurs along the entire bulk liquid–vapour coexistence curve and no wetting transition is seen. The system allows a detailed study of the approach to complete wetting, starting in the bulk liquid region and reducing the density by moving the walls apart. As the bulk coexistence curve is approached, with the bulk density varying smoothly, the density profile near the walls drops dramatically, indicating complete wetting by vapour: a vapour film develops between solid and liquid. It was shown that this is associated with capillary wave fluctuations in the liquid–vapour interface, and a diverging correlation length, parallel to the wall. In the subsequent study incorporating attractive molecule–wall contributions, the wetting transition was observed. A square-well fluid was confined between one square-well wall and one hard wall. This arrangement made it possible to maintain liquid–vapour coexistence, by having a free interface in the system, with the vapour phase being next to the hard wall. This interface was sufficiently far from the square-well wall not to interfere with the wetting behaviour of interest. Instead of moving along the bulk coexistence curve by varying the temperature, van Swol and Henderson chose a fixed point on the curve and varied the magnitude of the wall–molecule attraction, thus moving the wetting temperature $T_w$. Strong evidence in favour of a first order change from partial wetting to complete wetting by vapour was obtained.

These simulations went hand-in-hand with a detailed theoretical analysis, and there were also some interesting technical features. Standard Metropolis Monte Carlo was used for preliminary simulations and equilibration, but the production runs employed the molecular dynamics method. This was partly because evidence emerged of bottleneck problems with MC, and partly because it is more straightforward to obtain the pressure in a dynamic simulation. In fact, a check that the normal component $P^N$ was constant throughout the system was an essential indicator of equilibration. This quantity was determined via the usual virial expression and compared with the value at the walls given by the average momentum transferred in wall collisions and also by extrapolating the density profile to contact. To maintain the desired temperature, velocity rescaling was used at each square-well collision, thus simulating the constant-$NV\mathcal{T}$ ensemble (Section 7.4.3). Finally, grand canonical Monte Carlo was not used. This would correspond to simulation of an open capillary, in which case the wetting transition would most likely be pre-empted by capillary condensation (or in this case, evaporation). In other words, if the total number of molecules is allowed to vary, the simulation box will tend to completely fill up (or in this case completely empty) rather than show the desired surface adsorption. As is common in surface work, large simulation boxes were required here, both in the longitudinal direction (to minimize interference between the walls and allow several interfaces to exist) and transverse (to allow structural correlations to develop). With system sizes varying up to 8192 atoms, the link-list method (Section 5.3.2) was essential in speeding up the program.

More exotic liquids can be studied in the presence of hard, structureless walls. Heyes and Clarke [1981] have studied a Tosi–Fumi potential model of molten KCl near a charged hard wall, representing an electrode surface. Valleau, Torrie and co-workers [Torrie and Valleau 1980, 1982; Torrie, Valleau, and Patey 1982; Valleau and Torrie, 1984; Carnie and Torrie, 1984] have examined the behaviour of the primitive model under similar conditions. In the study of KCl, using the MD method, the imposition of a realistic electric field ($10^9$ Vm$^{-1}$) between the walls polarizes the sample, with $K^+$ ions migrating to the negative electrode. However, this effect is not dramatic, and does not establish alternating planes of opposite charge near the electrode. The single-ion density function $\rho(z)$ is determined primarily by packing against the wall, and is highly structured because of this. It is also clear from this study that, even for a reasonably large sample size $N = 504$, making the walls typically 5 nm apart, the strength and range of Coulombic interactions causes interference between the structures at the two walls. The variation in ion densities across the simulation box partly motivated the choice of grand canonical Monte Carlo for the primitive model simulations. With a strongly varying single-particle density, we can no longer be certain of the density of the bulk fluid with which the system is in equilibrium. Specifying the chemical potential, rather than the number of particles, means that the state point of the system is no longer in doubt. In all this work, the shortcomings of the Ewald method, particularly its overemphasis of fluctuating dipole correlations, have been re-examined (see Section 5.5.5). Valleau and co-workers [Valleau and Whittington 1977a; Valleau 1980] use an image charge method, which is described in detail in the original publications. Heyes [1983d] has suggested a somewhat different formulation of the lattice sum for this application. A primary result of these studies is that the simplest theory of ionic liquids at a charged electrode, namely the modified Gouy–Chapman theory, is surprisingly successful at describing the single-particle distributions, for singly charged ions in a solvent of moderate relative permittivity $\varepsilon$. As the charges go up, or $\varepsilon$ goes down, more sophisticated theories are required. Clearly these simulations, with the need for large systems and the handling of long-range forces, are at the limits of available computing resources at present.

Real surfaces are not structureless. Consider an adatom at position $\mathbf{r}_i$ near to a graphite surface in the $x$–$y$ plane. The external potential $v_s(\mathbf{r}_i)$ acting on the adatom should depend on all components of $\mathbf{r}_i$, not just $r_{iz}$. One possibility is to write the potential as a sum of interactions with carbon atoms at positions $\mathbf{r}_c$ within the graphite

$$v_s(\mathbf{r}_i) = \sum_c v_{ic}(|\mathbf{r}_i - \mathbf{r}_c|). \tag{11.28}$$

This is an extension of the site–site model discussed in Chapter 1; $v_{ic}(r)$ might typically be a Lennard-Jones potential. To obtain an accurate representation of $v_s(\mathbf{r}_i)$ we might need to include hundreds of surface atoms. A useful simplification [Steele 1974] is to expand in the reciprocal lattice vectors of the surface

$$v_s(\mathbf{r}_i) = v_0(r_{iz}) + \sum_k v_k(r_{iz}) f_k(r_{ix}, r_{iy}) . \tag{11.29}$$

The coefficients $v_k$ depend only on the height of the adatom, and the basis set $f_k(r_{ix}, r_{iy})$ depends on the symmetry of the surface. The sum is over wave numbers $k$ corresponding to $k$-vectors in the surface. For graphite, the expansion is rapidly convergent and inclusion of $k = 0$ and $k = 1$ reproduces $v_s(r_i)$ to better than 1 per cent. If $v_{ic}(r)$ is a Lennard-Jones potential, then a closed form can be obtained for the coefficients. $v_0(r_{iz})$ still involves a sum over graphite planes, but this can be accurately approximated using an Euler–Maclaurin expansion. In calculating $v_1(r_{iz})$, it is usual to consider just the first plane of graphite atoms. For the hexagonally symmetrical basal planes of graphite,

$$f_1(r_{ix}, r_{iy}) = -2\left[\cos\frac{2\pi}{a}\left(r_{ix} + \frac{r_{iy}}{\sqrt{3}}\right) + \cos\frac{2\pi}{a}\left(r_{ix} - \frac{r_{iy}}{\sqrt{3}}\right) + \cos\frac{4\pi r_{iy}}{a\sqrt{3}}\right] \tag{11.30}$$

where $a$ is the magnitude of the graphite lattice vector. Further details of this model are given elsewhere [Joshi and Tildesley 1985]. The details of the expansion for a variety of different surface symmetries are also available [Steele 1974].

In MD simulations with a static external field, energy is conserved but momentum is not. The periodic boundary conditions must match the symmetry of the external field. This is not a problem when the adsorbed phase is essentially fluid-like, or when it is a solid in registry with the substrate, but clearly a conflict arises when a solid phase forms with a different periodicity from the underlying lattice (an incommensurate solid). In addition, simulation techniques involving a continuously changing box-shape or size cannot be conducted with a static external field of this kind, and so as yet we have no method of simulating at constant spreading pressure for these models.

As an example of the use of this technique, we show in Fig. 11.9 a snapshot from a simulation of $N_2$ on graphite [Talbot, Tildesley, and Steele 1985]. The molecules form a monolayer which behaves like a two-dimensional liquid. At this relatively high density, many molecules are forced to tilt out of the surface. At lower densities, molecules are found preferentially parallel to the surface.

In some simulations, we must allow the surface atoms complete freedom of movement. A typical example is the study of surface melting of a Lennard-Jones f.c.c. crystal in a vacuum [Broughton and Woodcock 1978]. The aim is to model a semi-infinite f.c.c. lattice, with a vacuum surface in the $z$-direction and periodic boundary conditions in the $x$- and $y$-directions. The four uppermost layers, comprising 1024 atoms, are free to move. The lower layers, which are not expected to be involved in the surface melting, are represented by a static external field. This model goes some of the way towards a completely realistic simulation, although the presence of the static field and the periodic boundaries both act to artificially stabilize the crystal.

A plot of internal energy as a function of temperature showed three clear discontinuities corresponding to the distinct melting of each of the first three

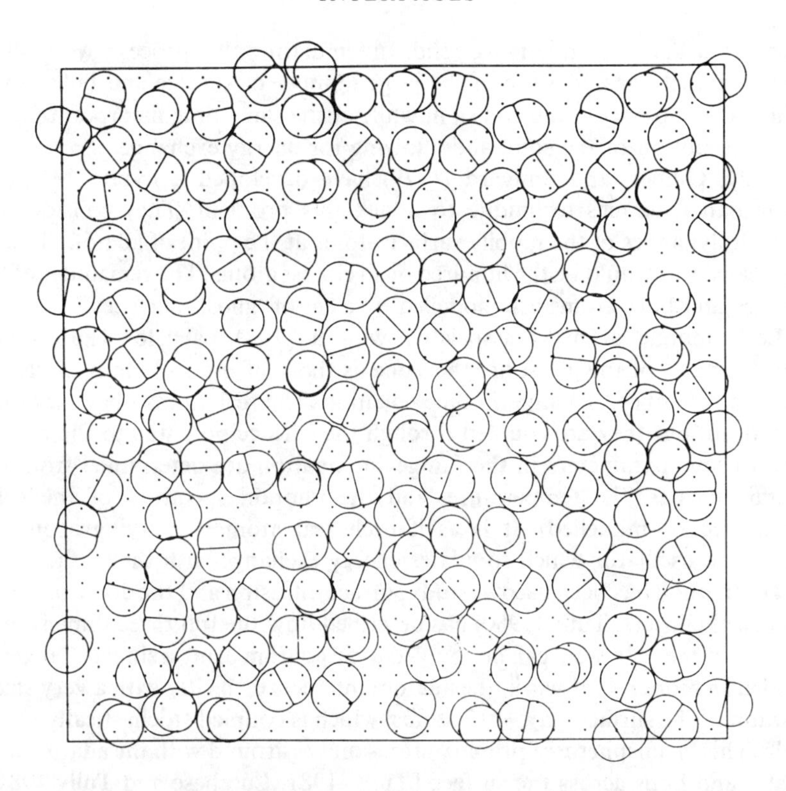

**Fig. 11.9** A snapshot from a simulation of $N_2$ adsorbed on graphite, showing the projection of the molecules on the surface. The coverage is 5.45 $nm^{-2}$ and the temperature is approximately 80 K. (This diagram was supplied by J. Talbot, Southampton).

layers. An interesting change in the density profile was observed on melting. The peaks corresponding to the four layers were broadened, and the average lattice spacing in the $z$-direction increased, with temperature. This change was marked for the top two layers. Striking time-dependent properties were also seen at a temperature just above melting of the top layer.

The mean-squared displacements for atoms in this layer increased linearly with time as in a liquid (see Fig. 6.5(a)). At the same time, in the second layer, a plateau value characteristic of the solid was quickly achieved. The overall conclusion was that the surface melted in a stepwise succession of intralayer transitions. The mechanism for each melting transition involved the promotion of atoms to the fluid layer above and an increased mobility in the depleted layer.

In some studies of surfaces, the use of any static external field is inappropriate. An important example is the scattering of atoms from surfaces. This process will involve a possible rearrangement of the surface atoms at the point of collision, surface diffusion of the adatom, and energy exchange between it and the underlying solid. In simulating this process, we wish to avoid explicit consideration of the large number of surface and bulk atoms which play no part in the collision, while at the same time incorporating the thermal motion of the solid and allowing for energy exchange.

Tully, Gilmer, and Shugard (1979) have developed a hybrid MD and Brownian dynamics technique to attack this problem. They consider the explicit dynamics of the mobile surface atoms at the point of adatom impact. Typically, this would be the hundred or so closest atoms. The remainder of the surface and bulk atoms are modelled by a set of 'ghost' particles.

Each mobile atom is associated with a ghost particle. The $x$- and $y$-coordinates of the ghost are the same as those of the real atom, while the $z$-coordinate obeys a Langevin-type equation. The real atoms obey Newton's equations, but they are coupled through the surface potential to the ghosts. The various parameters in the Langevin equation are determined from the atomic masses, the temperature, and the phonon spectrum of the solid (measured or modelled). It is as if each real atom is carrying around a Brownian oscillator which simulates energy exchange with the surface.

The method has been used to study surface diffusion and the dissociation of adsorbed clusters. Tully (1980) has described how the full three-dimensional motion of the bulk solid can be coupled to an adatom using a similar Langevin model. In more recent work, the adatom interacts explicitly with a very small portion of the surface (say $\sim$ 12 atoms) which is coupled stochastically to the bulk. This 'hemisphere' of primary atoms moves around with the adatom as it skates and hops across the surface [Tully, 1981; Lucchese and Tully 1984].

# APPENDICES

## APPENDIX A

# COMPUTERS AND COMPUTER SIMULATION

## A.1 Computer hardware

In this appendix, we attempt to describe the 'tools' of a computer simulator's trade: the experimental apparatus; the computer and its software and programming languages. Computer simulations require large quantities of computer time, but this does not mean that they are restricted to the most expensive and powerful machines. The ability to conduct uninterrupted runs for hours, perhaps days, on a dedicated small machine may make this a very attractive alternative.

Such is the speed of development of computer hardware that almost any statement made regarding current capabilities is liable to be out of date very rapidly. Nonetheless, we think it useful to give a summary of the types of machines that may be used for computer simulation. We discuss the hardware broadly in order of increasing power and expense.

Firstly, we have the smallest microcomputers, produced by a wide range of manufacturers, which are creating such an impact on the domestic and business fronts. These machines are, typically, based on a single microprocessor chip, usually an 8-bit or 16-bit processor; however, the 32-bit chips used in much larger machines are now beginning to make an appearance in micros. Microcomputers suffer from the disadvantages of slow speed, low memory ($\leqslant$ 128 kbyte) and, in the main, the restriction to simple interpretive languages, such as BASIC, or assembler code. Advantages are low cost, making the machines very accessible, and the great emphasis on interactive graphics. For serious computer simulation, the current generation of microcomputers are of limited power, although they clearly have a future in educational applications. These micros can also serve a purpose as 'intelligent terminals', allowing local program editing and graphics while acting as a means of communicating with a much larger machine. Here they begin to merge with the more powerful single-user workstation (exemplified by the Three Rivers/ICL PERQ of Fig. 1.1). These differ from the microcomputer in that they are faster, have more memory, support high level language compilers (such as FORTRAN, C, PASCAL), and provide more sophisticated software development tools (editors). With a continuing emphasis on graphics, these machines are ideal for the 'molecular mechanics' applications of computer simulation (user-controlled enzyme docking manoeuvres for example). As dedicated machines, they can also be useful for simple liquid state simulations, with the advantages of sophisticated graphical output of results.

The workstations are essentially single-user machines. Here they differ from the mini- and midi- ranges (typified by the PDP-11 and VAX-11 series from DEC), which are multi-user computers allowing interactive use simultaneously with background ('batch') jobs. These machines are characterized by memory in the Mbyte range; in the more modern computers, however, there is almost no memory limitation, due to the introduction of the virtual machine approach: rapid transfers to and from disk drives allow a trade-off of speed against memory requirements, which is almost transparent to the user. Computers in this range are ideal for computer simulation, and provide excellent interactive facilities for program development together with a choice of languages and standard packages and substantial computing power at modest cost. Moreover, a range of add-on processors (of the pipeline type, see below) has appeared (the Floating Point Systems array processors are examples), which greatly increase the power and make large-scale simulations possible.

For large-scale computations, a mainframe computer may be needed. Here, the manufacturers (IBM and CDC are well known) provide a powerful 'number cruncher' with a very wide range of languages, libraries, and packages. Although these machines are not always the most suitable for developing programs, they have been the mainstay of research in computer simulation for years. Their only real disadvantage is cost: mainframes are usually found only in the larger universities or at regional computing centres. Even more expensive are the latest supercomputers. The machines discussed above are serial machines, i.e. one instruction is carried out after another in the central processor. Ultimately, there are limits on the speed of such machines, imposed firstly by the amount of heat which must be dissipated in a rapidly operating unit, and secondly by the finite speed of light, which restricts the speed at which information can be passed around the system. Supercomputers use changes in the architecture to improve throughput. Pipeline machines or vector processors (such as the CRAY-1 and the CYBER-205) attempt to optimize the speed with which a sequence of elementary operations is carried out on a large set of data. Suppose that each item of data 1, 2, 3 . . . is to be subjected to the same set of operations $A \rightarrow B \rightarrow C \rightarrow D$. A might be fetching a number from store, B taking its modulus, C taking its square root, and D replacing it in store. Then overall throughput is maximized by feeding the data consecutively into an operation 'pipe'. Schematically:

```
                                   |       OPERATION PIPE         |
                                   |                              |
              INPUT STREAM      -> |   A -> B -> C -> D           | -> OUTPUT STREAM
                                   |                              |
STEP 1      5,    4,    3,    2    |   1                          |
STEP 2      6,    5,    4,    3    |   2    1                     |
STEP 3      7,    6,    5,    4    |   3    2    1                |
STEP 4      8,    7,    6,    5    |   4    3    2    1           |
STEP 5      9,    8,    7,    6    |   5    4    3    2           |  1
STEP 6     10,    9,    8,    7    |   6    5    4    3           |  2, 1
  :                     :          |   :    :    :    :           |  :
  :                     :          |   :    :    :    :           |  :
STEP 104  108,  107,  106,  105    |  104  103  102  101          |  100, 99, 98 ... 3, 2, 1
```

To process 100 numbers requires (here) 104 steps, the time taken being dictated by the slowest of the operations A, B, C, D. On a serial machine, 400 steps would be needed. These vector operations are treated very efficiently on a pipeline machine. There are some restrictions on the possible tasks that can be performed this way: for example, no decisions based on the values of the data can be taken within the pipeline operation (i.e. no IF statements). We return to this shortly; suffice it to say here that most molecular dynamics programs can be 'vectorized', whereas most Monte Carlo algorithms cannot. The second class of supercomputer currently available is the genuine array processor (typified by the ICL DAP). This is simply a set of processors working in parallel; the processors each operate on their own set of data, although there may be inter-processor connections to allow more complex operations. Also, logical masks may be used to switch some of the processors off. This approach is simple and very powerful, although it does require a new approach to many programming problems. The DAP, for instance, uses a parallel language adaptation of FORTRAN. Molecular dynamics may be approached in this way, but Monte Carlo is not intrinsically a 'parallel' technique, so only limited applications in this area exist.

Finally, we turn to machines constructed especially for the purpose of computer simulation. Monte Carlo simulations of Ising-type models are obvious candidates for purpose-built computers [Pearson, Richardson, and Toussaint 1983; Hoogland, Spaa, Selman, and Compagner 1983] but one molecular dynamics machine, for a general atomic system with a tabulated potential, has been described [Bakker, Bruin, van Dieren, and Hilhorst 1982]. Although these machines are outstandingly fast, they are of course restricted in application and required substantial input of effort and expertise to construct. Recently, this situation has changed. Developments in integrated circuit technology mean that a complicated parallel processor can be contained on one chip. An example of this technology is the INMOS transputer. Each transputer contains a 32-bit processor, its own memory and some fast communications links. Using a large number of transputers in parallel, it may be possible to construct a machine comparable to a supercomputer at a fraction of the cost. Although transputers work in parallel, different processors can perform different instructions on different sets of data at the same time. This multiple instruction, multiple data architecture makes the transputer array more versatile than the DAP. In the future, computer simulations may well be performed on microcomputers with transputer boards owned by individual research groups. Current developments in computer architecture have recently been surveyed from the viewpoint of computer simulation [Berendsen 1984; Abraham 1985].

## A.2 Programming languages

Historically, FORTRAN (FORmula TRANslation) has been the most common programming language for computer simulation, in view of its suitability for scientific applications, with a variety of mathematical functions, and the availability of FORTRAN compilers producing efficient code. The importance of extracting the most from the computer in the early days of simulation is emphasized by the use of assembler coding, instead of FORTRAN, for the most time-consuming program sections (the 'inner loop'). With some exceptions, the situation has changed somewhat since the early days. Now, computer time is more plentiful, and the gaps in speed between assembler and compiled FORTRAN, and between FORTRAN compilers and other languages, have narrowed. Rather more significant now is the programmer effort required to translate an idea into a working program. A scientist's time is valuable, and more structured programming languages (C, PASCAL, ALGOL-68) may be preferable to FORTRAN (at least to the older FORTRAN-66 or FORTRAN-IV dialects) in this respect. Having said this, the FORTRAN-77 standard incorporates many of the elements of these languages, notably the IF THEN . . . ELSE . . . ENDIF construct, which greatly simplify the writing, and final appearance, of programs. There are still some drawbacks to FORTRAN-77 (restriction to six-character variable names, absence of DO WHILE . . . ENDDO and CASE statements, and recursion, although these may appear in extended versions of the language on many machines) but the widespread implementation of FORTRAN-77, the features built into it, and its historical position, still make it the natural choice for the examples in this book.

We have attempted to adopt certain features of FORTRAN-77 consistently throughout. We use block-IF constructions instead of jumps to labelled statements, and avoid GOTO statements, because this clarifies the program structure. We make little use of COMMON blocks, preferring to pass information via subprogram argument lists, again for clarity, although the storage of the major arrays in a program (the positions, velocities, forces etc.) in COMMON may be the most efficient. In general, we have tried to follow consistent rules of style, indenting DO-loops and IF-blocks etc., to make the program easier to follow, in accord with recommended practice [Balfour and Marwick 1979; Ledgard and Chmura 1978].

Some features of FORTRAN-77 have been deliberately avoided. We have made no use of the FORTRAN initial letter convention (according to which, variables beginning with the letters I, J, K, L, M, N are assumed to be of INTEGER type, and all others are assumed to be REAL, unless otherwise stated). This allows us to adopt more mnemonic names for variables, and it is in any case recommended practice to declare all variables at the start of a program segment. In our examples, we either declare the type of each variable, or assume it to be obvious from the context. In the same way, we do not use the IMPLICIT statement to set our own initial letter convention. We should mention one non-FORTRAN-77 feature which is implemented on some compilers, namely the IMPLICIT NONE statement: this forces the user to declare all variables (flagging those which are not explicitly declared) and is extremely useful in program development, for catching misspelt variables and possible clashes of variable use. The FORTRAN-77 statement IMPLICIT CHARACTER* 99 (A–Z) can be used to have almost the same effect: any variables not explicitly declared are almost certain to show up sooner rather than later!

In our program examples, we have used REAL variables for floating-point operations. On some machines (e.g. the CDC 7600) REAL variables consist of 60 bits, and are therefore quite accurate enough for most computer simulation work. On 32-bit machines, however, it is more common to use DOUBLE PRECISION variables for accuracy, at least in molecular dynamics. There is not the same need for numerical precision in Monte Carlo. It is not even clear that DOUBLE PRECISION is necessary for MD simulations; very similar transport coefficients result from simulations conducted with both 32-bit and 64-bit variables, although there may be more doubt about the accuracy of time correlation functions computed from single-precision MD trajectories. DOUBLE PRECISION calculations are, of course, slower than REAL ones, so some experimentation in this field may be worthwhile.

The exceptions to the use of FORTRAN mentioned above relate to the non-standard programming required on vector (pipeline) and array processors. Perhaps the most obvious feature of these machines is that functions are provided that refer to entire vectors or matrices at once. In DAP-FORTRAN, for example, MATRIX variables consisting, typically, of $64 \times 64 = 4096$ elements may be defined, and 4096 additions or multiplications carried out simultaneously by a simple statement like $A = B + C$; a serial machine would use a lengthy DO-loop. Many pipeline machines have functions for manipulating vectors of any length in a convenient and efficient fashion. In terms of program structure, pipeline and array machines tend to require the replacement of IF statements by logical masks. For example, to implement a spherical cutoff, DAP-FORTRAN might use the following statements:

```
TOO_FAR_APART = ( RIJSQ .GT. RCUTSQ )
VIJ ( TOO_FAR_APART ) = 0.0
WIJ ( TOO_FAR_APART ) = 0.0
```

Here, RCUTSQ contains the squared potential cutoff distance, RIJSQ is a MATRIX variable containing all the squared pair separations, and VIJ, WIJ similarly contain all the pair potential and pair virial terms. The LOGICAL MATRIX TOO_FAR_APART (note that DAP-FORTRAN allows very descriptive variable names!) picks out all those pairs which are out of range, and is used as a mask in two statements which set the relevant potential and virial elements to zero. Note that, on a parallel machine, there is nothing to be gained by avoiding the calculation of some of the pair interactions, if we approach the problem this way. In the same way, we see in Section 1.5 a function which performs the same task of spherical cutoff on the CRAY-1S. Operations of the kind described above are almost certain to be incorporated into the next internationally agreed FORTRAN standard.

Finally, although some low-cost add-on vector processors will compile high-level 'vectorized' languages, for some applications on some machines it becomes essential to program in the low-level assembler code unique to the machine. Typically, this involves considering a step-by-step implementation of various machine functions, integer additions, floating-point multiplies, memory fetches etc., which can be carried out simultaneously. If these operations take different amounts of time (for example, two machine cycles for an integer addition, three cycles for a floating-point multiply, one cycle for a memory fetch) then the whole process becomes very much like the composition of a musical 'round', with the different operations taking the parts of different voices.

## A.3 Efficient programming in FORTRAN-77

We considered some possible general ways of improving the efficiency of computer simulation programs in Chapter 5; here, we draw the reader's attention to some machine-dependent points which may have a bearing on execution speed.

In the text, we have assumed that exponentiation (**) is the slowest FORTRAN arithmetic operation, followed by division (/), multiplication (*), and finally addition and subtraction (+, −). Operations on integers may be slower than floating-point operations; operations on double-precision numbers are always slower. Much depends on whether all or some of the operations are performed by hardware or by (slower) software. On most machines, which are designed to manipulate 4-byte, 32-bit numbers, it is not usually the case that INTEGER*2 or LOGICAL*1 operations are significantly faster than INTEGER*4 or LOGICAL*4. FORTRAN mathematical functions, such as SQRT, EXP, and SIN, are almost invariably slower than any of the arithmetic operations.

These factors have dictated the form of some of the statements in the text and on microfiche. If several divisions by the same number are to be performed, we frequently compute the inverse number first and carry out several multiplications by the inverse instead. Also, we take care to avoid square roots if possible. There are some points to watch out for. It may be quicker to compute X + X than to calculate 2.0 *X. On the other hand, if X is an INTEGER variable, the compiler may recognize 2 *X and X/2 as simple bit-shifting operations, which can be performed very rapidly. In the same way, X *X may be faster than X**2, and X * X * X may be faster than X **3. This will be true if exponentiation is performed, in effect, by taking logs. On the other hand, the compiler may recognize squaring, cubing etc. as operations that may be optimized, so X**2 might be faster. The rule, as always, is to try it on your machine and see.

In rare instances, the above hierarchy is completely changed. The DAP, for instance, is built around bitwise processors. For these, operations on short variables (INTEGER*1, REAL*3) are faster than operations on standard length ones, while LOGICAL operations (on one bit) are exceptionally fast. All floating-point operations are performed by software, and so are comparatively slow. Functions such as SQRT and EXP are no longer to be avoided at all costs.

The second main point concerns the use of arrays. Array references are time consuming, hence the use of statements such as RXI = RX(I) for frequently used variables that appear inside an inner loop. Often, references to multi-dimensional arrays are especially slow. In the text, therefore, we have used three separate arrays, RX(N), RY(N), RZ(N), rather than a single array R(3, N), for position vectors. For the same reason, some workers prefer to use a large, one-dimensional array R of length 3N, with successive elements storing $(x, y, z)$ values. When we come to molecular systems, particularly non-rigid ones, when we may wish to store several site positions for each molecule, then the alternative to the use of arrays such as RX(I, A), RY(I, A), RZ(I, A) for $\mathbf{r}_{ia}$ becomes rather complicated, and we have stuck to this usage for simplicity. When multi-dimensional arrays are used, some attention should be paid to the order in which the elements are referenced: the default order of storage in FORTRAN has the first index varying most rapidly and the last index most slowly, and this may affect the speed of access. It is frequently recommended that constructions such as

```
          DO 100 I = 1, IMAX
             DO 90 J = 1, JMAX
                ... refer to ARRAY(I,J) ...
90           CONTINUE
100       CONTINUE
```

should be avoided; the alternative

```
          DO 100 J = 1, JMAX
             DO 90 I = 1, IMAX
                ... refer to ARRAY(I,J) ...
90           CONTINUE
100       CONTINUE
```

is often much faster.

Our final point concerns the use of FORTRAN functions such as ANINT in the implementation of the minimum image convention and periodic boundary correction. There are many ways of coding the minimum image calculation, some involving functions, some involving tests. On pipeline machines we should avoid IF statements, but on serial machines they are worth considering. Suppose we have calculated the vector between two particles $i$ and $j$ by statements such as

```
RXIJ = RX(I) - RX(J)
```

and similar statements in the $y$ and $z$ directions. Then the statement

```
RXIJ = RXIJ - BOXL * ANINT ( RXIJ / BOXL )
```

or

```
RXIJ = RXIJ - BOXL * ANINT ( RXIJ * BOXI )
```

where BOXL is the box length, and BOXI its inverse, with corresponding statements for $y$ and $z$, will carry out minimum imaging in a cubic box for any initial pair of images of particles $i$ and $j$. These statements become more efficient when working in a box of unit length:

```
RXIJ = RXIJ - ANINT ( RXIJ )
```

Curiously, on some vector processors, the ANINT function may inhibit vectorization. Some workers have used the AINT function instead

```
RXIJ = RXIJ - BOX * AINT ( RXIJ / BOXL2 )
```

where BOXL2 is half the box length. Here it is easiest to use a box length of two, but this method requires that $i$ and $j$ be in the same box. With the same proviso, the following statements will also perform the minimum image correction

```
IF ( RXIJ .GT. BOXL2 ) THEN
   RXIJ = RXIJ - BOXL
ELSEIF ( RXIJ .LT. -BOXL2 ) THEN
   RXIJ = RXIJ + BOXL
ENDIF
```

Two tests may be avoided if we put

```
IF ( ABS ( RXIJ ) .GT. BOXL2 ) THEN
   RXIJ = RXIJ - SIGN ( BOXL, RXIJ )
ENDIF
```

In the case that we only wish to find the distance between $i$ and $j$, not the components of the separation vector, as in a Monte Carlo simulation or in a calculation of the tables for $g(r)$, then we may use

```
DXIJ = ABS ( RX(I) - RX(J) )
IF ( DXIJ .GT. BOXL2 ) DXIJ = BOXL - DXIJ
```

together with similar statements for $y$ and $z$, to compute $r_{ij}^2$.

# APPENDIX B
# REDUCED UNITS

## B.1 Reduced units

For systems consisting of just one type of molecule, it is sensible to use the mass of the molecule as a fundamental unit, i.e. set $m_i = 1$. As a consequence, the particle momenta $\mathbf{p}_i$ and velocities $\mathbf{v}_i$ become numerically identical, as do the forces $\mathbf{f}_i$ and acceleraions $\mathbf{a}_i$. This approach can be extended further. If the molecules interact by pair potentials of a simple form, i.e. like the Lennard-Jones potential (eqn (1.6)) they are completely specified by a few parameters such as $\varepsilon$ and $\sigma$, then further fundamental units of energy, length etc. may be defined. From these definitions, units of other quantities (pressure, time, momentum etc.) follow directly. Static and dynamic properties of the Lennard-Jones system are invariably quoted in reduced units

| | | |
|---|---|---|
| density | $\rho^* = \rho\sigma^3$ | (B.1) |
| temperature | $T^* = k_B T/\varepsilon$ | (B.2) |
| energy | $E^* = E/\varepsilon$ | (B.3) |
| pressure | $P^* = P\sigma^3/\varepsilon$ | (B.4) |
| time | $t^* = (\varepsilon/m\sigma^2)^{1/2}t$ | (B.5) |
| force | $\mathbf{f}^* = \mathbf{f}\sigma/\varepsilon$ | (B.6) |
| torque | $\boldsymbol{\tau}^* = \boldsymbol{\tau}/\varepsilon$ | (B.7) |
| surface tension | $\gamma^* = \gamma\sigma^2/\varepsilon$ | (B.8) |

and so on. The reduced thermodynamic variables determine the state point or, to be precise, a set of corresponding states with closely related properties. Quite generally, if the potential takes the form $v(r) = \varepsilon f(r/\sigma)$, there is a principle of corresponding states which applies to thermodynamic, structural, and dynamic properties [Helfand and Rice 1960]. Thus, the Lennard-Jones potential may be used to fit the equation of state for a large number of systems [Rahman 1964; McDonald and Singer 1972]. For the even simpler soft-sphere potential of eqn (1.9), a single reduced variable $(\rho\sigma^3)(\varepsilon/k_B T)^{3/\nu}$ defines the excess (i.e. non-ideal) properties [see e.g. Hoover, Ross, Johnson, Henderson, Barker, and Brown 1970; Hoover *et al.* 1971]. In the limit of the hard sphere potential (formally corresponding to $\nu \to \infty$) the temperature becomes a totally redundant variable so far as static quantities are concerned, and enters the dynamic properties only through the definition of a reduced time

$$t^* = (k_B T/m\sigma^2)^{1/2}t\,. \tag{B.9}$$

The use of reduced units avoids the possible embarrassment of conducting essentially duplicate simulations. There are also technical advantages in the use of reduced units. If parameters such as $\varepsilon$ and $\sigma$ have been given a value of unity, they need not appear in a computer simulation program at all; consequently some time will be saved in the calculation of potential energies, forces etc. Of course, the program then becomes unique to the particular functional form of the chosen potential. For complicated potentials, with many adjustable parameters, or in the case of mixtures of species, there is only a slight technical advantage to be gained by choosing one particular energy parameter, one characteristic length, and one molecular mass, to be unity.

In SI units Coulomb's law is

$$v^{zz}(r_{ij}) = z_i z_j/(4\pi\varepsilon_0 r_{ij}) \tag{B.10}$$

where $z_i$ and $z_j$ are charges in Coulombs, and $r_{ij}$ is their separation in metres. $\varepsilon_0 = 8.8542 \times 10^{-12}\ \mathrm{C^2N^{-1}m^{-2}}$ is the permittivity of free space. In reduced units based on the Lennard-Jones energy and length parameters, the charge, dipole, and quadrupole are

$$\begin{aligned} z^* &= z/(4\pi\varepsilon_0\sigma\varepsilon)^{1/2} \\ \mu^* &= \mu/(4\pi\varepsilon_0\sigma^3\varepsilon)^{1/2} \\ Q^* &= Q/(4\pi\varepsilon_0\sigma^5\varepsilon)^{1/2}. \end{aligned} \tag{B.11}$$

Many older papers give the moments in electrostatic units. Useful conversion factors are

$$\begin{aligned} &\text{Charge:} && 1\ \mathrm{C} = 2.9979 \times 10^9\ \text{e.s.u.} \\ &\text{Dipole:} && 1\ \mathrm{C\,m} = 2.9979 \times 10^{11}\ \text{e.s.u. cm} \\ &\text{Quadrupole:} && 1\ \mathrm{C\,m^2} = 2.9979 \times 10^{13}\ \text{e.s.u. cm}^2. \end{aligned} \tag{B.12}$$

It is convenient to use an alternative definition of the unit of charge, whether or not other reduced units are employed. In most of this book eqn (B.10) is used without the factor $4\pi\varepsilon_0$. In this case the charge $z_i$ is divided by $(4\pi\varepsilon_0)^{1/2}$ and has units of m $\mathrm{N}^{1/2}$.

# APPENDIX C

# CALCULATION OF FORCES AND TORQUES

## C.1 Introduction

The correct calculation of the forces and torques resulting from a given potential model is essential in the construction of a properly functioning molecular dynamics program. In this appendix, we consider forces, and, where appropriate, torques, arising from three complicated potential models:

(a) a polymer chain with constrained bond lengths, but realistic bond angle bending and torsional potentials;
(b) a molecular fluid of linear molecules, where the permanent electrostatic interactions are handled using a multipole expansion;
(c) a fluid of atoms with three-body interactions modelled using the Axilrod–Teller triple–dipole potential.

The formulae given here will be useful to anyone constructing simulation programs containing these potential models. In addition, the methods of derivation may assist the reader in handling a range of more complicated potentials.

## C.2 The polymer chain

The model of a polymer consists of $n_a$ atoms linked by rigid bonds. The angle between successive bonds, $\theta_a$, and the torsional angle $\phi_a$ defined by three successive bond vectors, are both allowed to vary. The way in which the atoms and angles are labelled is shown in Fig. C.1. If the bond vector between atoms $a$–1 and $a$ is $\mathbf{d}_a = \mathbf{r}_a - \mathbf{r}_{a-1}$, then $\theta_a$ may be calculated from

$$\cos\theta_a = \frac{\mathbf{d}_a \cdot \mathbf{d}_{a-1}}{|\mathbf{d}_a|\,|\mathbf{d}_{a-1}|} \tag{C.1}$$

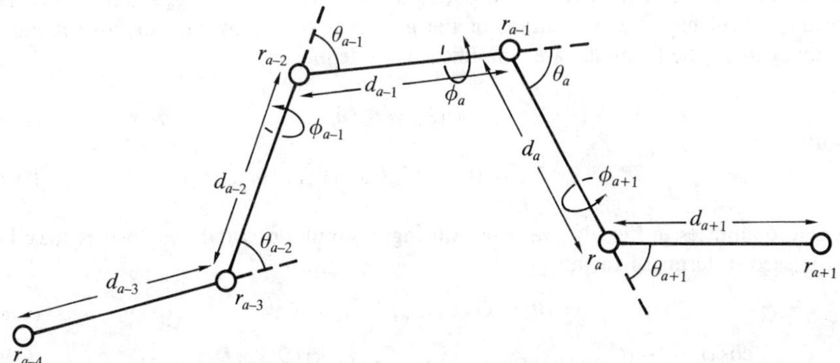


**Fig. C.1** A polymer chain. The bending angle $\theta_a$ is the angle between the bond vectors $\mathbf{d}_a$ and $\mathbf{d}_{a-1}$. The torsional angle $\phi_a$ is the angle between the plane defined by $\mathbf{d}_a$ and $\mathbf{d}_{a-1}$ and that defined by $\mathbf{d}_{a-1}$ and $\mathbf{d}_{a-2}$.

and $\phi_a$ may be obtained from

$$\cos\phi_a = -\frac{(\mathbf{d}_a \times \mathbf{d}_{a-1})\cdot(\mathbf{d}_{a-1}\times\mathbf{d}_{a-2})}{|\mathbf{d}_a \times \mathbf{d}_{a-1}|\,|\mathbf{d}_{a-1}\times\mathbf{d}_{a-2}|}. \tag{C.2}$$

Associated with each $\theta_a$ will be a bond bending potential term, which might take the quadratic form

$$v_\theta(\theta_a) = \tfrac{1}{2}k_\theta(\theta_a - \theta_a^0)^2 \tag{C.3}$$

where $\theta_a^0$ is the equilibrium value. Associated with the torsional angles will be potential terms of the form exemplified in Fig. 1.8, i.e. a sum of trigonometric functions such as

$$v_\phi(\phi_a) = \sum_k c_k \cos^k\phi_a \tag{C.4}$$

where the coefficients $c_k$ define the potential and the summation is truncated at (say) $k = 6$. The precise forms of eqns (C.3) and (C.4) are not necessary for the following discussion, however. We assume that the dynamics of this model will be solved by a constraint algorithm such as SHAKE or RATTLE (see Chapter 3), which is based on atomic motion. The key quantities to be calculated, then, are the forces on each atom due to intermolecular potentials (typically straightforward atom–atom) and due to potentials such as those of eqns (C.3), (C.4). Following the general approach of Pear and Weiner [1979], it is possible to obtain expressions for these forces. The position coordinate of atom $a$ will appear in the potential energy expressions for bending $v_\theta(\theta_a)$, $v_\theta(\theta_{a+1})$, and $v_\theta(\theta_{a+2})$, and also in the potential energies for torsion, $v_\phi(\phi_a)$, $v_\phi(\phi_{a+1})$, $v_\phi(\phi_{a+2})$, and $v_\phi(\phi_{a+3})$. Hence there will be contributions to the force on atom $a$ from all these sources. These contributions are evaluated by simple differentiation:

$$\begin{aligned}\mathbf{f}_a &= \sum_{c=a}^{a+2} -\nabla_{\mathbf{r}_a} v_\theta(\theta_c) + \sum_{c=a}^{a+3} -\nabla_{\mathbf{r}_a} v_\phi(\phi_c)\\ &= \sum_{c=a}^{a+2} -\left(\frac{\mathrm{d}v_\theta(\theta_c)}{\mathrm{d}\cos\theta_c}\right)\nabla_{\mathbf{r}_a}\cos\theta_c + \sum_{c=a}^{a+3} -\left(\frac{\mathrm{d}v_\phi(\phi_c)}{\mathrm{d}\cos\phi_c}\right)\nabla_{\mathbf{r}_a}\cos\phi_c.\end{aligned} \tag{C.5}$$

We assume that the derivatives of the potentials with respect to $\cos\phi_a$ and $\cos\theta_a$ may be readily calculated. The evaluation of the gradients of the cosine functions is more complicated. The formulae are simplified if we define

$$C_{ab} = C_{ba} = \mathbf{d}_a\cdot\mathbf{d}_b \tag{C.6}$$

and

$$D_{ab} = D_{ba} = C_{aa}C_{bb} - C_{ab}^2. \tag{C.7}$$

These quantities are easily evaluated during a simulation, and the cosines may be expressed in terms of them:

$$\cos\theta_a = C_{aa-1}(C_{aa}C_{a-1a-1})^{-1/2} \tag{C.8}$$

$$\cos\phi_a = -(C_{aa-1}C_{a-1a-2} - C_{aa-2}C_{a-1a-1})(D_{aa-1}D_{a-1a-2})^{-1/2}. \tag{C.9}$$

There are a simple set of rules governing the vector differentiation of the $C$ and $D$ functions with respect to the position of atom $a$:

$$
\begin{aligned}
\nabla_{\mathbf{r}_a} C_{aa} &= 2\mathbf{d}_a \\
\nabla_{\mathbf{r}_a} C_{aa+1} &= \mathbf{d}_{a+1} - \mathbf{d}_a \\
\nabla_{\mathbf{r}_a} C_{a+1a+1} &= -2\mathbf{d}_{a+1} \\
\nabla_{\mathbf{r}_a} C_{ab} &= \mathbf{d}_b \qquad (b \neq a, a+1) \\
\nabla_{\mathbf{r}_a} C_{a+1b} &= -\mathbf{d}_b \qquad (b \neq a, a+1) \\
\nabla_{\mathbf{r}_a} C_{bc} &= 0 \qquad (b, c \neq a, a+1)
\end{aligned} \tag{C.10}
$$

and

$$
\begin{aligned}
\nabla_{\mathbf{r}_a} D_{aa+1} &= 2C_{a+1a+1}\mathbf{d}_a - 2C_{aa}\mathbf{d}_{a+1} - 2C_{aa+1}\mathbf{d}_{a+1} + 2C_{aa+1}\mathbf{d}_a \\
\nabla_{\mathbf{r}_a} D_{ab} &= 2C_{bb}\mathbf{d}_a - 2C_{ab}\mathbf{d}_b \qquad (b \neq a, a+1) \\
\nabla_{\mathbf{r}_a} D_{a+1b} &= -2C_{bb}\mathbf{d}_{a+1} + 2C_{a+1b}\mathbf{d}_b \qquad (b \neq a, a+1) \\
\nabla_{\mathbf{r}_a} D_{bc} &= 0 \qquad (b, c \neq a, a+1).
\end{aligned} \tag{C.11}
$$

These relations are derived with the help of the identity

$$
\nabla(\mathbf{A}\cdot\mathbf{B}) = (\mathbf{B}\cdot\nabla)\mathbf{A} + (\mathbf{A}\cdot\nabla)\mathbf{B} + \mathbf{B}\times(\nabla\times\mathbf{A}) + \mathbf{A}\times(\nabla\times\mathbf{B}) \tag{C.12}
$$

in which the two terms involving $\nabla\times$ vanish for the vectors involved here. The expressions required in eqn (C.5) are then

$$
\begin{aligned}
\nabla_{\mathbf{r}_a}\cos\theta_a &= -(C_{aa}C_{a-1a-1})^{-1/2}((C_{aa-1}/C_{aa})\,\mathbf{d}_a - \mathbf{d}_{a-1}) \\
\nabla_{\mathbf{r}_a}\cos\theta_{a+1} &= (C_{a+1a+1}C_{aa})^{-1/2}((C_{aa+1}/C_{a+1a+1})\mathbf{d}_{a+1} - (C_{aa+1}/C_{aa})\mathbf{d}_a \\
&\qquad + \mathbf{d}_{a+1} - \mathbf{d}_a) \\
\nabla_{\mathbf{r}_a}\cos\theta_{a+2} &= (C_{a+2a+2}C_{a+1a+1})^{-1/2}((C_{a+2a+1}/C_{a+1a+1})\mathbf{d}_{a+1} - \mathbf{d}_{a+2})
\end{aligned} \tag{C.13}
$$

and

$$
\begin{aligned}
\nabla_{\mathbf{r}_a}\cos\phi_a = &-(D_{aa-1}D_{a-1a-2})^{-1/2}(C_{a-1a-2}\mathbf{d}_{a-1} - C_{a-1a-1}\mathbf{d}_{a-2} \\
&- D_{aa-1}^{-1}(C_{aa-1}C_{a-1a-2} - C_{aa-2}C_{a-1a-1})(C_{a-1a-1}\mathbf{d}_a - C_{aa-1}\mathbf{d}_{a-1})) \\
\nabla_{\mathbf{r}_a}\cos\phi_{a+1} = &-(D_{a+1a}D_{aa-1})^{-1/2}(C_{aa-1}\mathbf{d}_{a+1} - C_{aa-1}\mathbf{d}_a \\
&+ C_{aa+1}\mathbf{d}_{a-1} + C_{aa}\mathbf{d}_{a-1} - 2C_{a+1a-1}\mathbf{d}_a \\
&- D_{aa-1}^{-1}(C_{aa+1}C_{aa-1} - C_{a+1a-1}C_{aa})(C_{a-1a-1}\mathbf{d}_a - C_{aa-1}\mathbf{d}_{a-1}) \\
&- D_{aa+1}^{-1}(C_{aa+1}C_{aa-1} - C_{a-1a+1}C_{aa}) \\
&\times (C_{a+1a+1}\mathbf{d}_a - C_{aa}\mathbf{d}_{a+1} - C_{aa+1}\mathbf{d}_{a+1} + C_{aa+1}\mathbf{d}_a)) \\
\nabla_{\mathbf{r}_a}\cos\phi_{a+2} = &-(D_{a+2a+1}D_{a+1a})^{-1/2}(-C_{aa+1}\mathbf{d}_{a+2} + C_{a+1a+2}\mathbf{d}_{a+1} \\
&- C_{a+1a+2}\mathbf{d}_a - C_{a+1a+1}\mathbf{d}_{a+2} + 2C_{aa+2}\mathbf{d}_{a+1} \\
&- D_{aa+1}^{-1}(C_{a+1a+2}C_{aa+1} - C_{aa+2}C_{a+1a+1}) \\
&\times (C_{a+1a+1}\mathbf{d}_a - C_{aa}\mathbf{d}_{a+1} - C_{aa+1}\mathbf{d}_{a+1} + C_{aa+1}\mathbf{d}_a) \\
&- D_{a+1a+2}^{-1}(C_{a+1a+2}C_{aa+1} - C_{aa+2}C_{a+1a+1}) \\
&\times (-C_{a+2a+2}\mathbf{d}_{a+1} + C_{a+1a+2}\mathbf{d}_{a+2}))
\end{aligned}
$$

$$\nabla_{\mathbf{r}_a} \cos \phi_{a+3} = -(D_{a+3a+2} D_{a+2a+1})^{-1/2} (-C_{a+2a+3} \mathbf{d}_{a+2} + C_{a+2a+2} \mathbf{d}_{a+3}$$
$$- D_{a+1a+2}^{-1} (C_{a+2a+3} C_{a+1a+2} - C_{a+1a+3} C_{a+2a+2})$$
$$\times (-C_{a+2a+2} \mathbf{d}_{a+1} + C_{a+1a+2} \mathbf{d}_{a+2})). \tag{C.14}$$

In this way the force on each atom $a$ may be calculated. We note that some of the terms in these equations will vanish if $a$ is close to the end of the polymer chain.

## C.3 The molecular fluid with multipoles

The methods for calculating the force and torque in an interaction site model fluid are described in Chapters 1 and 5. Here, we discuss the forces and torques which arise from the permanent electrostatic interactions within the framework of the multipole expansion. For simplicity, we take the example of linear molecules. The configuration of a pair of linear molecules is shown in Fig. C.2.

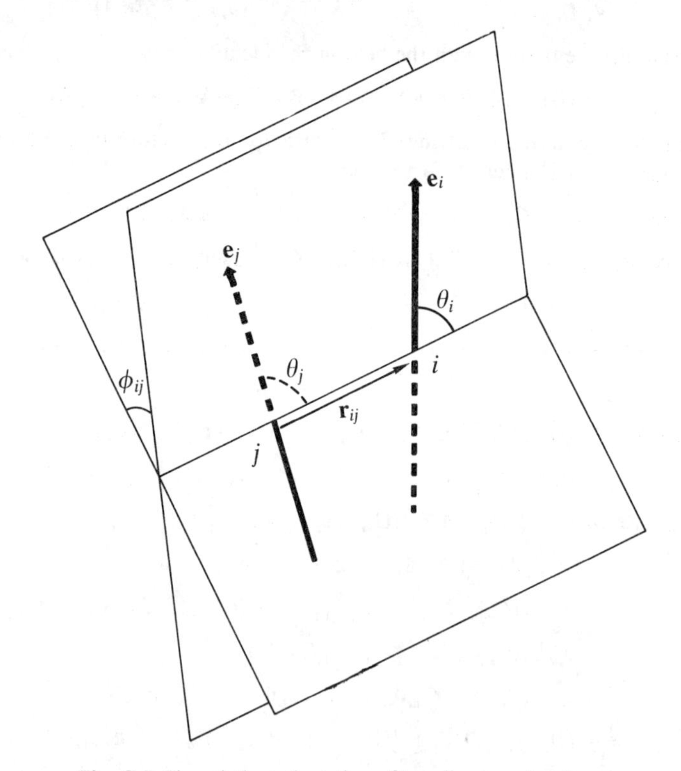


**Fig. C.2** The relative orientation of two linear molecules.

The centres of the molecules are separated by a vector $\mathbf{r}_{ij}$. $\theta_i$ and $\theta_j$ are the angles between $\mathbf{r}_{ij}$ and the unit vectors directed along the molecular axes, $\mathbf{e}_i$ and $\mathbf{e}_j$. $\phi_{ij}$ is the angle between the plane containing $\mathbf{e}_i$ and $\mathbf{r}_{ij}$ and that containing $\mathbf{e}_j$ and $\mathbf{r}_{ij}$. It is convenient to define an angle $\gamma_{ij}$ such that

$$\cos \gamma_{ij} = \mathbf{e}_i \cdot \mathbf{e}_j = \cos \theta_i \cos \theta_j + \sin \theta_i \sin \theta_j \cos \phi_{ij} \tag{C.15}$$

If molecules $i$ and $j$ each have a dipole moment of size $\mu$ and a quadrupole moment of size $Q$, then the electrostatic interaction energies are:

(a) dipole–dipole

$$v_{ij}^{\mu\mu} = (\mu^2/r_{ij}^3)(\cos\gamma_{ij} - 3\cos\theta_i \cos\theta_j)\,; \tag{C.16}$$

(b) dipole–quadrupole + quadrupole–dipole

$$v_{ij}^{\mu Q} = (3\mu Q/2r_{ij}^4)(\cos\theta_i - \cos\theta_j)(1 + 5\cos\theta_i\cos\theta_j - 2\cos\gamma_{ij})\,; \tag{C.17}$$

(c) quadrupole–quadrupole

$$\begin{aligned} v_{ij}^{QQ} = (3Q^2/4r_{ij}^5)(1 &- 5\cos^2\theta_i - 5\cos^2\theta_j \\ &- 15\cos^2\theta_i\cos^2\theta_j + 2(\cos\gamma_{ij} - 5\cos\theta_i\cos\theta_j)^2). \end{aligned} \tag{C.18}$$

The force on molecule $i$ due to $j$ is

$$\mathbf{f}_{ij} = -\nabla_{\mathbf{r}_{ij}} v_{ij} \tag{C.19}$$

where the arguments ($r_{ij}$, $\cos\theta_i$, $\cos\theta_j$, $\cos\gamma_{ij}$) are understood. Using the chain rule,

$$\begin{aligned} \mathbf{f}_{ij} = &-\left(\frac{\partial v_{ij}}{\partial r_{ij}}\right)\nabla_{\mathbf{r}_{ij}} r_{ij} - \left(\frac{\partial v_{ij}}{\partial \cos\theta_i}\right)\nabla_{\mathbf{r}_{ij}}\cos\theta_i \\ &-\left(\frac{\partial v_{ij}}{\partial \cos\theta_j}\right)\nabla_{\mathbf{r}_{ij}}\cos\theta_j - \left(\frac{\partial v_{ij}}{\partial \cos\gamma_{ij}}\right)\nabla_{\mathbf{r}_{ij}}\cos\gamma_{ij}\,. \end{aligned} \tag{C.20}$$

The angle $\gamma_{ij}$ is independent of $\mathbf{r}_{ij}$, so the last term vanishes. Using $\cos\theta_i = (\mathbf{e}_i \cdot \mathbf{r}_{ij})/r_{ij}$, we obtain

$$\nabla_{\mathbf{r}_{ij}}\cos\theta_i = -\cos\theta_i \mathbf{r}_{ij}/r_{ij}^2 + \mathbf{e}_i/r_{ij} \tag{C.21}$$

and a similar result for $\nabla_{\mathbf{r}_{ij}}\cos\theta_j$. Combining these results gives

$$\begin{aligned} \mathbf{f}_{ij} = &-\left(\frac{\mathbf{r}_{ij}}{r_{ij}}\right)\left(\frac{\partial v_{ij}}{\partial r_{ij}}\right) - \left(\frac{\mathbf{e}_i}{r_{ij}} - \mathbf{r}_{ij}\frac{\cos\theta_i}{r_{ij}^2}\right)\left(\frac{\partial v_{ij}}{\partial \cos\theta_i}\right) \\ &-\left(\frac{\mathbf{e}_j}{r_{ij}} - \mathbf{r}_{ij}\frac{\cos\theta_j}{r_{ij}^2}\right)\left(\frac{\partial v_{ij}}{\partial \cos\theta_j}\right). \end{aligned} \tag{C.22}$$

Now we turn to the evaluation of the torque on molecule $i$ due to molecule $j$, which is defined by

$$\boldsymbol{\tau}_{ij} = -\mathbf{e}_i \times \nabla_{\mathbf{e}_i} v_{ij}\,. \tag{C.23}$$

We should only consider the component of the gradient tangential to the vector $\mathbf{e}_i$, but in fact any non-physical radial component will disappear on taking the vector product, and so we can ignore this complication. Again applying the chain rule,

$$\begin{aligned} \nabla_{\mathbf{e}_i} v_{ij} = &\left(\frac{\partial v_{ij}}{\partial r_{ij}}\right)\nabla_{\mathbf{e}_i} r_{ij} + \left(\frac{\partial v_{ij}}{\partial \cos\theta_i}\right)\nabla_{\mathbf{e}_i}\cos\theta_i \\ &+\left(\frac{\partial v_{ij}}{\partial \cos\theta_j}\right)\nabla_{\mathbf{e}_i}\cos\theta_j + \left(\frac{\partial v_{ij}}{\partial \cos\gamma_{ij}}\right)\nabla_{\mathbf{e}_i}\cos\gamma_{ij}\,. \end{aligned} \tag{C.24}$$

The first and third terms vanish, and we obtain finally

$$\boldsymbol{\tau}_{ij} = -\mathbf{e}_i \times \left[ \frac{\mathbf{r}_{ij}}{r_{ij}} \left( \frac{\partial v_{ij}}{\partial \cos\theta_i} \right) + \mathbf{e}_j \left( \frac{\partial v_{ij}}{\partial \cos\gamma_{ij}} \right) \right]. \tag{C.25}$$

Note that the force and torque on molecule $j$ due to $i$ can be obtained by interchanging the labels and changing the signs of $\cos\theta_i$ and $\cos\theta_j$. From eqns (C.22) and (C.25) we see that $\mathbf{f}_{ij} = -\mathbf{f}_{ji}$ but that $\boldsymbol{\tau}_{ij} \neq -\boldsymbol{\tau}_{ji}$. As an example of the use of these equations, the force and torque between a pair of dipoles are

$$\mathbf{f}_{ij} = -\mathbf{f}_{ji} = \frac{3\mu^2}{r_{ij}^4} \Big( (\cos\gamma_{ij} - 5\cos\theta_i \cos\theta_j)(\mathbf{r}_{ij}/r_{ij}) + \cos\theta_j \mathbf{e}_i + \cos\theta_i \mathbf{e}_j \Big) \tag{C.26a}$$

$$\boldsymbol{\tau}_{ij} = -\frac{\mu^2}{r_{ij}^3} (\mathbf{e}_i \times \mathbf{e}_j - 3\cos\theta_j (\mathbf{e}_i \times \mathbf{r}_{ij})/r_{ij}) \tag{C.26b}$$

$$\boldsymbol{\tau}_{ji} = -\frac{\mu^2}{r_{ij}^3} (\mathbf{e}_j \times \mathbf{e}_i - 3\cos\theta_i (\mathbf{e}_j \times \mathbf{r}_{ij})/r_{ij}). \tag{C.26c}$$

The development in this section is based on a paper by Cheung [1976]. Price, Stone, and Alderton [1984] have given a more formal and thorough development which includes the electrostatic interactions for non-linear molecules. In both these papers, the convention employed is $\mathbf{r}_{ij} = \mathbf{r}_j - \mathbf{r}_i$, which is opposite to that adopted in this book.

## C.4 The triple-dipole potential

In this section, we consider the interaction between triplets of atoms through a potential of the Axilrod–Teller form

$$\begin{aligned} v^{\mathrm{AT}}(\mathbf{r}_i, \mathbf{r}_j, \mathbf{r}_k) &= \frac{\nu(1 + 3\cos\theta_i \cos\theta_j \cos\theta_k)}{r_{ij}^3 r_{jk}^3 r_{ik}^3} \\ &= \frac{\nu(r_{ij}^2 r_{jk}^2 r_{ik}^2 - 3(\mathbf{r}_{ik}\cdot\mathbf{r}_{jk})(\mathbf{r}_{ik}\cdot\mathbf{r}_{ij})(\mathbf{r}_{ij}\cdot\mathbf{r}_{jk}))}{r_{ij}^5 r_{jk}^5 r_{ik}^5} \end{aligned} \tag{C.27}$$

where $\nu$ is a constant, and the geometry is defined in Fig. C.3.

For acute-angled triangles this energy term is positive, but if one of the angles is obtuse it can become negative: thus near-linear configurations are slightly favoured. The net contribution in a liquid, however, is positive, and may amount to approximately 10–15 per cent of the total energy in, for example, argon.

The forces are readily calculated by straightforward differentiation.

$$\begin{aligned} \mathbf{f}_i &= -\nabla_{\mathbf{r}_i} v^{\mathrm{AT}}(\mathbf{r}_i, \mathbf{r}_j, \mathbf{r}_k) \\ &= \nu r_{ij}^{-5} r_{ik}^{-5} r_{jk}^{-5} [5(r_{ij}^2 r_{jk}^2 r_{ik}^2 - 3(\mathbf{r}_{ik}\cdot\mathbf{r}_{jk})(\mathbf{r}_{ik}\cdot\mathbf{r}_{ij})(\mathbf{r}_{ij}\cdot\mathbf{r}_{jk}))(r_{ij}^{-2}\mathbf{r}_{ij} + r_{ik}^{-2}\mathbf{r}_{ik}) \\ &\quad + 3((\mathbf{r}_{ij}\cdot\mathbf{r}_{jk})(\mathbf{r}_{ik}\cdot\mathbf{r}_{ij}) + (\mathbf{r}_{ik}\cdot\mathbf{r}_{ij})(\mathbf{r}_{ik}\cdot\mathbf{r}_{jk}))\mathbf{r}_{jk} \\ &\quad + 3((\mathbf{r}_{ij}\cdot\mathbf{r}_{jk})(\mathbf{r}_{ik}\cdot\mathbf{r}_{jk}))(\mathbf{r}_{ij} + \mathbf{r}_{ik}) \\ &\quad - 2(r_{jk}^2 r_{ik}^2 \mathbf{r}_{ij} + r_{jk}^2 r_{ij}^2 \mathbf{r}_{ik})] \end{aligned} \tag{C.28}$$

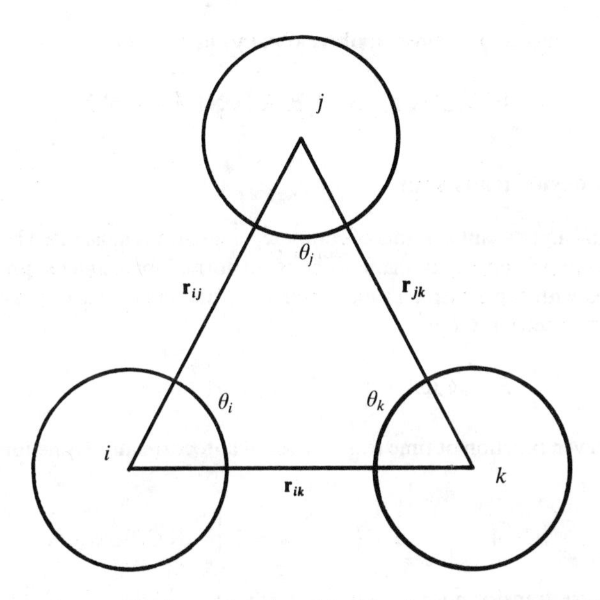


**Fig. C.3.** A triplet of atoms. The internal angles of the triangle are used in calculating the triple-dipole potential.

$$
\begin{aligned}
\mathbf{f}_j &= -\nabla_{\mathbf{r}_j} v^{\mathrm{AT}}(\mathbf{r}_i, \mathbf{r}_j, \mathbf{r}_k) \\
&= \nu r_{ij}^{-5} r_{ik}^{-5} r_{jk}^{-5} [5(r_{ij}^2 r_{jk}^2 r_{ik}^2 - 3(\mathbf{r}_{ik}\cdot\mathbf{r}_{jk})(\mathbf{r}_{ik}\cdot\mathbf{r}_{ij})(\mathbf{r}_{ij}\cdot\mathbf{r}_{jk}))(r_{jk}^{-2}\mathbf{r}_{jk} - r_{ij}^{-2}\mathbf{r}_{ij}) \\
&\quad + 3((\mathbf{r}_{ij}\cdot\mathbf{r}_{jk})(\mathbf{r}_{ik}\cdot\mathbf{r}_{ij}) - (\mathbf{r}_{ik}\cdot\mathbf{r}_{jk})(\mathbf{r}_{ij}\cdot\mathbf{r}_{jk}))\mathbf{r}_{ik} \\
&\quad + 3((\mathbf{r}_{ik}\cdot\mathbf{r}_{jk})(\mathbf{r}_{ik}\cdot\mathbf{r}_{ij}))(\mathbf{r}_{ij} - \mathbf{r}_{jk}) \\
&\quad - 2(r_{ij}^2 r_{ik}^2 \mathbf{r}_{jk} - r_{jk}^2 r_{ik}^2 \mathbf{r}_{ij})]
\end{aligned} \tag{C.29}
$$

$$
\begin{aligned}
\mathbf{f}_k &= -\nabla_{\mathbf{r}_k} v^{\mathrm{AT}}(\mathbf{r}_i, \mathbf{r}_j, \mathbf{r}_k) \\
&= \nu r_{ij}^{-5} r_{ik}^{-5} r_{jk}^{-5} [5(r_{ij}^2 r_{jk}^2 r_{ik}^2 - 3(\mathbf{r}_{ik}\cdot\mathbf{r}_{jk})(\mathbf{r}_{ik}\cdot\mathbf{r}_{ij})(\mathbf{r}_{ij}\cdot\mathbf{r}_{jk}))(-r_{ik}^{-2}\mathbf{r}_{ik} - r_{jk}^{-2}\mathbf{r}_{jk}) \\
&\quad - 3((\mathbf{r}_{ik}\cdot\mathbf{r}_{jk})(\mathbf{r}_{ik}\cdot\mathbf{r}_{ij}) + (\mathbf{r}_{ik}\cdot\mathbf{r}_{jk})(\mathbf{r}_{ij}\cdot\mathbf{r}_{jk}))\mathbf{r}_{ij} \\
&\quad - 3((\mathbf{r}_{ik}\cdot\mathbf{r}_{ij})(\mathbf{r}_{ij}\cdot\mathbf{r}_{jk}))(\mathbf{r}_{jk} + \mathbf{r}_{ik}) \\
&\quad + 2(r_{ij}^2 r_{ik}^2 \mathbf{r}_{jk} + r_{ij}^2 r_{jk}^2 \mathbf{r}_{ik})]
\end{aligned} \tag{C.30}
$$

The expressions for the potential and force both contain odd powers of the intermolecular separations, and an evaluation of these distances in the triplet will inevitably involve taking square roots (see Chapter 5). The forces will be evaluated in a triple loop as described in Chapter 1.

# APPENDIX D
# FOURIER TRANSFORMS

## D.1 The Fourier transform

The structural and dynamic results of computer simulations must often be transformed between time and frequency domains or between normal space and reciprocal space. To be compared with experiment, a time correlation function $C(t)$ is usually transformed to produce a spectrum $\hat{C}(\omega)$

$$\hat{C}(\omega) = \int_{-\infty}^{+\infty} \mathrm{d}t C(t) \exp(-i\omega t) . \tag{D.1}$$

If $C(t)$ is an even function of time (e.g. a classical autocorrelation function) this may be written

$$\hat{C}(\omega) = \int_{-\infty}^{+\infty} \mathrm{d}t C(t) \cos \omega t = 2 \int_{0}^{\infty} \mathrm{d}t\, C(t) \cos \omega t , \tag{D.2}$$

and the inverse transform is

$$C(t) = \int_{-\infty}^{\infty} \frac{\mathrm{d}\omega}{2\pi} \hat{C}(\omega) \exp(i\omega t) = \frac{1}{\pi} \int_{0}^{\infty} \mathrm{d}\omega \hat{C}(\omega) \cos \omega t . \tag{D.3}$$

It should be noted that true (i.e. quantum-mechanical) autocorrelation functions are not even in time, obeying instead the detailed balance condition (see Section 2.9). The time-reversal symmetry of classical cross-correlation functions is discussed in many places [e.g. Berne and Harp 1970].

A variety of combinations of numerical prefactors may appear in the above definitions. Also, it is sometimes convenient to use, instead of $\omega$, the variable $\nu = \omega/2\pi$, when the definitions become symmetrical. We shall stick to $\omega$, reserving $\nu$ for use as a discrete frequency index in the following section.

In practice, the time correlation function is known to some finite maximum time, $t_{max}$, which is determined by the method of analysis and the simulation run time. This value of $t_{max}$ replaces the upper limit in eqn (D.2). Useful formulae may be derived from the orthogonality relations

$$\int_{-\infty}^{\infty} \frac{\mathrm{d}\omega}{2\pi} \exp(i\omega t) \exp(-i\omega t') = \delta(t - t') \tag{D.4a}$$

$$\int_{0}^{\infty} \mathrm{d}\omega \cos \omega t \cos \omega t' = \frac{1}{2} \pi \delta(t - t') \tag{D.4b}$$

The convolution/correlation theorem states that if

$$C(t) = \int_{-\infty}^{\infty} \mathrm{d}t' A(t') B(t - t') \tag{D.5}$$

then

$$\hat{C}(\omega) = \hat{A}(\omega) \hat{B}(\omega) \tag{D.6}$$

while if

$$C(t) = \int_{-\infty}^{\infty} \mathrm{d}t' A(t')B(t+t') \tag{D.7}$$

then

$$\hat{C}(\omega) = \hat{A}(-\omega)\hat{B}(\omega) = \hat{A}^*(\omega)\hat{B}(\omega) \tag{D.8}$$

where* denotes the complex conjugate, and we take $A$ and $B$ to be real.

Structural quantities such as $g(r)$ may be related to quantities observed in (say) scattering experiments by a three-dimensional Fourier transform, such as

$$S(k) - 1 = \int \mathrm{d}\mathbf{r} \exp(i\mathbf{k}\cdot\mathbf{r})\rho g(r)\,. \tag{D.9}$$

These may be treated in a manner analogous to the one-dimensional case. In practice, when the functions depend only upon the magnitude of their arguments, it is sensible to integrate over the angular variables to obtain an equation such as

$$S(k) - 1 = 4\pi \int_0^{\infty} \mathrm{d}r\, r^2 \frac{\sin kr}{kr} \rho g(r) \tag{D.10}$$

with the inverse transform being

$$\rho g(r) = \frac{1}{2\pi^2} \int_0^{\infty} \mathrm{d}k\, k^2 \frac{\sin kr}{kr} (S(k) - 1)\,. \tag{D.11}$$

## D.2 The discrete Fourier transform

A discrete Fourier transform pair is defined

$$\hat{C}(\nu) = \sum_{\tau=0}^{n-1} C(\tau) \exp(-2\pi i \nu\tau/n) \qquad \nu = 0, 1, \ldots, n-1\,, \tag{D.12a}$$

$$C(\tau) = \frac{1}{n} \sum_{\nu=0}^{n-1} \hat{C}(\nu) \exp(2\pi i \nu\tau/n) \qquad \tau = 0, 1, \ldots, n-1\,. \tag{D.12b}$$

This is a relationship between a function $C(t)$ of time tabulated at $n$ points, $\delta t$ apart, so that $C(\tau) = C(\tau\delta t)$, and a function $\hat{C}(\omega)$ of frequency $\omega$, also tabulated at $n$ points, $\delta\omega$ apart, so that $\hat{C}(\nu) = \hat{C}(\nu\delta\omega)$. The intervals in time and frequency are related by

$$n\delta t\delta\omega = 2\pi\,. \tag{D.13}$$

The orthogonality relation is

$$\frac{1}{n} \sum_{\nu=0}^{n-1} \exp(2\pi i\nu\tau/n) \exp(-2\pi i\nu\tau'/n) = \delta_{\tau\tau'}\,. \tag{D.14}$$

The analogy between these equations and eqns (D.1)–(D.4) is obvious. There are, however, some subtleties involved in the use of the discrete Fourier transforms [Brigham 1974; Smith 1982b, 1982c]. In particular, when we use eqns (D.12), (D.14), we must understand $C(t)$ to be periodic in time with period $n\delta t$, and $\hat{C}(\omega)$ to be periodic in frequency with period $n\delta\omega$. Note that the indices above run from 0 to $n-1$, rather than taking positive and negative values (contrast eqns (D.12) with (D.1)–(D.3)). This is not serious, since a shift of time origin simply implies multiplication of the transform by a complex number. These transforms may be calculated very rapidly on a digital computer [Cooley and Tukey 1965].

The discrete convolution/correlation theorem is that if

$$C(\tau) = \sum_{\tau'=0}^{n-1} A(\tau')B(\tau-\tau') \tag{D.15}$$

then

$$\hat{C}(\nu) = \hat{A}(\nu)\hat{B}(\nu) \tag{D.16}$$

while if

$$C(\tau) = \sum_{\tau'=0}^{n-1} A(\tau')B(\tau+\tau') \tag{D.17}$$

then

$$\hat{C}(\nu) = \hat{A}^*(\nu)\hat{B}(\nu) \tag{D.18}$$

These equations are used in the computation of correlation functions by the FFT method (Section 6.3). In this application, the $\tau_{run}$ data items generated in a run are supplemented by $\tau_{run}$ zeroes, so that $n = 2\tau_{run}$ in this case. This avoids the introduction of spurious correlations due to the implied periodicity of the functions mentioned above.

## D.3 Numerical Fourier transforms

Most functions have to be Fourier transformed numerically. For large values of the frequency, $\omega$, the integrand in the transform oscillates rapidly and methods such as Simpson's rule are inadequate. An accurate method due to Filon [1928] fits a quadratic polynomial between discrete function points and evaluates the resulting integral analytically. For an integral of the form

$$\hat{C}(\omega) = 2\int_0^{t_{max}} \mathrm{d}t C(t) \cos\omega t\,, \tag{D.19}$$

the range is divided into $2n$ equal intervals, so that

$$t_{max} = 2n\delta t\,. \tag{D.20}$$

If we define

$$\theta = \omega\delta t\,, \tag{D.21}$$

then

$$\hat{C}(\omega) = 2\delta t(\alpha C(t_{max}) \sin\omega t_{max} + \beta C_e + \gamma C_o)\,, \tag{D.22}$$

where

$$\begin{aligned} \alpha &= (1/\theta^3)(\theta^2 + \theta\sin\theta\cos\theta - 2\sin^2\theta) \\ \beta &= (2/\theta^3)(\theta(1+\cos^2\theta) - 2\sin\theta\cos\theta) \\ \gamma &= (4/\theta^3)(\sin\theta - \theta\cos\theta)\,. \end{aligned} \tag{D.23}$$

$C_e$ is the sum of all the even ordinates of the curve, $C(t)\cos\omega t$, less one-half of the first and last ones. $C_o$ is the sum of all the odd ordinates. This algorithm, though accurate, does not preserve the orthogonality of the transform: transformation from $t$-space to $\omega$-space and back again will not, in general, regenerate the initial correlation function exactly. Lado [1971] has suggested a simple algorithm which preserves the orthogonality of the transform and can be used with the fast Fourier method of Cooley and Tukey [1965]. The integral is replaced by a discrete sum

$$\hat{C}(\nu-\tfrac{1}{2}) = 2\delta t \sum_{\tau=1}^{n} C(\tau-\tfrac{1}{2})\cos\left((\tau-\tfrac{1}{2})(\nu-\tfrac{1}{2})\pi/(n-\tfrac{1}{2})\right). \tag{D.24}$$

The back transform is

$$C(\tau-\tfrac{1}{2}) = \frac{\delta\omega}{\pi} \sum_{\nu=1}^{n} \hat{C}(\nu-\tfrac{1}{2})\cos\left((\tau-\tfrac{1}{2})(\nu-\tfrac{1}{2})\pi/(n-\tfrac{1}{2})\right). \tag{D.25}$$

The upper limit $n$ of the summations can be replaced by $n-1$, since the last term vanishes in each case. The points at which the function is evaluated are fixed in this method. $C(\tau-\frac{1}{2})$ means $C((\tau-\frac{1}{2})\delta t)$ where $\delta t = t_{max}/(n-\frac{1}{2})$ and $\hat{C}(\nu-\frac{1}{2})$ means $\hat{C}((\nu-\frac{1}{2})\delta\omega)$ with $\delta\omega=\pi/t_{max}$. These 'half-integer' values would usually be calculated by interpolation from the simulation data. Apart from the trivial 'half-integer' phase shifts, these are straightforward discrete Fourier transforms, and they may be computed by the efficient FFT method. This method is less accurate than that of Filon, being essentially a trapezoidal rule, but it can be made more accurate by decreasing $\delta t$ (i.e. calculating the correlation function at finer intervals). A program to calculate the one-dimensional transform using Lado's method and Filon's method is given in F.37. Sine transforms are tackled in a way analogous to cosine transforms as shown in the microfiche.

The one-dimensional transform of eqn (D.10) may be calculated by Filon's method, if an extra factor of $r$ is incorporated into the function being transformed. Lado [1971] also discusses in detail the calculation of two- and three-dimensional Fourier transforms.

## APPENDIX E

# THE GEAR PREDICTOR–CORRECTOR

### E.1 The Gear predictor–corrector

Both translational and rotational equations of motion have been numerically solved by applying the predictor–corrector method. If a predictor–corrector is used, the most common form has been that in which the molecular position (or orientation) and several time derivatives, all evaluated at the same time, are stored. This is the so-called Nordsieck representation. Alternatives would involve storing 'old' velocities, accelerations, etc. The corrector coefficients for various predictor–corrector methods have been discussed by Gear [1966, 1971]. It is most convenient to discuss the Nordsieck–Gear predictor–corrector in terms of time step scaled velocities, accelerations etc. Quite generally, suppose that $\mathbf{r}_0$ represents a set of particle positions, or perhaps quaternion parameters specifying the orientations. Let us define the successive scaled time derivatives of $\mathbf{r}_0$ to be $\mathbf{r}_1 = \delta t(\mathrm{d}\mathbf{r}_0/\mathrm{d}t)$, $\mathbf{r}_2 = \frac{1}{2}\delta t^2(\mathrm{d}^2\mathbf{r}_0/\mathrm{d}t^2)$, $\mathbf{r}_3 = \frac{1}{6}\delta t^3(\mathrm{d}^3\mathbf{r}_0/\mathrm{d}t^3)$, etc. Taking, as an example, the four-value Gear algorithm, the Taylor series predictor becomes, in matrix form,

$$\begin{pmatrix} \mathbf{r}_0^{\mathrm{p}}(t+\delta t) \\ \mathbf{r}_1^{\mathrm{p}}(t+\delta t) \\ \mathbf{r}_2^{\mathrm{p}}(t+\delta t) \\ \mathbf{r}_3^{\mathrm{p}}(t+\delta t) \end{pmatrix} = \begin{pmatrix} 1 & 1 & 1 & 1 \\ 0 & 1 & 2 & 3 \\ 0 & 0 & 1 & 3 \\ 0 & 0 & 0 & 1 \end{pmatrix} \begin{pmatrix} \mathbf{r}_0(t) \\ \mathbf{r}_1(t) \\ \mathbf{r}_2(t) \\ \mathbf{r}_3(t) \end{pmatrix} \tag{E.1}$$

The matrix is the Pascal triangle matrix. The form of the matrix makes the predictor easy to apply; if desired, the predictor can be coded so as to involve additions of the variables, not multiplications, and there may be some numerical advantages in this. This form of the predictor is easily extended to different orders of algorithm. The corrector coefficients due to Gear [1966, 1971] are always quoted on the assumption that time step scaled variables are being used; note that the time step factors make these coefficients different from the ones introduced in Section 3.2. The corrector takes the form

$$\begin{pmatrix} \mathbf{r}_0^{\mathrm{c}}(t+\delta t) \\ \mathbf{r}_1^{\mathrm{c}}(t+\delta t) \\ \mathbf{r}_2^{\mathrm{c}}(t+\delta t) \\ \mathbf{r}_3^{\mathrm{c}}(t+\delta t) \end{pmatrix} = \begin{pmatrix} \mathbf{r}_0^{\mathrm{p}}(t+\delta t) \\ \mathbf{r}_1^{\mathrm{p}}(t+\delta t) \\ \mathbf{r}_2^{\mathrm{p}}(t+\delta t) \\ \mathbf{r}_3^{\mathrm{p}}(t+\delta t) \end{pmatrix} + \begin{pmatrix} c_0 \\ c_1 \\ c_2 \\ c_3 \end{pmatrix} \Delta\mathbf{r} \tag{E.2}$$

The values of the corrector coefficients depend upon the order of the differential equation being solved. For a first-order equation of motion of the form

$$\dot{\mathbf{r}} = f(\mathbf{r}) \tag{E.3}$$

we have $\Delta\mathbf{r} = \mathbf{r}_1^{\mathrm{c}} - \mathbf{r}_1^{\mathrm{p}}$ where $\mathbf{r}_1^{\mathrm{p}}$ is the predicted first derivative according to eqn (E.1), and $\mathbf{r}_1^{\mathrm{c}}$ is the corrected first derivative obtained by substituting $\mathbf{r}_0^{\mathrm{p}}$ into the equation of motion. The Gear corrector coefficients for a first-order equation are given in Table E.1.

**Table E.1** *Gear corrector coefficients for a first-order equation*

| Values | $c_0$ | $c_1$ | $c_2$ | $c_3$ | $c_4$ | $c_5$ |
|---|---|---|---|---|---|---|
| 3 | 5/12 | 1 | 1/2 | | | |
| 4 | 3/8 | 1 | 3/4 | 1/6 | | |
| 5 | 251/720 | 1 | 11/12 | 1/3 | 1/24 | |
| 6 | 95/288 | 1 | 25/24 | 35/72 | 5/48 | 1/120 |

**Table E.2** *Gear corrector coefficients for a second-order equation*

| Values | $c_0$ | $c_1$ | $c_2$ | $c_3$ | $c_4$ | $c_5$ |
|---|---|---|---|---|---|---|
| 3 | 0 | 1 | 1 | | | |
| 4 | 1/6 | 5/6 | 1 | 1/3 | | |
| 5 | 19/120 | 3/4 | 1 | 1/2 | 1/12 | |
| 6 | 3/20 | 251/360 | 1 | 11/18 | 1/6 | 1/60 |

For a second-order equation

$$\ddot{\mathbf{r}} = f(\mathbf{r}) \tag{E.4}$$

we have $\Delta\mathbf{r} = \mathbf{r}_2^c - \mathbf{r}_2^p$ where $\mathbf{r}_2^p$ is the predicted value and $\mathbf{r}_2^c$ is obtained by substituting $\mathbf{r}_0^p$ into the equation of motion. The coefficients for a second-order equation are given in Table E.2.

For second-order equations of the form

$$\ddot{\mathbf{r}} = f(\mathbf{r},\dot{\mathbf{r}}) \tag{E.5}$$

in which the first derivatives also appear on the right, the coefficients $c_0$ should be replaced by 19/90 in the five-value method and by 3/16 in the six-value method, respectively. The coefficients given above for the second-order three-value method are actually those corresponding to the velocity Verlet algorithm discussed in Section 3.3.1, which we have formally fitted into the Gear scheme.

In Chapter 3 we have argued that the Verlet methods are generally simpler and exhibit better energy conservation than the Gear algorithms, for straightforward MD simulation of atomic systems. The situation is slightly different when rotational motion is involved, or when some of the special techniques described in Chapters 7–9 are used. In these cases, the velocities often appear on the right of the equations of motion. It is frequently possible to adapt one of the Verlet/leap-frog methods to solve these differential equations (and some examples appear in the text) but it is probably more convenient to use one of the general-purpose Gear algorithms. A first-order Gear method based on all six vectors (positions and momenta) may be used. In this case, time derivatives of all these quantities should be stored. Alternatively, the momenta may be eliminated from the equations of motion. It is then possible to apply the Gear method for a second-order differential equation based on $\mathbf{r}$ and its derivatives. If the values of $\dot{\mathbf{r}}$ appear on the right, the alternative coefficients mentioned above (e.g. 3/16 not 3/20, 19/90 not 19/120) should be used. If the momenta are required (they may not be equivalent to the velocities in some of the techniques described in Chapters 7 and 8) they may be obtained from $\mathbf{r}$, $\dot{\mathbf{r}}$ etc.

## APPENDIX F

# PROGRAM AVAILABILITY

The programs referred to in the text are available from the World Wide Web server maintained by the Engineering and Physical Sciences Research Council Collaborative Computational Project CCP5, which covers the computer simulation by molecular dynamics, Monte Carlo, and molecular mechanics of liquid and solid phases. The Web server is located at CCLRC Daresbury Laboratory, Warrington, WA4 4AD, UK, and the URL is:

http://www.dl.ac.uk/CCP/CCP5/main.html

The Web server also provides access to a library of other simulation programs, and links to groups carrying out activities in the field.

Although a few complete programs are provided, our aim has been to offer building blocks rather than black boxes. As far as we are aware, the programs work correctly, but we can accept no responsibility for the consequences of any errors, and would be grateful to hear from you if you find any. You should always check out a routine for your particular application. The programs contain some explanatory comments, and are written, in the main, in FORTRAN-77. One or two routines are written in BASIC, for use on microcomputers. In the absence of any universally agreed standard for BASIC, we have chosen a very rudimentary dialect. These programs have been run on an Acorn model B computer. Hopefully, the translation of these programs into more sophisticated languages such as PASCAL or C should not be difficult.

A package of simulation programs built around the SHAKE algorithm, is available from the authors, W. F. van Gunsteren and H. J. C. Berendsen, at the University of Groningen. Note that, in this case, copyright resides with the original authors, who should be consulted regarding any possible commercial applications.

*List of programs*

F.1 Periodic boundary conditions in various geometries.
F.2 5-Value Gear predictor–corrector algorithm.
F.3 Low-storage MD programs using leap-frog Verlet algorithm.
F.4 Velocity version of Verlet algorithm.
F.5 Quaternion parameter predictor–corrector algorithm.
F.6 Leap-frog algorithms for rotational motion.
F.7 Constraint dynamics for a non-linear triatomic molecule.
F.8 SHAKE algorithm for constraint dynamics of a chain molecule.
F.9 RATTLE algorithm for constraint dynamics of a chain molecule.
F.10 Hard sphere molecular dynamics program.
F.11 Constant-$NVT$ Monte Carlo for Lennard-Jones atoms.
F.12 Constant-$NPT$ Monte Carlo algorithm.
F.13 The heart of a constant-$\mu VT$ Monte Carlo program.
F.14 Algorithm to handle indices in constant-$\mu VT$ Monte Carlo.
F.15 Routines to randomly rotate molecules.

| | |
|---|---|
| F.16 | Hard dumb-bell Monte Carlo program. |
| F.17 | A simple Lennard-Jones force routine. |
| F.18 | Algorithm for avoiding the square root operation. |
| F.19 | The Verlet neighbour list. |
| F.20 | Routines to construct and use cell linked-list method |
| F.21 | Multiple time-step molecular dynamics. |
| F.22 | Routines to perform the Ewald sum. |
| F.23 | Routine to set up $\alpha$-f.c.c. lattice of linear molecules. |
| F.24 | Initial velocity distribution. |
| F.25 | Routine to calculate translational order parameter. |
| F.26 | Routines to fold/unfold trajectories in periodic boundaries. |
| F.27 | Program to compute time correlation functions. |
| F.28 | Constant-$NVT$ molecular dynamics (extended system method). |
| F.29 | Constant-$NVT$ molecular dynamics (constraint method). |
| F.30 | Constant-$NPH$ molecular dynamics (extended system method). |
| F.31 | Constant-$NPT$ molecular dynamics (constraint method). |
| F.32 | Cell linked lists in sheared boundaries. |
| F.33 | Brownian dynamics for a Lennard-Jones fliuid. |
| F.34 | An efficient clustering routine. |
| F.35 | The Voronoi construction in 2D and 3D. |
| F.36 | Monte Carlo simulation of hard lines in 2D. |
| F.37 | Routines to calculate Fourier transforms. |

APPENDIX G

# RANDOM NUMBERS

## G.1 Random number generators

Before the development of computers, random sequences of numbers had to be generated by physical methods such as rolling dice, tossing coins, picking numbers from an urn, or analysing noise generated in an electronic valve. These processes are slow, sometimes unreliable, and generate sequences which cannot be easily reproduced. To assist workers in this field, large tables of pre-calculated random sequences were published [Rand Corporation, 1955]. Many applications, including Monte Carlo and stochastic dynamics simulation, require random sequences of millions of numbers, for which the above approaches are inadequate. A solution is to generate numbers with the desired properties purely by arithmetic manipulation on a computer. For example, an early suggestion of von Neumann [1951] was to start with a four-digit number (e.g. 9876), square it (to give 97535376), and extract the middle four digits (5353). This process is repeated, to generate a sequence of four digit numbers (9876, 5353, 6546, 8501, 2670, . . . ). Methods of this type are repeatable and, of course, completely deterministic, so the numbers cannot be called 'random'; the terms 'pseudo-random' or 'quasi-random' are used instead. This leads us to consider carefully what properties of 'randomness' are required in any particular application. In fact, the 'mid-square' method described above performs rather poorly by most criteria: it tends to produce cyclical sequences of numbers and it terminates whenever a zero is generated. The trick of pseudo-random number generation is to produce a repeatable sequence that passes a wide range of statistical tests for independence, and that is sampled from the desired distribution.

## G.2 Random numbers uniform on (0, 1)

Most random number generators have at their heart a means of generating numbers that are uniformly distributed in the range (0, 1). This is commonly done via a sequence of large positive integers $X_i$, each generated from the previous one by some operation (e.g. multiplication) conducted in finite modulus arithmetic. A typical multiplicative generator is of the form

$$X_{i+1} = aX_i \bmod (M) \tag{G.1}$$

where $a$ and $M$ are both large positive integers. The modulo operation simply means (as for the MOD function in FORTRAN)

$$X \bmod (M) = X - M \operatorname{int}(X/M) \tag{G.2}$$

where int ( . . . ) stands for 'integer part of . . .'. Thus each integer in the sequence lies in the range $(0, M-1)$ and the desired random numbers $\xi$ are

$$\xi_i = X_i/M\,. \tag{G.3}$$

In this book, $\xi$ will denote a random number chosen in this way, uniformly on the range (0, 1). Since $X_i$ can never exceed $M-1$ the sequence must repeat after $M-1$ integers have been generated, and it may cycle round more often. Knuth [1973, Chapter 3] has outlined a number of principles for generating uniform random variates. For generators such as eqn (G.1) the important rules are:

(a) Choose a large value of $M$ to maximize the period of the generator. In many generators $M$ is the largest integer that will fit into the machine, i.e. $2^l$ where $l$ is the computer's word-length.
(b) If $M$ is a power of 2, then choose $a$ so that $a$ mod (8) = 5 or 3.
(c) The right-hand bits of $X_i$ may not be very random. This problem can be avoided by always basing decisions on the normalized random number, $\xi_i$, in which the non-random bits must be insignificant, rather than on $X_i$ itself.
(d) The seed $X_0$ should be relatively prime to $M$.

If $M$ is equal to $2^l$, then the maximum period of a simple multiplicative generator is $M/4$; if $M$ is chosen to be prime the maximum possible period is $M-1$ ($X = 0$ should never be generated for obvious reasons). The actual period may depend critically on the value of the initial seed $X_0$.

The generator $(M, a) = (2^{31}-1, 7^5)$ is used in the assembly language programs GGL1 and GGL2, and in the FORTRAN routine GGL, on the IBM 360 mainframe. The value of $M$ is the largest prime that can be represented in a 32-bit word with one bit reserved for the sign. In the NAG library routine G05CAF [NAG 1984] the generator $(M, a) = (2^{59}, 13^{13})$ is used. This generator can be coded efficiently on a machine with a 60-bit word length (e.g. the CDC 7600), but is also available as a slower multiprecision assembler routine on other machines. However, in computer simulations the raw speed of a random number generator is seldom important, since most of the computing time is spent elsewhere (for example in computing energies and forces). The generation of a sequence with good statistical properties and a long period is more important than the timing of the routine. A random number generator should allow the user to initialize the seed repeatably (good for program testing) or non-repeatably (essential for independent production runs).

Random number sequences should be thoroughly tested. Rubinstein [1981, Chapter 2] describes seven important statistical tests which can be applied to a sequence of random numbers. To illustrate the idea we mention the serial test. If the random numbers are taken successively in $(x, y)$ pairs, they should describe points which are randomly distributed in the unit square; taken in $(x, y, z)$ triplets, they should be randomly distributed in the unit cube etc. The once popular generators $(M, a) = (2^{35}, 2^{18}+3)$ and $(M, a) = (2^{32}, 2^{16}+3)$ fail the pair and triplet tests respectively [Greenberger 1965]. By contrast, the NAG random number generator mentioned above satisfies the serial test for hypercubes of dimension eight and lower. In addition to a serial test a spectral analysis of the output sequence can be a discriminating tool in testing and designing new generators [Conveyou and MacPherson 1967].

Purpose-built random number generators are available on mainframe and minicomputers as part of the manufacturers' software or in the common applications libraries. They are best written in assembler by an expert and it is probably not advisable to write one for your own MC program. A serious user should check with the author of the routine about the estimated period of the generator and the statistical tests to which it has been subjected. Generating long random number sequences on microcomputers (word-length 16 or less), is currently an area of active research.

## G.3 Generating non-uniform distributions

Using the random number $\xi_i$ generated uniformly on (0, 1) it is possible to construct random numbers taken from a variety of distributions. There are many distributions which are of interest to statisticians but only a limited number which are required in liquid-state simulation. In this section we discuss generating random variables on the normal (Gaussian), and exponential distributions. The interested reader is referred to Rubinstein [1981] for a comprehensive discussion of other distributions, and proof of the results quoted in this section.

The normal distribution, with mean $\langle x \rangle$, and variance $\sigma^2$ is defined as

$$\rho(x) = \frac{1}{\sigma(2\pi)^{1/2}} \exp\left(\frac{-(x - \langle x \rangle)^2}{2\sigma^2}\right) \quad -\infty < x < +\infty . \tag{G.4}$$

A random number $\zeta'$ generated from this distribution is related to a number $\zeta$ generated from the normal distribution with zero mean and unit variance by

$$\zeta' = \langle x \rangle + \sigma\zeta . \tag{G.5}$$

The problem is reduced to sampling from a normal distribution with zero mean and unit variance, and we have chosen two methods from the many possibilities. The first method involves two steps and the generation of two uniform random variates [Box and Muller 1958]:

(a) generate uniform random variates $\xi_1$ and $\xi_2$ on $(0, 1)$;
(b) calculate $\zeta_1 = (-2\ln\xi_1)^{1/2}\cos 2\pi\xi_2$ and $\zeta_2 = (-2\ln\xi_1)^{1/2}\sin 2\pi\xi_2$ .

The numbers $\zeta_1$ and $\zeta_2$ are the desired (independent) normally distributed random numbers. The second method also involves two steps and the generation of 12 uniform random variates:

(a) generate 12 uniform random variates, $\xi_1 \ldots \xi_{12}$ in the range $(0, 1)$;
(b) calculate $\zeta = \sum_{i=1}^{12} \xi_i - 6$ .

This second method yields numbers $\zeta$ which are sampled from an approximately normal distribution (by virtue of the central limit theorem of probability). Clearly, random variates outside the range $(-6, 6)$ will never be generated using this method, but it is adequate for most purposes, and is quite fast. The speed relative to the Box–Muller technique will depend on the precise timings of the FORTRAN functions ALOG, SQRT, COS, SIN, and the random number generator itself. A slight improvement [Knuth 1973] is to calculate

$$R = \left(\sum_{i=1}^{12} \xi_i - 6\right) \Big/ 4 \tag{G.6}$$

and then form the polynomial

$$\zeta = ((((a_9R^2 + a_7)R^2 + a_5)R^2 + a_3)R^2 + a_1)R \tag{G.7}$$

where the coefficients are

$$
\begin{aligned}
a_1 &= 3.949846138 \\
a_3 &= 0.252408784 \\
a_5 &= 0.076542912 \\
a_7 &= 0.008355968 \\
a_9 &= 0.029899776
\end{aligned} \tag{G.8}
$$

In some applications (e.g. Brownian dynamics, Chapter 9) we need to generate correlated pairs of numbers that are normally distributed. Given two independent normal random deviates $\zeta_1$ and $\zeta_2$, with zero means and unit variances, obtained by the methods outlined above, the variables

$$
\begin{aligned}
\zeta_1' &= \sigma_1 \zeta_1 \\
\zeta_2' &= \sigma_2 (c_{12}\zeta_1 + (1 - c_{12}^2)^{1/2} \zeta_2)
\end{aligned} \tag{G.9}
$$

are sampled from the bivariate Gaussian distribution with zero means, variances $\sigma_1^2$ and $\sigma_2^2$, and correlation coefficient $c_{12}$.

In the Brownian dynamics simulations described in Section 9.4, we need to sample a large number of correlated random numbers from a multivariate Gaussian distribution. Suppose we require $n$ correlated normal random deviates $\zeta_i'$ with zero mean and specified covariance matrix $C_{ij} = \langle \zeta_i' \zeta_j' \rangle$. The distribution is

$$
\rho(\mathbf{x}) = (|\mathbf{C}^{-1}|/(2\pi)^n)^{1/2} \exp(-\mathbf{x} \cdot \mathbf{C}^{-1} \cdot \mathbf{x}) . \tag{G.10}
$$

Some numerical packages include routines for sampling from this distribution (e.g. GGNRM in the IMSL library, G05EAF/G05EZF in the NAG library), and it is also possible to write your own. Suppose that $n$ independent Gaussian random numbers $\zeta_i$ are generated, using the methods described above, with zero means and unit variances. The method relies on the existence of a lower triangular matrix $\mathbf{L}$ which satisfies $\mathbf{L}\mathbf{L}^{\mathrm{t}} = \mathbf{C}$. The elements of this matrix are determined by [Ermak and McCammon 1978]

$$
L_{11} = C_{11}^{\frac{1}{2}} \tag{G.11a}
$$

$$
L_{i1} = C_{i1}/L_{11} \tag{G.11b}
$$

$$
L_{ii} = \left[ C_{ii} - \sum_{k=1}^{i-1} L_{ik}^2 \right]^{\frac{1}{2}} \qquad i > 1 \tag{G.11c}
$$

$$
L_{ij} = \left[ C_{ij} - \sum_{k=1}^{j-1} L_{ik} L_{jk} \right] \Big/ L_{jj} \quad i > j > 1 \tag{G.11d}
$$

and the desired random variables are just

$$
\zeta_i' = \sum_{j=1}^{i} L_{ij} \zeta_j . \tag{G.12}
$$

Note the limits of summation in eqns (G.11c, d). These make it possible to evaluate the first row of $\mathbf{L}$, (i.e. just $L_{11}$), followed by the second row ($L_{21}$, $L_{22}$), and so on. Note also that eqn (G.9) is a special case of this result, for two variables with $C_{ij} = \sigma_i \sigma_j c_{ij}$, $i,j = 1, 2$. In Brownian dynamics $n = 3N$ and $\mathbf{C} = \langle \delta\mathbf{r} \delta\mathbf{r} \rangle = 2\mathbf{D}\delta t$.

The exponential distribution is

$$
\begin{aligned}
\rho(x) &= \langle x \rangle^{-1} \exp(-x/\langle x \rangle) \quad & 0 < x < \infty \\
&= 0 & \text{otherwise}
\end{aligned} \tag{G.13}
$$

where $\langle x \rangle$ is a positive parameter. A method of generating a random variate $\zeta$ on this distribution is

(a) Generate a uniform random variate, $\xi$, on (0, 1);
(b) Calculate $\zeta = -\langle x \rangle \ln \xi$.

An example of the use of such a distribution is in the selection of random angular velocities $\boldsymbol{\omega}$ for a set of linear molecules. This may be accomplished by choosing the direction of $\boldsymbol{\omega}$ randomly in the plane perpendicular to the molecular axis (see next section), and then selecting the value of $\omega^2$ from the distribution $I/2k_BT \exp(-I\omega^2/2k_BT)$, i.e. from the exponential distribution with $\langle x \rangle = 2k_BT/I$.

## G.4 Random vectors on the surface of a sphere

There are a number of suitable methods for generating a vector on the surface of a unit sphere. The simplest of these uses the acceptance–rejection technique of von Neumann [1951]. The procedure is iterative:

(a) Generate three uniform random variates, $\xi_1$, $\xi_2$, and $\xi_3$, on (0, 1);
(b) Calculate $\zeta_i = 1 - 2\xi_i$ for $i = 1, 3$ so that the vector $\boldsymbol{\zeta} = (\zeta_1, \zeta_2, \zeta_3)$ is distributed uniformly in a cube of side 2, centred at the origin;
(c) Form the sum $\zeta^2 = \zeta_1^2 + \zeta_2^2 + \zeta_3^2$;
(d) For $\zeta^2 < 1$ (i.e. inside the inscribed sphere) take $\hat{\boldsymbol{\zeta}} = (\zeta_1/\zeta, \zeta_2/\zeta, \zeta_3/\zeta)$ as the vector;
(e) For $\zeta^2 > 1$ reject the vector and return to step (a).

The efficiency of this method tends to $\pi/6$ so that the algorithm requires 5.73 uniform variates on average.

Marsaglia [1972] has suggested an interesting improvement:

(a) Generate two uniform random variates, $\xi_1$, $\xi_2$ on (0, 1);
(b) Calculate $\zeta_i = 1 - 2\xi_i$ for $i = 1, 2$;
(c) Form the sum $\zeta^2 = \zeta_1^2 + \zeta_2^2$;
(d) For $\zeta^2 < 1$ take the vector, $\hat{\boldsymbol{\zeta}} = (2\zeta_1(1-\zeta^2)^{1/2},\ 2\zeta_2(1-\zeta^2)^{1/2},\ 1 - 2\zeta^2)$;
(e) For $\zeta^2 > 1$, reject and return to step (a).

This method requires on average 2.55 uniform variates and a square root. The method can be readily extended to choosing points on a four-sphere and Marsaglia gives an appropriate algorithm. To obtain random vectors in a plane normal to a given unit vector $\mathbf{e}$, simply subtract that part of $\hat{\boldsymbol{\zeta}}$ parallel to $\mathbf{e}$ i.e. form $\hat{\boldsymbol{\zeta}} - (\hat{\boldsymbol{\zeta}} \cdot \mathbf{e})\mathbf{e}$ and renormalize.

## G.5 Choosing randomly and uniformly from complicated regions

von Neumann [1951] suggested the following algorithm for generating from an arbitrary distribution $\rho(x)$. The distribution function is split in the following way

$$\rho(x) = C a(x) b(x) \tag{G.14}$$

where $a(x)$ is a simpler distribution function, from which it is easy to generate a random variate, $b(x)$ is a function which lies between zero and one and $C\ (\geqslant 1)$ is a constant. The following steps generate a random number on $\rho(x)$:

(a) generate a uniform random variate, $\xi$, on (0, 1);
(b) generate $\zeta$ randomly on the distribution $a(x)$;
(c) if $\xi \leqslant b(\zeta)$ then $\zeta$ is random on $\rho(x)$;
(d) if not, go to step (a).

A trivial example illustrates this technique. The probem is to generate a number uniformly on $(x_1, x_2)$ assuming we can generate randomly on the range $(x_1, x_3)$

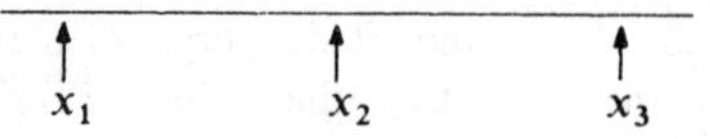


We require to sample from the distribution,

$$\rho(x) = \frac{1}{x_2 - x_1} \qquad x_1 < x \leqslant x_2 . \tag{G.15}$$

The distribution $\rho(x)$ is split up as follows:

$$a(x) = \frac{1}{x_3 - x_1} \qquad x_1 < x \leqslant x_3,$$

$$C = \frac{x_3 - x_1}{x_2 - x_1}$$

and $b(x)$ is the unit step function $\theta(x_2 - x)$. This separation is easily checked by substituting $a(x)$, $b(x)$, and $C$ into eqn (G.14). A little reflection shows that in this case the von Neumann algorithm simplifies considerably:

(a) generate $\zeta$ uniformly on $(x_1, x_3)$
(b) if $\zeta < x_2$ then take $\zeta$ as the random number on $\rho(x)$;
(c) if not, go to (a).

This simple one-dimensional example can be readily extended to higher dimensions and more complicated shapes. It will be noticed that the basic Monte Carlo sampling method is based on this approach.

To generate a vector which lies uniformly on the surface of a sphere but within a solid angle $\delta\boldsymbol{\Omega}$ of a given direction (see Fig. G.1):

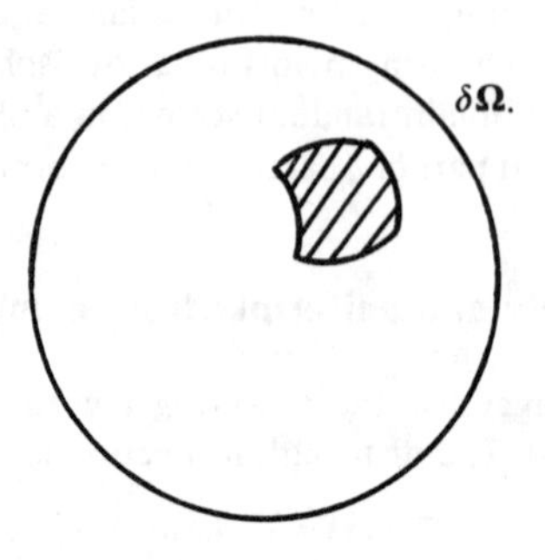


**Fig. G.1** Choosing a random vector on the surface of a sphere, with its end inside the shaded region $\delta\boldsymbol{\Omega}$.

(a) generate $\boldsymbol{\zeta}$ uniformly on the sphere;
(b) if $\boldsymbol{\zeta} \in \delta\boldsymbol{\Omega}$, then accept $\boldsymbol{\zeta}$;
(c) if not go to (a).

This method is the basis of choosing 'random' rotations restricted in magnitude, as used in Monte Carlo simulations of molecules. Clearly the secret of success in this method, in general, is to choose $a(x)$ so that it is convenient for calculating random variates, (a hypersphere, hyperellipsoid etc.), but which approximately covers the same range as the required region $\delta\boldsymbol{\Omega}$. The closer the region $\delta\boldsymbol{\Omega}$ to the region covered by $a(x)$, the fewer random variates are required.

## G.6 Sampling from an arbitrary distribution

In a number of applications it is necessary to sample from an arbitrary distribution, $\rho(x)$ which is bounded and of finite range, $x_1 \leqslant x \leqslant x_2$. Von Neumann's algorithm can be used by writing

$$\rho(x) = A\rho(x)\ \theta(x_1, x, x_2)/A \tag{G.13}$$

Where $A \geqslant 1$ is a normalizing constant which ensures that $A^{-1}\rho(x)$ is always $\leqslant 1$ and $\theta(x_1,x,x_2) = 1$ for $x_1 \leqslant x \leqslant x_2$ and zero otherwise. To generate randomly on $\rho(x)$ [Rubinstein 1981, p. 45]:

(a) generate two uniform random variates $\xi_1$ and $\xi_2$;
(b) if $\xi_1 \leqslant A^{-1}\rho(x_1 + (x_2 - x_1)\xi_2)$, then $x_1 + (x_2 - x_1)\xi_2$ is random on $\rho(x)$;
(c) if not, go to (a).

# REFERENCES

As well as primary literature references, we have included work reported in the 'Rapport d'activité scientifique du CECAM' (CECAM Workshop reports) and the 'Daresbury Laboratory Information Quarterly for MD and MC simulations', which we abbreviate 'CCP5 Quarterly'. It should be noted that these are unrefereed journals, which are used as an informal, but very useful, medium of information exchange. Some of the papers listed below are included in a reprint collection *Simulation of liquids and solids. Molecular dynamics and Monte Carlo methods in statistical mechanics* (eds. G. Ciccotti, D. Frenkel, and I. R. McDonald). North Holland (in press).

The number in square brackets following each entry gives the section number in which the reference is made.


Abraham, F. F. (1982). Statistical surface physics: a perspective via computer simulation of microclusters, interfaces and simple films. *Rep. Prog. Phys.* **45**, 1113–61. [4.5, 11.3]

Abraham, F. F. (1985). Computational statistical mechanics. Methodology, applications and supercomputing. *Adv. Phys.* **35**, 1–111 [A.1]

Ackerson, B. J. and Clark, N. A. (1983). Sheared colloidal suspensions. *Physica* **118A**, 221–49. [11.7]

Adams, D. J. (1974). Chemical potential of hard sphere fluids by Monte Carlo methods. *Mol. Phys.* **28**, 1241–52. [4.6]

Adams, D. J. (1975). Grand canonical ensemble Monte Carlo for a Lennard-Jones fluid. *Mol. Phys.* **29**, 307–11. [2.5, 4.6]

Adams, D. J. (1979). Computer simulation of ionic systems: the distorting effects of the boundary conditions. *Chem. Phys. Lett.* **62**, 329–32. [1.5.2]

Adams, D. J. (1980). Periodic truncated octahedral boundary conditions. *The problem of long-range forces in the computer simulation of condensed media* (ed. D. Ceperley). NRCC Workshop Proceedings Vol. 9, p. 13. [1.5.2]

Adams, D. J. (1983a). Alternatives to the periodic cube in computer simulation. *CCP5 Quarterly* **10**, 30–6. [1.5.4]

Adams, D. J. (1983b). On the use of the Ewald summation in computer simulation. *J. chem. Phys.* **78**, 2585–90. [5.5.1, 5.5.5]

Adams, D. J. and McDonald, I. R. (1976). Thermodynamic and dielectric properties of polar lattices. *Mol. Phys.* **32**, 931–47. [5.5.2, 5.5.3]

Adams, D. J., Adams, E. H., and Hills, G. J. (1979). The computer simulation of polar liquids. *Mol. Phys.* **38**, 387–400. [5.5.3, 5.5.5]

Adelman, S. A. (1979). Generalized Langevin theory for many-body problems in chemical dynamics: general formulation and the equivalent harmonic chain representation. *J. chem. Phys.* **71**, 4471–86. [9.3, 9.4]

Alder, B. J. (1983). Comments on nonequilibrium molecular dynamics. *Physica* **118A**, 413–18. [11.7]

Alder, B. J. (1986) Molecular dynamics simulations. *Molecular dynamics simulation of statistical mechanical systems.* Proceedings of the Enrico Fermi Summer School, Varenna, 1985, pp 66–80. Soc. Italiana di Fisica, Bologna. [11.7]

Alder, B. J. and Wainwright, T. E. (1957). Phase transition for a hard sphere system. *J. chem. Phys.* **27**, 1208–9. [1.1, 11.3]

Alder, B. J. and Wainwright, T. E. (1959). Studies in molecular dynamics. I. General method. *J. chem. Phys.* **31**, 459–66. [1.1, 3.6]

Alder, B. J. and Wainwright, T. E. (1960). Studies in molecular dynamics. II. Behaviour of a small number of elastic spheres. *J. chem. Phys.* **33**, 1439–51. [3.6, 5.6].

Alder, B. J. and Wainwright, T. E. (1962). Phase transition in elastic disks. *Phys. Rev.* **127**, 359–61. [11.3]

Alder, B. J. and Wainwright, T. E. (1969). Enhancement of diffusion by vortex-like motion of classical hard particles. Proceedings of the International Conference on Statistical Mechanics, Kyoto, 1968. *J. phys. Soc. Japan Suppl.* **26**, 267–9, [11.7]

Alder, B. J. and Wainwright, T. E. (1970). Decay of the velocity autocorrelation function, *Phys. Rev.* **A1**, 18–21. [6.4.4, 11.7]

Alder, B. J., Frankel, S. P., and Lewinson, V. A. (1955). Radial distribution function calculated by the Monte Carlo method for a hard sphere fluid. *J. chem. Phys.* **23**, 417–19. [4.2.2]

Alder, B. J., Gass, D. M., and Wainwright, T. E. (1970). Studies in molecular dynamics. VIII. The transport coefficients for a hard sphere fluid. *J. chem. Phys.* **53**, 3813–26. [6.5.5, 11.7]

Allen, M. P. (1980). Brownian dynamics simulation of a chemical reaction in solution. *Mol. Phys.* **40**, 1073–87. [9.3]

Allen, M. P. (1982). Algorithms for Brownian dynamics. *Mol. Phys.* **47**, 599–601. [9.3]

Allen, M. P. (1984a). A molecular dynamics simulation study of octopoles in the field of a planar surface. *Mol. Phys.* **52**, 717–32. [3.3.1]

Allen, M. P. (1984b). Atomic and molecular representations of molecular hydrodynamic variables. *Mol. Phys.* **52**, 705–16. [8.6]

Allen, M. P. and Cunningham, I. C. H. (1986). A computer simulation study of idealized model tetrahedral molecules. *Mol. Phys.* **58**, 615–25. [3.6.2]

Allen, M. P. and Imbierski, A. A. (1987). A molecular dynamics study of the hard dumb-bell system. *Mol. Phys.* **60**, 453–73. [3.6.2]

Allen, M. P. and Kivelson, D. (1981). Non-equilibrium molecular dynamics simulation and generalized hydrodynamics of transverse modes in molecular fluids. *Mol. Phys.* **44**, 945–65. [8.6]

Allen, M. P. and Maréchal, G. (1986). Nonequilibrium molecular dynamics simulation of a molecular fluid subjected to an oscillatory shear perturbation. *Mol. Phys.* **57**, 7–19. [8.6]

Allen, M. P., Frenkel, D., Gignac, W., and McTague, J. P. (1983). A Monte Carlo simulation study of the two-dimensional melting mechanism. *J. chem. Phys.* **78**, 4206–22. [11.3]

Andersen, H. C. (1980). Molecular dynamics simulations at constant pressure and/or temperature. *J. chem. Phys.* **72**, 2384–93. [7.4.1, 7.5.1, 7.5.4, 7.6, 8.3]

Andersen, H. C. (1983). Rattle: a 'velocity' version of the shake algorithm for molecular dynamics calculations, *J. comput. Phys.* **52**, 24–34. [3.4]

Andersen, H. C., Allen, M. P., Bellemans, A., Board, J., Clarke, J. H. R., Ferrario, M., Haile, J. M., Nosé, S., Opheusden, J. V., and Ryckaert, J. P. (1984). New molecular dynamics methods for various ensembles, Rapport d'activité scientifique du CECAM, pp. 82–115. [7.4, 7.4.3]

Anderson, J. B. (1975). A random walk simulation of the Schrödinger equation: $H_3^+$. *J. chem. Phys.* **63**, 1499–1503. [10.4]

Anderson, J. B. (1976). Quantum chemistry by random walk. $H\,^2P$, $H_3^+\,D_{3h}\,^1A_1'$, $H_2\,^3\Sigma_u^+$, $H_4\,^1\Sigma_g^+$, $Be\,^1S$. *J. chem. Phys.* **65**, 4121–27. [10.4]

Anderson, J. B. (1980). Quantum chemistry by random walk: higher accuracy. *J. chem. Phys.* **73**, 3897–9. [10.4]

Andrea, T. A., Swope, W. C., and Andersen, H. C. (1983). The role of long ranged forces in determining the structure and properties of liquid water. *J. chem. Phys.* **79**, 4576–84. [5.2.3, 5.5.3, 7.4.1]

Ashurst, W. T. and Hoover, W. G. (1972). Nonequilibrium molecular dynamics: shear viscosity and thermal conductivity. *Bull. Am. phys. Soc.* **17**, 1196. [8.1]

Ashurst, W. T. and Hoover, W. G. (1973). Argon shear viscosity via a Lennard-Jones potential with equilibrium and nonequilibrium molecular dynamics. *Phys. Rev. Lett.* **31**, 206–8. [8.1]

Ashurst, W. T. and Hoover, W. G. (1975). Dense fluid shear viscosity via non-equilibrium molecular dynamics. *Phys. Rev.* **A11**, 658–78. [8.1]

Ashurst, W. T. and Hoover, W. G. (1977). Shear viscosity via periodic nonequilibrium molecular dynamics. *Phys. Lett.* **61A**, 175–7. [8.7]

Aviram, I., Tildesley, D. J., and Streett, W. B. (1977). The virial pressure in a fluid of hard polyatomic molecules, *Mol. Phys.* **34**, 881–5. [6.5.3]

Axilrod, B. M. and Teller, E. (1943). Interaction of the van der Waals' type between three atoms. *J. chem. Phys.* **11**, 299–300. [1.3.2]

Bakker, A. F., Bruin, C., van Dieren, F., and Hilhorst, H. J. (1982). Molecular dynamics of 16000 Lennard-Jones particles. *Phys. Lett.* **93A**, 67–9. [A.1]

Balfour, A. and Marwick, D. H. (1979). *Programming in standard FORTRAN-77*. Heinemann, London. [3.5, A.2]

Barker, A. A. (1965). Monte Carlo calculations of the radial distribution functions for a proton–electron plasma. *Aust. J. Phys.* **18**, 119–33. [4.3]

Barker, J. A. (1976). Interatomic potentials for inert gases from experimental data. In *Rare gas solids* (eds. M. L. Klein and J. A. Venables) Vol. I, pp. 212–64. Academic Press, London. [1.3.2]

Barker, J. A. (1979). A quantum statistical Monte Carlo method; path integrals with boundary conditions. *J. chem. Phys.* **70**, 2914–18. [10.2]

Barker, J. A. (1980). Reaction field method for polar fluids. *The problem of long-range forces in the computer simulation of condensed matter* (ed. D. Ceperley). NRCC Workshop Proceedings Vol. 9, pp. 45–6. [5.5.3]

Barker, J. A. and Henderson, D. (1976). What is 'liquid'? Understanding the states of matter. *Rev. Mod. Phys.* **48**, 587–671. [1.1, 4.6]

Barker, J. A. and Watts, R. O. (1969). Structure of water: a Monte Carlo calculation. *Chem. Phys. Lett.* **3**, 144–145. [1.1, 4.7.1]

Barker, J. A. and Watts, R. O. (1973). Monte Carlo studies of the dielectric properties of water-like models. *Mol. Phys.* **26**, 789–92. [5.5.3]

Barker, J. A., Fisher, R. A., and Watts, R. O. (1971). Liquid argon: Monte Carlo and molecular dynamics calculations. *Mol. Phys.* **21**, 657–73. [1.3.2, 5.2.3]

Barker, J. A., Henderson, D., and Abraham, F. F. (1981). Phase diagram of the two-dimensional Lennard-Jones system: evidence for first order transitions. Physica **A106**, 226–38. [11.3]

Barojas, J., Levesque, D., and Quentrec, B. (1973). Simulation of diatomic homonuclear liquids. *Phys. Rev.* **A7**, 1092–105. [1.1, 1.3.3, 3.3]

Baxter, R. J. (1970). Ornstein–Zernike relation and Percus–Yevick approximation for fluid mixtures. *J. chem. Phys.* **52**, 4559–62. [6.5.2]

Baxter, R. J. (1982). *Exactly solved models in statistical mechanics*. Academic Press, London. [1.2]

Beeman, D. (1976). Some multistep methods for use in molecular dynamics calculations. *J. comput. Phys.* **20**, 130–9. [3.2.1]

Beenakker, C. W. J. (1986). Ewald sum of the Rotne–Prager tensor. *J. chem. Phys.* **85**, 1581–2. [9,4]

Bellemans, A., Orban, J., and van Belle, D. (1980). Molecular dynamics of rigid and non-rigid necklaces of hard discs. *Mol. Phys.* **39**, 781–82. [3.6.2]

Bennett, C. H. (1976). Efficient estimation of free energy differences from Monte Carlo data. *J. comput. Phys.* **22**, 245–68. [7.2.2]

Berendsen, H. J. C. (1984). Special purpose computational machines. Rapport d'activité scientifique du CECAM, pp. 166–75. [A.1]

Berendsen, H. J. C. and Van Gunsteren, W. F. (1984). Molecular dynamics simulations: techniques and approaches. In *Molecular liquids, dynamics and interaction* (eds. A. J. Barnes, W. J. Orville-Thomas, and J. Yarwood). NATO ASI series C135, pp. 475–500, Reidel, New York. [2.10]

Berendsen, H. J. C. and Van Gunsteren, W. F. (1986). Practical algorithms for dynamic simulations. *Molecular dynamics simulation of statistical mechanical systems.* Proceedings of the Enrico Fermi Summer School. Varenna, 1985 pp 43–65. Soc. Italiana di Fisica, Bologna. [3.2, 3.2.2]

Berendsen, H. J. C., Postma, J. P. M., Van Gunsteren, W. F., Di Nola, A., and Haak, J. R. (1984). Molecular dynamics with coupling to an external bath. *J. chem. Phys.* **81**, 3684–90. [7.4.4, 7.5.3, 7.6]

Berens, P. H. and Wilson, K. R. (1981). Molecular dynamics and spectra. I. Diatomic rotation and vibration. *J. chem. Phys.* **74**, 4872–82. [6.5.6]

Berens, P. H., Mackay, D. H. J., White, G. M., and Wilson, K. R. (1983). Thermodynamics and quantum corrections from molecular dynamics for liquid water. *J. chem. Phys.* **79**, 2375–89. [2.9, 3.5, 5.5.3]

Bernal, J. D. and King, S. V. (1968). Experimental studies of a simple liquid model. In *Physics of simple liquids.* (eds. H. N. V. Temperley, J. S. Rowlinson, and G. S. Rushbrooke) pp. 231–52. North-Holland, Amsterdam. [1.1]

Berne, B. J. (1977). Molecular dynamics of the rough sphere fluid. II. Kinetic models of partially sticky spheres, structured spheres, and rough screwballs. *J. chem. Phys.* **66**, 2821–30. [3.6, 3.6.1]

Berne, B. J. and Harp, G. D. (1970). On the calculation of time correlation functions. *Adv. Chem. Phys.* **17**, 63–227. [1.1, 2.7, 2.9, 3.2.3, 3.3, 11.6, D.1]

Berne, B. J. and Pechukas, P. (1972). Gaussian model potentials for molecular interactions. *J. chem. Phys.* **56**, 4213–16. [1.3.3, 11.5]

Berne, B. J. and Pecora, R. (1976). *Dynamic light scattering.* Wiley, New York. [9.2]

Berthaut, F. (1952). L'énergie électrostatique de réseaux ioniques. *J. Phys. Radium* **13**, 499–505. [5.5.2]

Beysens, D. and Gbadamassi, M. (1979). Shear-induced transition to mean-field critical behaviour. *J. Phy. Lett.* **40**, L565–7. [11.7]

Binder, K. (1984). *Applications of the Monte Carlo method in statistical physics.* Topics in current Physics Vol. 36, Springer, Berlin. [1.3.4, 4.7.2]

Binder, K. (1986). *Monte Carlo methods in statistical physics* (2nd edn). Topics in Current Physics Vol. 7. Springer, Berlin. [1.3.4]

Bishop, M., Ceperley, D., Frisch, H. L., and Kalos, M. H. (1980). Investigations of static properties of model bulk polymer fluids. *J. chem. Phys.* **72**, 3228–35. [4.7.2]

Bobetic, M. V. and Barker, J. A. (1970). Lattice dynamics with three-body forces: argon. *Phys. Rev.* **B2**, 4169–75. [1.3.2]

Booth, A. D. (1972). *Numerical methods* (3rd edn). Butterworth, London [5.2.3]

Born, M. and Von Karman, Th. (1912). Über Schwingungen in Raumgittern. *Physik. Z.* **13**, 297–309. [1.5.2]

Bossis, G., Quentrec, B., and Boon, J. P. (1982). Brownian dynamics and the fluctuation–dissipation theorem. *Mol. Phys.* **45**, 191–6. [9.4]

Box, G. E. P. and Muller, M. E. (1958). A note on the generation of random normal deviates. *Ann. math. Stat.* **29**, 610–11. [G.3]

Bradbury, T. C. (1968). *Theoretical mechanics.* Wiley, New York. [3.4]

Brender, C. and Lax, M. (1977). A Monte Carlo off-lattice method: the slithering snake in a continuum. *J. chem. Phys.* **79**, 2423–5. [4.7.2]

Brigham, E. O. (1974). *The fast Fourier transform.* Prentice-Hall, Englewood Cliffs, NJ. [D.2]

Broughton, J. Q. and Woodcock, L. V. (1978). A molecular dynamics study of surface melting. *J. Phys.* **C11**, 2743–62. [11.8]

Broughton, J. Q., Gilmer, G. H., and Weeks, J. D. (1981). Constant-pressure molecular dynamics simulations of the 2D $r^{-12}$ system: comparison with isochores and isotherms. *J. chem. Phys.* **75**, 5128–32. [11.3]

Broughton, J. Q., Gilmer, G. H., and Weeks, J. D. (1982). Molecular dynamics study of melting in two dimensions. Inverse twelfth-power interaction. *Phys. Rev.* **B25**, 4651–69. [11.3]

Brown, D. and Clarke, J. H. R. (1983). The rheological properties of model liquid n-hexane determined by nonequilibrium molecular dynamics. *Chem. Phys. Lett.* **98**, 579–83. [8.6]

Brown, D. and Clarke, J. H. R. (1984). A comparison of constant energy, constant temperature, and constant pressure ensembles in molecular dynamics simulations of atomic liquids. *Mol. Phys.* **51**, 1243–52. [7.4.3, 7.5.1]

Burgos, E., Murthy, C. S., and Righini, R. (1982). Crystal structure and lattice dynamics of chlorine. The role of electrostatic and anisotropic atom–atom potentials. *Mol. Phys.* **47**, 1391–1403. [1.4.2]

C. R. C. (1984). *CRC handbook of chemistry and physics.* Chemical Rubber Company, Florida. [1.4.2]

Car, R. and Parrinello, M. (1985). Unified approach for molecular dynamics and density functional theory. *Phys. Rev. Lett.* **55**, 2471–4. [1.3.1]

Carnie, S. L. and Torrie, G. M. (1984). The statistical mechanics of the electrical double layer. *Adv. Chem. Phys.* **56**, 141–253. [4.6, 11.8]

Ceperley, D. M. and Kalos, M. H. (1986). Quantum many-body problems. In *Monte Carlo methods in statistical physics* (2nd edn) (1986) (ed. K. Binder). Topics in Current Physics Vol. 7, pp. 145–94. Springer, Berlin. [1.1]

Ceperley, D. M., Chester, G. V., and Kalos, M. H. (1977). Monte Carlo simulation of a many-fermion study. *Phys. Rev.* **B16**, 3081–99. [4.4]

Chandler, D. (1982). Equilibrium theory of polyatomic fluids. In *The liquid state of matter: fluids simple and complex* (eds. E. W. Montroll and J. L. Lebowitz). Studies in Statistical Mechanics Vol. VIII, 275–340. North Holland, Amsterdam. [1.3.3]

Chandler, D. (1987). *Introduction to modern statistical mechanics.* Oxford University Press, New York. [2.1]

Chandler, D. and Berne, B. J. (1979). Comment on the role of constraints on the structure of n-butane in liquid solvents. *J. chem. Phys.* **71**, 5386–7. [2.10]

Chandler, D. and Pratt, L. R. (1976). Statistical mechanics of chemical equilibria and intramolecular structures of nonrigid molecules in condensed phases. *J. chem. Phys.* **65**, 2925–40. [9.4]

Chandler, D. and Wolynes, P. G. (1981). Exploiting the isomorphism between quantum theory and classical statistical mechanics of polyatomic fluids. *J. chem. Phys.* **74**, 4078–95. [10.2]

Chandrasekhar, S. (1943). Stochastic problems in physics and astronomy. *Rev. Mod. Phys.* **15**, 1–89. Also in *Selected papers on noise and stochastic processes* (ed. N. Wax). Dover, New York, 1954. [9.3]

Chapela, G. A., Martinez-Casas, S. E., and Alejandre, J. (1984). Molecular dynamics for discontinuous potentials. I. General method and simulation of hard polyatomic molecules. *Mol. Phys.* **53**, 139–59. [3.6.1, 3.6.2]

Chapela, G. A., Saville, G., Thompson, S. M., and Rowlinson, J. S. (1977). Computer simulation of a gas-liquid surface. Part 1. *J. chem. Soc. Faraday II* **73**, 1133–44. [1.1, 1.5.2]

Chapman, W. and Quirke, N. (1985). Metropolis Monte Carlo simulation of fluids with multiparticle moves. Proceedings of the NATO Workshop on the Simulation of Condensed Matter, 1984. *Physica* **B131**, 34–40. [4.4]

Chatfield, C. (1984). The analysis of time series. An introduction (3rd edn). Chapman and Hall. [6.4.1]

Chesnut, D. A. (1963). Monte Carlo calculations for the two-dimensional triangular lattice gas: supercritical region. *J. chem. Phys.* **39**, 2081–84. [4.6]

Cheung, P. S. Y. (1976). On the efficient evaluation of torques and forces for anisotropic potentials in computer simulation of liquids composed of linear molecules. *Chem. Phys. Lett.* **40**, 19–22. [C. 3]

Cheung, P. S. Y. (1977). On the calculation of specific heats, thermal pressure coefficients, and compressibilities in molecular dynamics simulations. *Mol. Phys.* **33**, 519–26. [2.4, 2.5]

Cheung, P. S. Y. and Powles, J. G. (1975). The properties of liquid nitrogen. IV. A computer simulation. *Mol. Phys.* **30**, 921–49. [1.3.3, 1.4.2, 3.3.2]

Chung, K. L. (1960). *Markov chains with stationary state probabilities* Vol. 1. Springer, Heidelberg. [4.3]

Ciccotti, G. and Jacucci, G. (1975). Direct computation of dynamical response by molecular dynamics: the mobility of a charged Lennard-Jones particle. *Phys. Rev. Lett.* **35**, 789–92. [8.5, 8.7]

Ciccotti, G. and Ryckaert, J. P. (1980). Computer simulation of the generalized Brownian motion. I. The scalar case. *Mol. Phys.* **40**, 141–59. [9.4]

Ciccotti, G. and Ryckaert, J. P. (1981). On the derivation of the generalized Langevin equation for interacting Brownian particles. *J. stat. Phys.* **26**, 73–82. [9.4]

Ciccotti, G. and Ryckaert, J. P. (1986). Molecular dynamics simulation of rigid molecules. *Comput. Phys. Rep.* **4**, 345–92 [3.4]

Ciccotti, G. and Tenenbaum, A. (1980). Canonical ensemble and nonequilibrium states by molecular dynamics. *J. stat. Phys.* **23**, 767–72. [7.4.4]

Ciccotti, G., Ferrario, M., and Ryckaert, J. P. (1982). Molecular dynamics of rigid systems in cartesian coordinates. A general formulation, *Mol. Phys.* **47**, 1253–64. [3.4]

Ciccotti, G., Jacucci, G., and McDonald, I. R. (1975). Transport properties of molten alkali halides. *Phys. Rev.* **A13**, 426–36. [8.2]

Ciccotti, G., Jacucci, G., and McDonald, I. R. (1978). Thermal response to a weak external field. *J. Phys.* **C11**, L509–13. [8.4]

Ciccotti, G., Jacucci, G., and McDonald, I. R. (1979). 'Thought experiments' by molecular dynamics. *J. stat. Phys.* **21**, 1–22. [1.1, 8.1, 8.2, 8.3, 8.4, 8.5, 8.7]

Ciccotti, G., Orban, J., and Ryckaert, J. P. (1976). Stochastic approach to the dynamics of large molecules in a solvent. Rapport d'activité scientifique du CECAM, 'Models for protein dynamics'. [9.4]

Clark, N. A. and Ackerson, B. J. (1980). Observation of the coupling of concentration fluctuations to steady-state shear flow. *Phys. Rev. Lett.* **44**, 1005–8. [11.7]

Conveyou, R. R. and Macpherson, R. D. (1967). Fourier analysis of uniform random number generators. *J. Ass. Comput. Mach.* **14**, 100–19. [G.2]

Cooley, J. N. and Tukey, J. W. (1965). An algorithm for the machine calculation of complex Fourier series. *Math. Comput.* **19**, 297–301. [D.2, D.3]

Corbin, N. and Singer, K. (1982). Semiclassical molecular dynamics of wave packets. *Mol. Phys.* **46**, 671–7. [1.1, 10.3]

Corner, J. (1948). The second virial coefficient of a gas of nonspherical molecules. *Proc. R. Soc. Lond.* **A192**, 275–92. [1.3.3]

Creutz, M. (1983). Microcanonical Monte Carlo simulation. *Phys. Rev. Lett.* **50**, 1411–14. [2.2]

Crowell, A. D. (1958). Potential energy functions for graphite. *J. chem. Phys.* **29**, 446–7. [1.4.2]

Cunningham, G. W. and Meijer, P. H. E. (1976). A comparison of two Monte Carlo methods for computations in statistical mechanics. *J. comput. Phys.* **20**, 50–63. [4.3]

Curro, J. (1974). Computer simulation of multiple chain systems: the effect of density on the average chain dimensions. *J. chem. Phys.* **61**, 1203–7. [4.7.2]

Dahlquist, G. and Björk, A. (1974). *Numerical methods.* Prentice-Hall. Englewood Cliffs, NJ [3.2.1]

Debye, P. (1929). *Polar molecules.* Dover Press, New York. [11.6]

De Gennes, P. (1971). Reptation of a polymer chain in the presence of fixed obstacles. *J. chem. Phys.* **55**, 572–9. [4.7.2]

De Leeuw, S. W., Perram, J. W., and Smith, E. R. (1980). Simulation of electrostatic systems in periodic boundary conditions. I. Lattice sums and dielectric constant. *Proc. R. Soc. Lond.* **A373**, 27–56. [5.5.2]

De Raedt, B., Sprik, M., and Klein, M. L. (1984). Computer simulation of muonium in water. *J. chem. Phys.* **80**, 5719–24. [10.2]

D'Evelyn, M. P. and Rice, S. A. (1981). Configuration space diffusion criterion for optimization of the force-bias Monte Carlo method. Comments. *Chem. Phys. Lett.* **77**, 630–3. [7.3.2]

Dixon, M. and Hutchinson, P. (1977). A method for the extrapolation of pair distribution functions. *Mol. Phys.* **33**, 1663–70. [6.5.2]

Dixon, M. and Sangster, M. J. L. (1976). The structure of molten NaCl from a simulation model which allows for the polarization of both ions. *J. Phys.* **C9**, L5–9. [1.3.2, 11.4]

Doll, J. D. and Dion, D. R. (1976). Generalized Langevin equation approach for atom/solid surface scattering: numerical techniques for Gaussian generalized Langevin dynamics. *J. chem. Phys.* **65**, 3762–6. [9.4]

Doob, J. L. (1942). The Brownian movement and stochastic equations. *Ann. Math.* **43**, 351–69. Also in *Selected papers on noise and stochastic processes* (ed. N. Wax). Dover, New York, 1954. [9.3]

Doran, M. B. and Zucker, I. J. (1971). Higher order multipole three-body Van der Waals' interactions and stability of rare gas solids. *J. Phys.* **C4**, 307–12. [1.3.2]

Dorfman, J. R. and Cohen, E. G. D. (1970). Velocity correlation functions in two and three dimensions. *Phys. Rev. Lett.* **25**, 1257–60. [11.7]

Dorfman, J. R. and Cohen, E. G. D. (1972). Velocity correlation functions in two and three dimensions: low density. *Phys. Rev.* **A6**, 776–90. [11.7]

Dorfman, J. R. and Cohen, E. G. D. (1975). Velocity correlation functions in two and three dimensions. II. Higher density. *Phys. Rev.* **A12**, 292–316. [11.7]

Dymond, J. H. and Smith, E. B. (1980). *The virial coefficients of pure gases and mixtures. A critical compilation.* Oxford Science Research Paper Series. Oxford University Press, Oxford. [1.4.3]

Eastwood, J. W., Hockney, R. W., and Lawrence, D. (1980). P3M3DP—the three dimensional periodic particle–particle/particle–mesh program. *Comput. Phys. Commun.* **19**, 215–61. [5.5.4]

Edwards, F. G., Enderby, J. E., Howe, R. A., and Page, D. I. (1975). The structure of molten sodium chloride. *J. Phys.* **C8**, 3483–90. [11.4]

Egelstaff, P. A. (1961). The theory of the thermal neutron scattering law. In *Inelastic scattering of neutrons in solids and liquids*, pp. 25–38. IAEA, Vienna. [2.9]

Enderby, J. E. and Neilson, G. W. (1980). Structural properties of ionic liquids. *Adv. Phys.* **29**, 323–65. [11.4]

English, C. A. and Venables, J. A. (1974). The structure of diatomic molecular solids. *Proc. R. Soc. Lond.* **A340**, 57–80. [1.4.2]

Eppenga, R. and Frenkel, D. (1984). Monte Carlo study of the isotropic and nematic phases of infinitely thin hard platelets. *Mol. Phys.* **52**, 1303–34. [4.5, 4.7.1, 5.6, 11.5]

Ermak, D. L. (1975). A computer simulation of charged particles in solution. I. Technique and equilibrium properties. *J. chem. Phys.* **62**, 4189–96. [9.3]

Ermak, D. L. (1976). Brownian dynamics techniques and their application to dilute solutions. Rapport d'activité scientifique du CECAM, pp. 66–81. [9.3]

Ermak, D. L. and Buckholtz, H. (1980). Numerical integration of the Langevin equation: Monte Carlo simulation. *J. comput. Phys.* **35**, 169–82. [9.3, 9.4]

Ermak, D. L. and McCammon, J. A. (1978). Brownian dynamics with hydrodynamic interactions. *J. chem. Phys.* **69**, 1352–60. [9.4, G.3]

Ermak, D. L. and Yeh, Y. (1974). Equilibrium electrostatic effects on behaviour of polyions in solution: polyion–mobile ion interaction. *Chem. Phys. Lett.* **24**, 243–8. [7.4.4, 9.3]

Ernst, M. H., Hauge, E. H., and Van Leeuwen, J. M. J. (1970). Asymptotic time behaviour of correlation functions. *Phys. Rev. Lett.* **25**, 1254–6. [11.7]

Ernst, M. H., Hauge, E. H., and Van Leeuwen, J. M. J. (1971). Asymptotic time behaviour of correlation functions. I. Kinetic terms. *Phys. Rev.* **A4**, 2055–65. [11.7]

Ernst, M. H., Hauge, E. H., and Van Leeuwen, J. M. J. (1976a). Asymptotic time behaviour of correlation functions. II. Kinetic and potential terms. *J. stat. Phys.* **15**, 7–22. [11.7]

Ernst, M. H., Hauge, E. H., and Van Leeuwen, J. M. J. (1976b). Asymptotic time behaviour of correlation functions. III. Local equilibrium and mode-coupling theory, *J. stat. Phys.* **15**, 23–58. [11.7]

Ernst, M. H., Cichocki, B., Dorfman, J. R., Sharma, J., and Van Beijeren, H. (1978). Kinetic theory of nonlinear viscous flow in two and three dimensions. *J. stat. Phys.* **18**, 237–70. [11.7]

Erpenbeck, J. J. (1983). Nonequilibrium molecular dynamics calculations of the shear viscosity of hard spheres. *Physica* **118A**, 144–56. [8.7, 11.7]

Erpenbeck, J. J. (1984). Shear viscosity of the hard sphere fluid via nonequilibrium molecular dynamics. *Phys. Rev. Lett.* **52**, 1333–5. [11.7]

Erpenbeck, J. J. and Wood, W. W. (1977). Molecular dynamics techniques for hard core systems. In *Statistical mechanics B. Modern theoretical chemistry* (ed. B. J. Berne) Vol. 6, pp. 1–40. Plenum, New York. [2.4, 3.2, 5.3, 5.6, 8.5]

Erpenbeck, J. J. and Wood, W. W. (1981). Molecular dynamics calculations of shear viscosity time correlation functions for hard spheres. *J. stat. Phys.* **24**, 455–68. [11.7]

Evans, D. J. (1977). On the representation of orientation space. *Mol. Phys.* **34**, 317–25. [3.3.1]

Evans, D. J. (1979a). The frequency dependent shear viscosity of methane. *Mol. Phys.* **37**, 1745–54. [8.2, 8.7]

Evans, D. J. (1979b). Nonlinear viscous flow in the Lennard-Jones fluid. *Phys. Lett.* **74A**, 229–32. [8.2, 8.7, 11.7]

Evans, D. J. (1979c). The nonsymmetric pressure tensor in polyatomic fluids. *J. stat. Phys.* **20**, 547–55. [8.6]

Evans, D. J. (1980). Enhanced $t^{-3/2}$ long-time tail for the stress–stress time correlation function. *J. stat. Phys.* **22**, 81–90. [11.7]

Evans, D. J. (1981a). Equilibrium fluctuation expressions for the wavevector- and frequency-dependent shear viscosity. *Phys. Rev.* **A23**, 2622–6. [8.2]

Evans, D. J. (1981b). Nonequilibrium molecular dynamics study of the rheological properties of diatomic liquids. *Mol. Phys.* **42**, 1355–65. [8.6]

Evans, D. J. (1981c). Rheological properties of simple fluids by computer simulation. *Phys. Rev.* **A23**, 1988–97. [11.7]

Evans, D. J. (1982a). Homogeneous NEMD algorithm for thermal conductivity: application of non-canonical linear response theory. *Phys. Lett.* **91A**, 457–60. [8.4]

Evans, D. J. (1982b). Shear-induced melting of soft sphere crystals. *Phys. Rev.* **A25**, 2788–92. [11.7]

Evans, D. J. (1983a). Computer experiment for nonlinear thermodynamics of Couette flow. *J. chem. Phys.* **78**, 3297–302. [7.4.3, 7.6]

Evans, D. J. (1983b). Molecular dynamics simulations of the rheological properties of simple fluids. *Physica* **118A**, 51–68. [11.7]

Evans, D. J. and Gaylor, K. (1983). NEMD algorithm for calculating the Raman spectra of dense fluids. *Mol. Phys.* **49**, 963–72. [8.6]

Evans, D. J. and Hanley, H. J. M. (1982). Fluctuation expressions for fast thermal transport processes: vortex viscosity. *Phys. Rev.* **A25**, 1771–4. [8.6]

Evans, D. J. and Morriss, G. P. (1983a). The isothermal isobaric molecular dynamics ensemble. *Phys. Lett.* **98A**, 433–6. [7.4.3, 7.5.2]

Evans, D. J. and Morriss, G. P. (1983b). Isothermal isobaric molecular dynamics. *Chem. Phys.* **77**, 63–6. [7.5.2, 7.6]

Evans, D. J. and Morriss, G. P. (1984a). Non-Newtonian molecular dynamics. *Comput. Phys. Rep.* **1**, 297–344. [7.4.1, 7.4.3, 7.5.2, 8.1, 8.2, 8.4, 8.5]

Evans, D. J. and Morriss, G. P. (1984b). Nonlinear response theory for steady planar Couette flow. *Phys. Rev.* **A30**, 1528–30. [8.2]

Evans D. J. and Morriss G. P. (1986). Shear thickening and turbulence in simple fluids. *Phys. Rev. Lett.* **56**, 2172–5 [11.7]

Evans, D. J. and Murad, S. (1977). Singularity-free algorithm for molecular dynamics simulation of rigid polyatomics. *Mol. Phys.* **34**, 327–31. [3.3.1]

Evans, D. J. and Streett, W. B. (1978). Transport properties of homonuclear diatomics. II. Dense fluids. *Mol. Phys.* **36**, 161–76. [8.6]

Evans, D. J. and Watts, R. O. (1976). On the structure of liquid benzene. *Mol. Phys.* **32**, 93–100. [1.4.2]

Evans, D. J., Hoover, W. G., Failor, B. H., Moran, B., and Ladd, A. J. C. (1983). Nonequilibrium molecular dynamics via Gauss's principle of least constraint. *Phys. Rev.* **A28**, 1016–21. [8.5]

Evans, H., Tildesley, D. J., and Sluckin, T. (1984). Boundary effects in the orientational ordering of adsorbed nitrogen. *J. Phys.* **C17**, 4907–26. [4.3]

Evans, W. A. B. and Powles, J. G. (1982). The computer simulation of the dielectric properties of polar liquids. The dielectric constant and relaxation of liquid hydrogen chloride. *Mol. Phys.* **45**, 695–707. [8.6, 8.7]

Ewald, P. (1921). Die Berechnung optischer und elektrostatischer Gitterpotentiale. *Ann. Phys.* **64**, 253–87. [5.5.2]

Eyring, H. (1932). Steric hindrance and collision diameters. *J. Am. chem. Soc.* **54**, 3191–203. [1.3.3]

Fabbri, U. and Zannoni, C. (1986). A Monte Carlo investigation of the Lebwohl–Lasher lattice model in the vicinity of the orientational phase transition. *Mol. Phys.* **58**, 763–88. [11.5]

Felderhof, B. U. (1977). Hydrodynamic interaction between two spheres. *Physica* **A89**, 373–84. [9.4]

Felderhof, B. U. (1980). Fluctuation theorems for dielectrics with periodic boundary conditions. *Physica* **101A**, 275–82. [5.5.3]

Feller, W. (1957). *An introduction to probability theory and its applications* (2nd edn). Wiley, New York. [4.3]

Ferrario, M. and Ryckaert, J. P. (1985). Constant-pressure, constant-temperature molecular dynamics for rigid and partially rigid molecular systems. *Mol. Phys.* **54**, 587–604. [7.5.4]

Feynman, R. P. and Hibbs, A. R. (1965). *Quantum mechanics and path integrals*. McGraw-Hill, New York. [10.1, 10.2]

Filon, L. N. G. (1928). On a quadrature formula for trigonometric integrals. *Proc. R. Soc. Edinburgh* **A49**, 38–47. [D.3]

Fincham, D. (1981). An algorithm for the rotational motion of rigid molecules. *CCP5 Quarterly* **2**, 6–10. [3.3.1, 3.4]

Fincham, D. (1983). RDF on the DAP: easier than you think. *CCP5 Quarterly* **8**, 45–6. [6.2]

Fincham, D. (1984). More on rotational motion of linear molecules. *CCP5 Quarterly* **12**, 47–8. [3.3.2]

Fincham, D. (1985). How big is a timestep? *CCP5 Quarterly* **17**, 43–50. [3.5]

Fincham, D. and Heyes, D. M. (1982). Integration algorithms in molecular dynamics. *CCP5 Quarterly* **6**, 4–10. [3.2.1, 3.2.2]

Fincham, D. and Ralston, B. J. (1981). Molecular dynamics simulation using the CRAY-1 vector processing computer. *Comput. Phys. Commun.* **23**, 127–34. [1.5.4, 5.3.1]

Fincham, D., Quirke, N., and Tildesley, D. J. (1986). Computer simulation of molecular liquid mixtures. I. A diatomic Lennard-Jones model mixture for $CO_2/C_2H_6$. *J. chem. Phys.* **84**, 4535–46 [5.6, 6.4.1, 7.4.3]

Fisher, M. E. (1964). The free energy of a macroscopic system. *Arch. Rat. Mech. Anal.* **17**, 377–410. [2.3]

Fixman, M. (1974). Classical statistical mechanics of constraints: a theorem and application to polymers. *Proc. nat. Acad. Sci.* **71**, 3050–3. [2.10, 3.4]

Fixman, M. (1978a). Simulation of polymer dynamics. I. General theory. *J. chem. Phys.* **69**, 1527–37. [3.4]

Fixman, M. (1978b). Simulation of polymer dynamics. II. Relaxation rates and dynamic viscosity. *J. chem. Phys.* **69**, 1538–45. [3.4]

Flinn, P. A. and McManus, G. M. (1961). Monte Carlo calculation of the order–disorder transformation in the body-centred cubic lattice. *Phys. Rev.* **124**, 54–9. [4.3]

Fosdick, L. D. and Jordan, H. F. (1966). Path integral calculation of the two particle Slater sum for $He^4$. *Phys. Rev.* **143**, 58–66. [10.2]

Fowler, R. F. (1984). The identification of a droplet in equilibrium with its vapour. *CCP5 Quarterly* **13**, 58–62. [11.2]

Fox, J. R. and Andersen, H. C. (1984). Molecular dynamics simulations of a supercooled monatomic liquid and glass. *J. phys. Chem.* **88**, 4019–27. [7.5.1]

Freasier, B. C. (1980). Equation of state of fused hard spheres revisited. *Mol. Phys.* **39**, 1273–80. [6.5.3]

Freasier, B. C., Jolly, D., and Bearman, R. J. (1976). Hard dumbells: Monte Carlo pressures and virial coefficients. *Mol. Phys.* **31**, 255–63. [6.5.3]

Freasier, B. C., Jolly, D. L., Hamer, N. D., and Nordholm, S. (1979). Effective vibrational potentials of bromine in argon. Monte Carlo simulation in a mixed ensemble. *Chem. Phys.* **38**, 293–300. [2.2]

Frenkel, D. (1980). Intermolecular spectroscopy and computer simulations. *Intermolecular spectroscopy and dynamical properties of dense systems.* Proceedings of the Enrico Fermi Summer School, Vol. 75, pp. 156–201. Soc. Italiana di Fisica, Bologna. [6.4.1]

Frenkel, D. (1986). Free energy computation and first order phase transitions, *Molecular dynamics simulation of statistical mechanical systems.* Proceedings of the Enrico Fermi Summer School, Varenna 1985, pp 151–88. [2.4, 5.6, 7.2.3, 7.2.4]

Frenkel, D. and Eppenga, R. (1982). Monte Carlo study of the isotropic–nematic transition in a fluid of thin hard disks. *Phys. Rev. Lett.* **49**, 1089–92. [11.5]

Frenkel, D. and Ladd, A. J. C. (1984). New Monte Carlo method to compute the free energy of arbitrary solids. Application to the f.c.c. and h.c.p. phases of hard spheres. *J. chem. Phys.* **81**, 3188–93. [7.2.4]

Frenkel, D. and Maguire, J. F. (1983). Molecular dynamics study of the dynamical properties of an assembly of infinitely thin hard rods. *Mol. Phys.* **49**, 503–41. [3.6.2]

Frenkel, D. and McTague, J. P. (1979). Evidence for an orientationally ordered two dimensional fluid phase from molecular dynamics calculations. *Phys. Rev. Lett.* **42**, 1632–5. [11.3]

Frenkel, D. and McTague, J. P. (1980). Computer simulations of freezing and supercooled liquids. *A. Rev. Phys. Chem.* **31**, 491–521. [1.1, 11.3]

Frenkel, D. and Mulder, B. M. (1985). The hard ellipsoid-of-revolution fluid. I. Monte Carlo simulations, *Mol. Phys.* **55**, 1171–92. [11.5]

Frenkel, D., Mulder, B. M., and McTague, J. P. (1984). Phase diagram of a system of hard ellipsoids. *Phys. Rev. Lett.* **52**, 287–90. [11.5]

Frenkel, D., Mulder, B. M. and McTague, J. P. (1985). Phase diagram of hard ellipsoids of revolution. *Mol. Cryst. Liq. Cryst.* **123**, 119–28. [7.2.4, 11.5]

Friedberg, R. and Cameron, J. E. (1970). Test of the Monte Carlo method: fast simulation of a small Ising lattice. *J. chem. Phys.* **52**, 6049–58. [6.4.1]

Friedman, H. L. (1975). Image approximation to the reaction field. *Mol. Phys.* **29**, 1533–43. [5.5.3]

Friedman, H. L. (1985). *A course in statistical mechanics.* Prentice-Hall, Englewood Cliffs, NJ. [2.1]

Fröhlich, H. (1949). *Theory of dielectrics; dielectric constant and dielectric loss.* Oxford University Press, Oxford. [5.5.1, 5.5.3]

Fumi, F. G. and Tosi, M. P. (1964). Ionic sizes and Born repulsive parameters in the NaCl-type alkali halides. I. The Huggins–Mayer and Pauling forms. *J. Phys. Chem. Solids* **25**, 31–43. [11.4]

Futrelle, R. P. and McGinty, D. J. (1971). Calculation of spectra and correlation functions from molecular dynamics data using the fast Fourier transform. *Chem. Phys. Lett.* **12**, 285–7. [6.3.2]

Gay, J. G. and Berne, B. J. (1981). Modification of the overlap potential to mimic a linear site–site potential. *J. chem. Phys.* **74**, 3316–19. [1.3.3, 11.5]

Gear, C. W. (1966). The numerical integration of ordinary differential equations of various orders. Report ANL 7126, Argonne National Laboratory. [3.2, 8.7, E.1]

Gear, C. W. (1971). *Numerical initial value problems in ordinary differential equations.* Prentice-Hall, Englewood Cliffs, NJ. [3.2, 3.2.1, 8.7, E.1]

Gibson, W. G. (1974). Quantum corrections to the radial distribution function of a fluid. *Mol. Phys.* **28**, 793–800. [2.9]

Gibson, W. G. (1975a). Quantum corrections to the properties of a dense fluid with nonanalytic intermolecular potential function. I. The general case. *Mol. Phys.* **30**, 1–11. [2.9]

Gibson, W. G. (1975b). Quantum corrections to the properties of a dense fluid with nonanalytic intermolecular potential function. II. Hard spheres. *Mol. Phys.* **30**, 13–30. [2.9]

Gillan, M. J. and Dixon, M. (1983). The calculation of thermal conductivities by perturbed molecular dynamics simulation. *J. Phys.* **C16**, 869–78. [8.4]

Go, N. and Scheraga, H. A. (1976). On the use of classical statistical mechanics in the treatment of polymer chain conformation. *Macromolecules* **9**, 535–42. [3.4]

Goddard, J. D., Klein, M. L., and Ozaki, Y. (1983). *Ab initio* potential curves for sodium nitrite and the simulation of the molten salt. *J. Phys. Soc. Japan* **52**, 1168–72. [11.4]

Goldstein, H. (1980). *Classical mechanics* (2nd edn). Addison-Wesley, Reading, MA. [1.3.1, 3.1, 3.3, 3.3.1]

Gordon, R. (1968). Correlation functions for molecular motion. *Adv. Magn. Reson.* **3**, 1–42. [2.7]

Gosling, E. M., McDonald, I. R., and Singer, K. (1973). On the calculation by molecular dynamics of the shear viscosity of a simple fluid. *Mol. Phys.* **26**, 1475–84. [8.2, 8.7]

Gray, C. G. and Gubbins, K. E. (1984). *The theory of molecular fluids. I. Fundamentals.* Clarendon Press, Oxford. [1.4.2, 2.6, 6.2]

Green, H. S. (1951). The quantum mechanics of assemblies of interacting particles. *J. chem. Phys.* **19**, 955–62. [10.1]

Greenberger, M. (1965). Method in randomness. *Commun. Ass. Comput. Mach.* **8**, 177–9. [G.2]

Guiasu, S. and Shenitzer, A. (1985). The principle of maximum entropy. *Math. Intell.* **7**, 42–8. [6.5.6]

Guldbrand, L., Jönsson, B., Wennerström, H., and Linse, P. (1984). Electrical double layer forces. A Monte Carlo study. *J. chem. Phys.* **80**, 2221–8. [4.6]

Gull, S. F. and Daniell, G. J. (1978). Image reconstruction from incomplete and noisy data. *Nature* **272**, 686–90. [6.5.6]

Haile, J. M. (1978). Molecular dynamics simulation of simple fluids with three-body interactions included. In *Computer modelling of matter* (ed. P. Lykos). ACS Symposium series Vol. 86, pp. 172–90. Washington. [5.4]

Haile, J. M. and Graben, H. W. (1980). Molecular dynamics simulations extended to various ensembles. I. Equilibrium properties in the isoenthalpic–isobaric ensemble. *J. chem. Phys.* **73**, 2412–9 [7.5.1]

Haile, J. M. and Gray, C. G. (1980). Spherical harmonic expansions of the angular pair correlation function in molecular fluids. *Chem. Phys. Lett.* **76**, 583–8. [2.6]

Haile, J. M. and Gupta, S. (1983). Extensions of molecular dynamics simulation method. II. Isothermal systems. *J. chem. Phys.* **79**, 3067–76. [7.4.1]

Haken, H. (1975). Cooperative phenomena in systems far from thermal equilibrium and in non-physical systems. *Rev. Mod. Phys.* **47**, 67–121. [9.3]

Halperin, B. I. and Nelson, D. R. (1978). Theory of two-dimensional melting. *Phys. Rev. Lett.* **41**, 121–4, 519(E). [11.3]

Hammersley, J. M. and Handscomb, D. C. (1964). *Monte Carlo methods.* Wiley, New York. [4.2.2]

Hanley, H. J. M. (1983). Nonlinear fluid behaviour. Proceedings of a Conference in Boulder, Colorado. *Physica* **118A**, 1–454. [8.7]

Hansen, J. P. and McDonald, I. R. (1986). *Theory of simple liquids.* (2nd ed.). Academic Press, New York. [1.1, 1.2, 2.1, 2.6, 2.7, 6.5.2, 8.2, 8.3, 8.4, 9.2, 9.3, 11.4]

Hansen, J. P. and Verlet, L. (1969). Phase transitions of the Lennard-Jones system. *Phys. Rev.* **184**, 151–61. [7.2.4, 11.3]

Hansen, J. P. and Weis, J. J. (1969). Quantum corrections to the coexistence curve of neon near the triple point. *Phys. Rev.* **188**, 314–18. [2.9]

Hansen, J. P., Levesque, D., and Weis, J. J. (1979). Self diffusion in the two-dimensional classical electron gas. *Phys. Rev. Lett.* **43**, 979–82. [1.5.5]

Harp, G. D. and Berne, B. J. (1968). Linear and angular momentum autocorrelation functions in diatomic liquids. *J. chem. Phys.* **49**, 1249–54. [1.1, 3.3]

Harp, G. D. and Berne, B. J. (1970). Time correlation functions, memory functions, and molecular dynamics. *Phys. Rev.* **A2**, 975–96. [3.3]

Harris, F. J. (1978). On the use of windows for harmonic analysis with discrete Fourier transforms. *Proc. IEEE* **66**, 51–83. [6.5.6]

Hastings, W. K. (1970). Monte Carlo sampling methods using Markov chains, and their applications. *Biometrika* **57**, 97–109. [4.4]

Helfand, E. (1979). Flexible vs rigid constraints in statistical mechanics. *J. chem. Phys.* **71**, 5000–7. [3.4]

Helfand, E. and Rice, S. A. (1960). Principle of corresponding states for transport properties. *J. chem. Phys.* **32**, 1642–4. [B.1]

Heller, E. J. (1975). Time-dependent approach to semiclassical dynamics. *J. chem. Phys.* **62**, 1544–55. [10.3]

Heller, E. J. (1976). Time-dependent variational approach to semiclassical dynamics. *J. chem. Phys.* **64**, 63–73. [10.3]

Hemmer, P. C. (1968). The hard core quantum gas at high temperatures. *Phys. Lett.* **A27**, 377–8. [2.9]

Henderson, J. R. (1983). Statistical mechanics of fluids at spherical structureless walls. *Mol. Phys.* **50**, 741–61. [2.4]

Henderson, J. R. and Van Swol, F. (1984). On the interface between a fluid and a planar wall. Theory and simulations of a hard sphere fluid at a hard wall. *Mol. Phys.* **51**, 991–1010. [11.8]

Henderson, J. R. and Van Swol, F. (1985). On the approach to complete wetting by gas at a liquid–wall interface. Exact sum rules, fluctuation theory and the verification by computer simulation of the presence of long-range pair correlations at the wall. *Mol. Phys.* **56**, 1313–56. [11.8]

Herman, M. F., Bruskin, E. J., and Berne, B. J. (1982). On path integral Monte Carlo simulations. *J. chem. Phys.* **76**, 5150–5. [10.2]

Hess, W. and Klein, R. (1978). Dynamical properties of colloidal systems. I. Derivation of stochastic transport equations. *Physica* **94A**, 71–90. [9.4]

Heyes, D. M. (1981). Electrostatic potentials and fields in infinite point charge lattices. *J. chem. Phys.* **74**, 1924–9. [5.5.2]

Heyes, D. M. (1983a). MD incorporating Ewald summations on partial charge polyatomic systems. *CCP5 Quarterly* **8**, 29–36. [5.5.2]

Heyes, D. M. (1983b). Molecular dynamics at constant pressure and temperature. *Chem. Phys.* **82**, 285–301. [7.4.1]

Heyes, D. M. (1983c). Self-diffusion and shear viscosity of simple fluids. A molecular dynamics study. *J. chem. Soc. Faraday* II, **79**, 1741–58. [8.2]

Heyes, D. M. (1983d). A new method of performing MD and MC on point charge systems. I. The lamina. *CCP5 Quarterly* **9**, 20–7. [11.8]

Heyes, D. M. (1986). Shear thinning and thickening of the Lennard-Jones fluid. A molecular dynamics study. *J. chem. Soc. Faraday* II, **82**, 1365–83. [11.7]

Heyes, D. M. and Clarke, J. H. R. (1981). Computer simulation of molten salt interphases. Effect of a rigid boundary and an applied electric field. *J. chem. Soc. Faraday II* **77**, 1089–100. [11.8]

Heyes, D. M. and Singer, K. (1982). A very accurate molecular dynamics algorithm. *CCP5 Quarterly* **6**, 11–23. [3.2.1, 3.2.3]

Heyes, D. M., Montrose, C. J., and Litovitz, T. A. (1983). Viscoelastic shear thinning of liquids: a molecular dynamics study. *J. chem. Soc. Faraday II* **79**, 611–35. [11.7]

Heyes, D. M., Kim, J. J., Montrose, C. J., and Litovitz, T. A. (1980). Time dependent nonlinear shear stress effects in simple liquids: a molecular dynamics study. *J. chem. Phys.* **73**, 3987–96. [11.7]

Hill, T. L. (1956). *Statistical mechanics.* McGraw-Hill, New York.. [2.1, 2.2, 2.3]

Hockney, R. W. (1970). The potential calculation and some applications. *Methods comput. Phys.* **9**, 136–211. [3.2.1]

Hockney, R. W. and Eastwood, J. W. (1981). *Computer simulation using particles.* McGraw-Hill, New York. [1.2, 3.2, 3.2.1, 5.3.2, 5.5.4]

Holian, B. L. and Evans, D. J. (1983). Shear viscosities away from the melting line: a comparison of equilibrium and nonequilibrium molecular dynamics. *J. chem. Phys.* **78**, 5147–50. [8.7]

Honeycutt, J. P. and Andersen, H. C. (1984). The effect of periodic boundary conditions on homogeneous nucleation observed in computer simulations. *Chem. Phys. Lett.* **108**, 535–8. [1.5.2]

Hoogland, A., Spaa, J., Selman, B., and Compagner, A. (1983). A special-purpose processor for the Monte Carlo simulation of Ising spin systems. *J. comput. Phys.* **51**, 250–60. [A.1]

Hoover, W. G. (1983a). Nonequilibrium molecular dynamics. *A. Rev. phys. Chem.* **34**, 103–27. [2.2, 7.4.3, 8.1]

Hoover, W. G. (1983b). Atomistic nonequilibrium computer simulations. *Physica* **118A**, 111–22. [2.2, 8.1]

Hoover, W. G. (1985). Canonical dynamics: equilibrium phase-space distributions. *Phys. Rev.* **A31**, 1695–7. [7.4.4, 7.5.3, 7.6]

Hoover, W. G. and Alder, B. J. (1967). Studies in molecular dynamics. IV. The pressure, collision rate, and their number dependence for hard disks. *J. chem. Phys.* **46**, 686–91. [5.6]

Hoover, W. G. and Ashurst, W. T. (1975). Nonequilibrium molecular dynamics. In *Theoretical chemistry: Advances and perspectives* (ed. H. Eyring and D. Henderson). Vol. 1, pp.1–51. Academic Press, New York. [1.1, 8.1, 8.5]

Hoover, W. G. and Ree, F. H. (1968). Melting transition and communal entropy for hard spheres. *J. chem. Phys.* **49**, 3609–17. [7.2.4, 11.3]

Hoover, W. G., Gray, S. G., and Johnson, K. W. (1971). Thermodynamic properties of the fluid and solid phases for inverse power potentials. *J. chem. Phys.* **55**, 1128–36. [11.3, B.1]

Hoover, W. G., Ladd, A. J. C., and Moran, B. (1982). High strain rate plastic flow studied via nonequilibrium molecular dynamics. *Phys. Rev. Lett.* **48**, 1818–20. [7.4.3, 7.6]

Hoover, W. G., Ladd, A. J. C., Hickman, R. B., and Holian, B. L. (1980a). Bulk viscosity via nonequilibrium and equilibrium molecular dynamics. *Phys. Rev.* **A21**, 1756–60. [8.3, 8.7, 11.7]

Hoover, W. G., Evans, D. J., Hickman, R. B., Ladd, A. J. C., Ashurst, W. T., and Moran, B. (1980b). Lennard-Jones triple-point bulk and shear viscosities. Green–Kubo theory, Hamiltonian mechanics, and non-equilibrium molecular dynamics. *Phys. Rev.* **A22**, 1690–7. [6.5.5, 8.2, 8.3, 8.7, 11.7]

Hoover, W. G., Ross, M., Johnson, K. W., Henderson, D., Barker, J. A., and Brown, B. C. (1970). Soft sphere equation of state. *J. chem. Phys.* **52**, 4931–41. [B.1]

Hsu, C. S. and Rahman, A., (1979). Crystal nucleation and growth in liquid rubidium. *J. chem. Phys.* **70**, 5234–40. [11.3]

Impey, R. W., Madden, P. A., and McDonald, I. R. (1982). Spectroscopic and transport properties of water. Model calculations and the interpretation of experimental results. *Mol. Phys.* **46**, 513–39. [11.6]

Impey, R. W., Madden, P. A., and Tildesley, D. J. (1981). On the calculation of the orientational correlation parameter $g_2$. *Mol. Phys.* **44**, 1319–34. [1.5.2]

Jackson, J. L. and Mazur, P. (1964). On the statistical mechanical derivation of the correlation formula for the viscosity. *Physica* **30**, 2295–304. [8.1]

Jacucci, G. (1984). Path integral Monte Carlo. In *Monte Carlo methods in quantum problems* (ed. M. H. Kalos). NATO ASI Series Vol. C125, pp. 117–44. Reidel, New York. [10.2]

Jacucci, G. and McDonald, I. R. (1975). Structure and diffusion in mixtures of rare gas liquids. *Physica* **80A**, 607–25. [8.5]

Jacucci, G. and Omerti, E. (1983). Monte Carlo calculation of the radial distribution function of quantum hard spheres at finite temperatures using path integrals with boundary conditions. *J. chem. Phys.* **79**, 3051–4. [10.2]

Jacucci, G. and Quirke, N. (1980a). Structural correlation functions in the coexistence region from molecular dynamics. *Nuovo Cimento* **58B**, 317–22. [5.7.3]

Jacucci, G. and Quirke, N. (1980b). Monte Carlo calculation of the free energy differences between hard- and soft-core fluids. *Mol. Phys.* **40**, 1005–9. [7.2.2]

Jacucci, G. and Rahman, A. (1984). Comparing the efficiency of Metropolis Monte Carlo and molecular dynamics methods for configuration space sampling. *Nuovo Cimento* **D4**, 341–56. [6.4.1]

Jancovici, B. (1969). Quantum mechanical equation of state of a hard sphere gas at high temperature. *Phys. Rev.* **178**, 295–7. [2.9]

Jansoone, V. M. (1974). Dielectric properties of a model fluid with the Monte Carlo method. *Chem. Phys.* **3**, 78–86. [4.7.1]

Jolly, D. L., Freasier, B. C., and Bearman, R. J. (1976). The extension of simulation radial distribution functions to an arbitrary range by Baxter's factorization technique. *Chem. Phys.* **15**, 237–42. [6.5.2]

Jordan, H. F. and Fosdick, L. D. (1968). Three-particle effects in the pair distribution function for $He^4$ gas. *Phys. Rev.* **171**, 128–49. [10.2]

Jorgensen, W. L. (1981). Transferable intermolecular potential functions for water, alcohols, and esters. Application to liquid water. *J. Am. chem. Soc.* **103**, 335–40. [1.4.2]

Jorgensen, W. L. (1982). Revised TIPS model for simulations of liquid water and aqueous solutions. *J. chem. Phys.* **77**, 4156–63. [4.5]

Joshi, Y. P. and Tildesley, D. J. (1985). A simulation study of the melting of patches of $N_2$ adsorbed on graphite. *Mol. Phys.* **55**, 999–1016. [11.8]

Kalos, M. H. (1984). Monte Carlo methods in quantum problems. *NATO Advanced Science Institute Series*, Vol. C125, Reidel, New York. [10.1]

Kalos, M. H., Levesque, D., and Verlet, L. (1974). Helium at zero temperature with hard sphere and other forces. *Phys. Rev.* **A9**, 2178–95. [10.4]

Kawasaki, K. and Gunton, J. D. (1973). Theory of nonlinear transport processes: nonlinear shear viscosity and norml stress effects. *Phys. Rev.* **A8**, 2048–64. [11.7]

Kelvin (Lord) (1901). Nineteenth century clouds over the dynamical theory of heat and light. *Phil. Mag.* **2**, 1–40. [4.1]

Kestemont, E. and Van Craen, J. (1976). On the computation of correlation functions in molecular dynamics experiments. *J. comput. Phys.* **22**, 451–8. [6.3.2]

Kirkwood, J. (1933). Quantum statistics of almost classical assemblies. *Phys. Rev.* **44**, 31–7. [2.9]

Klein, M. L. and McDonald, I. R. (1982). Properties of the paraelectric-solid and molten phases of sodium nitrite. *Proc. R. Soc. Lond.* **A382**, 471–82. [6.5.6]

Klein, M. L. and Weis, J. J. (1977). The dynamical structure factor $S(Q,\omega)$ of solid $\beta$-$N_2$. *J. chem. Phys.* **67**, 217–224. [1.5.2]

Knuth, D. (1973). *The art of computer programming*. (2nd edn). Addison-Wesley, Reading, MA. [5.3.2, G.2]

Kornfeld, H. (1924). Die Berechnung elektrostatischer Potentiale und der Energie von Dipol- und Quadrupolgittern. *Z. Phys.* **22**, 27–43. [5.5.2]

Kratky, K. W. (1980). New boundary condition for computer experiments of thermodynamic systems. *J. comput. Phys.* **37**, 205–17. [1.5.5]

Kratky, K. W. and Schreiner, W. (1982). Computational techniques for spherical boundary conditions. *J. comput. Phys.* **47**, 313–320. [1.5.5]

Kubo, R. (1957). Statistical mechanical theory of irreversible processes. I. General theory and simple applications to magnetic and conduction problems. *J. phys. Soc. Japan* **12**, 570–86. [8.5]

Kubo, R. (1966). The fluctuation dissipation theorem. *Rep. Prog. Phys.* **29**, 255–284. [8.5]

Kuharski, R. A. and Rossky, P. J. (1984). Quantum mechanical contributions to the structure of liquid water. *Chem. Phys. Lett.* **103**, 357–62. [10.2]

Kushick, J. and Berne, B. J. (1976). Computer simulation of anisotropic molecular fluids. *J. chem. Phys.* **64**, 1362–7. [11.5]

Ladd, A. J. C. (1977). Monte Carlo simulation of water. *Mol. Phys.* **33**, 1039–50. [5.5.4]

Ladd, A. J. C. (1978). Long-range dipolar interactions in computer simulations of polar liquids. *Mol. Phys.* **36**, 463–74. [5.5.4]

Ladd, A. J. C. (1984). Equations of motion for nonequilibrium molecular dynamics simulations of viscous flow in molecular liquids. *Mol. Phys.* **53**, 459–63. [8.2, 8.6]

Ladd, A. J. C. and Woodcock, L. V. (1977) Triple-point coexistence properties of the Lennard-Jones system. *Chem. Phys. Lett.* **51**, 155–9. [11.3]

Ladd, A. J. C. and Woodcock, L. V. (1978). Interfacial and coexistence properties of the Lennard-Jones system at the triple point. *Mol. Phys.* **36**, 611–619. [11.3]

Lado, F. (1971). Numerical Fourier transforms in one, two, and three dimensions for liquid state calculations. *J. comput. Phys.* **8**, 417–33. [D.3]

Lal, M. and Spencer, D. (1971). Monte Carlo computer simulation of chain molecules. III. Simulation of n-alkane molecules. *Mol. Phys.* **22**, 649–59. [4.7.2]

Landau, L. D. and Lifshitz, E. M. (1980). *Statistical physics* (Course of Theoretical Physics, **5**) (3rd edn, revised E. M. Lifshitz and L. P. Pitaevskii). Pergamon Press, Oxford. [2.1, 2.3, 2.5, 2.9]

Larsen, B. (1978). Studies in statistical mechanics of coulombic systems. III. Numerical solution of the HNC and RHNC equations for the restricted primitive model. *J. chem. Phys.* **68**, 4511–23. [11.4]

Lax, M. (1966). Classical noise. IV. Langevin methods. *Rev. Mod. Phys.* **38**, 541–66. [9.3, 9.4]

Lebowitz, J. L., Percus, J. K., and Verlet, L. (1967). Ensemble dependence of fluctuations with application to machine computations. *Phys. Rev.* **153**, 250–4. [2.3, 2.5]

Ledgard, H. F. and Chmura, L. J. (1978). *FORTRAN with style. Programming proverbs.* Hayden New Jersey. [3.5, A.2]

Lee, C. Y. and Scott, H. L. (1980). The surface tension of water: a Monte Carlo calculation using an umbrella sampling algorithm. *J. chem. Phys.* **73**, 4591–6. [7.2.1]

Lee, J. K., Barker, J. A., and Pound, G. M. (1974). Surface structure and surface tension: perturbation theory and Monte Carlo calculation. *J. chem. Phys.* **60**, 1976–80. [1.1, 4.3, 11.2]

Lees, A. W. and Edwards, S. F. (1972). The computer study of transport processes under extreme conditions. *J. Phys.* **C5**, 1921–9. [1.1, 8.2]

Lennard-Jones, J. E. and Devonshire, A. F. (1939a). Critical and cooperative phenomena. III. A theory of melting and the structure of liquids. *Proc. R. Soc. Lond.* **A169**, 317–38. [1.2]

Lennard-Jones, J. E. and Devonshire, A. F. (1939b). Critical and cooperative phenomena. IV. A theory of disorder in solids and liquids and the process of melting. *Proc. R. Soc. Lond.* **A170**, 464–84. [1.2]

LeSar, R. (1984). Improved electron-gas model calculations of solid $N_2$ to 10 GPa. *J. chem. Phys.* **81**, 5104–08. [1.3.1]

LeSar, R. and Gordon, R. G. (1982). Density functional theory for the solid alkali cyanides. *J. chem. Phys.* **77**, 3682–92. [1.3.1]

LeSar, R. and Gordon, R. G. (1983). Density functional theory for solid nitrogen and carbon dioxide at high pressure. *J. chem. Phys.* **78**, 4991–6. [1.3.1]

Levesque, D., Verlet, L., and Kurkijarvi, J. (1973). Computer experiments on classical fluids. IV. Transport properties and time correlation functions of the Lennard-Jones liquid near its triple point. *Phys. Rev.* **A7**, 1690–700. [6.5.5, 11.7]

Lewis, J. W. E. and Singer, K. (1975). Thermodynamic properties and self-diffusion of molten sodium chloride. *J. chem. Soc. Faraday II* **71**, 41–53. [11.4]

Lowden, L. J. and Chandler, D. (1974). Theory of intermolecular pair correlations for molecular liquids. Applications to liquid carbon disulfide, carbon diselenide and benzene. *J. chem. Phys.* **61**, 5228–41. [2.6]

Lucchese, R. R. and Tully, J. C. (1984). Trajectory studies of rainbow scattering from the reconstructed Si(100) surface., *Surf. Sci.* **137**, 570–98. [11.8]

Luckhurst, G. R. and Gray, G. W. (1979). *The molecular physics of liquid crystals.* Academic Press, New York. [11.5]

Luckhurst, G. R. and Romano S. (1980). Computer simulation of anisotropic systems. IV. The effect of translational freedom. *Proc. R. Soc. Lond.* **A373**, 111–30. [11.5]

Luckhurst, G. R. and Simpson, P. (1982). Computer simulation studies of anisotropic systems.VIII. The Lebwohl–Lasher model of nematogens revisited. *Mol. Phys.* **47**, 251–65. [1.5.2, 4.7.1, 11.5]

Luttinger, J. M. (1964). Theory of thermal transport coefficients. *Phys. Rev.* **135A** 1505–14. [8.5]

Lyklema, J. W. (1979a). Computer simulation of a rough sphere fluid. I. *Physica* **A96**, 573–93. [3.6.1]

Lyklema, J. W. (1979b). Computer simulation of a rough sphere fluid. II. Comparison with stochastic models. *Physica* **A96**, 594–605. [3.6.1]

Lynden-Bell, R. M. (1984). Comparison of the results from simulations with the predictions of models for molecular reorientation. In *Molecular liquids, dynamics and interactions.* (eds A. J. Barnes, W. J. Orville-Thomas, and J. Yarwood) NATO ASI Series Vol. 135, pp. 501–518, Reidel, New York. [11.6]

Lynden-Bell, R. M. and McDonald, I. R. (1981). Reorientational correlation functions for computer simulated liquids of tetrahedral molecules. *Mol. Phys.* **43**, 1429–40. [11.6]

McCammon, J. A., Gelin, B. R., and Karplus, M. (1977). Dynamics of folded proteins. *Nature* **267**, 585–90. [1.1]

McDonald, I. R. (1969). Monte Carlo calculations for one- and two-component fluids in the isothermal–isobaric ensemble. *Chem. Phys. Lett.* **3**, 241–3. [4.5]

McDonald, I. R. (1972). *NpT*-ensemble Monte Carlo calculations for binary liquid mixtures. *Mol. Phys.* **23**, 41–58. [4.5]

McDonald, I. R. (1986). Molecular liquids: orientational order and dielectric properties. *Molecular dynamics simulation of statistical mechanical systems.* Proceedings of the Enrico Fermi Summer School, Varenna, 1985, pp 341–70. Soc. Italiana di Fisica, Bologna. [5.5.2]

McDonald, I. R. and Singer, K. (1972). An equation of state for simple liquids. *Mol. Phys.* **23**, 29–40. [B.1]

McLachlan, A. D. (1964). A variational solution of the time-dependent Schrödinger equation. *Mol. Phys.* **8**, 39–44. [10.3]

McMillan, W. L. (1965) Ground state of liquid $He^4$. *Phys. Rev.* **A138**, 442–51. [10.4]

McNeill, W. J. and Madden, W. G. (1982). A new method for the molecular dynamics simulation of hard core molecules. *J. chem. Phys.* **76**, 6221–6. [3.6, 3.6.2]

McQuarrie, D. A. (1976). *Statistical mechanics.* Harper and Row, New York. [2.1, 2.6, 2.7, 2.9, 9.2]

McTague, J. P., Frenkel, D., and Allen, M. P. (1980). Simulation studies of the 2-D melting mechanism. In *Ordering in two dimensions*, (ed. S. K. Sinha). pp. 147–53. North Holland, Amsterdam. [11.3]

Madden, P. A. and Kivelson, D. (1984). A consistent molecular treatment of dielectric phenomena. *Adv. Chem. Phys.* **56**, 467–566. [5.5.2]

Madelung, E. (1918). Das elektrische Feld in Systemen von regelmässig angeordneten Punktladungen. *Phys. Z.* **19**, 524–32. [5.5.2]

Maitland, G. C. and Smith, E. B. (1971). The intermolecular pair potential of argon. *Mol. Phys.* **22**, 861–8. [1.3.2]

Maitland, G. C., Rigby, M., Smith, E. B., and Wakeham, W. A. (1981). *Intermolecular forces: their origin and determination.* Clarendon Press, Oxford. [1.3.2, 1.4.2, 5.4]

Mandell, M. J. (1976). On the properties of a periodic fluid. *J. stat. Phys.* **15**, 299–305. [1.5.2]

Maréchal, G. and Ryckaert, J. P. (1983). Atomic vs molecular description of transport properties in polyatomic fluids: n-butane as an example. *Chem. Phys. Lett.* **101**, 548–54. [1.3.3]

Marsaglia, G. (1972). Choosing a point from the surface of a sphere. *Ann. math. Stat.* **43**, 645–6. [G.4]

Mayer, J. E. and Wood, W. W. (1965). Interfacial tension effects in finite periodic two-dimensional systems. *J. chem. Phys.* **42**, 4268–74. [11.3]

Mazur, P. (1982). On the motion and Brownian motion of *n* spheres in a viscous fluid. *Physica* **A110**, 128–46. [9.4]

Mazur, P. and Van Saarloos, W. (1982). Many-sphere hydrodynamic interactions and mobilities in a suspension. *Physica* **A115**, 21–57. [9.4]

Mehrotra, P. K., Mezei, M., and Beveridge, D. L. (1983). Convergence acceleration in Monte Carlo simulation on water and aqueous solutions. *J. chem. Phys.* **78**, 3156–66. [7.3.1, 7.3.2]

Mentch, F. and Anderson, J. B. (1981). Quantum chemistry by random walk: importance sampling for $H_3^+$. *J. chem. Phys.* **74**, 6307–11. [10.4]

Metropolis, N. and Ulam, S. (1949). The Monte Carlo method. *J. Am. stat. Ass.* **44**, 335–41. [4.1, 10.4]

Metropolis, N., Rosenbluth, A. W., Rosenbluth, M. N., Teller, A. H., and Teller, E. (1953). Equation of state calculations by fast computing machines. *J. chem. Phys.* **21**, 1087–92. [1.1, 1.5.3, 4.1, 4.3, 11.3]

Mezei, M. (1980). A cavity-biased ($TV\mu$) Monte Carlo method for the computer simulation of fluids. *Mol. Phys.* **40**, 901–6. [4.6, 7.3.1]

Mezei, M. (1983). Virial bias Monte Carlo methods: efficient sampling in the ($T$, $P$, $N$) ensemble. *Mol. Phys.* **48**, 1075–82. [7.3.4]

Mohling, F. (1982). *Statistical mechanics. Methods and applications.* Halsted, New York. [11.3]

Mon, K. K. and Griffiths, R. B. (1985). Chemical potential by gradual insertion of a particle in Monte Carlo simulation. *Phys. Rev.* **A31**, 956–9. [7.2.1]

Monson, P. A., Rigby, M., and Steele, W. A. (1983). Non-additive energy effects in molecular liquids. *Mol. Phys.* **49**, 893–8. [1.3.2]

Morf, R. and Stoll, E. (1977). Numerical methods for the calculation of molecular dynamics. International Series of Numerical Mathematics (eds. J. Descloux and J. Marti) Vol. 37, pp. 139–52. Proceedings of the Colloquium on Numerical Analysis. Birkhauser, Basel. [9.3]

Mori, H. (1965a). Transport, collective motion and Brownian motion. *Prog. theor. Phys.* **33**, 423–55. [9.2]

Mori, H. (1965b). A continued fraction representation of the time correlation functions. *Prog. theor. Phys.* **34**, 399–416. [9.2]

Morrell, W. E. and Hildebrand, J. H. (1936). The distribution of molecules in a model liquid. *J. chem. Phys.* **4**, 224–27. [1.1]

Morse, M. D. and Rice, S. A. (1982). Tests of effective pair potentials for water: predicted ice structures. *J. chem. Phys.* **76**, 650–60. [1.1]

Mouritsen, O. G. (1984). *Computer studies of phase transitions and critical phenomena.* Springer Series in Computational Physics. Springer, Berlin. [11.3, 11.5]

Mouritsen, O. G. and Berlinsky, A. J. (1982). Fluctuation-induced first order phase transition in an anisotropic planar model of $N_2$ on graphite. *Phys. Rev. Lett.* **48**, 181–4. [1.5.2]

Müller-Krumbhaar, H. and Binder, K. (1973). Dynamic properties of the Monte Carlo method in statistical mechanics. *J. stat. Phys.* **8**, 1–24. [6.4.1]

Münster, A. (1969). *Statistical thermodynamics.* Springer/Academic Press, Berlin/New York. [2.3, 2.4]

Murad, S. (1978). LINEAR and NONLINEAR. Quantum chemistry program exchange. *QCPE* **12**, 357. (Indiana University). [1.4.3]

Murad, S. and Gubbins, K. E. (1978). Molecular dynamics simulation of methane using a singularity free algorithm. In *Computer modelling of matter* (ed. P. Lykos). ACS Symposium Series Vol. 86, pp. 62–71. American Chemical Society, Washington. [1.4.2]

Murphy, T. J. and Aguirre, J. L. (1972). Brownian motion of $N$ interacting particles. I. Extension of the Einstein diffusion relation to the $N$-particle case. *J. chem. Phys.* **57**, 2098–104. [9.4]

Murthy, C. S., O'Shea, S. F., and McDonald, I. R. (1983). Electrostatic interactions in molecular crystals. Lattice dynamics of solid nitrogen and carbon dioxide. *Mol. Phys.* **50**, 531–41. [1.3.3]

Murthy, C. S., Singer, K., Klein, M. L., and McDonald, I. R. (1980). Pairwise effective potentials for nitrogen. *Mol. Phys.* **41**, 1387–99. [1.4.2]

NAG (1984). Routine G05CAF. NAG FORTRAN Library Manual Vol. 4, mark 9. [G.2]

Naitoh, T. and Ono, S. (1976). The shear viscosity of hard sphere fluid via nonequilibrium molecular dynamics. *Phys. Lett.* **57A**, 448–50. [8.2, 8.7, 11.7]

Naitoh, T. and Ono, S. (1978). The shear viscosity of 500 hard spheres via nonequilibrium molecular dynamics. *Phys. Lett.* **69A**, 125–6. [11.7]

Naitoh, T. and Ono, S. (1979). The shear viscosity of a hard sphere fluid via nonequilibrium molecular dynamics. *J. chem. Phys.* **70**, 4515–23. [8.2, 11.7]

Neto, N., Righini, R., Califano, S., and Walmsley, S. H. (1978). Lattice dynamics of molecular crystals using atom–atom and multipole–multipole potentials. *Chem. Phys.* **29**, 167–79. [1.4.3]

Neumann, M. (1983). Dipole moment fluctuation formulas in computer simulations of polar systems. *Mol. Phys.* **50**, 841–58. [5.5.3]

Neumann, M. and Steinhauser, O. (1980). The influence of boundary conditions used in machine simulations on the structure of polar systems. *Mol. Phys.* **39**, 437–54. [5.5.3]

Neumann, M. and Steinhauser. O. (1983a). On the calculation of the dielectric constant using the Ewald–Kornfeld tensor. *Chem. Phys. Lett.* **95**, 417–22. [5.5.2]

Neumann, M. and Steinhauser, O. (1983b). On the calculation of the frequency-dependent dielectric constant in computer simulations. *Chem. Phys. Lett.* **102**, 508–513. [5.5.2]

Neumann, M., Steinhauser, O., and Pawley, G. S. (1984). Consistent calculation of the static and frequency-dependent dielectric constant in computer simulations. *Mol. Phys.* **52**, 97–113. [5.5.2, 5.5.5]

Nezbeda, I. (1977). Statistical thermodynamics of interaction site molecules. The theory of hard dumb-bells. *Mol. Phys.* **33**, 1287–99. [6.5.3]

Nicholson, D. (1984). Grand ensemble Monte Carlo. *CCP5 Quarterly* **11**, 19–24. [4.6]

Nicholson, D. and Parsonage, N. (1982). *Computer simulation and the statistical mechanics of adsorption*. Academic Press, New York. [4.6]

Nicolas, J. J., Gubbins, K. E., Streett, W. B., and Tildesley, D. J. (1979) Equation of state for the Lennard-Jones fluid. *Mol. Phys.* **37**, 1429–54. [1.1, 5.2.4, 5.4]

Nordholm, S. and Zwanzig, R. (1975). A systematic derivation of exact generalized Brownian motion theory. *J. stat. Phys.* **13**, 347–71. [9.2]

Norman, G. E. and Filinov, V. S. (1969). Investigations of phase transitions by a Monte Carlo method. *High Temp. (USSR)* **7**, 216–22. [4.6]

Northrup, S. H. and McCammon, J. A. (1980). Simulation methods for protein structure fluctuations. *Biopolymers* **19**, 1001–16. [7.3.3]

Nosé, S. (1984). A molecular dynamics method for simulations in the canonical ensemble. *Mol. Phys.* **52**, 255–68. [7.4.2]

Nosé, S. and Klein, M. L. (1983). Constant pressure dynamics for molecular systems. *Mol. Phys.* **50**, 1055–76. [7.5.4]

Nosé, S., Kataoka, Y., Okada, K., and Yamamoto, T. (1981). A Monte Carlo study of phase transitions in a f.c.c. octopolar array. *J. chem. Phys.* **75**, 985–92. [1.3.4]

O'Dell, J. and Berne, B. J. (1975). Molecular dynamics of the rough sphere fluid. I. Rotational relaxation. *J. chem. Phys.* **63**, 2376–94. [3.6, 3.6.1]

Onsager, L. (1936). Electric moments of molecules in liquids. *J. Am. chem. Soc.* **58**, 1486–93. [5.5.3]

Onsager, L. (1949). The effect of shape on the interaction of colloidal particles. *Ann. NY Acad. Sci.* **51**, 627–59. [11.5]

Oppenheim, I. and Ross, J. (1957). Temperature dependence of distribution functions in quantum statistical mechanics. *Phys. Rev.* **107**, 28–32. [10.1]

Orban, J. and Ryckaert, J. P. (1974). Methods in molecular dynamics. Rapport d'activite scientifique du CECAM. [3.4]
O'Shea, S. F. (1978). A Monte Carlo study of the classical octopolar solid, *J. chem. Phys.* **68**, 5435–41. [1.3.4, 4.7.1]
O'Shea, S. F. (1983). Neighbour lists again. *CCP5 Quarterly* **9**, 41–6. [5.3.1]
Owicki, J. C. (1978). Optimization of sampling algorithms in Monte Carlo calculations on fluids. In *Computer modelling of matter* (ed. P. Lykos). ACS Symposium series Vol. 86, pp. 159–71. American Chemical Society, Washington. [7.3.1]
Owicki, J. C. and Scheraga, H. A. (1977a). Preferential sampling near solutes in Monte Carlo calculations on dilute solutions. *Chem. Phys. Lett.* **47**, 600–2. [4.3, 7.3.1]
Owicki, J. C. and Scheraga, H. A. (1977b). Monte Carlo calculations in the isothermal–isobaric ensemble. 2. Dilute aqueous solutions of methane. *J. Am. chem. Soc.* **99**, 7413–18. [4.5, 4.7.1]
Panagiotopoulos, A. Z. (1987a). Direct determination of phase coexistence properties of fluids by Monte Carlo simulation in a new ensemble. *Mol. Phys.* **61**, 813–26. [7.7]
Panagiotopoulos, A. Z. (1987b). Adsorption and capillary condensation of fluids in cylindrical pores by Monte Carlo simulation in the Gibbs ensemble. *Mol. Phys.* **62**, 701–20. [7.7].
Panagiotopoulos, A. Z., Quirke, N., Stapleton, M., and Tildesley, D. J. (1988). Phase equilibria by simulation in the Gibbs ensemble. Alternative derivation, generalization and application to mixture and membrane equilibria. *Mol. Phys.* **63**, 527–46. [7.7].
Pangali, C., Rao, M., and Berne, B. J. (1978). On a novel Monte Carlo scheme for simulating water and aqueous solutions. *Chem. Phys. Lett.* **55**, 413–17. [7.3.2]
Papoulis, A. (1965). *Probability, random variables, and stochastic processes.* McGraw-Hill, New York. [6.4.1]
Parrinello, M. and Rahman, A. (1980). Crystal structure and pair potentials: a molecular dynamics study. *Phys. Rev. Lett.* **45**, 1196–99. [7.5.4]
Parrinello, M. and Rahman, A. (1981). Polymorphic transitions in single crystals: a new molecular dynamics method. *J. appl. Phys.* **52**, 7182–90. [7.5.4]
Parrinello, M. and Rahman, A. (1982). Strain fluctuations and elastic constants. *J. chem. Phys.* **76**, 2662–6. [7.5.4]
Parrinello, M. and Rahman, A. (1984). Study of an F center in molten KC1. *J. chem. Phys.* **80**, 860–7. [10.2]
Patey, G. N., Levesque, D., and Weis, J. J. (1982). On the theory and computer simulation of dipolar fluids. *Mol. Phys.* **45**, 733–46. [5.5.3]
Pear, M. R. and Weiner, J. H. (1979). Brownian dynamics study of a polymer chain of linked rigid bodies. *J. chem. Phys.* **71**, 212–24. [2.10, 3.4, 4.7.2, C.2]
Pearson, R. B., Richardson, J. L., and Toussaint, D. (1983). A fast processor for Monte Carlo simulation. *J. comput. Phys.* **51**, 241–9. [A.1]
Pedoe, D. S. (1958). *The gentle art of mathematics.* Penguin, Harmondsworth. [4.1]
Perram, J. W. Wertheim, M. S., Lebowitz, J. L., and Williams, G. O. (1984). Monte Carlo simulation of hard spheroids. *Chem. Phys. Lett.* **105**, 277–80. [11.5]
Peskun, P. H. (1973). Optimum Monte Carlo sampling using, Markov chains. *Biometrika* **60**, 607–12. [4.3]
Pieranski, P., Malecki, J., Kuczynski, W., and Wojciechowski, K. (1978). A hard disc system, an experimental model. *Phil. Mag.* **A37**, 107–15. [1.1]

Pitzer, K. S. and Gwinn, W. D. (1942). Energy levels and thermodynamic functions for molecules with internal rotation. I. Rigid frame with attached tops. *J. chem. Phys.* **10**, 428–40. [4.7.2]

Potter, D. (1972). *Computational physics.* Wiley, New York. [3.2.1, 3.3.1]

Powles, J. G. (1984a). Singer on the root. *CCP5 Quarterly* **11**, 39–41. [5.2.2]

Powles, J. G. (1984b). The liquid–vapour coexistence line for Lennard-Jones-type fluids. *Physica* **A126**, 289–99. [5.2.4]

Powles, J. G. and Rickayzen, G. (1979). Quantum corrections and the computer simulation of molecular fluids. *Mol. Phys.* **38**, 1875–92. [2.9]

Powles, J. G., Evans, W. A. B., and Quirke, N. (1982). Non-destructive molecular dynamics simulation of the chemical potential of a fluid. *Mol. Phys.* **46**, 1347–70. [2.4, 5.2.4, 5.6, 6.5.1]

Powles, J. G., Fowler, R. F., and Evans, W. A. B. (1983a). A new method for computing surface tension using a drop of liquid. *Chem. Phys. Lett.* **96**, 289–92. [11.2]

Powles, J. G., Fowler, R. F., and Evans, W. A. B. (1983b). The surface thickness of simulated microscopic liquid drops. *Phys. Lett.* **98A**, 421–5. [11.2]

Powles, J. G., Fowler, R. F., and Evans, W. A. B. (1984). The dielectric constant of a polar liquid by the simulation of microscopic drops. *Chem. Phys. Lett.* **107**, 280–3. [11.2]

Powles, J. G., Rickayzen, G., and Williams, M. L. (1985). A curious result concerning the Clausius virial. *J. chem. Phys.* **83**, 293–6. [11.2]

Powles, J. G., Evans, W. A. B., McGrath, E., Gubbins, K. E., and Murad, S. (1979). A computer simulation for a simple model of liquid hydrogen chloride. *Mol. Phys.* **38**, 893–908. [3.3.1]

Pratt, L. R. and Haan, S. W. (1981). Effects of periodic boundary conditions on equilibrium properties of computer simulated fluids. I. Theory. *J. chem. Phys.* **74**, 1864–72. [1.5.2]

Price, S. L., Stone, A. J., and Alderton, M. (1984). Explicit formulae for the electrostatic energy, forces and torques between a pair of molecules of arbitrary symmetry. *Mol. Phys.* **52**, 987–1001. [1.3.3, 3.3, C.3]

Quentrec, B. and Brot, C. (1975). New method for searching for neighbors in molecular dynamics computations. *J. comput. Phys.* **13**, 430–2. [5.3.2]

Rahman, A. (1964). Correlations in the motion of atoms in liquid argon. *Phys. Rev.* **136A**, 405–11. [1.1, 3.2.3, B.1]

Rahman, A. and Stillinger, F. H. (1971). Molecular dynamics study of liquid water. *J. chem. Phys.*, **55**, 3336–59. [1.1]

Rand Corporation (1955). *A million random digits with* 100 000 *normal deviates.* The Free Press, Glencoe, Ill. [G.1]

Rao, M. and Berne, B. J. (1979). On the force-bias Monte Carlo simulation of simple liquids. *J. chem. Phys.* **71**, 129–32. [7.3.2, 7.3.3]

Rao, M. Pangali, C., and Berne, B. J. (1979). On the force-bias Monte Carlo simulation of water: methodology, optimization and comparison with molecular dynamics. *Mol. Phys.* **37**, 1773–98. [7.3.2]

Rapaport, D. C. (1978). Molecular dynamics simulation of polymer chains with excluded volume. *J. Phys.* **A11**, L213–17. [3.6.2]

Rapaport, D. C. (1979). Molecular dynamics study of a polymer chain in solution. *J. chem. Phys.* **71**, 3299–303. [3.6.2]

Rapaport, D. C. (1980). The event scheduling problem in molecular dynamics simulation. *J. comput. Phys.* **34**, 184–201. [5.3]

Ray, J. R. and Graben, H. W. (1981). Direct calculation of fluctuation formulae in the microcanonical ensemble. *Mol. Phys.* **43**, 1293–7. [2.5]

Ray, J. R., Graben, H. W., and Haile, J. M. (1981). Statistical mechanics of the isoenthalpic–isobaric ensemble. *Nuovo Cimento* **64B**, 191–206. [7.5.1]

Rebertus, D. W. and Sando, K. M. (1977). Molecular dynamics simulation of a fluid of hard spherocylinders. *J. chem. Phys.* **67**, 2585–90. [3.6.2]

Ree, F. H. (1970). Statistical mechanics of single-occupancy systems of spheres, disks and rods. *J. chem. Phys.* **53**, 920–31. [4.4]

Reimers, J. R., Wilson, K. R., and Heller, E. J. (1983). Complex time dependent wave packet technique for thermal equilibrium systems: electronic spectra. *J. chem. Phys.* **79**, 4749–57. [10.3]

Reynolds, P. J., Ceperley, D. M., Alder, B. J., and Lester, W. A. (1982). Fixed-node Monte Carlo for molecules. *J. chem. Phys.* **77**, 5593–603. [10.4]

Righini, R., Maki, K., and Klein, M. L. (1981). An intermolecular potential for methane. *Chem. Phys. Lett.* **80**, 301–5. [1.3.3]

Romano, S. and Singer, K. (1979). Calculation of the entropy of liquid chlorine and bromine by computer simulation. *Mol. Phys.* **37**, 1765–72. [4.7.1, 5.6]

Rosenbluth, M. N. and Rosenbluth, A. W. (1954). Further results on Monte Carlo equations of state, *J. chem. Phys.* **22**, 881–4. [11.3]

Rossky, P. J., Doll, J. D., and Friedman, H. L. (1978). Brownian dynamics as smart Monte Carlo simulation, *J. chem. Phys.* **69**, 4628–33. [7.3.3]

Rowley, L. A., Nicholson, D., and Parsonage, N. G. (1975). Monte Carlo grand canonical ensemble calculation in a gas–liquid transition region for 12–6 argon. *J. comput. Phys.* **17**, 401–14. [4.6]

Rowley, L. A., Nicholson, D., and Parsonage, N. G. (1978). Long range corrections to grand canonical ensemble Monte Carlo calculations for adsorption systems. *J. comput. Phys.* **26**, 66–79. [4.6]

Rowlinson, J. S. (1963). *The perfect gas.* Pergamon Press, Oxford. [2.2]

Rowlinson, J. S. (1969). *Liquids and liquid mixtures.* (2nd edn). Butterworth, London. [2.5]

Rowlinson, J. S. and Swinton, F. L. (1982). *Liquids and liquid mixtures* (3rd edn). Butterworth, London. [1.1, 1.4.3]

Rowlinson, J. S. and Widom, B. (1982). *Molecular theory of capillarity.* Clarendon Press, Oxford. [11.2]

Rubinstein, R. Y. (1981). *Simulation and Monte Carlo methods.* Wiley, New York. [4.1, 4.2.2, 4.3, G.2, G.3, G.6]

Rusanov, A. I. and Brodskaya, E. N. (1977). A molecular dynamics simulation of a small drop. *J. Colloid Interface Sci.* **62**, 542–55. [11.2]

Ryckaert, J. P. (1985). Special geometrical constraints in the molecular dynamics of chain molecules. *Mol. Phys.* **55**, 549–56. [3.4]

Ryckaert, J. P. and Bellemans, A. (1975). Molecular dynamics of liquid n-butane near its boiling point. *Chem. Phys. Lett.* **30**, 123–5. [1.1, 1.3.3, 3.4, 4.7.2]

Ryckaert, J. P. and Bellemans, A. (1978). Molecular dynamics of liquid alkanes. *Chem. Soc. Faraday Discuss.* **66**, 95–106. [3.4]

Ryckaert, J. P., Ciccotti, G., and Berendsen, H. J. C. (1977). Numerical integration of the cartesian equations of motion of a system with constraints: molecular dynamics of n-alkanes. *J. comput. Phys.* **23**, 327–41. [3.4]

St. Pierre, A. G. and Steele, W. A. (1969). The rotational Wigner function. *Ann. Phys.* **52**, 251–92. [2.9]

Saito, Y. (1982). Melting of dislocation vector systems in two dimensions. *Phys. Rev. Lett.* **48**, 1114–17. [11.3]

Saito, Y. (1983). Two-dimensional melting of dislocation vector systems. *Surf. Sci.* **125**, 285–90. [11.3]

Salsburg, Z. W., Jacobson, J. D., Fickett, W., and Wood, W. W. (1959). Application of the Monte Carlo method to the lattice gas model. I. Two dimensional triangular lattice. *J. chem. Phys.* **30**, 65–72. [4.6]

Sangster, M. J. L. and Dixon, M. (1976). Interionic potentials in alkali halides and their use in simulation of molten salts. *Adv. Phys.* **25**, 247–342. [11.4]

Schlichting, H. (1979). *Boundary layer theory* (7th edn). McGraw-Hill, New York. [8.2]

Schmidt, K. E. and Kalos, M. H. (1984). Few- and many-fermion problems. In *Applications of the Monte Carlo method in statistical physics* (ed. K. Binder). Topics in Current Physics Vol. 36, pp. 125–43. Springer, Berlin. [10.1, 10.4]

Schmitz, R. and Felderhof, B. U. (1982). Mobility matrix for two spherical particles with hydrodynamic interactions. *Physica* **A116**, 163–77. [9.4].

Schneider, T. and Stoll, E. (1978). Molecular dynamics study of a three-dimensional one-component model for distortive phase transitions. *Phys. Rev.* **B17**, 1302–22. [7.4.4, 9.3]

Schoen, M., Vogelsang, R., and Hoheisel, C. (1984). The recurrence time in molecular dynamics ensembles. *CCP5 Quarterly* **13**, 27–37. [6.4.4]

Schofield, P. (1960). Space–time correlation function formalism for slow neutron scattering. *Phys. Rev. Lett.* **4**, 239–40. [2.9]

Schofield, P. (1973). Computer simulation studies of the liquid state. *Comput. Phys. Commun.* **5**, 17–23. [5.7.1]

Schofield, P. and Henderson, J. R. (1982). Statistical mechanics of inhomogeneous fluids. *Proc. R. Soc. Lond.* **A379**, 231–46. [11.2]

Schweizer, K. S., Stratt, R. M., Chandler, D., and Wolynes, P. G. (1981). Convenient and accurate discretized path integral methods for equilibrium quantum mechanical calculations. *J. chem. Phys.* **75**, 1347–64. [10.2]

Severin, E. S. and Tildesley, D. J. (1980). A methane molecule adsorbed on a graphite surface. *Mol. Phys.* **41**, 1401–18. [1.5.2]

Severin, E. S., Freasier, B. C., Hamer, N. D., Jolly, D. L., and Nordholm, S. (1978). An efficient microcanonical sampling method. *Chem. Phys. Lett.* **57**, 117–20. [2.2]

Shing, K. S. and Chung, S. T. (1987). Computer simulation methods for the calculation of solubility in supercritical extraction systems. *J. Phys. Chem.* **91**, 1674–81. [2.4]

Shing, K. S. and Gubbins, K. E. (1981). The chemical potential from computer simulation. Test particle method with umbrella sampling. *Mol. Phys.* **43**, 717–21. [7.2.1].

Shing, K. S. and Gubbins, K. E. (1982). The chemical potential in dense fluids and fluid mixtures via computer simulation. *Mol. Phys.* **46**, 1109–28. [7.2.1]

Singer, J. V. L. and Singer, K. (1984). Molecular dynamics based on the first order correction in the Wigner–Kirkwood expansion. *CCP5 Quarterly* **14**, 24–6. [10.1]

Singer, K. (1983). A fast method for the evaluation of square roots in computer simulation. *CCP5 Quarterly* **8**, 47–8. [5.2.2]

Singer, K. and Smith, W. (1986). Semiclassical many-particle dynamics with Gaussian wave packets. *Mol. Phys.* **57**, 761–75. [10.3]

Singer, K., Singer, J. V. L., and Fincham, D. (1980). Determination of the shear viscosity of atomic liquids by nonequilibrium molecular dynamics. *Mol. Phys.* **40**, 515–19. [8.7]

Singer, K., Taylor, A., and Singer, J. V. L. (1977). Thermodynamic and structural properties of liquids modelled by 'two-Lennard-Jones centres' pair potentials. *Mol. Phys.* **33**, 1757–95. [1.3.3, 1.4.2, 3.3.2]

Smith, E. B. and Wells, B. H. (1984). Estimating errors in molecular simulation calculations. *Mol. Phys.* **52**, 701–4. [6.4.1]

Smith, E. R. (1987). Boundary conditions on hydrodynamics in simulations of dense suspensions. *Chem. Soc. Faraday Discuss.* **83**, to appear. [9.4]

Smith, W. (1982a). Point multipoles in the Ewald summation. *CCP5 Quarterly* **4**, 13–25. [5.5.2]

Smith, W. (1982b). An introduction to the discrete Fourier transform. *CCP5 Quarterly* **5**, 34–41. [6.3.2, D.2]

Smith, W. (1982c). Correlation functions and the fast Fourier transform. *CCP5 Quarterly* **7**, 12–24. [6.3.2, D.2]

Smith, W. (1983). The periodic boundary condition in non-cubic MD cells. Wigner–Seitz cells with reflection symmetry. *CCP5 Quarterly* **10**, 37–42. [1.5.4]

Snook, I., Van Megen, W., and Tough, R. J. A. (1983). Diffusion in concentrated hard sphere dispersions: effective two-particle mobility tensors. *J. chem. Phys.* **78**, 5825–36. [9.4]

Sprik, M., Klein, M. L., and Chandler, D. (1985). Staging: a sampling technique for the Monte Carlo evaluation of path integrals. *Phys. Rev.* **B31**, 4234–44 [10.2]

Stark, P. A. (1970). *Introduction to numerical methods.* Macmillan, Toronto. [6.5.4]

Steele, W. A. (1969). Time correlation functions. In *Transport phenomena in fluids* (ed. H. J. M. Hanley) pp. 209–312. Dekker, New York. [2.7]

Steele, W. A. (1974). *Interaction of gases with solid surfaces.* Pergamon Press, Oxford. [11.8]

Steele, W. A. (1980). Molecular reorientation in dense systems. *Intermolecular spectroscopy and the dynamical properties of dense systems.* Proceedings of the Enrico Fermi Summer School Vol. 75, pp. 325–374. Soc. Italiana di Fisica, Bologna. [2.7, 11.6]

Stillinger, F. H. (1975). Theory and molecular models for water. *Adv. Chem. Phys.* **31**, 1–101. [1.1]

Stillinger, F. H. (1980). Water revisited. *Science* **209**, 451–7. [1.1]

Stillinger, F. H. and Weber, T. A. (1981). Gaussian core model in two dimensions. II. Solid and fluid phase topological distribution functions. *J. chem. Phys.* **74**, 4020–8. [11.3]

Stoddard, S. D. (1978). Identifying clusters in computer experiments on systems of particles. *J. comput. Phys.* **27**, 291–3. [11.2]

Stoddard, S. D. and Ford, J. (1973). Numerical experiments on the stochastic behavior of a Lennard-Jones gas system. *Phys. Rev.* **A8**, 1504–12. [3.2, 5.2.4]

Stratonovich, R. L. (1963). *Topics in the theory of random noise I.* Gordon and Breach, New York. [9.3]

Stratonovich, R. L. (1967). *Topics in the theory of random noise II.* Gordon and Breach, New York. [9.3]

Stratt, R. M. and Miller, W. H. (1977). A phase space sampling approach to equilibrium semiclassical statistical mechanics. *J. chem. Phys.* **67**, 5894–903. [10.2]

Stratt, R. M., Holmgren, S. L., and Chandler, D. (1981). Constrained impulsive molecular dynamics. *Mol. Phys.* **42**, 1233–43. [3.6, 3.6.2]

Streett, W. B. and Tildesley, D. J. (1976). Computer simulations of polyatomic molecules. I. Monte Carlo studies of hard diatomics. *Proc. R. Soc. Lond.* **A348**, 485–510. [1.3.3, 2.6, 6.2]

Streett, W. B. and Tildesley, D. J. (1977). Computer simulations of polyatomic molecules. II. Molecular dynamics studies of diatomic liquids with atom–atom and quadrupole–quadrupole potentials. *Proc. R. Soc. Lond.* **A355**, 239–66. [1.3.3]

Streett, W. B. and Tildesley, D. J. (1978). Computer simulations of polyatomic molecules. III. Monte Carlo simulation studies of heteronuclear and homonuclear hard diatomics. *J. chem. Phys.* **68**, 1275–84. [4.7.1]

Streett, W. B., Tildesley, D. J., and Saville, G. (1978a). Multiple time step methods and an improved potential function for molecular dynamics simulations of molecular liquids. In *Computer modelling of matter*. (ed. P. Lykos). ACS Symposium Series Vol. 86, pp. 144–58. American Chemical Society, Washington. [5.2.4, 5.4]

Streett, W. B., Tildesley, D. J., and Saville, G. (1978b), Multiple time step methods in molecular dynamics. *Mol. Phys.* **35**, 639–48. [5.4]

Student (nom de plume) (1908). The probable error of a correlation coefficient. *Biometrika* **6**, 302–10. [4.1]

Subramanian, G. and Davis, H. T. (1975). Molecular dynamical studies of rough sphere fluids. *Phys. Rev.* **A11**, 1430–9. [3.6.1]

Subramanian, G. and Davis, H. T. (1979). Molecular dynamics of a hard sphere fluid in small pores. *Mol. Phys.* **38**, 1061–66. [1.5.2]

Suzuki, K. and Nakata, Y. (1970). The three-dimensional structure of macromolecules. I. The conformation of ethylene polymers by the Monte Carlo method. *Bull. chem. Soc. Japan* **43**, 1006–10. [4.7.2]

Swindoll, R. D. and Haile, J. M. (1984). A multiple time step method for molecular dynamics simulations of fluids of chain molecules. *J. comput. Phys.* **53**, 289–98. [5.4]

Swope, W. C. and Andersen, H. C. (1984). A molecular dynamics method for calculating the solubility of gases in liquids and the hydrophobic hydration of inert gas atoms in aqueous solution. *J. phys. Chem.* **88**, 6548–56. [7.2.1]

Swope, W. C., Andersen, H. C., Berens, P. H., and Wilson, K. R. (1982). A computer simulation method for the calculation of equilibrium constants for the formation of physical clusters of molecules: application to small water clusters. *J. chem. Phys.* **76**, 637–49. [3.2.1, 6.4.1]

Talbot, J., Tildesley, D. J., and Steele, W. A. (1985). Molecular dynamics simulation of fluid $N_2$ adsorbed on a graphite surface. *Faraday Discuss. Chem. Soc.* **80**, 91–105. [11.8]

Tanemura, M., Hiwatari, Y., Matsuda, H., Ogawa, T., Ogita, N., and Ueda, A. (1977). Geometrical analysis of crystallization of the soft-core model. *Prog. theor. Phys.* **58**, 1079–95. [11.3]

Tanemura, M., Hiwatari, Y., Matsuda, H., Ogawa, T., Ogita, N., and Ueda, A. (1978). Geometrical analysis of crystallization of the soft-core model in an FCC crystal formation. *Prog. theor. Phys.* **59**, 323–4. [11.3]

Tenenbaum, A., Ciccotti, G., and Gallico, R. (1982). Stationary non-equilibrium states by molecular dynamics. Fourier's law. *Phys. Rev.* **A25**, 2778–87. [8.1]

Thirumalai, D. and Berne, B. J. (1983). On the calculation of time correlation functions in quantum systems: path integral techniques. *J. chem. Phys.* **79**, 5029–33. [10.2]

Thirumalai, D. and Berne, B. J. (1984). Time correlation functions in quantum systems. *J. chem. Phys.* **81**, 2512–13. [10.2]

Thompson, C. J. (1972). *Mathematical statistical mechanics.* Macmillan, London. [11.3]

Thompson, S. M. (1983). Use of neighbour lists in molecular dynamics. *CCP5 Quarterly* **8**, 20–8. [5.3.1]

Thompson, S. M., Gubbins, K. E., Walton, J. P. R. B., Chantry, R. A. R., and Rowlinson, J. S. (1984). A molecular dynamics study of liquid drops. *J. chem. Phys.* **81**, 530–42. [11.2]

Tildesley, D. J. and Madden, P. A. (1981). An effective pair potential for liquid carbon disulphide. *Mol. Phys.* **42**, 1137–56. [1.4.2, 3.4]

Tildesley, D. J. and Madden, P. A. (1983). Time correlation functions for a model of liquid carbon disulphide. *Mol. Phys.* **48**, 129–52. [6.5.5, 11.6]

Tildesley, D. J. and Streett, W. B. (1980). An equation of state for hard-dumb-bell fluids. *Mol. Phys.* **41**, 85–94. [6.5.3]

Tobochnik, J. and Chester, G. V. (1980). The melting of two-dimensional solids. In *Ordering in two dimensions* (ed. S. K. Sinha). pp. 339–40. North Holland, Amsterdam. [11.3]

Tobochnik, J. and Chester, G. V. (1982). Monte Carlo study of melting in two dimensions. *Phys. Rev.* **B25**, 6778–98. [11.3]

Toda, M., Kubo, R., and Saito, N. (1983). *Statistical physics I: Equilibrium statistical mechanics.* Springer, Berlin. [1.3.4]

Tolman, R. C. (1938). *The principles of statistical mechanics.* Clarendon Press, Oxford. [2.1, 7.4.1]

Torrie, G. M. and Valleau, J. P. (1974). Monte Carlo free energy estimates using non-Boltzmann sampling: application to the sub-critical Lennard-Jones fluid. *Chem. Phys. Lett.* **28**, 578–81. [7.2.1]

Torrie, G. M. and Valleau, J. P. (1977a). Nonphysical sampling distributions in Monte Carlo free energy estimation: umbrella sampling. *J. comput. Phys.* **23**, 187–99. [7.2.1]

Torrie, G. M. and Valleau, J. P. (1977b). Monte Carlo study of a phase separating liquid mixture by umbrella sampling. *J. chem. Phys.* **66**, 1402–8. [7.2.1]

Torrie, G. M. and Valleau, J. P. (1979). A Monte Carlo study of an electrical double layer. *Chem. Phys. Lett.* **65**, 343–6. [1.5.2]

Torrie, G. M. and Valleau, J. P. (1980). Electrical double layers. I. Monte Carlo study of a uniformly charged surface. *J. chem. Phys.* **73**, 5807–16. [11.8]

Torrie, G. M. and Valleau, J. P. (1982). Electrical double layers. IV. Limitations of the Gouy–Chapman theory. *J. phys. Chem.* **86**, 3251–7. [11.8]

Torrie, G. M., Valleau, J. P., and Patey, G. N. (1982). Electrical double layers. II. Monte Carlo and HNC studies of image effects. *J. chem. Phys.* **76**, 4615–22. [11.8]

Tough, R. J. A., Pusey, P. N., Lekkerkerker, H. N. W., and Van den Broeck, C. (1986). Stochastic descriptions of the dynamics of interacting Brownian particles. *Mol. Phys.* **59**, 595–619. [9.4]

Toxvaerd, S. (1982). A new algorithm for molecular dynamics calculations. *J. comput. Phys.* **47**, 444–51. [3.2.3]

Tully, J. C. (1980). Dynamics of gas surface interactions: 3D generalized Langevin model applied to f.c.c. and b.c.c. surfaces. *J. chem. Phys.* **74**, 1975–85. [11.8]

Tully, J. C. (1981). Dynamics of gas surface interactions: thermal desorption of Ar and Xe from platinum. *Surf. Sci.* **111**, 461–78. [11.8]

Tully, J. C., Gilmer, G. H., and Shugard, M. (1979). Molecular dynamics of surface diffusion. I. The motion of adatoms and clusters. *J. chem. Phys.* **71**, 1630–42. [11.8]

Turq, P., Lantelme, F., and Friedman, H. L. (1977). Brownian dynamics: its application to ionic solutions. *J. chem. Phys.* **66**, 3039–44. [1.1, 9.3]

Valleau, J. P. (1980). The problem of coulombic forces in computer simulation. *The problem of long-range forces in the computer simulation of condensed media* (ed. D. Ceperley). NRCC Workshop Proceedings Vol. 9, pp. 3–8. [5.5.5, 11.8]

Valleau, J. P. and Card, D. N. (1972). Monte Carlo estimation of the free energy by multistage sampling. *J. chem. Phys.* **57**, 5457–62. [7.2.1]

Valleau, J. P. and Torrie, G. M. (1977). A guide to Monte Carlo for statistical mechanics. 2. Byways. In *Statistical mechanics A. Modern theoretical chemistry* (ed. B. J. Berne). Vol. 5, pp. 169–94. Plenum Press, New York. [2.4]

Valleau, J. P. and Torrie, G. M. (1984). Electrical double layers. V. Asymmetric ion-wall interactions. *J. chem. Phys.* **81**, 6291–5. [11.8]

Valleau, J. P. and Whittington, S. G. (1977a). A guide to Monte Carlo for statistical mechanics. 1. Highways. In *Statistical mechanics A. Modern theoretical chemistry* (ed. B. J. Berne). Vol. 5, pp. 137–168. Plenum Press, New York. [4.3, 5.5.5, 11.8]

Valleau, J. P. and Whittington, S. G. (1977b). Monte Carlo in statistical mechanics: choosing between alternative transition matrices. *J. comput. Phys.* **24**, 150–7. [4.3]

Van Gunsteren, W. F. (1980). Constrained dynamics of flexible molecules. *Mol. Phys.* **40**, 1015–19. [2.10, 3.4, 9.3]

Van Gunsteren, W. F. and Berendsen, H. J. C. (1977). Algorithms for macromolecular dynamics and constraint dynamics. *Mol. Phys.* **34**, 1311–27. [3.2, 3.2.1, 3.2.2, 3.4]

Van Gunsteren, W. F. and Berendsen, H. J. C. (1982). Algorithms for Brownian dynamics. *Mol. Phys.* **45**, 637–47. [9.3]

Van Gunsteren, W. F. and Karplus, M. (1982). Effect of constraints on the dynamics of macromolecules. *Macromolecules* **15**, 1528–44. [3.4, 9.3]

Van Gunsteren, W. F., Berendsen, H. J. C., and Rullmann, J. A. C. (1978). Inclusion of reaction fields in molecular dynamics: application to liquid water. *Faraday Discuss Chem. Soc.* **66**, 58–70. [5.5.3]

Van Saarloos, W. and Mazur, P. (1983). Many-sphere hydrodynamic interactions. II. Mobilities at finite frequencies. *Physica* **A120**, 77–102. [9.4]

Van Swol, F. and Henderson, J. R. (1986). Wetting at a fluid–wall interface. Computer simulation and exact statistical sum rules. *J. chem Soc. Faraday* II, **82**, 1685–99. [11.8]

Verlet, L. (1967). Computer 'experiments' on classical fluids. I. Thermodynamical properties of Lennard-Jones molecules. *Phys. Rev.* **159**, 98–103. [1.1, 3.2.1, 5.3, 5.4, 5.7.3]

Verlet, L. (1968). Computer 'experiments' on classical fluids. II. Equilibrium correlation functions. *Phys. Rev.* **165**, 201–14. [1.1, 6.5.2]

Verwey, E. J. W. and Overbeek, J. Th. G. (1948). *Theory of the stability of lyophobic colloids.* Elsevier, Amsterdam. [9.4]

Vesely, F. J. (1982). Angular Monte Carlo integration using quaternion parameters: a spherical reference potential for carbon tetrachloride. *J. comput. Phys.* **47**, 291–6. [4.7.1].

Vieillard-Baron, J. E. (1972). Phase transitions of the classical hard ellipse system. *J. chem. Phys.* **56**, 4729–44. [5.7.3, 11.5]

Vieillard-Baron, J. E. (1974). The equation of state of a system of hard spherocylinders. *Mol. Phys.* **28**, 809–18. [11.5]

Von Neumann, J. (1951). Various techniques used in connection with random digits. *US Nat. Bur. stand. appl. Math.* Ser. **12**, 36–8. [G.1, G.4, G.5]

Von Neumann, J. and Ulam, S. (1945). Random ergodic theorems. *Bull. Am. math. Soc.* **51**, (9), 660 (No. 165). [4.1]

Voronstov-Vel'Yaminov, P. N., El'y-Ashevich, A. M., Morgenshtern, L. A., and Chakovskikh, V. P. (1970). Investigations of phase transitions in argon and coulomb gas by the Monte Carlo method using an isothermically isobaric ensemble. *High Temp.* (*USSR*) **8**, 261–8. [4.5]

Wainwright, T. E., Alder, B. J., and Gass, D. M. (1971). Decay of time correlations in two dimensions. *Phys. Rev.* **A4**, 233–7. [11.7]

Wall, F. T. and Mandel, F. (1975). Macromolecular dimensions obtained by an efficient Monte Carlo method without sample attrition. *J. chem. Phys.* **63**, 4592–5. [4.7.2]

Wall, F. T., Chin, J. C., and Mandel, F. (1977). Configurations of macromolecular chains confined to strips or tubes. *J. chem. Phys.* **66**, 3066–9. [4.7.2]

Wang, M. C. and Uhlenbeck, G. E. (1945). On the theory of Brownian motion. II. *Rev. Mod. Phys.* **17**, 323–342. Also in *Selected papers on noise and stochastic processes* (ed. N. Wax). Dover, New York, 1954. [9.3]

Wang, S. S. and Krumhansl, J. A. (1972). Superposition approximation. II. High density fluid argon. *J. chem. Phys.* **56**, 4287–90. [1.5.2]

Weeks, J. D., Chandler, D. and Andersen, H. C. (1971). Role of repulsive forces in determining the equilibrium structure of simple liquids. *J. chem. Phys.* **54**, 5237–47. [1.3.2]

Whitehouse, J. S., Nicholson, D., and Parsonage, N. G. (1983). A grand ensemble Monte Carlo study of krypton adsorbed on graphite. *Mol. Phys.* **49**, 829–47. [4.6]

Widom, B. (1963). Some topics in the theory of fluids. *J. chem. Phys.* **39**, 2808–12. [2.4, 5.6]

Widom, B. (1982). Potential distribution theory and the statistical mechanics of fluids. *J. phys. Chem.* **86**, 869–872. [2.4]

Wigner, E. (1932). On the quantum correction for thermodynamic equilibrium. *Phys. Rev.* **40**, 749–59. [2.9]

Wilemski, G. (1976). On the derivation of Schmoluchowski equations with corrections in the classical theory of Brownian motion. *J. stat. Phys.* **14**, 153–69. [9.4]

Williams, D. E. (1965). Non-bonded potential parameters derived from aromatic hydrocarbons. *J. chem. Phys.* **45**, 3770–8. [1.4.2]

Williams, D. E. (1967). Non-bonded potential parameters derived from crystalline hydrocarbons. *J. chem. Phys.* **47**, 4680–4. [1.4.2]

Wojick, M. and Gubbins, K. E. (1983). Thermodynamics of hard dumbell mixtures. *Mol. Phys.* **49**, 1401–15. [4.7.1]

Wood, D. W. (1979). Computer simulation of water and aqueous solutions. In *Water: a comprehensive treatise* (ed. F. Franks). Vol. 6, pp. 279–409. Plenum Press, New York. [1.1]

Wood, W. W. (1968a). Monte Carlo studies of simple liquid models. In *Physics of simple liquids.* (ed. H. N. V. Temperley, J. S. Rowlinson, and G. S. Rushbrooke), pp. 115–230. North Holland, Amsterdam. [4.3, 4.5, 11.3]

Wood, W. W. (1968b). Monte Carlo calculations for hard disks in the isothermal-isobaric ensemble. *J. chem. Phys.* **48**, 415–34. [4.5]

Wood, W. W. (1970). *NpT*-ensemble Monte Carlo calculations for the hard disk fluid. *J. chem. Phys.* **52**, 729–41. [4.5]

Wood, W. W. (1986). Early history of computer simulations in statistical mechanics. *Molecular dynamics simulation of statistical mechanical systems.* Proceedings of the Enrico Fermi Summer School, Varenna 1985, pp 3–14. [4.1, 11.3]

Wood, W. W. and Jacobson, J. D. (1957). Preliminary results from a recalculation of the Monte Carlo equation of state of hard spheres. *J. chem. Phys.* **27**, 1207–8. [11.3]

Wood, W. W. and Jacobson, J. D. (1959). Monte Carlo calculations in statistical mechanics. Proceedings of the Western Joint Computer Conference, (San Francisco), pp. 261–9. [4.3, 4.4]

Wood, W. W. and Parker, F. R. (1957). Monte Carlo equation of state of molecules interacting with the Lennard-Jones potential. I. A supercritical isotherm at about twice the critical temperature. *J. chem. Phys.* **27**, 720–33. [1.1]

Woodcock, L. V. (1971). Isothermal molecular dynamics calculations for liquid salts. *Chem. Phys. Lett.* **10**, 257–61. [7.4.3]

Woodcock, L. V. and Singer, K. (1971). Thermodynamic and structural properties of liquid ionic salts obtained by Monte Carlo computation. *Trans. Faraday Soc.* **67**, 12–30. [5.5.2]

Yamada, T. and Kawasaki, K. (1975). Applications of mode-coupling theory to the nonlinear stress tensor in fluids. *Prog. theor. Phys.* **53**, 111–24. [11.7]

Yao, J. Greenkorn, R. A., and Chao, K. C. (1982), Monte Carlo simulation of the grand canonical ensemble. *Mol. Phys.* **46**, 587–94. [4.6]

Yarwood, J. (1984). Experimental determination of correlation functions from infra-red and Raman spectra. In *Molecular liquids, dynamics and interactions.* (eds. A. J. Barnes, W. J. Orville-Thomas, and J. Yarwood) NATO ASI series Vol. 135, pp. 357–82. Reidel, New York. [11.6].

Young, A. P. (1979). Melting and the vector coulomb gas in two dimensions. *Phys. Rev.* **B19**, 1855–66. [11.3]

Zannoni, C. (1979). Computer simulation. In *The molecular physics of liquid crystals.* (eds. G. R. Luckhurst and G. W. Gray) Chapter 9, pp. 191–220. Academic Press, London. [11.5]

Zippelius, A., Halperin, B. I., and Nelson, D. R. (1980). Dynamics of two dimensional melting. *Phys. Rev.* **B22**, 2514–41. [11.3]

Zwanzig, R. (1960). Ensemble method in the theory of irreversibility. *J. chem. Phys.* **33**, 1338–41. [9.2]

Zwanzig, R. (1961a). Statistical mechanics of irreversibility. *Lectures in theoretical physics* (eds. W. E. Brittin, B. W. Downs, and J. Downs). Vol. 3. pp. 106–41. Interscience, New York. [9.2]

Zwanzig, R. (1961b). Memory effects in irreversible thermodynamics. *Phys. Rev.* **124**, 983–92. [9.2]

Zwanzig, R. (1965). Time correlation functions and transport coefficients in statistical mechanics. *Ann. Rev. phys. Chem.* **16**, 67–102. [2.7, 8.5]

Zwanzig, R. (1969). Langevin theory of polymer dynamics in dilute solution. *Adv. Chem. Phys.* **15**, 325–31. [9.3, 9.4]

Zwanzig, R. (1981). Nonlinear shear viscosity and long time tails. *Proc. nat. Acad. Sci.* **78**, 3296–7. [11.7]

Zwanzig, R. and Ailawadi, N. K. (1969). Statistical error due to finite time averaging in computer experiments. *Phys. Rev.* **182**, 280–3. [6.4.1]

# INDEX